COLLINS
MINI
ATLAS
OF THE WORLD

Collins Mini Atlas of the World

Collins
An Imprint of HarperCollins*Publishers*
77-85 Fulham Palace Road
London W6 8JB

First published 1999
Reprinted 1999, 2000
Reprinted with changes 2001

Printed and bound in the UK

ISBN 0 00 448893 8 (hardback)
0 00 448909 8 (paperback)

OH11093 (hardback)
OH11092 (paperback)

COLLINS MINI ATLAS OF THE WORLD

HarperCollins*Publishers*

COLLINS MINI ATLAS OF THE WORLD

CONTENTS

HOW TO USE THE ATLAS *6-7*

THE WORLD

PHYSICAL FEATURES 8-9
CLIMATE 10-11
COUNTRIES AND TIME ZONES 12-13
POPULATION AND CITIES 14-15

OCEANIA

OCEANIA Countries 1:70M *16-17*
AUSTRALIA 1:25M *18-19*
Southeastern AUSTRALIA 1:10M *20-21*
NEW ZEALAND 1:10M *22*

ANTARCTICA 1:60M *23*

ASIA

ASIA Countries 1:70M *24-25*
SOUTHEAST ASIA 1:30M *26-27*
MALAYSIA and Western INDONESIA 1:15M *28-29*
CONTINENTAL SOUTHEAST ASIA 1:15M *30-31*
PHILIPPINES 1:15M *32*
NORTH KOREA and SOUTH KOREA 1:9M *33*
JAPAN 1:10M *34-35*
CHINA and MONGOLIA 1:30M *36-37*
Central CHINA 1:15M *38-39*
SOUTHERN ASIA 1:20M *40-41*
PAKISTAN, INDIA and BANGLADESH 1:15M *42-43*
Central ASIA 1:20M *44-45*
ARABIAN PENINSULA 1:15M *46-47*
Eastern MEDITERRANEAN 1:15M *48-49*
RUSSIAN FEDERATION 1:42M *50-51*

EUROPE

EUROPE Countries 1:40M *52-53*

EUROPEAN RUSSIAN FEDERATION 1:20M *54-55*

Northeastern EUROPE 1:8M *56-57*

UKRAINE and MOLDOVA 1:8M *58-59*

SCANDINAVIA and ICELAND 1:10M *60-61*

BRITISH ISLES 1:8M *62-63*

SCOTLAND 1:4M *64*

IRELAND 1:4M *65*

ENGLAND and WALES 1:4M *66-67*

Northwestern EUROPE 1:4M *68-69*

Central EUROPE 1:8M *70-71*

FRANCE and SWITZERLAND 1:8M *72-73*

SPAIN and PORTUGAL 1:8M *74-75*

ITALY and the BALKANS 1:8M *76-77*

GREECE, ROMANIA and BULGARIA 1:8M *78-79*

AFRICA

AFRICA Countries 1:60M *80-81*

Northwestern AFRICA 1:26M *82-83*

Northeastern AFRICA 1:26M *84-85*

Central AFRICA 1:20M *86-87*

Southern AFRICA 1:20M *88-89*

REPUBLIC OF SOUTH AFRICA 1:10M *90-91*

NORTH AMERICA

NORTH AMERICA Countries 1:70M *92-93*

CANADA 1:30M *94-95*

Western CANADA 1:15M *96-97*

Eastern CANADA 1:15M *98-99*

UNITED STATES OF AMERICA 1:25M *100-101*

Western U.S.A. 1:11M *102-103*

North Central U.S.A. 1:11M *104-105*

South Central U.S.A. 1:11M *106-107*

Northeastern U.S.A. 1:11M *108-109*

Southeastern U.S.A. 1:11M *110-111*

MEXICO 1:15M *112-113*

CARIBBEAN 1:20M *114-115*

SOUTH AMERICA

SOUTH AMERICA Countries 1:70M *116-117*

Northern SOUTH AMERICA 1:25M *118-119*

Southern SOUTH AMERICA 1:25M *120-121*

Southeastern BRAZIL 1:10M *122-123*

PACIFIC OCEAN 1:120M *124-125*

ATLANTIC OCEAN 1:120M *126*

INDIAN OCEAN 1:120M *127*

ARCTIC OCEAN 1:60M *128*

WORLD DATA *129-160*

INDEX *161-256*

These pages help the reader to interpret the reference maps in the atlas. They explain the main features shown on the mapping and the policies adopted in deciding what to show and how to show it.

The databases used to create the maps provide the freedom to select the best map coverage for each part of the world. Maps are arranged on a continental basis, with each continent being introduced by a map of the continent. Maps of Antarctica and the world's oceans complete the worldwide coverage.

SYMBOLS and GENERALIZATION

Maps show information by using signs, or symbols, which are designed to reflect the features on the earth that they represent. Symbols can be in the form of points - such as those used to show towns and airports; lines - used to represent roads and rivers; or areas - such as lakes. Variation in size, shape and colour of these types of symbol allow a great range of information to be shown. The symbols used in this atlas are explained here. Not all information can be shown, and much has to be generalized to be clearly shown on the maps. This generalization takes the form of selection - the inclusion of some features and the omission of others of less importance; and simplification - where lines are smoothed, areas combined, or symbols displaced slightly to add clarity. This is done in such a way that the overall character of the area mapped is retained. The degree of generalization varies, and is determined largely by the scale at which the map is drawn.

SCALE

Scale is the relationship between the size of an area shown on the map and the actual size of the area on the ground. It determines the amount of detail shown on a map - larger scales show more, smaller scales show less - and can be used to measure the distance between two points, though the projection of the map must also be taken into account when measuring distances.

GEOGRAPHICAL NAMES

The spelling of place names on maps is a complex problem for the cartographer. There is no single standard way of converting them from one alphabet, or symbol set, to another. Changes in official languages also have to be taken into account when creating maps and policies need to be established for the spelling of names on individual atlases and maps. Such policies must take account of the local official position, international conventions or traditions, and the purpose of the atlas or map. The policy in this atlas is to use local name forms which are officially recognized by the governments of the countries concerned. However, English conventional name forms are used for the most well-known places. In these cases, the local form is often included in brackets on the map and also appears as a cross-reference in the index. All country names and those for international features appear in their English forms.

BOUNDARIES

The status of nations and their boundaries are shown in this atlas as they are in reality at the time of going to press, as far as can be ascertained. Where international boundaries are the subject of disputes the aim is to take a strictly neutral viewpoint, based on advice from expert consultants.

Boundaries

International

International disputed

Administrative
(selected countries only)

Ceasefire line

Settlements

POPULATION	NATIONAL CAPITAL	ADMINISTRATIVE CAPITAL	CITY or TOWN
over 1 million	BEIJING	Tianjin	New York
500 000 - 1 million	BANGUI	Douala	Barranquilla
100 000 - 500 000	WELLINGTON	Mansa	Mara
50 000 - 100 000	PORT OF SPAIN	Lubango	Arecibo
under 50 000	MALABO	Chinhoyi	El Tigre

Communications

- Motorway
- Main road
- Track
- Railway
- Main Airport
- Canal

Physical features

- Freshwater lake
- Seasonal freshwater lake
- Salt lake
- Seasonal salt lake
- Dry salt lake
- Glacier / Ice cap
- River
- Seasonal river
- 123 Mountain pass
- 1234 Summit

Other features

- Site of special interest
- Wall

Styles of lettering

Country name	**FRANCE**	Island	*Gran Canaria*
Overseas Territory / Dependency	**Guadeloupe**	Lake	*Lake Erie*
Administrative name	**SCOTLAND**	Mountain	*Mt Blanc*
Area name	PATAGONIA	River	*Thames*

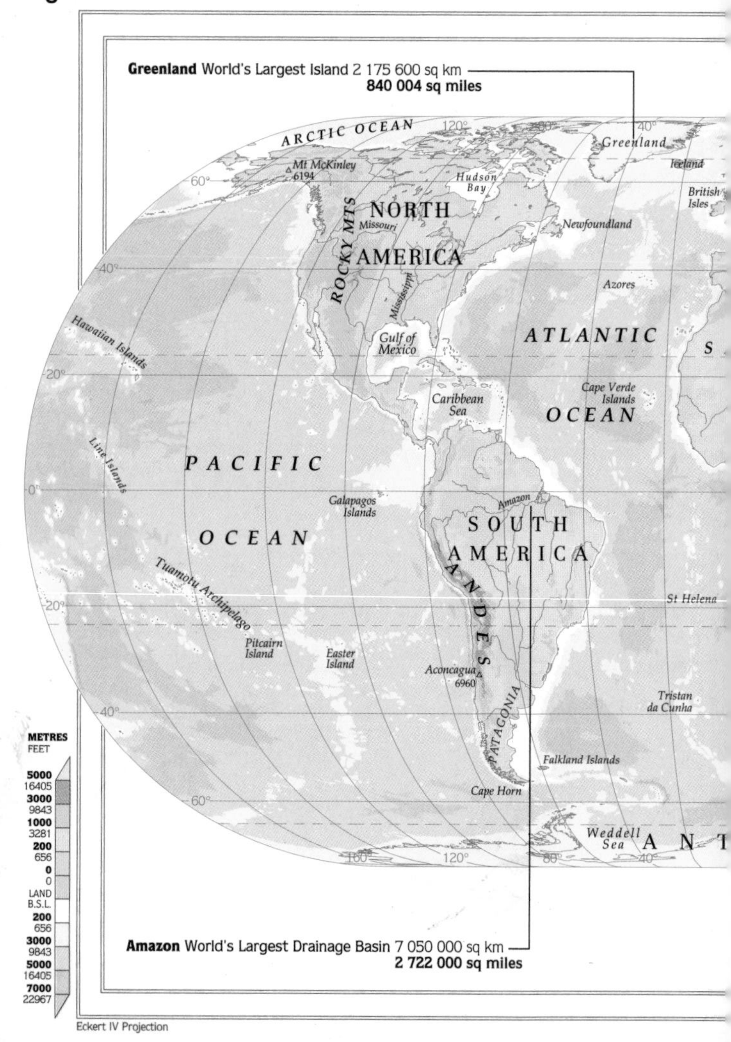

Eckert IV Projection

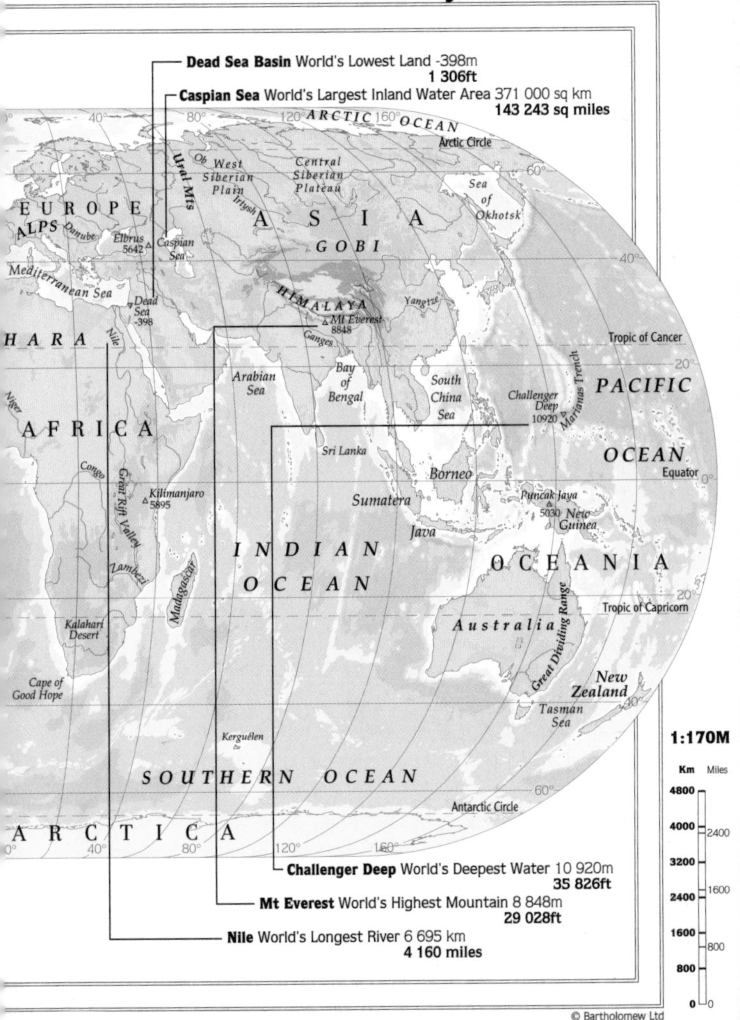
Dead Sea Basin World's Lowest Land -398m
1 306ft
Caspian Sea World's Largest Inland Water Area 371 000 sq km
143 243 sq miles
ARCTIC OCEAN
Arctic Circle
Ural Mts
Ob
West Siberian Plain
Central Siberian Plateau
Sea of Okhotsk
EUROPE
ALPS
Danube
Elbrus 5642
Caspian Sea
ASIA
Irtysh
GOBI
Mediterranean Sea
Dead Sea -398
HIMALAYA
Mt Everest 8848
Yangtze
Ganges
HARA
Nile
Tropic of Cancer
Arabian Sea
Bay of Bengal
South China Sea
Challenger Deep 10920
Marianas Trench
PACIFIC OCEAN
Niger
AFRICA
Sri Lanka
Borneo
Equator
Congo
Great Rift Valley
Kilimanjaro 5895
Sumatera
Java
Puncak Jaya 5030
New Guinea
INDIAN OCEAN
OCEANIA
Zambezi
Madagascar
Great Dividing Range
Tropic of Capricorn
Kalahari Desert
Australia
Cape of Good Hope
New Zealand
Tasman Sea
Kerguélen
SOUTHERN OCEAN
Antarctic Circle
ARCTICA
Challenger Deep World's Deepest Water 10 920m
35 826ft
Mt Everest World's Highest Mountain 8 848m
29 028ft
Nile World's Longest River 6 695 km
4 160 miles
1:170M
Km Miles
4800
4000 2400
3200
2400 1600
1600
800 800
0 0

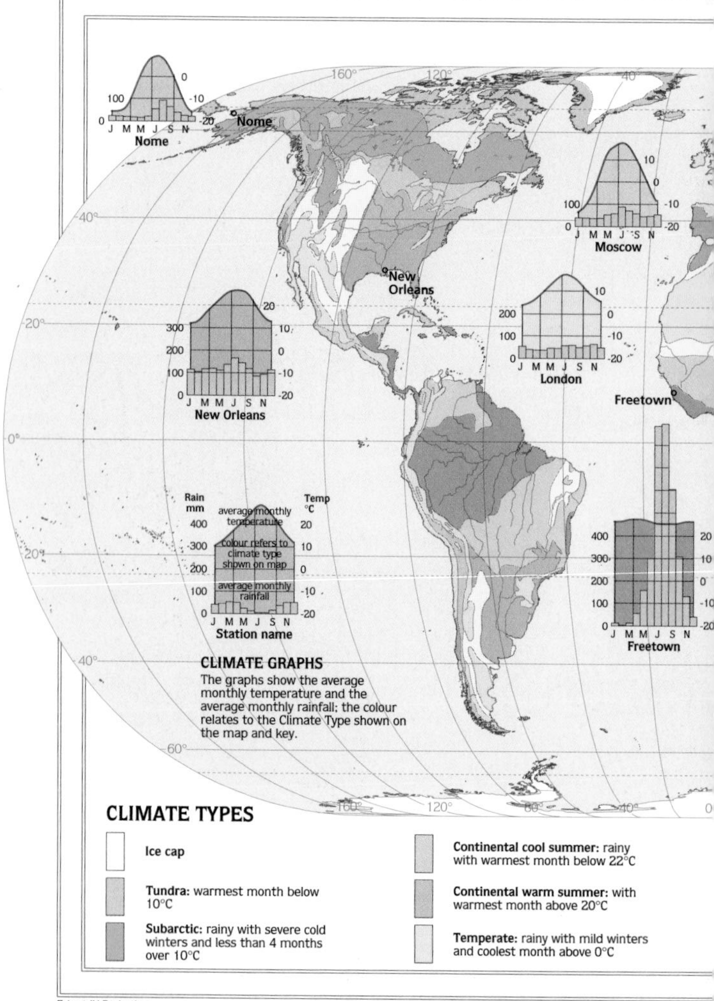

CLIMATE GRAPHS

The graphs show the average monthly temperature and the average monthly rainfall; the colour relates to the Climate Type shown on the map and key.

CLIMATE TYPES

- **Ice cap**
- **Tundra:** warmest month below 10°C
- **Subarctic:** rainy with severe cold winters and less than 4 months over 10°C
- **Continental cool summer:** rainy with warmest month below 22°C
- **Continental warm summer:** with warmest month above 20°C
- **Temperate:** rainy with mild winters and coolest month above 0°C

Eckert IV Projection

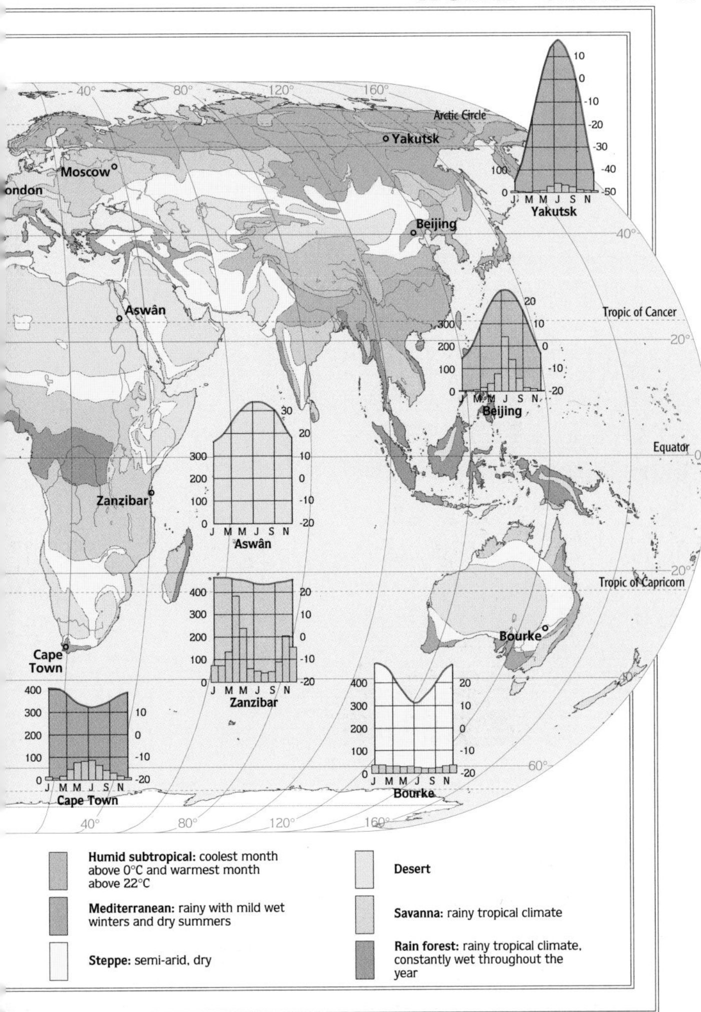
Arctic Circle
Yakutsk
Moscow
London
Beijing
Aswân
Tropic of Cancer
Equator
Zanzibar
Tropic of Capricorn
Bourke
Cape Town
40°
80°
120°
160°
J M M J S N
Humid subtropical: coolest month above 0°C and warmest month above 22°C
Mediterranean: rainy with mild wet winters and dry summers
Steppe: semi-arid, dry
Desert
Savanna: rainy tropical climate
Rain forest: rainy tropical climate, constantly wet throughout the year

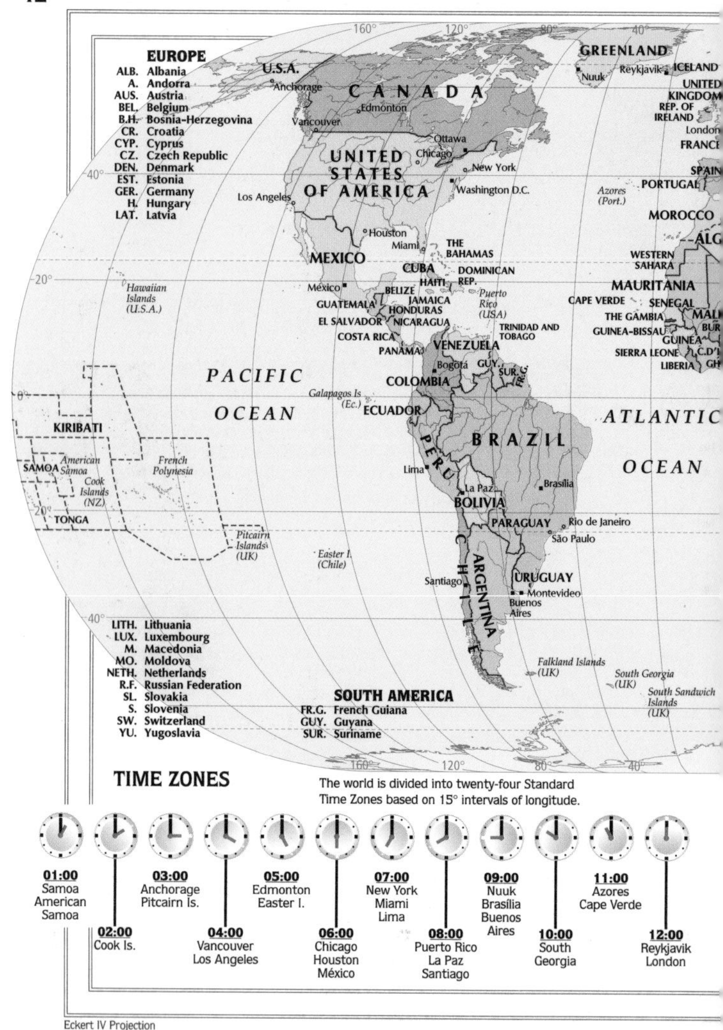
EUROPE
ALB. Albania
A. Andorra
AUS. Austria
BEL. Belgium
B.H. Bosnia-Herzegovina
CR. Croatia
CYP. Cyprus
CZ. Czech Republic
DEN. Denmark
EST. Estonia
GER. Germany
H. Hungary
LAT. Latvia
LITH. Lithuania
LUX. Luxembourg
M. Macedonia
MO. Moldova
NETH. Netherlands
R.F. Russian Federation
SL. Slovakia
S. Slovenia
SW. Switzerland
YU. Yugoslavia
SOUTH AMERICA
FR.G. French Guiana
GUY. Guyana
SUR. Suriname
160°
120°
80°
40°
20°
0°
U.S.A.
Anchorage
CANADA
Edmonton
Vancouver
Ottawa
Chicago
New York
Washington D.C.
UNITED STATES OF AMERICA
Los Angeles
Houston
Miami
MEXICO
México
THE BAHAMAS
CUBA
DOMINICAN REP.
HAITI
Puerto Rico (USA)
BELIZE
JAMAICA
GUATEMALA
HONDURAS
EL SALVADOR
NICARAGUA
COSTA RICA
PANAMA
TRINIDAD AND TOBAGO
VENEZUELA
Bogotá
GUY.
SUR.
FR.G.
COLOMBIA
Galapagos Is (Ec.)
ECUADOR
BRAZIL
PERU
Lima
La Paz
Brasília
BOLIVIA
PARAGUAY
Rio de Janeiro
São Paulo
CHILE
ARGENTINA
Santiago
URUGUAY
Montevideo
Buenos Aires
Falkland Islands (UK)
South Georgia (UK)
South Sandwich Islands (UK)
GREENLAND
Nuuk
Reykjavik
ICELAND
UNITED KINGDOM
REP. OF IRELAND
London
FRANCE
SPAIN
PORTUGAL
Azores (Port.)
MOROCCO
ALG
WESTERN SAHARA
MAURITANIA
CAPE VERDE
SENEGAL
THE GAMBIA
MALI
GUINEA-BISSAU
BUR
GUINEA
SIERRA LEONE
C.D'I
LIBERIA
GH
Hawaiian Islands (U.S.A.)
PACIFIC OCEAN
ATLANTIC OCEAN
KIRIBATI
SAMOA
American Samoa
Cook Islands (NZ)
TONGA
French Polynesia
Pitcairn Islands (UK)
Easter I. (Chile)
TIME ZONES
The world is divided into twenty-four Standard Time Zones based on 15° intervals of longitude.
01:00 Samoa American Samoa
02:00 Cook Is.
03:00 Anchorage Pitcairn Is.
04:00 Vancouver Los Angeles
05:00 Edmonton Easter I.
06:00 Chicago Houston México
07:00 New York Miami Lima
08:00 Puerto Rico La Paz Santiago
09:00 Nuuk Brasília Buenos Aires
10:00 South Georgia
11:00 Azores Cape Verde
12:00 Reykjavik London
Eckert IV Projection

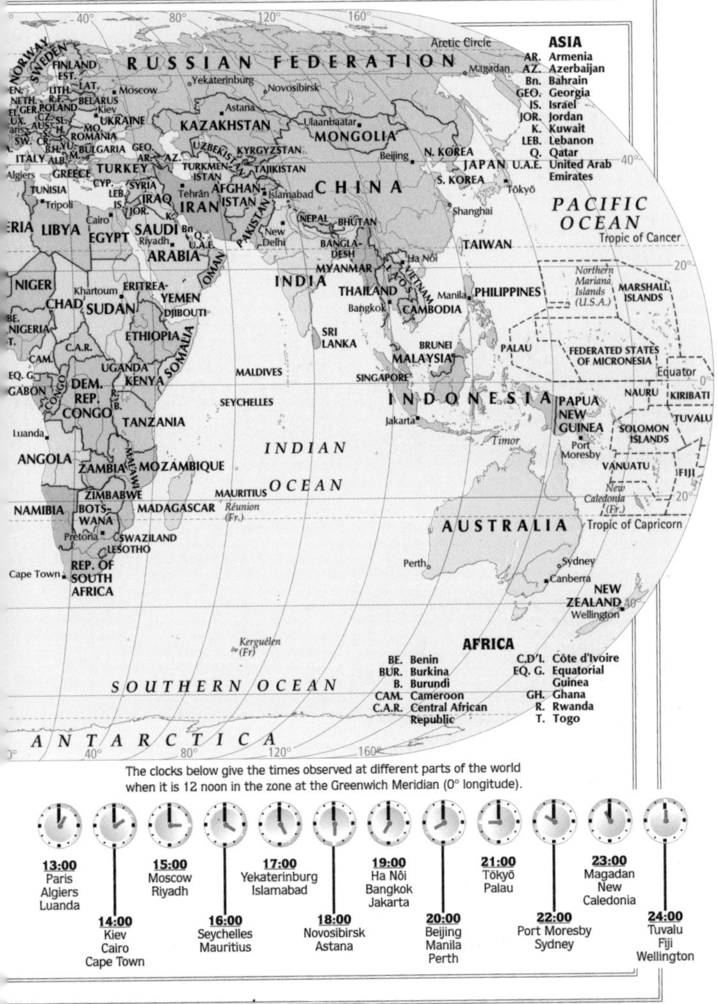
RUSSIAN FEDERATION
Arctic Circle
Magadan
Yekaterinburg
Novosibirsk
Moscow
FINLAND
SWEDEN
NORWAY
EST.
LAT.
LITH.
BELARUS
POLAND
Kiev
UKRAINE
MO.
ROMANIA
BULGARIA
ITALY
GREECE
TURKEY
GEO.
AR.
AZ.
Astana
KAZAKHSTAN
Ulaanbaatar
MONGOLIA
UZBEKISTAN
KYRGYZSTAN
TURKMENISTAN
TAJIKISTAN
Beijing
N. KOREA
JAPAN
S. KOREA
Tōkyō
CHINA
Islamabad
AFGHANISTAN
PAKISTAN
Tehrān
IRAN
IRAQ
SYRIA
CYP.
LEB.
IS.
JOR.
K.
Bn
Q.
U.A.E.
Algiers
TUNISIA
Tripoli
LIBYA
Cairo
EGYPT
SAUDI ARABIA
Riyadh
OMAN
NEPAL
BHUTAN
New Delhi
BANGLADESH
Shanghai
TAIWAN
Ha Nôi
MYANMAR
LAOS
VIETNAM
INDIA
THAILAND
Bangkok
CAMBODIA
Manila
PHILIPPINES
NIGER
Khartoum
ERITREA
YEMEN
DJIBOUTI
CHAD
SUDAN
NIGERIA
ETHIOPIA
SOMALIA
C.A.R.
CAM.
EQ. G.
GABON
CONGO
UGANDA
KENYA
DEM. REP. CONGO
R.
B.
TANZANIA
Luanda
ANGOLA
ZAMBIA
MALAWI
MOZAMBIQUE
ZIMBABWE
NAMIBIA
BOTSWANA
Pretoria
SWAZILAND
LESOTHO
MADAGASCAR
Cape Town
REP. OF SOUTH AFRICA
SRI LANKA
MALDIVES
SEYCHELLES
MAURITIUS
Réunion (Fr.)
INDIAN OCEAN
BRUNEI
MALAYSIA
SINGAPORE
INDONESIA
Jakarta
Timor
PALAU
PAPUA NEW GUINEA
Port Moresby
PACIFIC OCEAN
Tropic of Cancer
Northern Mariana Islands (U.S.A.)
MARSHALL ISLANDS
FEDERATED STATES OF MICRONESIA
Equator
NAURU
KIRIBATI
TUVALU
SOLOMON ISLANDS
VANUATU
FIJI
New Caledonia (Fr.)
Tropic of Capricorn
AUSTRALIA
Perth
Sydney
Canberra
NEW ZEALAND
Wellington
Kerguélen (Fr)
SOUTHERN OCEAN
ANTARCTICA
40°
80°
120°
160°
20°
0°
ASIA
AR. Armenia
AZ. Azerbaijan
Bn. Bahrain
GEO. Georgia
IS. Israel
JOR. Jordan
K. Kuwait
LEB. Lebanon
Q. Qatar
U.A.E. United Arab Emirates
AFRICA
BE. Benin
BUR. Burkina
B. Burundi
CAM. Cameroon
C.A.R. Central African Republic
C.D'I. Côte d'Ivoire
EQ. G. Equatorial Guinea
GH. Ghana
R. Rwanda
T. Togo
The clocks below give the times observed at different parts of the world when it is 12 noon in the zone at the Greenwich Meridian (0° longitude).
13:00 Paris Algiers Luanda
14:00 Kiev Cairo Cape Town
15:00 Moscow Riyadh
16:00 Seychelles Mauritius
17:00 Yekaterinburg Islamabad
18:00 Novosibirsk Astana
19:00 Ha Nôi Bangkok Jakarta
20:00 Beijing Manila Perth
21:00 Tōkyō Palau
22:00 Port Moresby Sydney
23:00 Magadan New Caledonia
24:00 Tuvalu Fiji Wellington

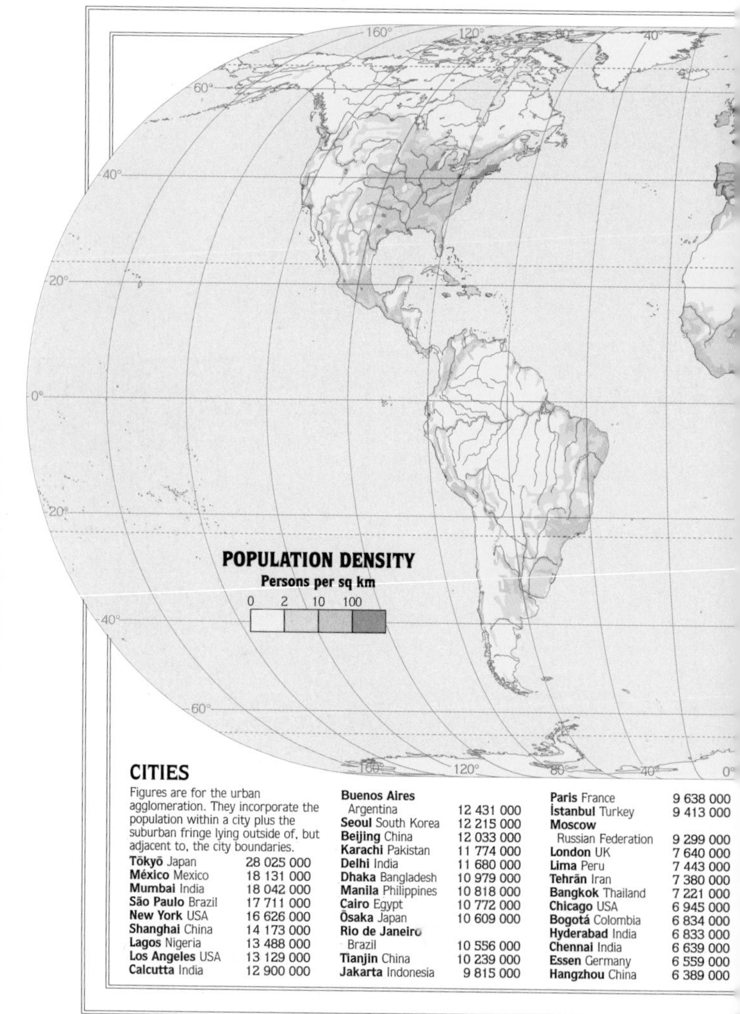

CITIES

Figures are for the urban agglomeration. They incorporate the population within a city plus the suburban fringe lying outside of, but adjacent to, the city boundaries.

Tōkyō Japan	28 025 000
México Mexico	18 131 000
Mumbai India	18 042 000
São Paulo Brazil	17 711 000
New York USA	16 626 000
Shanghai China	14 173 000
Lagos Nigeria	13 488 000
Los Angeles USA	13 129 000
Calcutta India	12 900 000
Buenos Aires Argentina	12 431 000
Seoul South Korea	12 215 000
Beijing China	12 033 000
Karachi Pakistan	11 774 000
Delhi India	11 680 000
Dhaka Bangladesh	10 979 000
Manila Philippines	10 818 000
Cairo Egypt	10 772 000
Ōsaka Japan	10 609 000
Rio de Janeiro Brazil	10 556 000
Tianjin China	10 239 000
Jakarta Indonesia	9 815 000
Paris France	9 638 000
İstanbul Turkey	9 413 000
Moscow Russian Federation	9 299 000
London UK	7 640 000
Lima Peru	7 443 000
Tehrān Iran	7 380 000
Bangkok Thailand	7 221 000
Chicago USA	6 945 000
Bogotá Colombia	6 834 000
Hyderabad India	6 833 000
Chennai India	6 639 000
Essen Germany	6 559 000
Hangzhou China	6 389 000

Eckert IV Projection

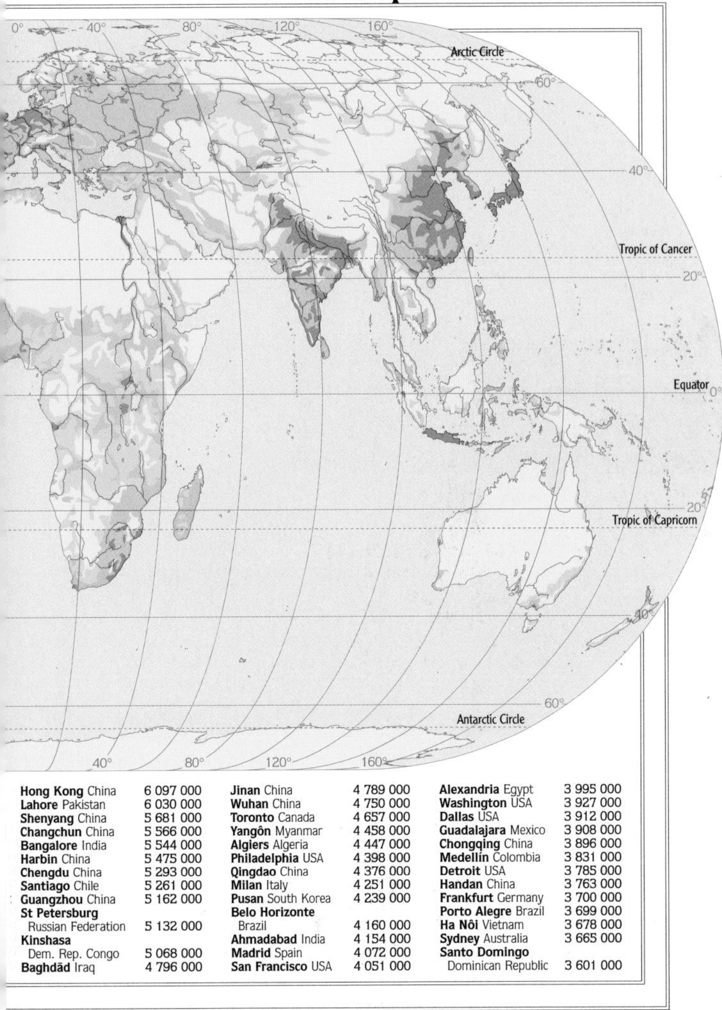

City	Population
Hong Kong China	6 097 000
Lahore Pakistan	6 030 000
Shenyang China	5 681 000
Changchun China	5 566 000
Bangalore India	5 544 000
Harbin China	5 475 000
Chengdu China	5 293 000
Santiago Chile	5 261 000
Guangzhou China	5 162 000
St Petersburg Russian Federation	5 132 000
Kinshasa Dem. Rep. Congo	5 068 000
Baghdād Iraq	4 796 000
Jinan China	4 789 000
Wuhan China	4 750 000
Toronto Canada	4 657 000
Yangôn Myanmar	4 458 000
Algiers Algeria	4 447 000
Philadelphia USA	4 398 000
Qingdao China	4 376 000
Milan Italy	4 251 000
Pusan South Korea	4 239 000
Belo Horizonte Brazil	4 160 000
Ahmadabad India	4 154 000
Madrid Spain	4 072 000
San Francisco USA	4 051 000
Alexandria Egypt	3 995 000
Washington USA	3 927 000
Dallas USA	3 912 000
Guadalajara Mexico	3 908 000
Chongqing China	3 896 000
Medellín Colombia	3 831 000
Detroit USA	3 785 000
Handan China	3 763 000
Frankfurt Germany	3 700 000
Porto Alegre Brazil	3 699 000
Ha Nôi Vietnam	3 678 000
Sydney Australia	3 665 000
Santo Domingo Dominican Republic	3 601 000

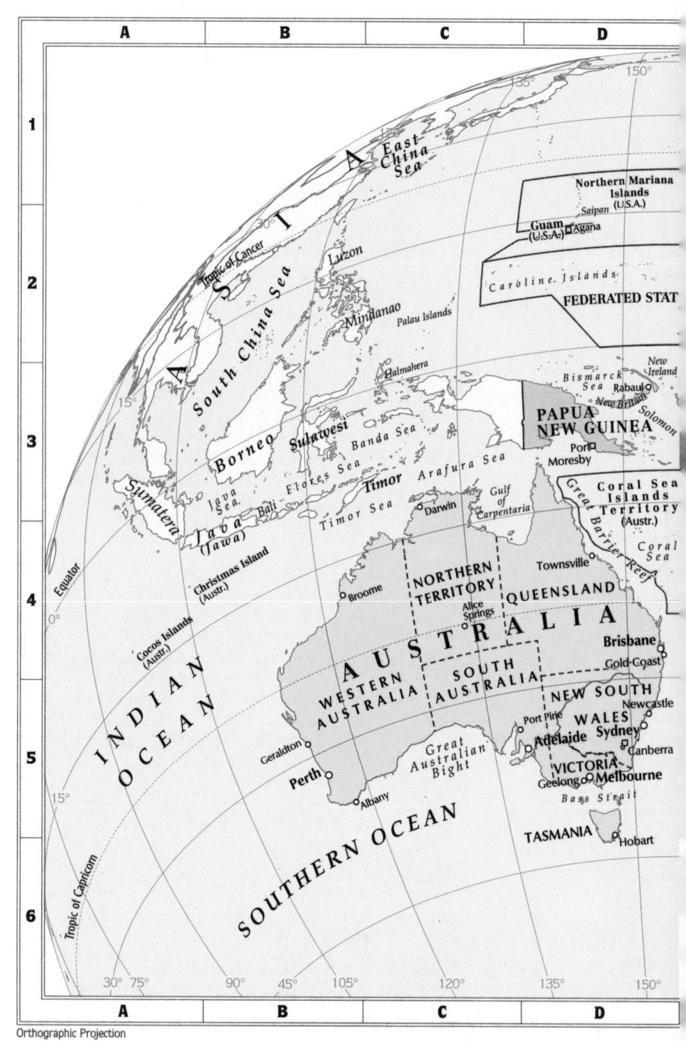

Orthographic Projection

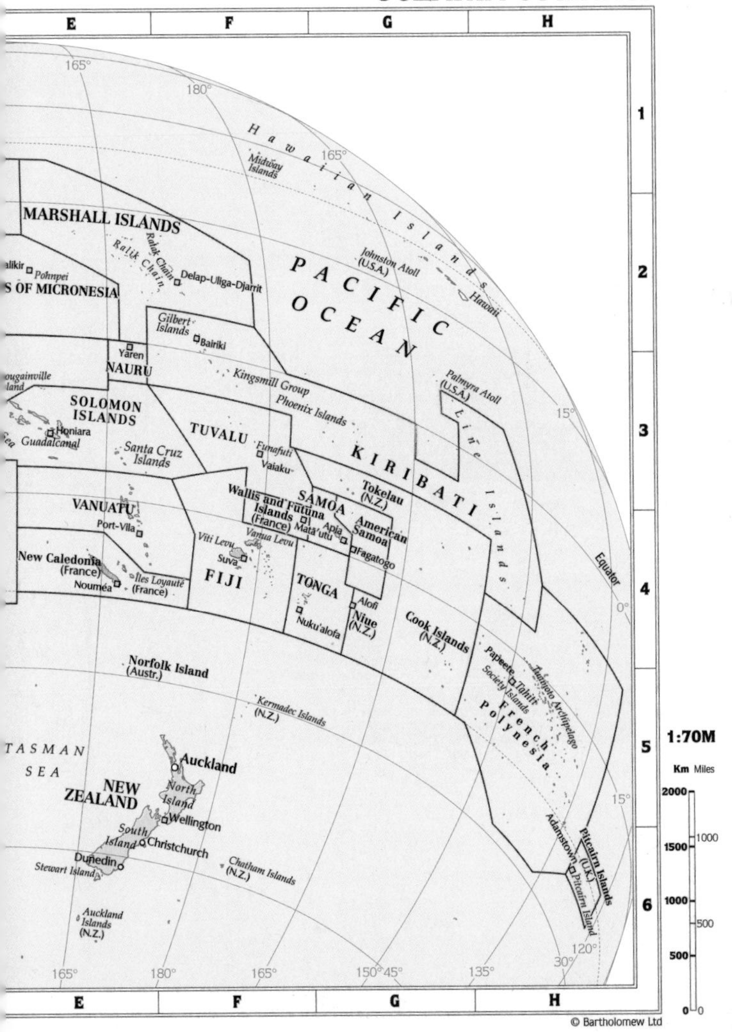
E
F
G
H
1
2
3
4
5
6
PACIFIC OCEAN
Hawaiian Islands
Midway Islands
Hawaii
Johnston Atoll (U.S.A.)
MARSHALL ISLANDS
Ralik Chain
Ratak Chain
Delap-Uliga-Djarrit
Palikir
Pohnpei
S OF MICRONESIA
Gilbert Islands
Bairiki
Yaren
NAURU
Kingsmill Group
Phoenix Islands
Palmyra Atoll (U.S.A.)
Line Islands
KIRIBATI
SOLOMON ISLANDS
Honiara
Guadalcanal
Santa Cruz Islands
TUVALU
Funafuti
Vaiaku
Tokelau (N.Z.)
SAMOA
Apia
American Samoa
Fagatogo
Wallis and Futuna Islands (France)
Matā'utu
VANUATU
Port-Vila
Viti Levu
Vanua Levu
Suva
FIJI
New Caledonia (France)
Nouméa
Îles Loyauté (France)
TONGA
Nuku'alofa
Alofi
Niue (N.Z.)
Cook Islands (N.Z.)
Equator
Papeete
Tahiti
Society Islands
Tuamotu Archipelago
French Polynesia
Norfolk Island (Austr.)
Kermadec Islands (N.Z.)
TASMAN SEA
NEW ZEALAND
Auckland
North Island
Wellington
South Island
Christchurch
Dunedin
Stewart Island
Chatham Islands (N.Z.)
Auckland Islands (N.Z.)
Adamstown
Pitcairn Islands (U.K.)
Pitcairn Island
1:70M
Km Miles
2000
1500
1000
500
0
1000
500
0

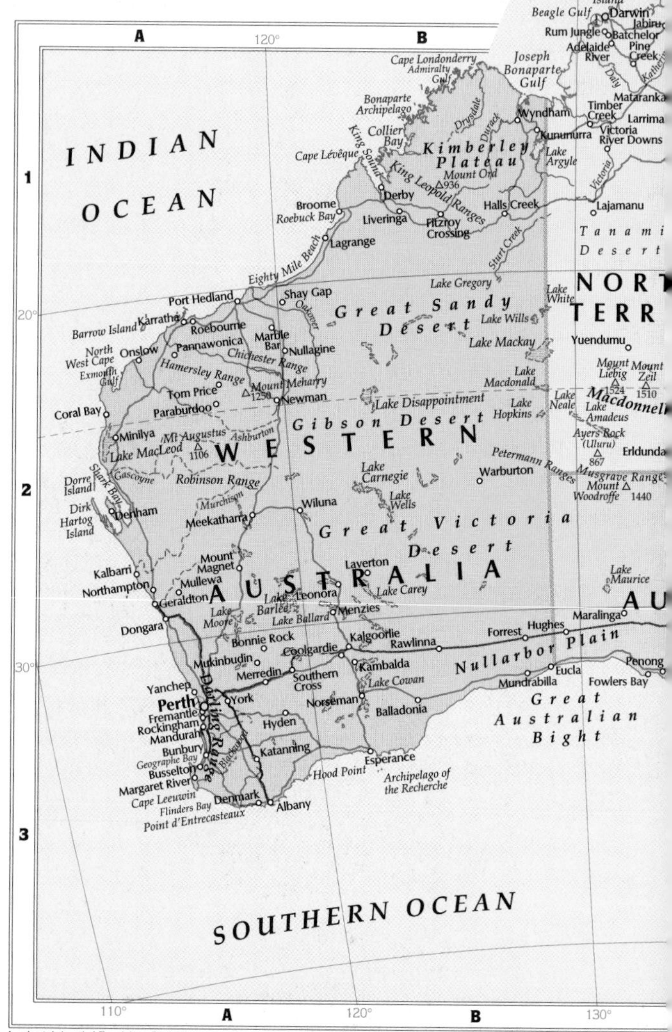

Lambert Azimuthal Equal Area Projection

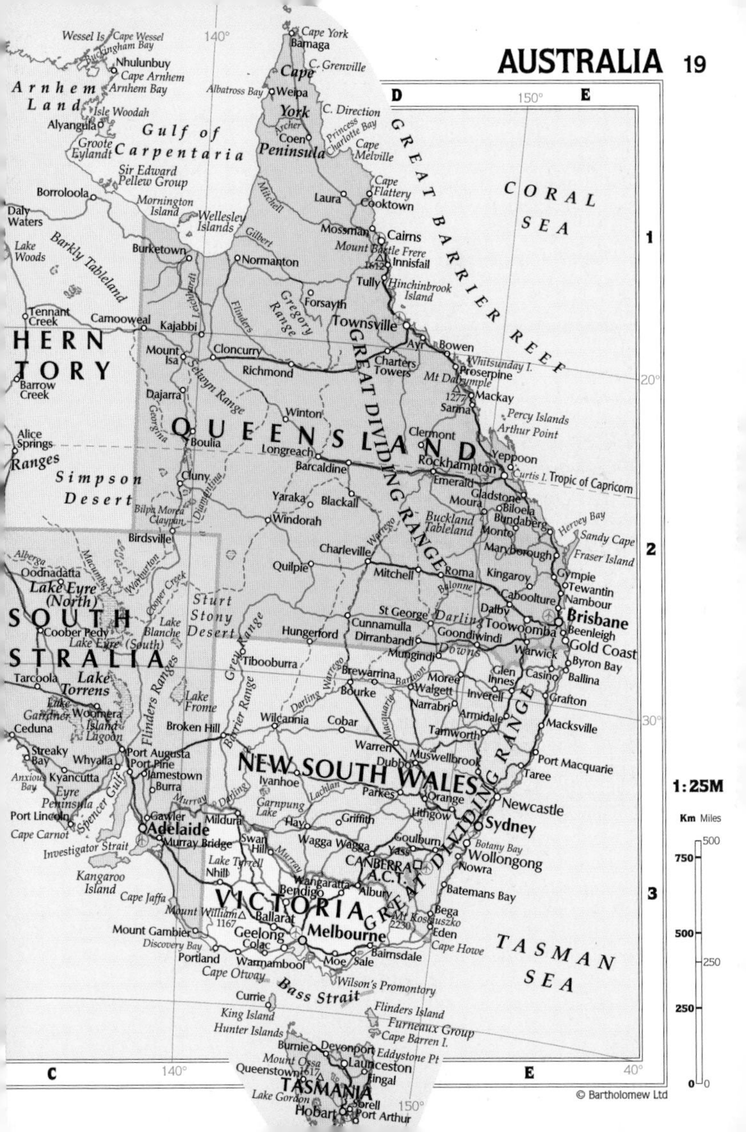
Wessel Is
Cape Wessel
Buckingham Bay
Nhulunbuy
Cape Arnhem
Arnhem Land
Arnhem Bay
Isle Woodah
Alyangula
Groote Eylandt
Gulf of Carpentaria
Sir Edward Pellew Group
Borroloola
Mornington Island
Wellesley Islands
Daly Waters
Lake Woods
Barkly Tableland
Burketown
Normanton
Tennant Creek
Camooweal
Kajabbi
HERN TORY
Mount Isa
Cloncurry
Richmond
Selwyn Range
Barrow Creek
Dajarra
Georgina
Winton
Alice Springs
Ranges
Boulia
QUEENSLAND
Longreach
Barcaldine
Simpson Desert
Cluny
Diamantina
Bilpa Morea Claypan
Yaraka
Blackall
Windorah
Birdsville
Alberga
Oodnadatta
Macumba
Warburton
Lake Eyre (North)
Cooper Creek
Sturt Stony Desert
Lake Blanche
SOUTH AUSTRALIA
Coober Pedy
Lake Eyre (South)
Tarcoola
Lake Torrens
Lake Gairdner
Woomera
Island Lagoon
Flinders Ranges
Lake Frome
Ceduna
Streaky Bay
Whyalla
Port Augusta
Port Pirie
Jamestown
Burra
Anxious Bay
Kyancutta
Eyre Peninsula
Spencer Gulf
Port Lincoln
Cape Carnot
Gawler
Adelaide
Murray Bridge
Investigator Strait
Kangaroo Island
Cape Jaffa
Cape York
Bamaga
C. Grenville
Cape York Peninsula
Albatross Bay
Weipa
Archer
Coen
C. Direction
Princess Charlotte Bay
Cape Melville
Mitchell
Laura
Cape Flattery
Cooktown
Gilbert
Mossman
Cairns
Mount Bartle Frere
1612
Innisfail
Tully
Hinchinbrook Island
Gregory Range
Forsayth
Flinders
Leichhardt
Townsville
Ayr
Bowen
Charters Towers
Whitsunday I.
Proserpine
Mt Dalrymple
1277
Mackay
Sarina
Percy Islands
Arthur Point
Clermont
Yeppoon
Rockhampton
Curtis I.
Tropic of Capricorn
Emerald
Gladstone
Moura
Biloela
Bundaberg
Hervey Bay
Sandy Cape
Fraser Island
Buckland Tableland
Monto
Maryborough
Warrego
Charleville
Quilpie
Mitchell
Roma
Kingaroy
Gympie
Tewantin
Nambour
Balonne
Caboolture
Brisbane
Dalby
St George
Darling Downs
Toowoomba
Beenleigh
Gold Coast
Cunnamulla
Hungerford
Dirranbandi
Goondiwindi
Mungindi
Warwick
Byron Bay
Ballina
Grey Range
Tibooburra
Brewarrina
Barwon
Moree
Glen Innes
Casino
Bourke
Walgett
Inverell
Grafton
Barrier Range
Darling
Narrabri
Armidale
Wilcannia
Cobar
Macquarie
Tamworth
Macksville
Broken Hill
Warren
Muswellbrook
Port Macquarie
Dubbo
Taree
NEW SOUTH WALES
Ivanhoe
Lachlan
Parkes
Orange
Newcastle
Garnpung Lake
Mildura
Murray
Lithgow
Sydney
Hay
Griffith
Swan Hill
Wagga Wagga
Goulburn
Yass
Botany Bay
Wollongong
Nowra
CANBERRA
A.C.T.
Lake Tyrrell
Nhill
Wangaratta
Albury
Batemans Bay
Bendigo
VICTORIA
GREAT DIVIDING RANGE
Mount William
1167
Ballarat
Mt Kosciuszko
2230
Bega
Eden
Cape Howe
Mount Gambier
Geelong
Colac
Melbourne
Bairnsdale
Discovery Bay
Portland
Warrnambool
Moe
Sale
Cape Otway
Bass Strait
Wilson's Promontory
Currie
King Island
Flinders Island
Hunter Islands
Furneaux Group
Cape Barren I.
Burnie
Devonport
Eddystone Pt
Mount Ossa
1617
Launceston
Queenstown
Fingal
TASMANIA
Lake Gordon
Sorell
Hobart
Port Arthur
GREAT BARRIER REEF
CORAL SEA
TASMAN SEA
GREAT DIVIDING RANGE
140°
150°
20°
30°
40°
C
D
E
1
2
3
1:25M
Km
Miles
750
500
250
0
500
250
0

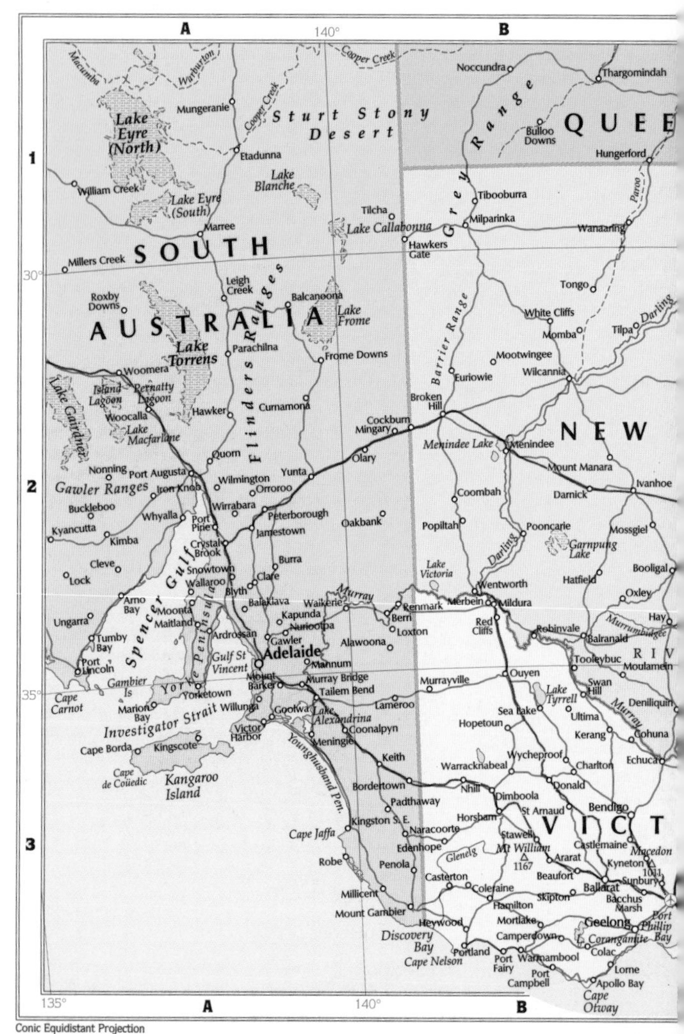

SOUTH AUSTRALIA
NEW
QUEE
VICT
Lake Eyre (North)
Lake Eyre (South)
Sturt Stony Desert
Lake Torrens
Lake Frome
Flinders Ranges
Grey Range
Barrier Range
Gawler Ranges
Spencer Gulf
Gulf St Vincent
Investigator Strait
Kangaroo Island
Yorke Peninsula
Adelaide
Broken Hill
Mildura
Bendigo
Geelong
Port Phillip Bay
Discovery Bay
Cape Otway
Conic Equidistant Projection

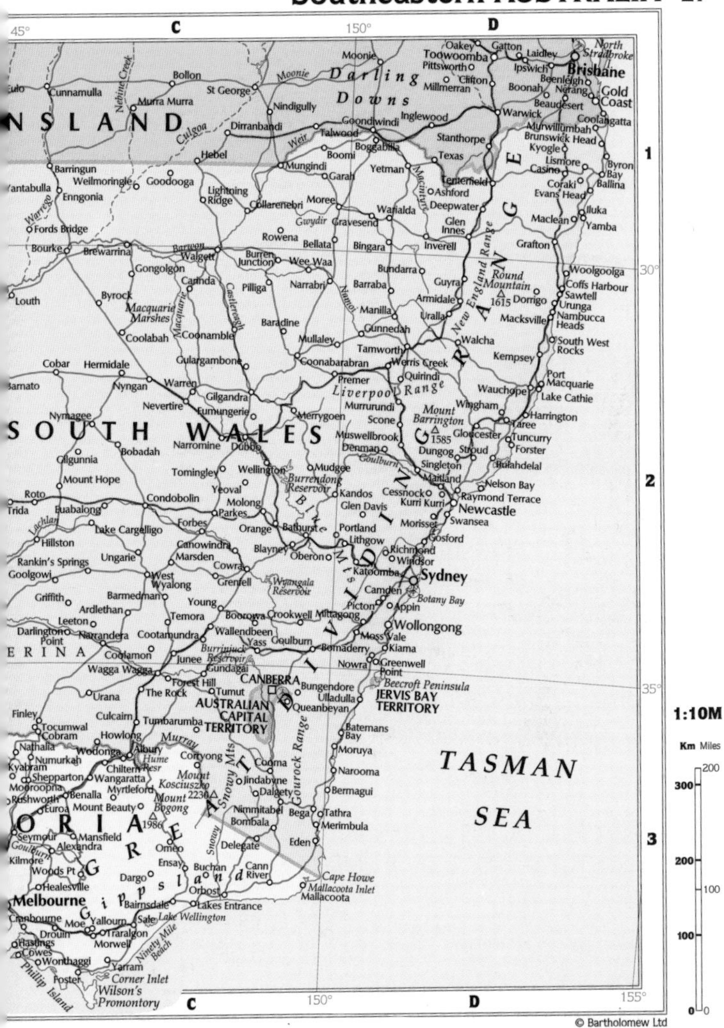
NEW SOUTH WALES
QUEENSLAND
VICTORIA
Darling Downs
GREAT DIVIDING RANGE
AUSTRALIAN CAPITAL TERRITORY
CANBERRA
JERVIS BAY TERRITORY
TASMAN SEA
Brisbane
Gold Coast
Toowoomba
Sydney
Newcastle
Wollongong
Melbourne
Dubbo
Bathurst
Orange
Tamworth
Armidale
Wagga Wagga
Albury
Goulburn
Moree
Bourke
Cobar
Cunnamulla
St George
Goondiwindi
Warwick
Lismore
Grafton
Coffs Harbour
Port Macquarie
Maitland
Gosford
Katoomba
Nowra
Cooma
Bega
Eden
Mount Kosciuszko 2230
Gippsland
Cape Howe
Wilson's Promontory
Phillip Island
Botany Bay
1:10M
Km Miles
© Bartholomew Ltd

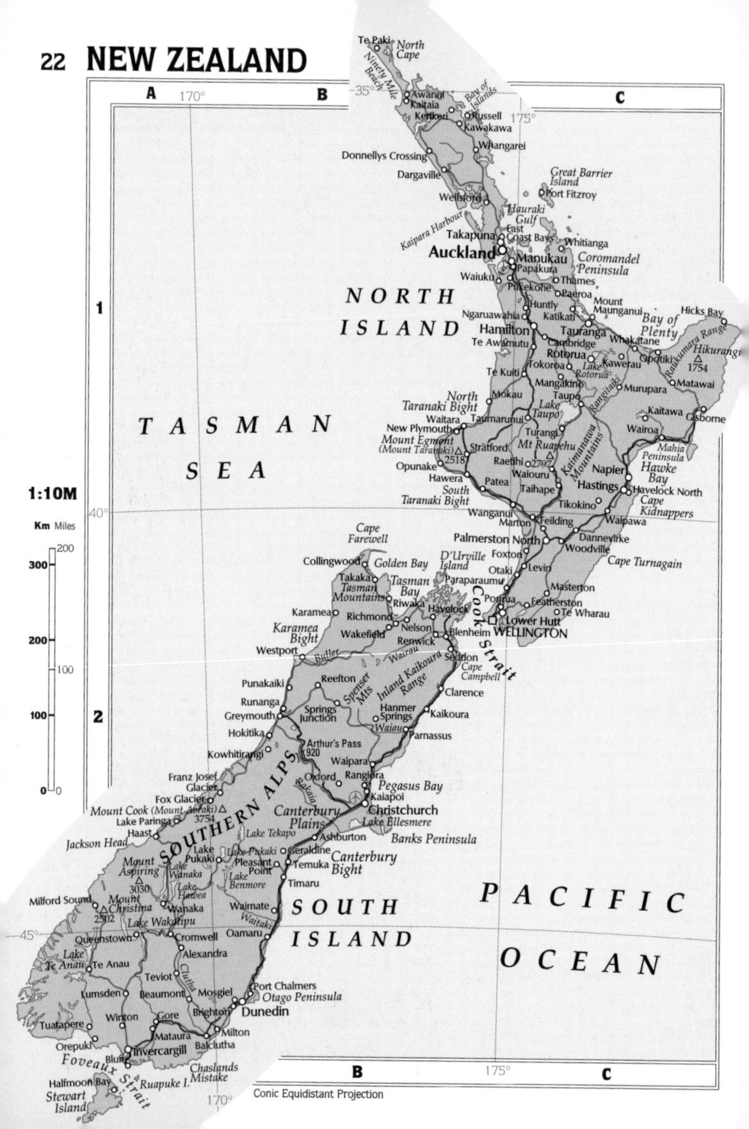
A
B
C
1
2
170°
175°
35°
40°
45°
1:10M
Km Miles
300
200
100
0
200
100
0
Te Paki
North Cape
Ninety Mile Beach
Awanui
Kaitaia
Bay of Islands
Kerikeri
Russell
Kawakawa
Whangarei
Donnellys Crossing
Dargaville
Great Barrier Island
Port Fitzroy
Wellsford
Hauraki Gulf
Kaipara Harbour
Takapuna
East Coast Bays
Whitianga
Auckland
Manukau
Coromandel Peninsula
Papakura
Waiuku
Thames
Pukekohe
Paeroa
NORTH ISLAND
Huntly
Katikati
Mount Maunganui
Bay of Plenty
Hicks Bay
Ngaruawahia
Hamilton
Tauranga
Whakatane
Raukumara Range
Te Awamutu
Cambridge
Hikurangi
1754
Rotorua
Lake Rotorua
Kawerau
Opotiki
Tokoroa
Te Kuiti
Mangakino
Murupara
Matawai
North Taranaki Bight
Mokau
Taupo
Rangitaiki
Lake Taupo
Kaitawa
Gisborne
Waitara
Taumarunui
TASMAN
SEA
New Plymouth
Turangi
Wairoa
Mount Egmont (Mount Taranaki)
2518
Stratford
Mt Ruapehu
Kaimanawa Mountains
Mahia Peninsula
Opunake
Raetihi
2797
Hawke Bay
Napier
Hawera
Waiouru
Hastings
South Taranaki Bight
Patea
Taihape
Havelock North
Cape Kidnappers
Tikokino
Wanganui
Marton
Feilding
Waipawa
Cape Farewell
Palmerston North
Dannevirke
Woodville
Collingwood
Golden Bay
D'Urville Island
Foxton
Cape Turnagain
Takaka
Tasman Mountains
Tasman Bay
Otaki
Levin
Paraparaumu
Masterton
Riwaka
Porirua
Featherston
Karamea
Richmond
Havelock
Te Wharau
Cook Strait
Lower Hutt
Karamea Bight
Nelson
WELLINGTON
Wakefield
Blenheim
Renwick
Westport
Buller
Wairau
Seddon
Cape Campbell
Inland Kaikoura Range
Reefton
Spenser Mts
Punakaiki
Clarence
Runanga
Hanmer Springs
Greymouth
Springs Junction
Kaikoura
Waiau
Hokitika
Parnassus
Arthur's Pass
920
Kowhitirangi
SOUTHERN ALPS
Waipara
Franz Josef Glacier
Oxford
Rangiora
Pegasus Bay
Fox Glacier
Rakaia
Kaiapoi
Mount Cook (Mount Aoraki)
3754
Christchurch
Lake Paringa
Canterbury Plains
Lake Ellesmere
Haast
Lake Tekapo
Ashburton
Banks Peninsula
Jackson Head
Lake Pukaki
Lake Pukaki
Geraldine
Canterbury Bight
Mount Aspiring
3030
Lake Wanaka
Pleasant Point
Temuka
Lake Benmore
Timaru
Lake Hawea
PACIFIC
Milford Sound
Mount Christina
2502
Wanaka
Waimate
SOUTH
Lake Wakatipu
Waitaki
Queenstown
Cromwell
Oamaru
ISLAND
Alexandra
OCEAN
Lake Te Anau
Te Anau
Teviot
Clutha
Port Chalmers
Lumsden
Beaumont
Mosgiel
Otago Peninsula
Brighton
Dunedin
Winton
Gore
Tuatapere
Milton
Mataura
Orepuki
Balclutha
Invercargill
Bluff
Foveaux Strait
Chaslands Mistake
Halfmoon Bay
Ruapuke I.
Stewart Island
Conic Equidistant Projection

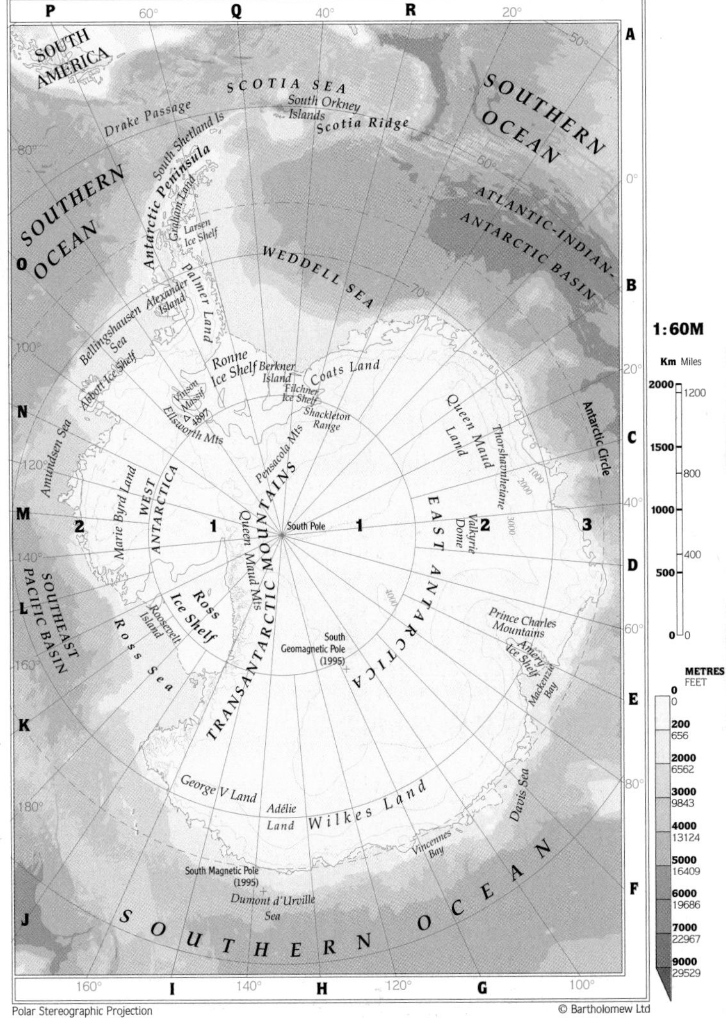

Polar Stereographic Projection

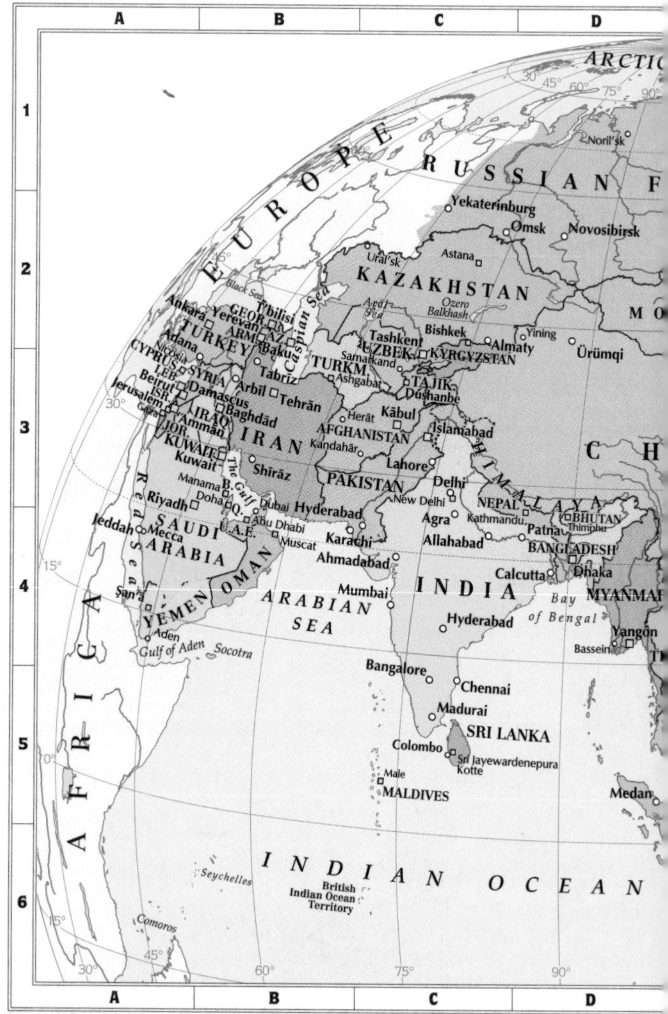

Orthographic Projection

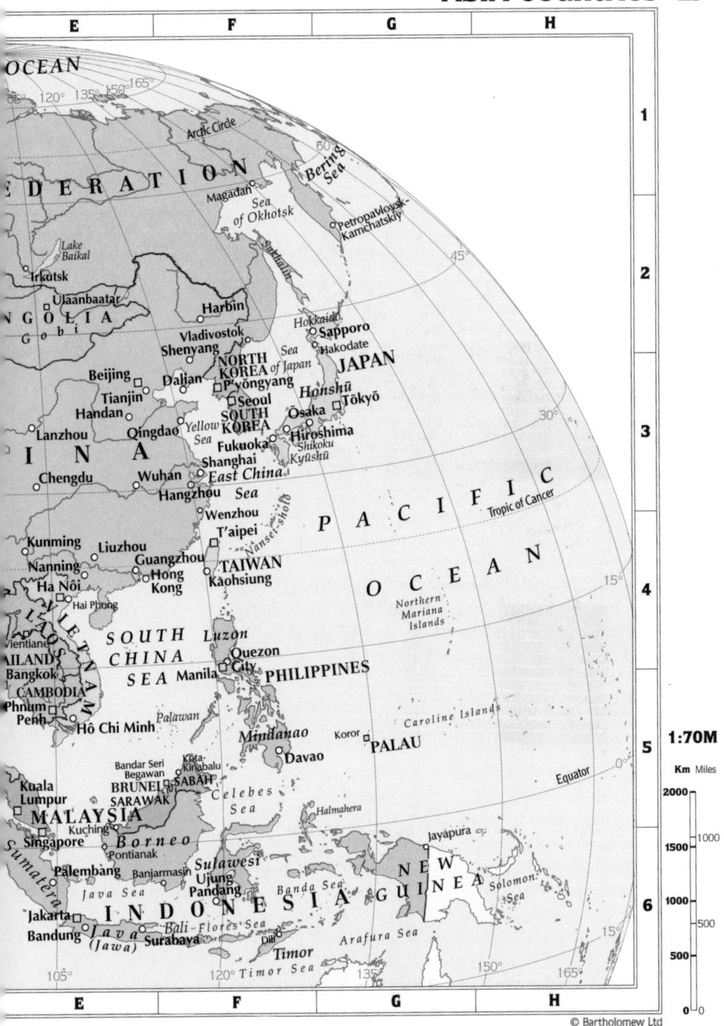

E
F
G
H
1
2
3
4
5
6
OCEAN
Arctic Circle
Bering Sea
Magadan
Sea of Okhotsk
Petropavlovsk-Kamchatskiy
Sakhalin
Lake Baikal
Irkutsk
Ulaanbaatar
Gobi
Harbin
Vladivostok
Shenyang
Hokkaido
Sapporo
Hakodate
Sea of Japan
NORTH KOREA
P'yŏngyang
JAPAN
Beijing
Tianjin
Dalian
Seoul
Honshū
Tōkyō
Handan
SOUTH KOREA
Ōsaka
Lanzhou
Qingdao
Yellow Sea
Hiroshima
Fukuoka
Shikoku
Kyūshū
Shanghai
East China Sea
Chengdu
Wuhan
Hangzhou
Wenzhou
Nansei-shotō
PACIFIC OCEAN
Tropic of Cancer
T'aipei
Kunming
Liuzhou
Guangzhou
TAIWAN
Nanning
Hong Kong
Kaohsiung
Ha Nôi
Hai Phong
Northern Mariana Islands
LAOS
VIETNAM
SOUTH CHINA SEA
Luzon
Vientiane
Quezon City
Bangkok
Manila
PHILIPPINES
CAMBODIA
Phnum Penh
Palawan
Caroline Islands
Hô Chi Minh
Mindanao
Koror
PALAU
Davao
Bandar Seri Begawan
Kota Kinabalu
BRUNEI
SABAH
SARAWAK
Kuala Lumpur
Celebes Sea
Equator
MALAYSIA
Halmahera
Kuching
Singapore
Borneo
Pontianak
Jayapura
Sumatera
Palembang
Banjarmasin
Sulawesi
NEW GUINEA
Ujung Pandang
Solomon Sea
Java Sea
Banda Sea
INDONESIA
Jakarta
Bandung
Java (Jawa)
Surabaya
Bali
Flores Sea
Dili
Timor
Arafura Sea
Timor Sea
1:70M
Km Miles

A
B
105°
120°
1
2
3
15°
THAILAND
LAOS
VIETNAM
CAMBODIA
MALAYSIA
INDONESIA
SOUTH CHINA SEA
Gulf of Tongking
Gulf of Thailand
Strait of Malacca
Java Sea
Bali Sea
Flores Sea
Sulu Sea
Luzon Strait
Macassar Strait
Selat Karimata
Selat Sunda
Selat Lombok
Balabac Strait
Mouths of the Mekong
INDIAN OCEAN
BORNEO
SUMATERA (SUMATRA)
JAVA (JAWA)
SULAWESI (CELEBES)
SARAWAK
SABAH
BRUNEI
Hainan (China)
Luzon
Palawan
Mindoro
Negros
Panay
Sumbawa
Flores
Sumba
Lombok
Bali
Madura
Timor
Sulu Archipelago
Babuyan Islands
Batan Islands
Calamian Group
Cuyo Islands
Natuna Besar (Indonesia)
Kepulauan Anambas (Indonesia)
Kepulauan Natuna (Indonesia)
Kepulauan Tambelan (Indonesia)
Kepulauan Riau
Kepulauan Lingga
Kepulauan Mentawai
Kepulauan Kangean
Christmas I. (Australia)
Simeulue
Nias
Pulau-pulau Batu
Siberut
Sipura
Pagai Utara
Pagai Selatan
Enggano
Bangka
Belitung
Pyinmana
Toungoo
Chiang Rai
Phayao
Chiang Mai
Nan
Phrae
Lampang
Pegu
Uttaradit
Thaton
Tak
Phitsanulok
Moulmein
Khon Kaen
Lop Buri
Ayutthaya
Tavoy
BANGKOK (Krung Thep)
Nakhon Ratchasima
Pattaya
Mergui
Palaw
Chanthaburi
Tenasserim
Prachuap Khiri Khan
Chumphon
Ranong
Surat Thani
Takua Pa
Nakhon Si Thammarat
Krabi
Phuket
Phatthalung
Songkhla
Hat Yai
Yala
Louangphrabang
Xiangkhoang
Thanh Hoa
Vinh
Ha Tinh
Nam Đinh
VIENTIANE (Viangchan)
Udon Thani
Savannakhet
Sarayan
Pakxé
Ubon Ratchathani
Surin
Đông Hôi
Huê
Đa Năng
Quang Ngai
Qui Nhơn
Buôn Mê Thuôt
Nha Trang
Đa Lat
Phan Rang
Phan Thiêt
Biên Hoa
Tây Ninh
Hô Chi Minh (Saigon)
Cân Thơ
Long Xuyên
Rach Gia
Ca Mau
Bac Liêu
Mui Ca Mau
Batdâmbâng
PHNUM PÉNH
Kâmpóng Spoe
Takêv
Sihanoukville
Haikou
Xuwen
Chengmai
Wenchang
Qionghai
Dongfang
Wanning
Aparri
Laoag
Tuguegarao
Vigan
Bontoc
Ilagan
San Fernando
Dagupan
Tarlac
Quezon City
MANILA
Lucena
Batangas
Romblon
Taytay
Iloilo
Puerto Princesa
Brooke's Point
Banggi
Kundat
Gunung Kinabalu
Kota Kinabalu
Sandakan
Lahad Datu
Semporna
Tawau
Zamboanga
Isabela
Basilan
Jolo
BANDAR SERI BEGAWAN
Miri
Bintulu
Mukah
Sibu
Kuching
Igan
Debak
Sri Aman
Lubok Antu
Serian
Liku
Sambas
Singkawang
Mempawah
Pontianak
Ketapang
Sukadana
Kendawangan
Pangkalanbuun
Sampit
Banjarmasin
Amuntai
Kotabaru
Martapura
Balikpapan
Samarinda
Sangkulirang
Tanjungredeb
Tanjungselor
Tarakan
Tg Selatan
Laut
Kota Bharu
Kangar
Alor Setar
Sungei Petani
George Town
Pasir Putih
Kuala Terengganu
Taiping
Kuala Lipis
Ipoh
KUALA LUMPUR
Putrajaya
Melaka
Keluang
Muar
Johor Bahru
Dumai
SINGAPORE
Banda Aceh
Sigli
Bireun
Langsa
Pangkalansusu
Medan
Prapat
Labuhanbilik
Gunungsitoli
Sibolga
Minas
Payakumbuh
Bukittinggi
Padang
Sijunjung
G. Kerinci
3805
Jambi
Pangkalpinang
Sungailiat
Belinyu
Manggar
Toboali
Bangko
Palembang
Sekayu
Tebingtinggi
Bengkulu
Lahat
G. Dempo
3159
Bintuhan
Krui
Tanjungkarang-Telukbetung
JAKARTA
Cirebon
Semarang
Sukabumi
Tk Palabuhanratu
Bandung
Cilacap
Surakarta
Surabaya
Jember
Malang
Denpasar
Mataram
Praya
Dompu
Raba
Reo
Ende
Waikabubak
Waingapu
Ujung Pandang
Bontosunggu
Bulukumba
Benteng
Salayar
Tanahjampea
Kep. Bonerate
Parepare
Watampone
Kolaka
Makale
Wotu
3074
Palu
Poso
Donggala
Teluk Tomini
Semenanjung
Tolitoli
Moutong
Malili
Uekuli
1:30M
Km
Miles
1000
750
500
250
0
600
400
200
Albers Equal Area Conic Projection

PACIFIC OCEAN
PHILIPPINE SEA
PHILIPPINES
Northern Mariana Islands (U.S.A.)
Pagan
Saipan
Tinian
Rota
Guam (U.S.A.)
FEDERATED STATES OF MICRONESIA
Ulithi
Fais
Yap
Ngulu
Sorol
Eauripik
Caroline Islands
PALAU
KOROR
Catanduanes
Sorsogon
Catarman
Samar
Catbalogan
Tacloban
Bacolod
Cebu
Surigao
Bohol Sea
Butuan
Cagayan de Oro
Oroquieta
Mindanao
Pagadian
Davao
Cotabato
Mati
General Santos
Kepulauan Talaud
Kepulauan Sangir
Morotai
Manado
Tobelo
Tondano
Ternate
Halmahera
Gorontalo
Sao-Siu
Molucca Sea
Labuna
Bacan
Obi
Maluku
Waigeo
Selat Dampir
Kwoka
Jazirah Doberai
3000
Sorong
Manokwari
Biak
Numfoor
Ransiki
Selat Yapen
Yapen
Tg d'Urville
Sarmi
Jayapura
Vanimo
Aitape
Salawati
Misool
Fafanlap
Inanwatan
Teluk Berau
Teluk Cenderawasih
Babo
Nabire
Fakfak
Kaimana
Seram
Seram Sea
G. Binaija
3019
Piru
Namlea
Buru
Ambon
Saparua
Kepulauan Banda
Kepulauan Watubela
Adi
Kepulauan Kai
Kai Kecil
Tual
Dobo
Wokam
Benjina
Kepulauan Aru
Kobroör
Trangan
Sia
Banda Sea
Kepulauan Barat Daya
Roma
Damar
Wuliaru
Babar
Larat
Kepulauan Tanimbar
Saumlakki
Selaru
Kepulauan Sermata
Leti
Alor
Kalabahi
Huaki
Kaiwatu
Wetar
Dili
Manatuto
2960
EAST TIMOR
Kefamenanu
Timor
Kupang
Rote
Arafura Sea
AUSTRALIA
Melville Island
Bathurst Island
Van Diemen Gulf
Beagle Gulf
Darwin
Jabiru
Croker I.
Wessel Is
C. Wessel
Nhulunbuy
C. Arnhem
Gulf of Carpentaria
Enarotali
Pegunungan Van Rees
Taritatu
Pk Jaya 5030
Pk Trikora 4730
Pk Mandala 4700
Pegunungan Maoke
Amamapare
NEW GUINEA
Digul
Tg Deyong
P. Dolak
Tg Vals
Merauke
Morehead
Daru
Schouten Islands
Wewak
Sepik
Manam I.
Madang
Long Island
Umboi
Huon Peninsula
PAPUA NEW GUINEA
Central Ra.
4000
4509
Mendi
Mount Hagen
Goroka
Lae
Wau
Kikori
Kerema
Morobe
Bereina
Mt Victoria 4073
Balimo
Gulf of Papua
PORT MORESBY
Equator
Pelleluhu Is
Hermit Is
C. York
Bamaga
Weipa
Coen
135°
15°
0°
C
D
1
2

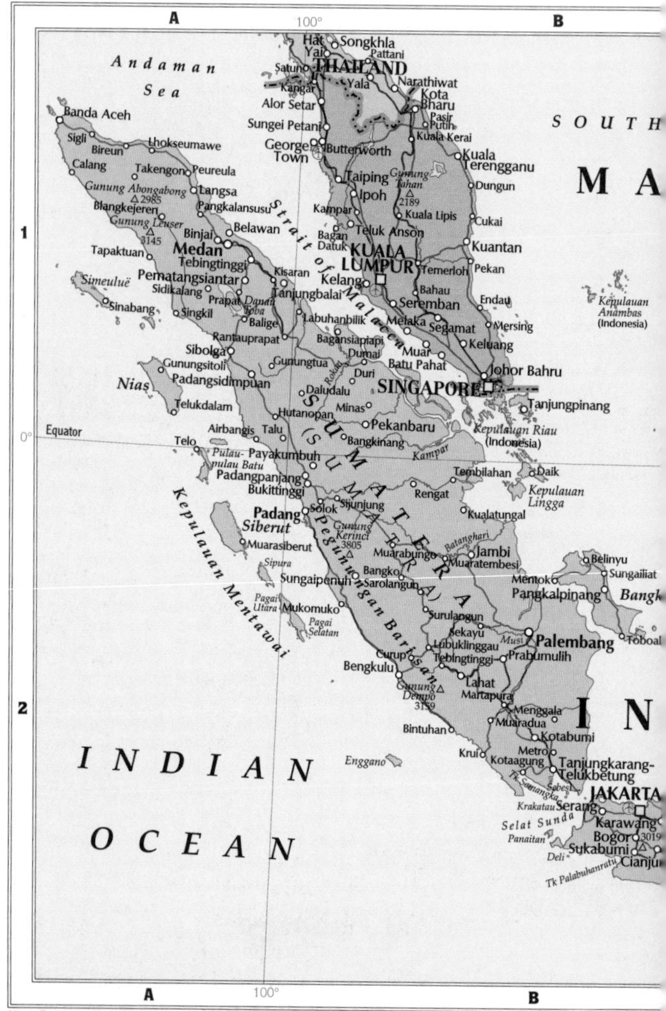

Albers Equal Area Conic Projection

C
110°
CHINA SEA
LAYSIA
Natuna Besar
(Indonesia)
Panarik
Kepulauan
Natuna
(Indonesia)
SULU
SEA
Kundat
Banggi
Kota Belud
Gunung
Kinabalu
4094
Kota
Kinabalu
Ranau
Sandakan
Beaufort
Labuan
Lamag
SABAH
Lahad
Datu
BANDAR SERI
BEGAWAN
BRUNEI
Lawas
Kuamut
Tumindao
Kuala Belait
Lutong
Miri
Seria
Banjaran Crocker
Pensiangan
Semporna
Lumbis
Tawau
CELEBES
SEA
1
Baram
SARAWAK
Long
Akah
Kubuang
Tarakan
Bintulu
Igan
Mukah
Belaga
Kayan
Tanjungselor
Sibu
Sarikei
Rajang
Kapit
Tanjungredeb
Pegunungan Iran
Datadian
Liku
Sematan
Kuching
Kota
Samarahan
Saratok
Debak
2988
Sambaliung
Sepinang
Sambas
Pemangkat
Singkawang
Serian
Sri Aman
Lubok
Antu
Putusibau
Pegunungan Muller
Mahakam
Sangkulirang
Kepulauan
Pulauan
Tembelan
Bengkayang
Semitau
BORNEO
Bontang
Mempawah
Ngabang
Sanggau
Kapuas
Sintang
0°
Pontianak
Longiram
Tenggarong
Nangahpinoh
Balaiberkuak
Pegunungan Schwaner
Muaralaung
Samarinda
Telukbatang
Muarateweh
KALIMANTAN
Pulau-pulau
Karimata
Sukadana
Nangatayap
Rantaupanjang
Balikpapan
Ketapang
Seruyan
Katingan
Barito
Tanahgrogot
Babana
Selat Karimata
Palangkaraya
Makassar Strait
Kendawangan
Belangiran
Sampit
Amuntai
Mamuju
Tanjungpandan
Pangkalanbuun
Kandangan
Bukit
Gandadiwata
3074
Manggar
Tanjung
Sambar
Kualapembuang
Martapura
Kotabaru
Polewali
Belitung
Banjarmasin
Pagatan
Majene
Laut
Tanjung
Puting
Tanjung
Selatan
INDONESIA
2
JAVA SEA
Kepulauan
Laut Kecil
Pulau-pulau
Karimunjawa
Kemujan
Bawean
Tanjung
Indramayu
Tanjung
Bugel
Kepulauan
Kangean
Sabalana
Purwakarta
Cirebon
Tegal
Semarang
Kudus
Pati
Tuban
Bangkalan
Madura
Sumenep
Bandung
Garut
3428
Pekalongan
Surabaya
Selat Madura
Raas
Kepulauan
Tengah
Temanggung
Surakarta
Jombang
Pasuruan
Situbondo
Bali Sea
Ciamis
Madiun
Semeru
3676
Banyuwangi
Cilacap
Kebumen
Yogyakarta
Malang
G. Raung
3332
Singaraja
3142
Selat Lombok
Mataram
Alas
Sumbawa
Dompu
Raba
JAVA
(JAWA)
Lumajang
Jember
Gianyar
Sumbawabesar
Barung
Bali
Denpasar
Praya
Taliwang
Lombok
110°
C
1:15M
Km Miles
450
300
300
150
150
0 0
© Bartholomew Ltd

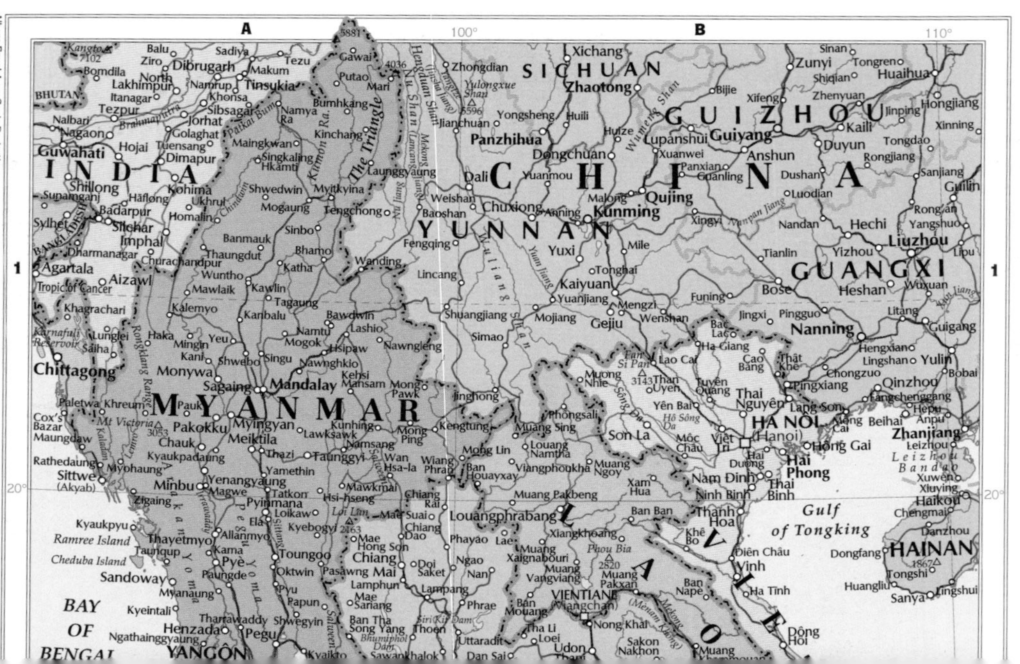

A
B
1
100°
110°
20°
INDIA
CHINA
MYANMAR
SICHUAN
YUNNAN
GUIZHOU
GUANGXI
HAINAN
BHUTAN
BANGLADESH
Gulf of Tongking
BAY OF BENGAL
Kunming
Guiyang
Nanning
HA NỘI
(Hanoi)
Hai Phong
Mandalay
YANGÔN
VIENTIANE
(Viangchan)
Chittagong
Guwahati
Imphal
Aizawl
Agartala
Shillong
Kohima
Dali
Zhaotong
Panzhihua
Liuzhou
Zhanjiang
Haikou
Louangphrabang
Chiang Mai
Sittwe
(Akyab)
Pegu
Myitkyina
Lashio
Kengtung
Hengduan Shan
Nu Shan
Wuliang Shan
Wumeng Shan
Mekong
Salween
Irrawaddy
Chindwin
Yuan Jiang
Rongklang Range
Tropic of Cancer
Pegu Yoma
Arakan Yoma
Leizhou Bandao
Ramree Island
Cheduba Island
The Triangle
Kumon Ra.
Patkai Bum
Brahmaputra
Yulongxue Shan
Mt Victoria
Fan Si Pan
Phou Bia
Kangto

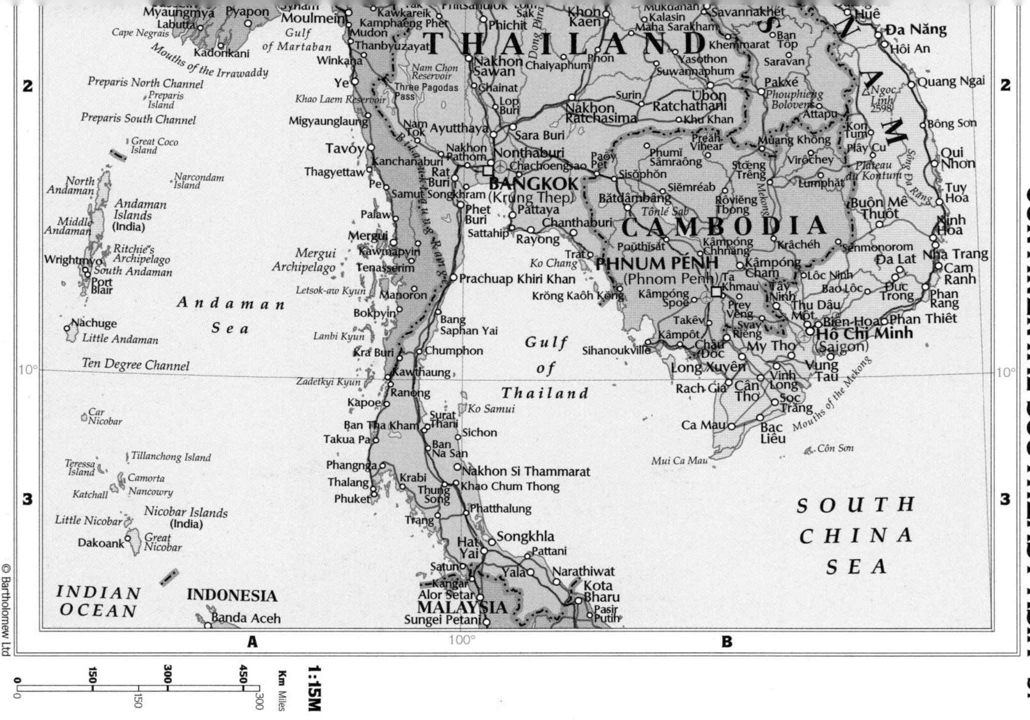
THAILAND
CAMBODIA
NAM
MALAYSIA
INDONESIA
SOUTH CHINA SEA
Gulf of Thailand
Andaman Sea
INDIAN OCEAN
Gulf of Martaban
Mouths of the Irrawaddy
Mouths of the Mekong
Mergui Archipelago
Andaman Islands (India)
Nicobar Islands (India)
Ten Degree Channel
Preparis North Channel
Preparis South Channel
Preparis Island
Great Coco Island
Narcondam Island
North Andaman
Middle Andaman
South Andaman
Ritchie's Archipelago
Little Andaman
Car Nicobar
Tillanchong Island
Teressa Island
Camorta
Nancowry
Katchall
Little Nicobar
Great Nicobar
Myaungmya
Pyapon
Labutta
Cape Negrais
Kadonkani
Moulmein
Kawkareik
Kamphaeng Phet
Mudon
Thanbyuzayat
Winkana
Ye
Khao Laem Reservoir
Three Pagodas Pass
Nam Chon Reservoir
Migyaunglaung
Tavoy
Thagyettaw
Pe
Palaw
Mergui
Kawmapyin
Tenasserim
Letsok-aw Kyun
Manoron
Bokpyin
Lanbi Kyun
Kra Buri
Kawthaung
Zadetkyi Kyun
Ranong
Kapoe
Ban Tha Kham
Takua Pa
Phangnga
Thalang
Phuket
Krabi
Thung Song
Trang
Satun
Kangar
Alor Setar
Sungei Petani
Banda Aceh
Wrightmyo
Port Blair
Nachuge
Dakoank
Bilauktaung Range
Phichit
Nakhon Sawan
Chainat
Lop Buri
Nam Tok
Ayutthaya
Kanchanaburi
Nakhon Pathom
Nonthaburi
Rat Buri
Samut Songkhram
Phet Buri
BANGKOK (Krung Thep)
Chachoengsao
Sara Buri
Pattaya
Sattahip
Rayong
Chanthaburi
Trat
Ko Chang
Prachuap Khiri Khan
Bang Saphan Yai
Chumphon
Ko Samui
Surat Thani
Sichon
Ban Na San
Nakhon Si Thammarat
Khao Chum Thong
Phatthalung
Songkhla
Hat Yai
Pattani
Yala
Narathiwat
Kota Bharu
Pasir Putih
Dong Phra
Khon Kaen
Chaiyaphum
Phon
Nakhon Ratchasima
Surin
Ubon Ratchathani
Khu Khan
Mukdahan
Kalasin
Maha Sarakham
Savannakhet
Khemmarat
Yasothon
Suwannaphum
Ban Tôp
Saravan
Pakxé
Phouphieng Bolovens
Attapu
Mùang Không
Paoy Pet
Sisôphôn
Phumĭ Sâmraông
Preăh Vihéar
Siĕmréab
Bătdâmbâng
Tônlé Sab
Stœ̆ng Trêng
Rôviĕng Tbong
Mekong
Pouthĭsăt
Kâmpóng Chhnăng
Krâchéh
PHNUM PÉNH (Phnom Penh)
Kâmpóng Cham
Ta Khmau
Kâmpóng Spœ
Krŏng Kaôh Kŏng
Sihanoukville
Takêv
Kâmpôt
Prey Vêng
Svay Riĕng
Virôchey
Lumphăt
Sênmonorom
Huê
Đa Năng
Hôi An
Quang Ngai
Ngoc Linh 2598
Bông Son
Kon Tum
Plây Cu
Plateau du Kontum
Sông Đa Răng
Qui Nhon
Tuy Hoa
Buôn Mê Thuôt
Ninh Hoa
Nha Trang
Cam Ranh
Đa Lat
Đuc Trong
Phan Rang
Phan Thiêt
Lôc Ninh
Bao Lôc
Tây Ninh
Thu Dâu Môt
Biên Hoa
Hô Chi Minh (Saigon)
Vung Tau
My Tho
Châu Đôc
Long Xuyên
Rach Gia
Cân Tho
Vinh Long
Soc Trang
Bac Liêu
Ca Mau
Mui Ca Mau
Côn Son
10°
100°
A
B
2
3
1:15M
Km Miles
450
300
150
0

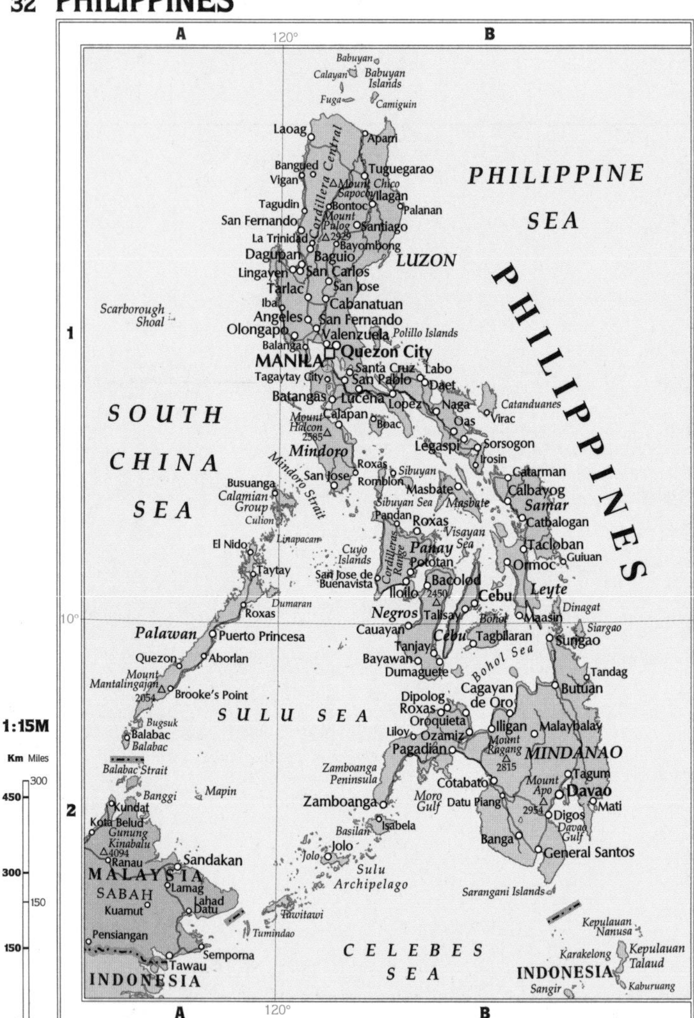

Albers Equal Area Conic Projection

CHINA
JILIN
LIAONING
NORTH KOREA
SOUTH KOREA
JAPAN
SEA OF JAPAN
YELLOW SEA (HUANG HAI)
Korea Bay
Korea Strait (Tsushima-kaikyō)
Cheju-haehyŏp
Tongjosŏn man
Haeju-man
P'YŎNGYANG
SEOUL (Sŏul)
Shenyang
Fushun
Anshan
Benxi
Liaoyang
Dandong
Siping
Liaoyuan
Tieling
Tonghua
Baishan
Yanji
Tumen
Ch'ŏngjin
Najin
Kimch'aek
Hamhŭng
Hŭngnam
Wŏnsan
Sinŭiju
Namp'o
Sariwŏn
Haeju
Kaesŏng
Inch'ŏn
Puch'ŏn
Suwŏn
Sŏngnam
Taejŏn
Taegu
Kwangju
Pusan
Ulsan
P'ohang
Kyŏngju
Masan
Chinju
Chŏnju
Kunsan
Mokp'o
Cheju
Cheju-do (S. Korea)
Ullŭng-do (S. Korea)
Paengnyŏng-do (S.Korea)
Tsushima
Shimonoseki
Kita-Kyūshū
Fukuoka
Kurume
Saga
Sasebo
Paekdu-san 2750
Halla-san 1950
Chiri-san 1915
Kwanmo-bong 2541
Puksubaek-san 2522
Changbai Shan
A B C
1 2 3
125° 130°
40° 35°
1:9M
Km Miles
300 200 100 0
150 100 50 0

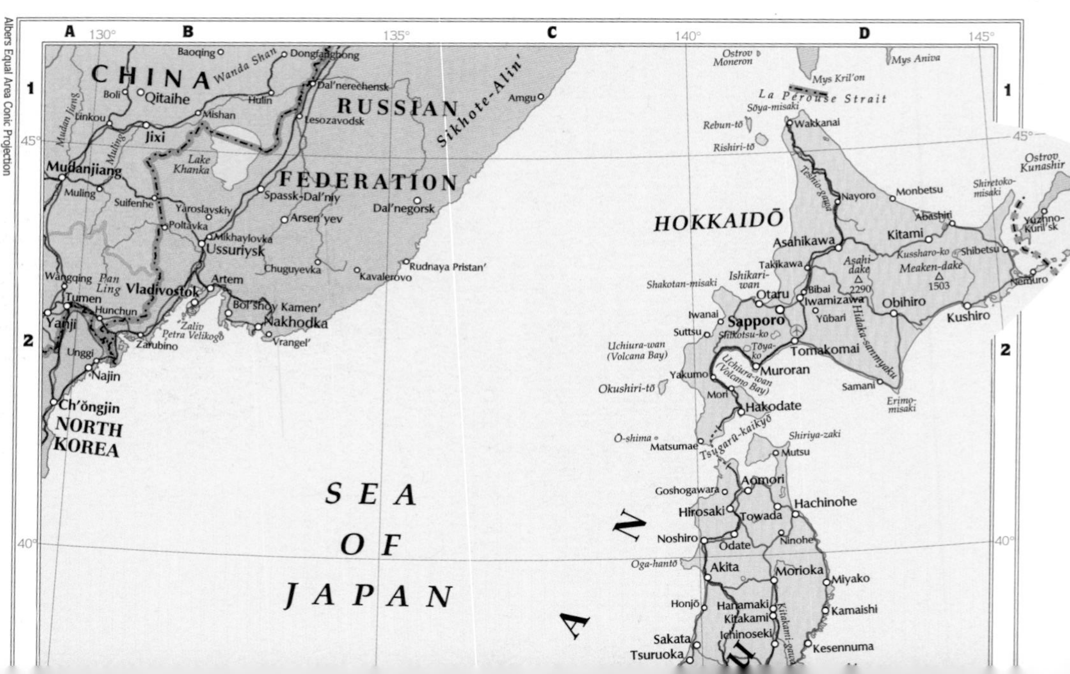

Albers Equal Area Conic Projection

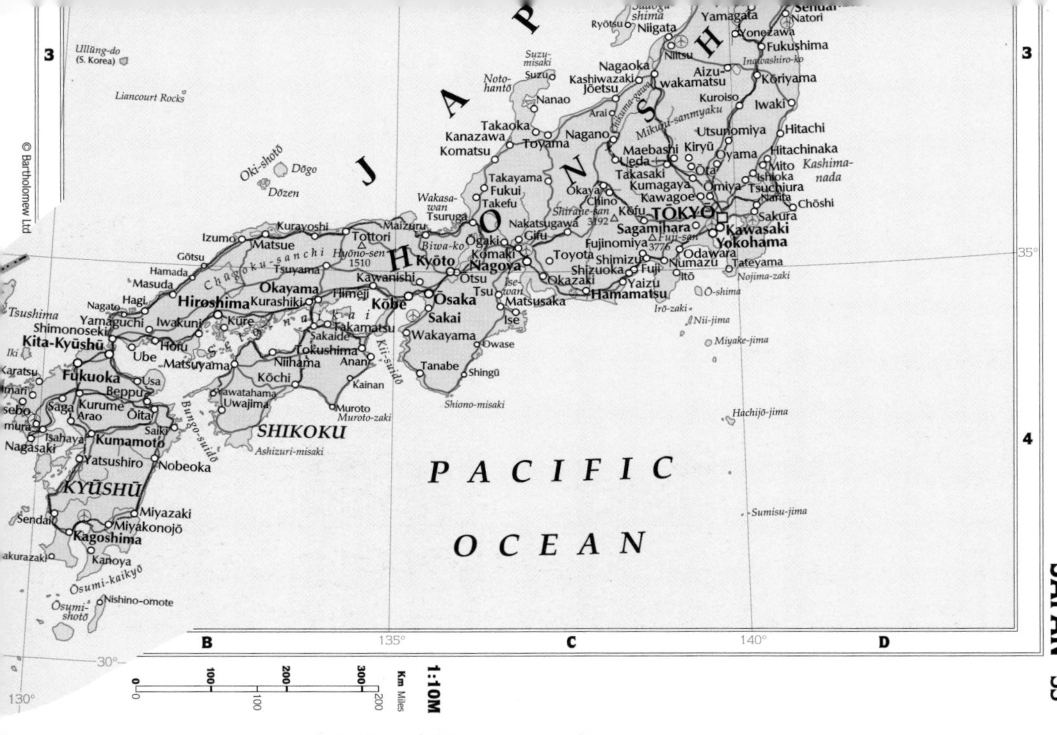
PACIFIC
OCEAN
JAPAN
HONSHU
SHIKOKU
KYŪSHŪ
TŌKYŌ
Yokohama
Kawasaki
Nagoya
Kyōto
Ōsaka
Kōbe
Sakai
Hiroshima
Fukuoka
Kita-Kyūshū
Kagoshima
Niigata
Fukushima
Kōriyama
Iwaki
Hitachi
Hitachinaka
Utsunomiya
Maebashi
Takasaki
Kōfu
Sagamihara
Shizuoka
Hamamatsu
Toyota
Gifu
Kanazawa
Toyama
Fukui
Nagano
Okayama
Kurashiki
Matsue
Tottori
Tokushima
Takamatsu
Matsuyama
Kōchi
Wakayama
Nagasaki
Kumamoto
Ōita
Miyazaki
Miyakonojō
Shimonoseki
Yamaguchi
Biwa-ko
Fuji-san
3776
Shirane-san
3192
Hyōno-sen
1510
Oki-shotō
Dōgo
Dōzen
Noto-hantō
Wakasa-wan
Ise-wan
Kii-suidō
Bungo-suidō
Ōsumi-kaikyō
Ōsumi-shotō
Kashima-nada
Nojima-zaki
Shiono-misaki
Muroto-zaki
Ashizuri-misaki
Ō-shima
Nii-jima
Miyake-jima
Hachijō-jima
Sumisu-jima
Liancourt Rocks
Ullŭng-do
(S. Korea)
Tsushima
Iki
Inawashiro-ko
35°
140°
135°
30°
130°
B
C
D
3
4
1:10M
Km Miles
300
200
100
0
200
100
0
© Bartholomew Ltd

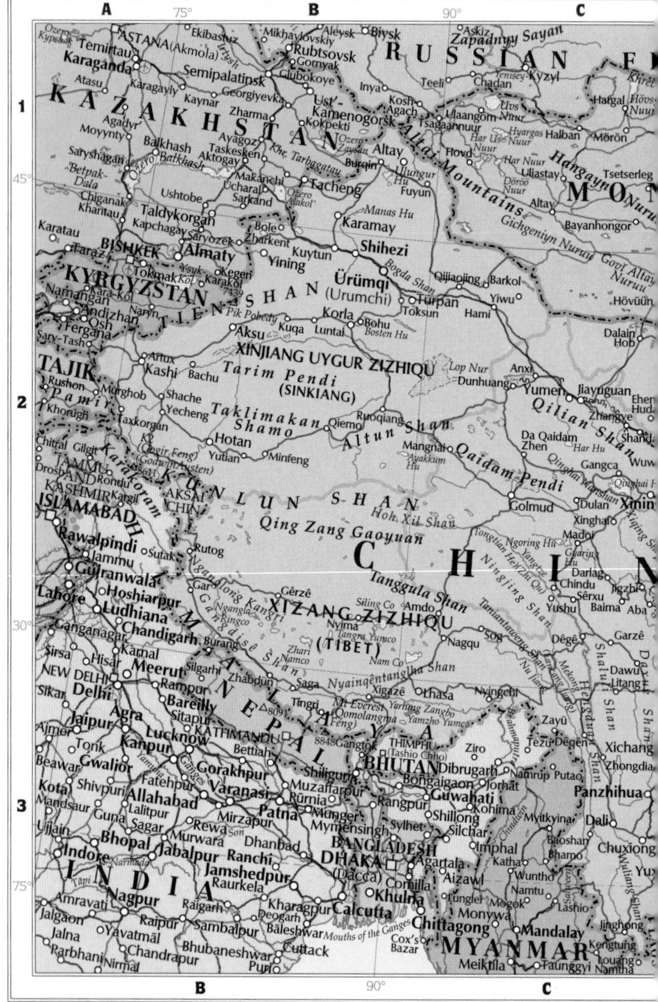

Albers Equal Area Conic Projection

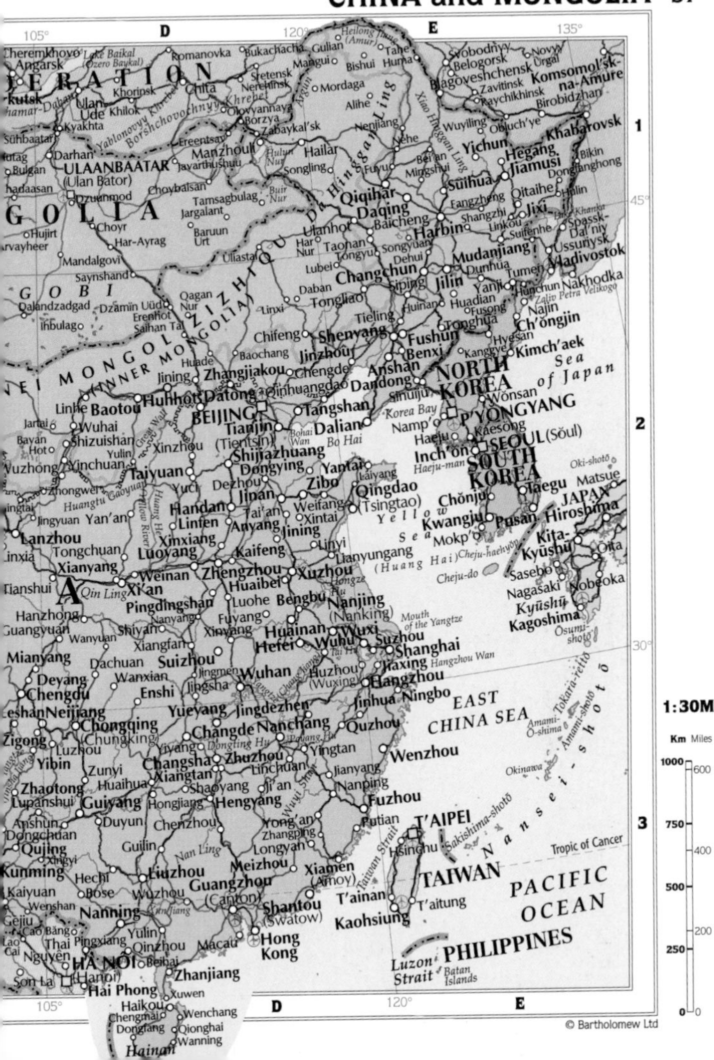
105°
D
120°
E
135°
Cheremkhovo
Angarsk
Lake Baikal
(Ozero Baykal)
Romanovka
Bukachacha
Gulian
Heilong Jiang
(Amur)
Tahe
Huma
Svobodnyy
Novyy Urgal
Belogorsk
Blagoveshchensk
Komsomol'sk-na-Amure
Zavitinsk
Raychikhinsk
Birobidzhan
Mangui
Bishui
Sretensk
Nerchinsk
Mordaga
Khorinsk
Chita
Ulan-Ude
Khilok
Olovyannaya
Borzya
Zabaykal'sk
Alihe
Nenjiang
Yablonovyy Khrebet
Borshchovochnyy Khrebet
Da Hinggan Ling
Xiao Hinggan Ling
Wuyiling
Obluch'ye
Khabarovsk
Kyakhta
Sühbaatar
Ereentsav
Manzhouli
Hulun Nur
Hailar
Nehe
Bei'an
Mingshui
Yichun
Hegang
Jiamusi
Bikin
Dongfanghong
Darhan
Bulgan
ULAANBAATAR
(Ulan Bator)
Javarthushuu
Songling
Fuyu
Suihua
Qitaihe
Hulin
Choybalsan
Tamsagbulag
Buir Nur
Qiqihar
Fangzheng
Dzuunmod
Jargalant
Daqing
Baicheng
Harbin
Shangzhi
Linkou
Jixi
Spassk-Dal'niy
Suifenhe
Choyr
Baruun Urt
Ulanhot
Taonan
Songyuan
Ussuriysk
Vladivostok
Hujirt
Har-Ayrag
Har Nur
Tongyu
Dehui
Mudanjiang
Mandalgovi
Uliastai
Lubei
Changchun
Jilin
Dunhua
Tumen
Nakhodka
Saynshand
Siping
Yanji
Hunchun
GOBI
Daban
Tongliao
Huadian
Zaliv Petra Velikogo
Dalandzadgad
Dzamin Uüd
Qagan Nur
Erenhot
Linxi
Huinan
Fusong
Najin
Tieling
Tonghua
Ch'ŏngjin
Saihan Tal
Chifeng
Shenyang
Fushun
Hyesan
Kimch'aek
Benxi
Kanggye
Baochang
Jinzhou
Sea of Japan
NEI MONGOL ZIZHIQU
(INNER MONGOLIA)
Huade
Anshan
NORTH KOREA
Jining
Zhangjiakou
Chengde
Dandong
Sinuiju
Wŏnsan
Huhhot
Datong
Qinhuangdao
Baotou
BEIJING
Tangshan
Korea Bay
P'YONGYANG
Linhe
Jartai
Wuhai
Tianjin
(Tientsin)
Bohai Wan
Dalian
Namp'o
Kaesŏng
Bayan Hot
Shizuishan
Yulin
Great Wall
Xinzhou
Bo Hai
Haeju
Inch'ŏn
SEOUL (Sŏul)
Shijiazhuang
Haeju-man
SOUTH KOREA
Oki-shotō
Yinchuan
Taiyuan
Dongying
Yantai
Laiyang
Matsue
Zhongwei
Yuci
Dezhou
Zibo
Qingdao
(Tsingtao)
Taegu
JAPAN
Huangtu Gaoyuan
Handan
Jinan
Weifang
Chŏnju
Hiroshima
Jingyuan
Yan'an
Yellow River
Huang He
Linfen
Tai'an
Xintai
Yellow Sea
Kwangju
Pusan
Lanzhou
Anyang
Jining
Mokp'o
Kita-Kyūshū
Tongchuan
Xinxiang
Linyi
Cheju-haehyŏp
Ōita
Luoyang
Kaifeng
Lianyungang
(Huang Hai)
Xianyang
Zhengzhou
Xuzhou
Cheju-do
Sasebo
Tianshui
Weinan
Hongze Hu
Nobeoka
Xi'an
Huaibei
Nagasaki
Qin Ling
Pingdingshan
Luohe
Bengbu
Nanjing
(Nanking)
Kyūshū
Hanzhong
Nanyang
Fuyang
Mouth of the Yangtze
Kagoshima
Guangyuan
Shiyan
Xinyang
Huainan
Wuxi
Ōsumi-shotō
Wanyuan
Hefei
Wuhu
Suzhou
Xiangfan
Tai Hu
Shanghai
30°
Mianyang
Dachuan
Suizhou
Jiaxing
Hangzhou Wan
Jingmen
Wuhan
Huzhou
(Wuxing)
Wanxian
Deyang
Jingsha
Hangzhou
Chengdu
Enshi
Jinhua
Ningbo
EAST CHINA SEA
Leshan
Neijiang
Yueyang
Jingdezhen
Chongqing
(Chungking)
Changde
Nanchang
Quzhou
Amami-Ō-shima
Zigong
Luzhou
Dongting Hu
Poyang Hu
Yingtan
Wenzhou
Nansei-shotō
Yibin
Changsha
Zhuzhou
Zunyi
Xiangtan
Linchuan
Jianyang
Okinawa
Zhaotong
Huaihua
Shaoyang
Ji'an
Nanping
Lupanshui
Guiyang
Hongjiang
Hengyang
Wuyi Shan
Fuzhou
Anshun
Duyun
Chenzhou
Yong'an
Putian
T'AIPEI
Dongchuan
Zhangping
Hsinchu
Sakishima-shotō
Qujing
Guilin
Nan Ling
Longyan
Tropic of Cancer
Xingyi
Taiwan Strait
Kunming
Hechi
Liuzhou
Meizhou
Xiamen
(Amoy)
TAIWAN
Kaiyuan
Bose
Wuzhou
Guangzhou
(Canton)
T'ainan
PACIFIC OCEAN
Wenshan
Nanning
Xi Jiang
Shantou
(Swatow)
T'aitung
Gejiu
Cao Bằng
Yulin
Kaohsiung
Lao Cai
Thai Nguyên
Pingxiang
Qinzhou
Macau
Hong Kong
PHILIPPINES
HÀ NÔI
(Hanoi)
Beihai
Zhanjiang
Luzon Strait
Batan Islands
Sơn La
Hai Phong
Xuwen
Haikou
Wenchang
Chengmai
Dongfang
Qionghai
Wanning
Hainan
1
2
3
1:30M
Km
Miles
1000
750
500
250
0
600
400
200
0

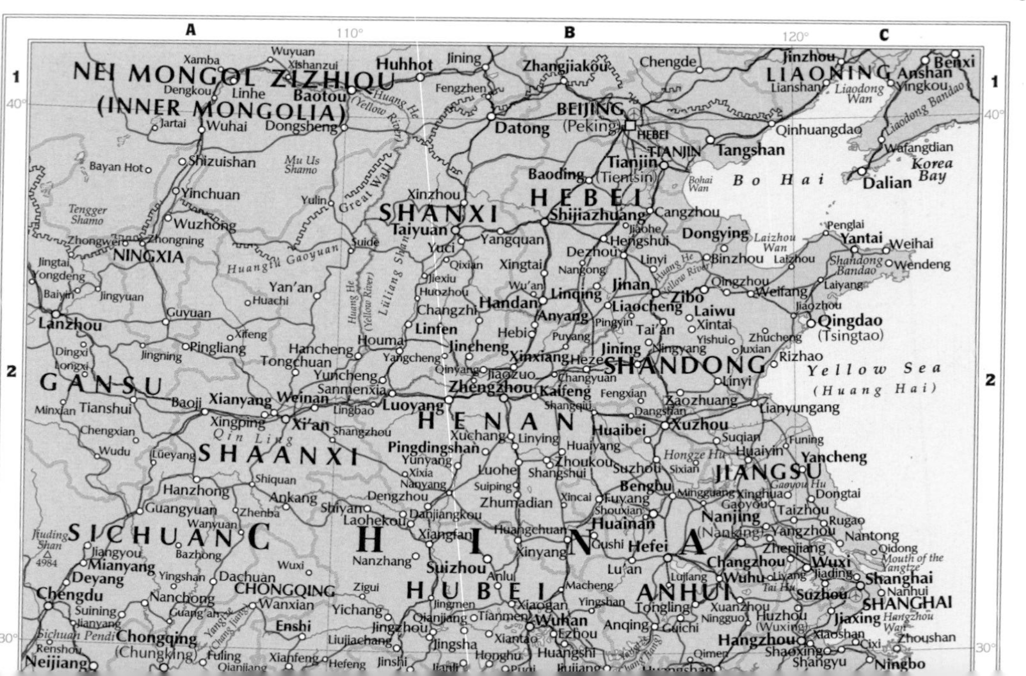
NEI MONGOL ZIZHIQU
(INNER MONGOLIA)
LIAONING
BEIJING
(Peking)
TIANJIN
HEBEI
SHANXI
SHANDONG
HENAN
JIANGSU
ANHUI
SHANGHAI
HUBEI
SHAANXI
GANSU
NINGXIA
SICHUAN
CHONGQING
CHINA
Bo Hai
Yellow Sea
(Huang Hai)
Korea Bay
Liaodong Wan
Bohai Wan
Laizhou Wan
Liaodong Bandao
Shandong Bandao
Huang He
(Yellow River)
Great Wall
Lüliang Shan
Qin Ling
Mu Us Shamo
Tengger Shamo
Sichuan Pendi
Hongze Hu
Gaoyou Hu
Tai Hu
Hangzhou Wan
Mouth of the Yangtze
Jiuding Shan
4984
Huhhot
Baotou
Datong
Zhangjiakou
Chengde
Tangshan
Qinhuangdao
Jinzhou
Anshan
Benxi
Yingkou
Dalian
Wafangdian
Tianjin
(Tientsin)
Baoding
Shijiazhuang
Taiyuan
Yangquan
Handan
Xingtai
Jinan
Zibo
Qingdao
(Tsingtao)
Yantai
Weihai
Penglai
Weifang
Rizhao
Lianyungang
Xuzhou
Zhengzhou
Luoyang
Kaifeng
Xinxiang
Anyang
Pingdingshan
Nanjing
(Nanking)
Shanghai
Suzhou
Wuxi
Changzhou
Hefei
Huainan
Bengbu
Wuhan
Hangzhou
Ningbo
Xi'an
Xianyang
Baoji
Lanzhou
Yinchuan
Wuhai
Tianshui
Chengdu
Chongqing
(Chungking)
Nanchong
Mianyang
Deyang
Neijiang
Yichang
Wanxian
Xiangfan
Nanyang
Hanzhong
Ankang
Yan'an
Yulin
Dongsheng
Linfen
Changzhi
Jincheng
Yuncheng
Sanmenxia
Weinan
Tongchuan
Guyuan
Pingliang
Zhongwei
Baiyin
Jingtai
Yongdeng
Dingxi
Minxian
40°
30°
110°
120°
A
B
C
1
2

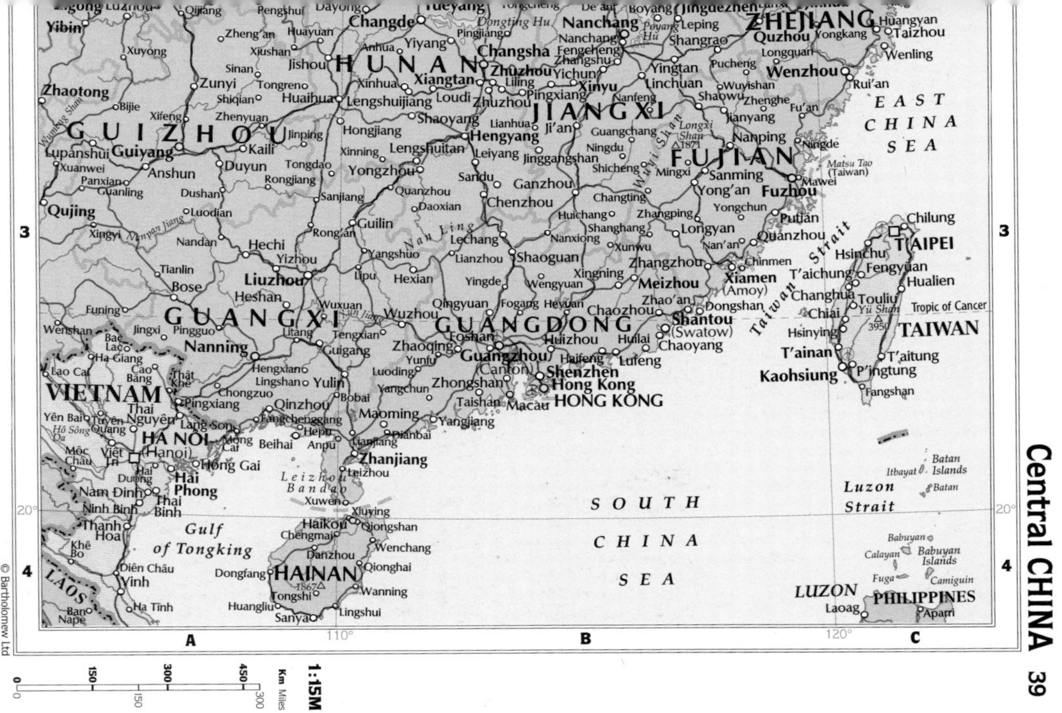

1:15M

Km Miles

0 150 300 450

0 150 300

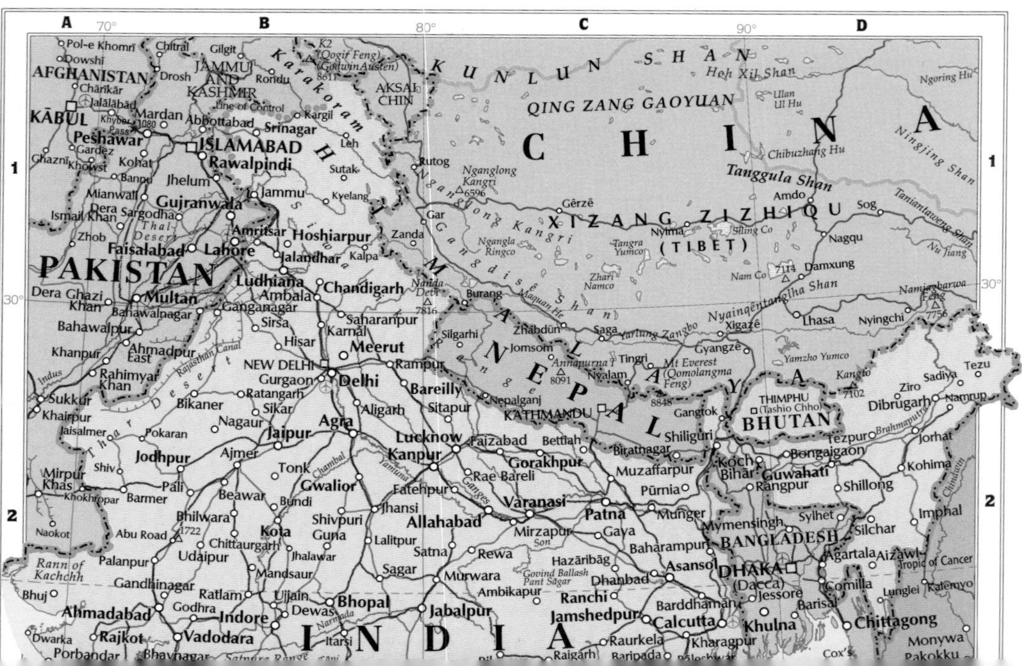

Albers Equal Area Conic Projection

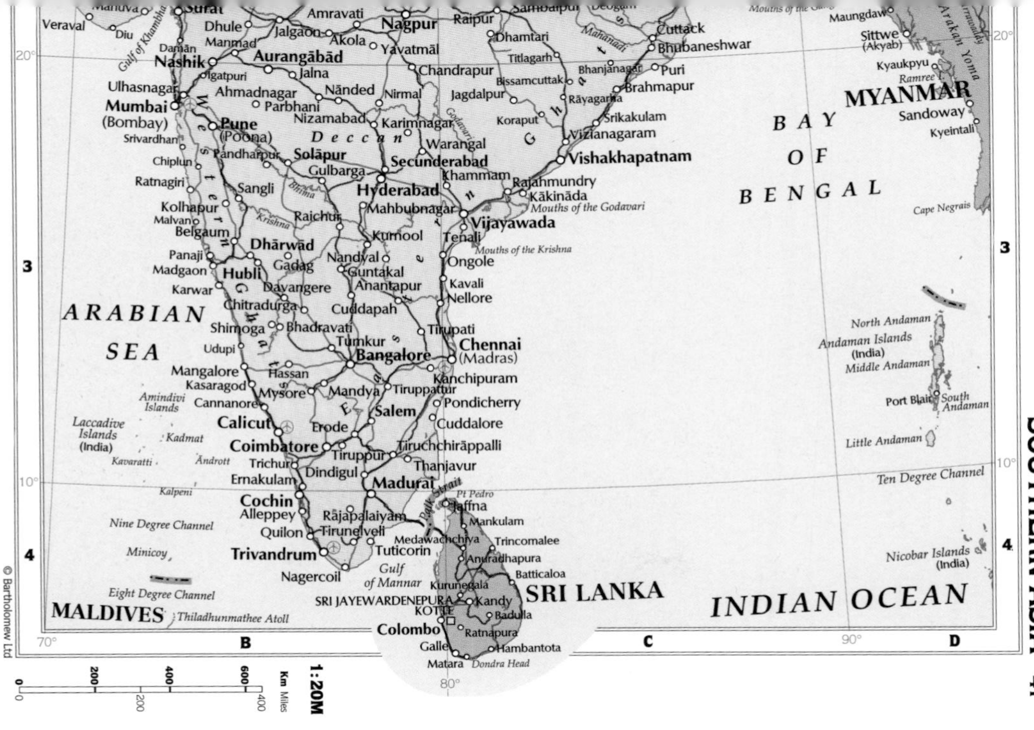

MYANMAR
BAY OF BENGAL
ARABIAN SEA
INDIAN OCEAN
SRI LANKA
MALDIVES
Arakan Yoma
Maungdaw
Sittwe (Akyab)
Kyaukpyu
Ramree I.
Sandoway
Kyeintali
Cape Negrais
North Andaman
Andaman Islands (India)
Middle Andaman
South Andaman
Port Blair
Little Andaman
Ten Degree Channel
Nicobar Islands (India)
Mouths of the Godavari
Mouths of the Krishna
Cuttack
Bhubaneshwar
Puri
Brahmapur
Srikakulam
Vizianagaram
Vishakhapatnam
Rajahmundry
Kākināda
Vijayawada
Tenali
Ongole
Kavali
Nellore
Tirupati
Chennai (Madras)
Kanchipuram
Pondicherry
Cuddalore
Tiruchchirāppalli
Thanjavur
Madurai
Tuticorin
Tirunelveli
Nagercoil
Trivandrum
Quilon
Alleppey
Cochin
Ernakulam
Trichur
Coimbatore
Tiruppur
Dindigul
Erode
Salem
Calicut
Cannanore
Kasaragod
Mangalore
Udupi
Shimoga
Bhadravati
Hassan
Mysore
Mandya
Tiruppattur
Bangalore
Tumkur
Chitradurga
Davangere
Hubli
Dhārwād
Gadag
Belgaum
Karwar
Panaji
Madgaon
Kolhapur
Malvan
Ratnagiri
Chiplun
Sangli
Pandharpur
Solāpur
Gulbarga
Raichur
Kurnool
Nandyal
Guntakal
Anantapur
Cuddapah
Hyderabad
Secunderabad
Mahbubnagar
Khammam
Warangal
Karimnagar
Nizamabad
Nirmal
Nānded
Parbhani
Jalna
Aurangābād
Ahmadnagar
Igatpuri
Pune (Poona)
Mumbai (Bombay)
Ulhasnagar
Srivardhan
Nashik
Damān
Diu
Veraval
Dhule
Manmad
Jalgaon
Akola
Amravati
Yavatmāl
Nagpur
Chandrapur
Raipur
Dhamtari
Jagdalpur
Koraput
Titlagarh
Bissamcuttak
Rāyagarha
Bhanjanagar
Godavari
Krishna
Bhima
Mahanadi
Deccan
Western Ghats
Eastern Ghats
Gulf of Khambhat
Laccadive Islands (India)
Amindivi Islands
Kadmat
Āndrott
Kavaratti
Kalpeni
Minicoy
Nine Degree Channel
Eight Degree Channel
Thiladhunmathee Atoll
Jaffna
Pt Pedro
Palk Strait
Mankulam
Trincomalee
Anuradhapura
Medawachchiya
Batticaloa
Kurunegala
Kandy
Badulla
SRI JAYEWARDENEPURA KOTTE
Colombo
Ratnapura
Hambantota
Galle
Matara
Dondra Head
Gulf of Mannar
Rājapālaiyam
1:20M
Km Miles

Albers Equal Area Conic Projection

XINJIANG UYGUR ZIZHIQU
(SINKIANG)
Hoh Xil Shan
CHINA
QING ZANG GAOYUAN
(PLATEAU OF TIBET)
XIZANG ZIZHIQU
(TIBET)
Tanggula Shan
AKSAI CHIN
Under Chinese administration, claimed by India
Nganglong Kangri
Gangdisê Shan
Nyainqêntanglha Shan
NEPAL
KATHMANDU
BHUTAN
THIMPHU
(Tashio Chho)
BANGLADESH
DHAKA
(Dacca)
Mt Everest
(Qomolangma Feng)
8848
Lhasa
Delhi
Lucknow
Kanpur
Varanasi
Allahabad
Patna
Calcutta
Jabalpur
Nagpur
Ranchi
Jamshedpur
Mouths of the Ganges
BAY OF BENGAL
1:15M
Km Miles

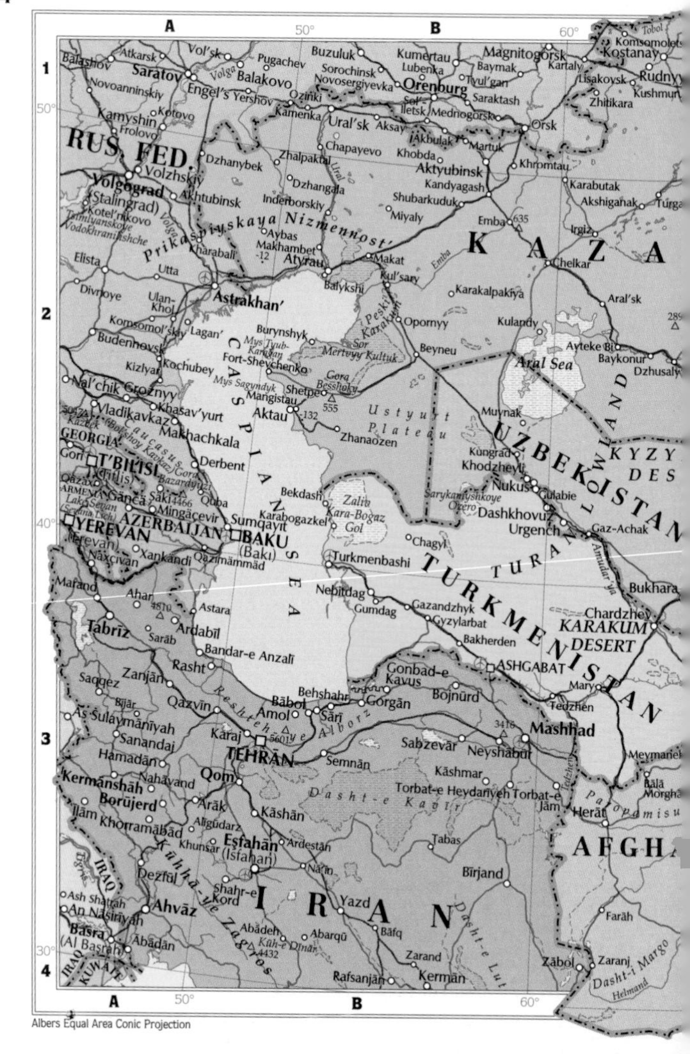
RUS. FED.
K A Z A
UZBEKISTAN
TURKMENISTAN
AZERBAIJAN
GEORGIA
ARMENIA
IRAN
IRAQ
KUWAIT
AFGHA
CASPIAN SEA
Aral Sea
KARAKUM DESERT
TURAN LOWLAND
Ustyurt Plateau
Prikaspiyskaya Nizmennost'
Dasht-e Kavir
Dasht-e Lut
Kühhā-ye Zagros
Reshteh-ye Alborz
Caucasus (Bol'shoy Kavkaz)
Saratov
Volgograd (Stalingrad)
Astrakhan'
Atyrau
Aktyubinsk
Orenburg
Ural'sk
Aktau
Makhachkala
T'BILISI (Tiflis)
YEREVAN (Jerevan)
BAKU (Bakı)
Tabrīz
Rasht
TEHRĀN
Qom
Esfahān (Isfahan)
Mashhad
ASHGABAT
Turkmenbashi
Nukus
Urgench
Chardzhev
Herāt
Ahvāz
Basra (Al Basrah)
Yazd
Kermān
Albers Equal Area Conic Projection

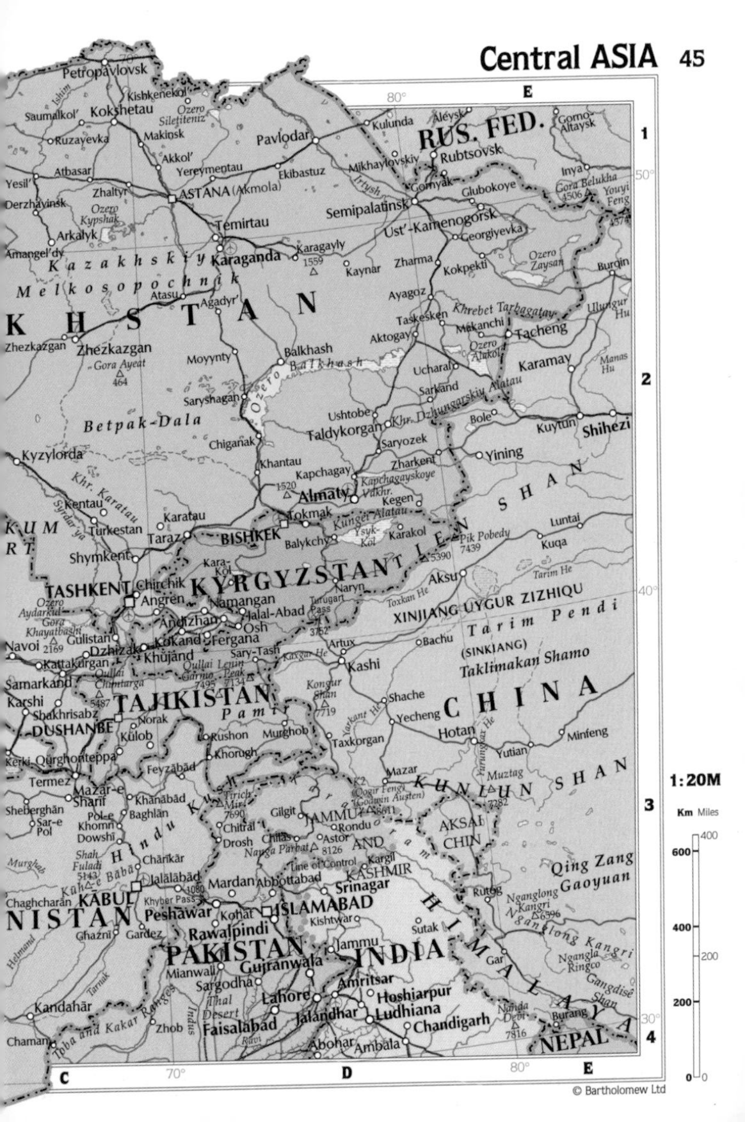
RUS. FED.
KAZAKHSTAN
KYRGYZSTAN
TAJIKISTAN
CHINA
PAKISTAN
INDIA
NEPAL
ASTANA (Akmola)
BISHKEK
TASHKENT
DUSHANBE
KABUL
ISLAMABAD
Petropavlovsk
Kokshetau
Pavlodar
Semipalatinsk
Ust'-Kamenogorsk
Karaganda
Temirtau
Zhezkazgan
Balkhash
Taldykorgan
Almaty
Tokmak
Taraz
Shymkent
Kyzylorda
Turkestan
Namangan
Andizhan
Osh
Jalal-Abad
Fergana
Kokand
Khujand
Samarkand
Karshi
Kashi
Aksu
Yining
Shihezi
Karamay
Tacheng
Hotan
Yecheng
Shache
Kuqa
Peshawar
Rawalpindi
Gujranwala
Lahore
Faisalabad
Srinagar
Jammu
Amritsar
Jalandhar
Ludhiana
Chandigarh
Kandahār
Tian Shan
Pamir
Hindu Kush
Karakoram
Kunlun Shan
Himalaya
Betpak-Dala
Kazakhskiy Melkosopochnik
Ozero Balkhash
Ysyk-Köl
Ozero Zaysan
Tarim Pendi
Taklimakan Shamo
XINJIANG UYGUR ZIZHIQU
(SINKIANG)
Qing Zang Gaoyuan
JAMMU AND KASHMIR
AKSAI CHIN
Line of Control
K2 (Qogir Feng) (Godwin Austen) 8611
Pik Pobedy 7439
Nanga Parbat 8126
Tirich Mir 7690
Thal Desert
Toba and Kakar Ranges
1:20M
Km Miles
© Bartholomew Ltd

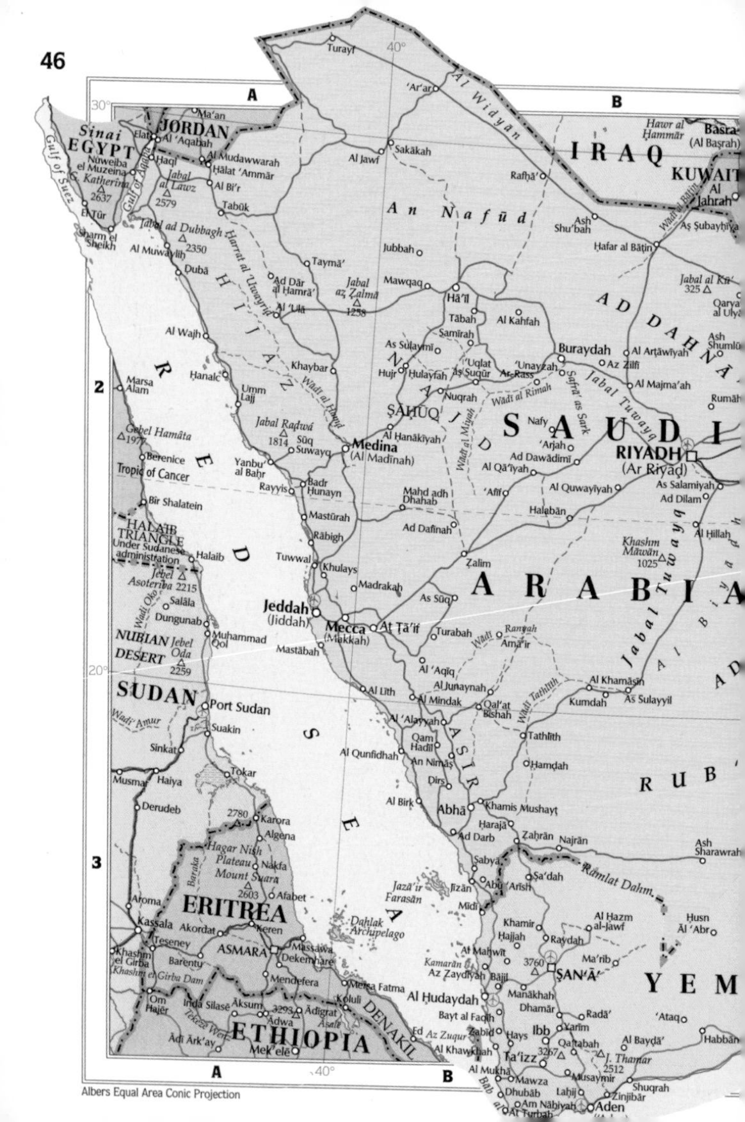
SAUDI ARABIA
IRAQ
KUWAIT
JORDAN
EGYPT
Sinai
SUDAN
ERITREA
ETHIOPIA
YEMEN
RED SEA
HIJAZ
NAJD
ASIR
An Nafūd
AD DAHNĀ'
RUB'
RIYADH
(Ar Riyāḍ)
Medina
(Al Madīnah)
Mecca
(Makkah)
Jeddah
(Jiddah)
ŞAN'Ā'
ASMARA
Port Sudan
Aden
Al Ḩudaydah
Ta'izz
Ibb
At Ṭā'if
Tabūk
Ḩā'il
Buraydah
'Unayzah
Ḩafar al Bāţin
Sakākah
Al Jawf
Turayf
'Ar'ar
Rafḩā'
Tropic of Cancer
HALAIB TRIANGLE
Under Sudanese administration
NUBIAN DESERT
Dahlak Archipelago
Jazā'ir Farasān
Gulf of Suez
Gulf of Aqaba
Wādī al Batin
Al Widyān
Jabal Ṭuwayq
Ramlat Dahm
Abhā
Khamis Mushayt
Najrān
Jīzān
Kassala
Keren
Massawa
Mek'elē
Tropic of Cancer
Albers Equal Area Conic Projection

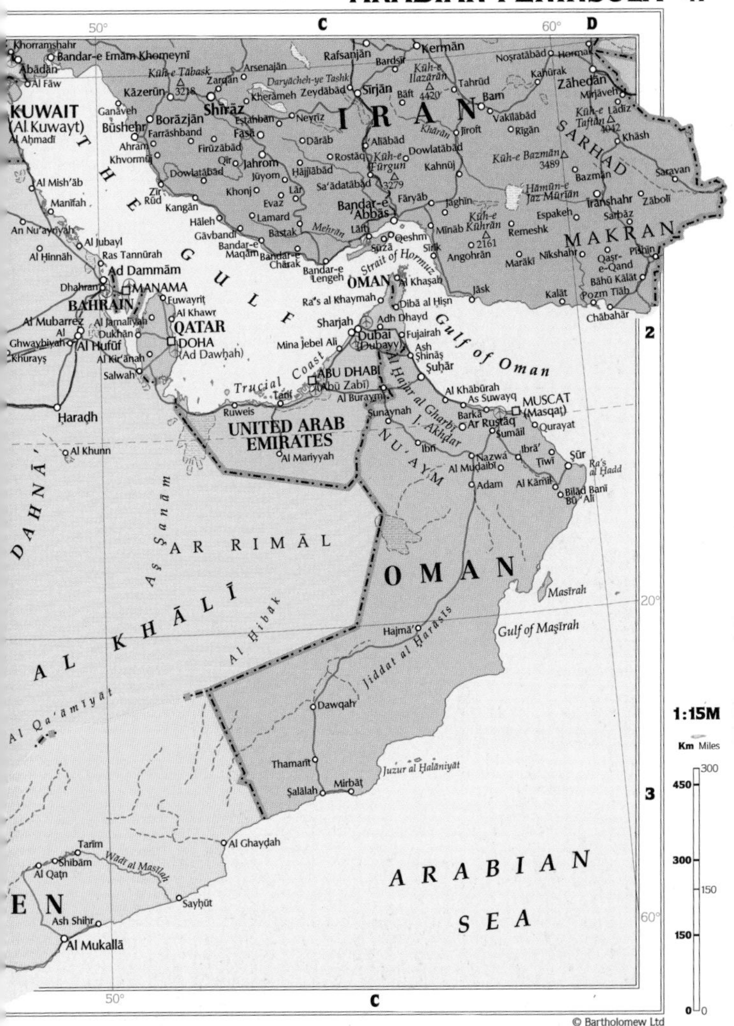
IRAN
KUWAIT (Al Kuwayt)
BAHRAIN
MANAMA
QATAR
DOHA (Ad Dawhah)
UNITED ARAB EMIRATES
ABU DHABI (Abū Zabī)
MUSCAT (Masqaţ)
OMAN
THE GULF
Gulf of Oman
Strait of Hormuz
Gulf of Maşīrah
ARABIAN SEA
AR RIMĀL
AL KHĀLĪ
DAHNĀ'
SARHAD
MAKRAN
Trucial Coast
Al Hajar al Gharbī
J. Akhdar
Shīrāz
Kermān
Bandar-e Abbās
Dubai (Dubayy)
Sharjah
Ad Dammām
Al Hufūf
Şalālah
Al Mukallā
Maşīrah
Zāhedān
Bam
Sīrjān
Rafsanjān
Ābādān
Khorramshahr
Bandar-e Emām Khomeynī
Būshehr
Ra's al Khaymah
Fujairah
Şuḩār
Nazwá
Şūr
Ra's al Ḩadd
Dawqah
Thamarīt
Mirbāţ
Al Ghayḑah
Sayḩūt
Ash Shiḩr
Tarīm
Shibām
Al Qaţn
Haradh
Chābahār
Jāsk
1:15M
Km Miles
© Bartholomew Ltd

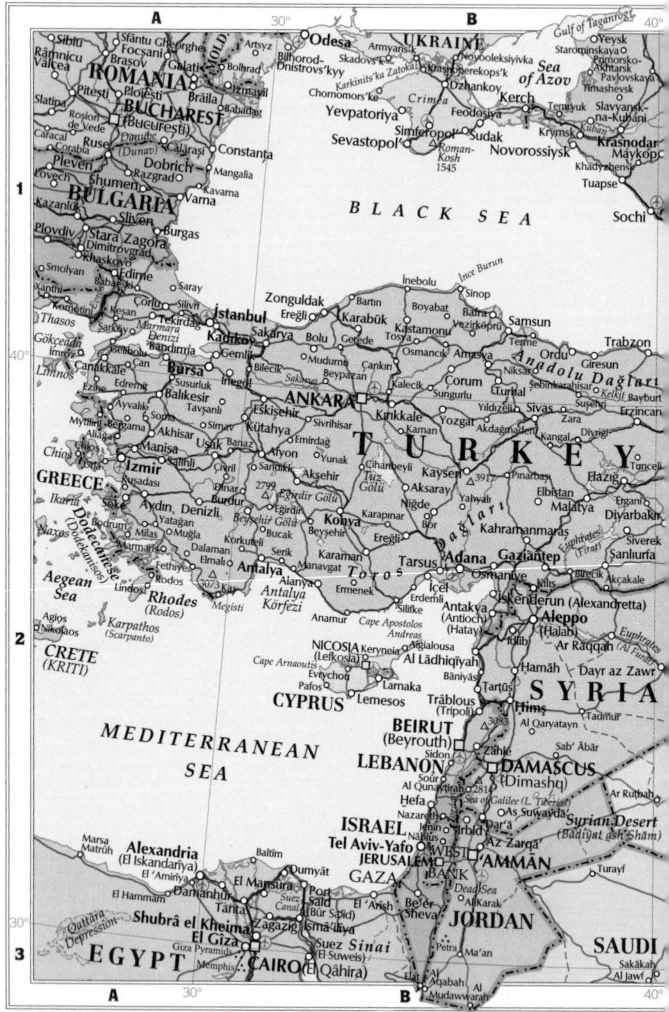

Albers Equal Area Conic Projection

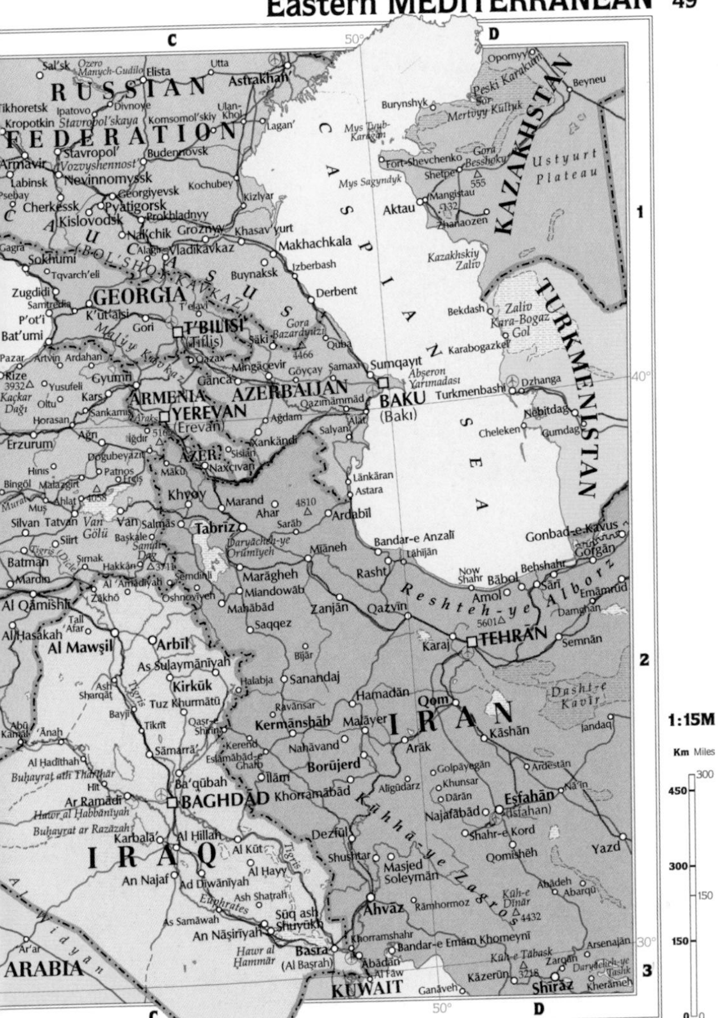
RUSSIAN FEDERATION
GEORGIA
ARMENIA
AZERBAIJAN
AZER.
KAZAKHSTAN
TURKMENISTAN
IRAN
IRAQ
KUWAIT
ARABIA
CASPIAN SEA
T'BILISI
(Tiflis)
YEREVAN
(Erevan)
BAKU
(Bakı)
TEHRĀN
BAGHDĀD
Astrakhan'
Elista
Stavropol'
Groznyy
Vladikavkaz
Makhachkala
Derbent
Sumqayıt
Gəncə
Tabrīz
Rasht
Qazvīn
Hamadān
Qom
Kāshān
Eşfahān
(Isfahan)
Yazd
Shīrāz
Ahvāz
Basra
(Al Başrah)
Kirkūk
Arbīl
Al Mawşil
Karbalā'
An Najaf
Turkmenbashi
Aktau
Beyneu
Ustyurt Plateau
Reshteh-ye Alborz
Kūhhā-ye Zāgros
Tigris
Euphrates
1
2
3
C
D
50°
40°
30°
1:15M
Km Miles
450
300
150
0
300
150
0

ARCTIC
BARENTS SEA
Kara Sea
(Karskoye More)
Norwegian Sea
Greenland Sea
Svalbard
(Norway)
RUSSIAN
KAZAKHSTAN
NORWAY
SWEDEN
FINLAND
Caspian Sea
Black Sea
Ural Mountains
Altai Mountains
Gulf of Bothnia
White Sea
MOSCOW
ST. Petersburg
Novosibirsk
Omsk
Yekaterinburg
Chelyabinsk
Murmansk
Archangel
Vorkuta
Noril'sk
Surgut
Tomsk
Barnaul
ASTANA
Karaganda
Volgograd
Samara
Kazan'
Perm'
Ufa
Orenburg
Astrakhan'
Conic Equidistant Projection

OCEAN
Laptev Sea
(More Laptevykh)
Chukchi Sea
Sea of Okhotsk
RUSSIAN FEDERATION
FEDERATION
CHINA
MONGOLIA
N. KOREA
S. KOREA
U.S.A.
Severnaya Zemlya
Arkhipelag Nordenshel'da
Poluostrov Taymyr
Novosibirskiye Ostrova
Ostrov Vrangelya
Sakhalin
Khrebet Cherskogo
Verkhoyanskiy Khr.
Stanovoy Khrebet
Sredinnyy Khrebet
Koryakskiy Khrebet
Khrebet Kolymskiy
Sredne-Sibirskoye Ploskogor'ye
Yakutsk
Magadan
Petropavlovsk-Kamchatskiy
Vladivostok
Khabarovsk
Irkutsk
Krasnoyarsk
Chita
Blagoveshchensk
Harbin
Qiqihar
Changchun
Shenyang
ULAANBAATAR
(Ulan Bator)
P'YONGYANG
SEOUL
(Soul)
Lake Baikal
GOBI
1:42M
Km
Miles
1200
900
600
300
0
600
300
0
© Bartholomew Ltd

Orthographic Projection

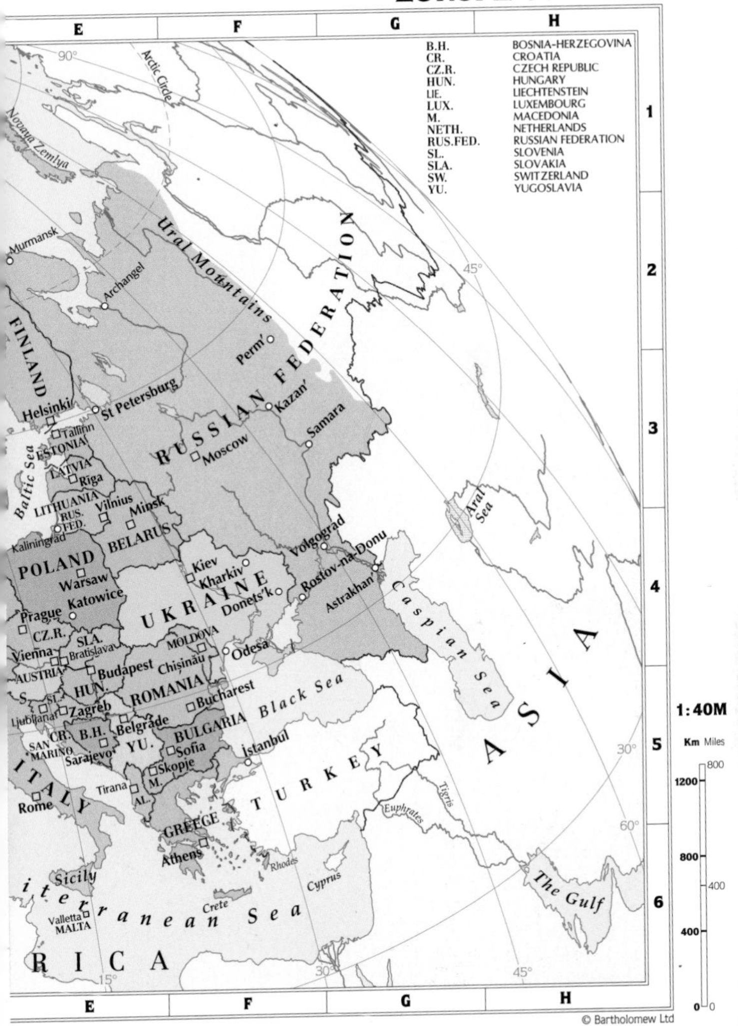

B.H. BOSNIA-HERZEGOVINA
CR. CROATIA
CZ.R. CZECH REPUBLIC
HUN. HUNGARY
LIE. LIECHTENSTEIN
LUX. LUXEMBOURG
M. MACEDONIA
NETH. NETHERLANDS
RUS.FED. RUSSIAN FEDERATION
SL. SLOVENIA
SLA. SLOVAKIA
SW. SWITZERLAND
YU. YUGOSLAVIA
E
F
G
H
1
2
3
4
5
6
Arctic Circle
Novaya Zemlya
Murmansk
Archangel
Ural Mountains
RUSSIAN FEDERATION
FINLAND
Perm'
Helsinki
St Petersburg
Kazan'
Tallinn
ESTONIA
Samara
Moscow
Baltic Sea
LATVIA
Riga
LITHUANIA
Vilnius
Minsk
RUS. FED.
Kaliningrad
BELARUS
Aral Sea
Volgograd
Rostov-na-Donu
POLAND
Warsaw
Kiev
Kharkiv
UKRAINE
Donets'k
Astrakhan'
Caspian Sea
Katowice
Prague
CZ.R.
SLA.
Bratislava
Vienna
AUSTRIA
MOLDOVA
Chişinău
Odesa
Budapest
HUN.
ROMANIA
Bucharest
Black Sea
ASIA
SL.
Ljubljana
Zagreb
CR.
B.H.
Belgrade
BULGARIA
SAN MARINO
Sarajevo
YU.
Sofia
Istanbul
TURKEY
Skopje
M.
Tirana
AL.
ITALY
Rome
Tigris
Euphrates
GREECE
Athens
Rhodes
Sicily
Mediterranean Sea
Cyprus
The Gulf
Crete
Valletta
MALTA
AFRICA
90°
45°
30°
60°
15°
1:40M
Km Miles
1200
800
400
0
800
400
0

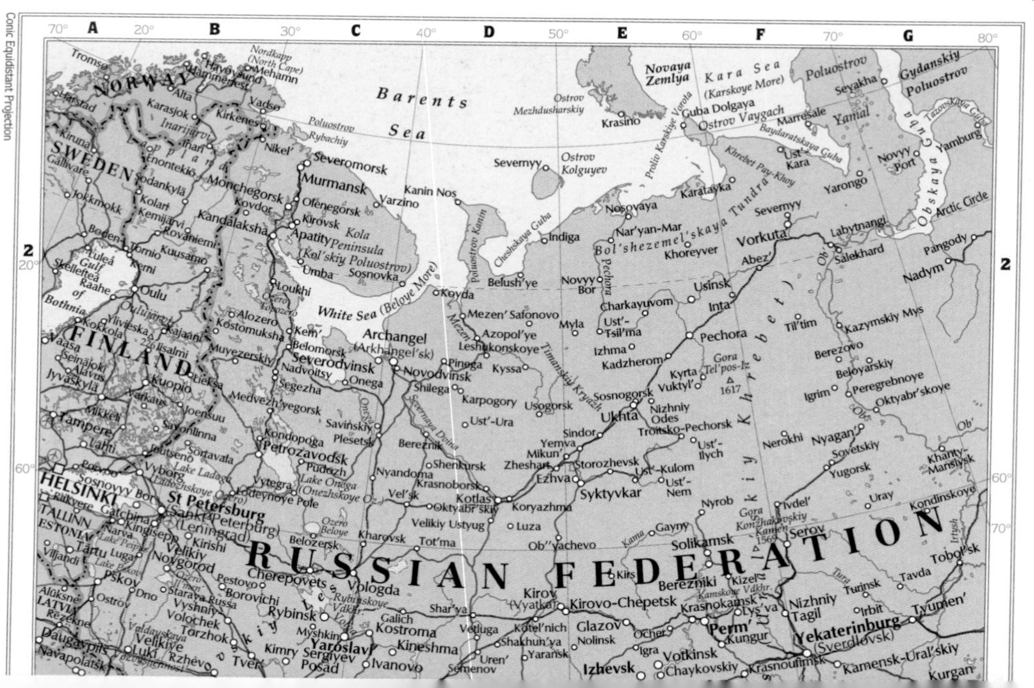
Barents Sea
Kara Sea
(Karskoye More)
White Sea (Beloye More)
Novaya Zemlya
Ostrov Kolguyev
Ostrov Vaygach
Kola Peninsula
(Kol'skiy Poluostrov)
Gydanskiy Poluostrov
Obskaya Guba
Arctic Circle
Gulf of Bothnia
Lake Ladoga
Lake Onega
NORWAY
SWEDEN
FINLAND
ESTONIA
LATVIA
RUSSIAN FEDERATION
HELSINKI
TALLINN
St Petersburg
(Sankt-Peterburg)
(Leningrad)
Murmansk
Severomorsk
Archangel
(Arkhangel'sk)
Severodvinsk
Petrozavodsk
Vorkuta
Salekhard
Nadym
Pechora
Ukhta
Syktyvkar
Kotlas
Vologda
Cherepovets
Yaroslavl'
Kostroma
Kirov
(Vyatka)
Kirovo-Chepetsk
Perm'
Berezniki
Solikamsk
Yekaterinburg
(Sverdlovsk)
Izhevsk
Tyumen'
Tobol'sk
Oulu
Tampere
Tver'
Ivanovo
Conic Equidistant Projection

MOSCOW
(Moskva)
Nizhniy Novgorod
Kazan'
Naberezhnyye Chelny
Chelyabinsk
Ufa
Orenburg
Samara
Tol'yatti
Ul'yanovsk
Penza
Saratov
Volgograd
(Stalingrad)
Astrakhan'
Voronezh
Ryazan'
Tula
Tambov
Lipetsk
Kursk
Belgorod
Orel
Bryansk
Smolensk
Rostov-na-Donu
Krasnodar
Stavropol'
Sochi
Groznyy
Makhachkala
Vladikavkaz
KIEV
(Kyyiv)
MINSK
BELARUS
UKRAINE
Kharkiv
Dnipropetrovs'k
Donets'k
Zaporizhzhya
Mariupol'
Odesa
Kherson
Mykolayiv
Sevastopol'
Simferopol'
Crimea
Sea of Azov
Black Sea
Caspian Sea
Aral Sea
KAZAKHSTAN
Ural'sk
Aktyubinsk
Atyrau
Aktau
UZBEKISTAN
TURKMENISTAN
TURAN LOWLAND
GEORGIA
T'BILISI
(Tiflis)
AZER.
ARM.
TURKEY
ANKARA
Trabzon
Samsun
Volga
1:20M
Km Miles
600
400
200
0
400
200
0

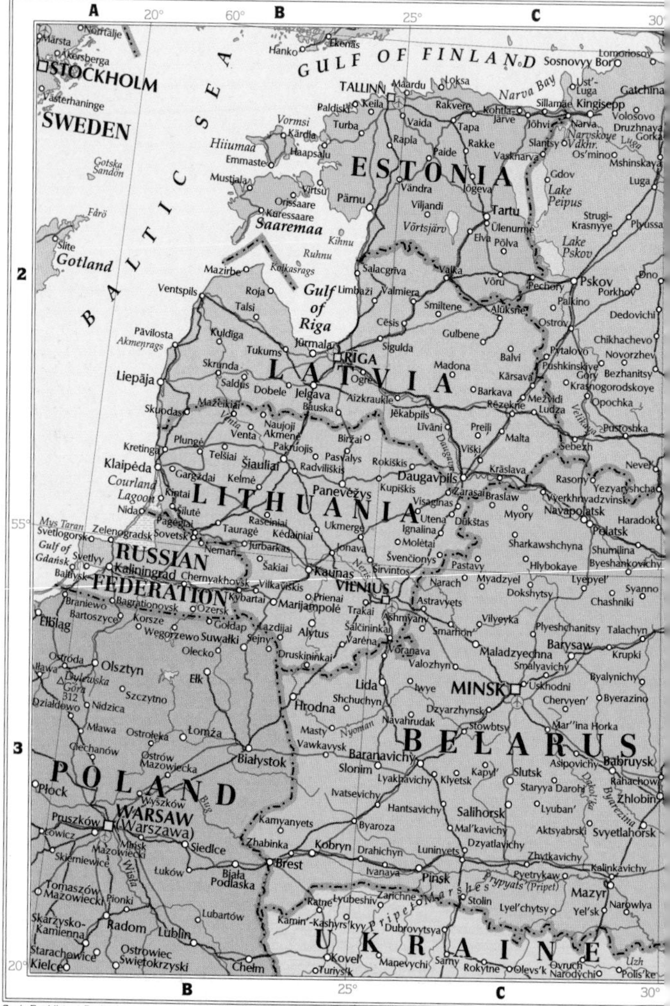

Conic Equidistant Projection

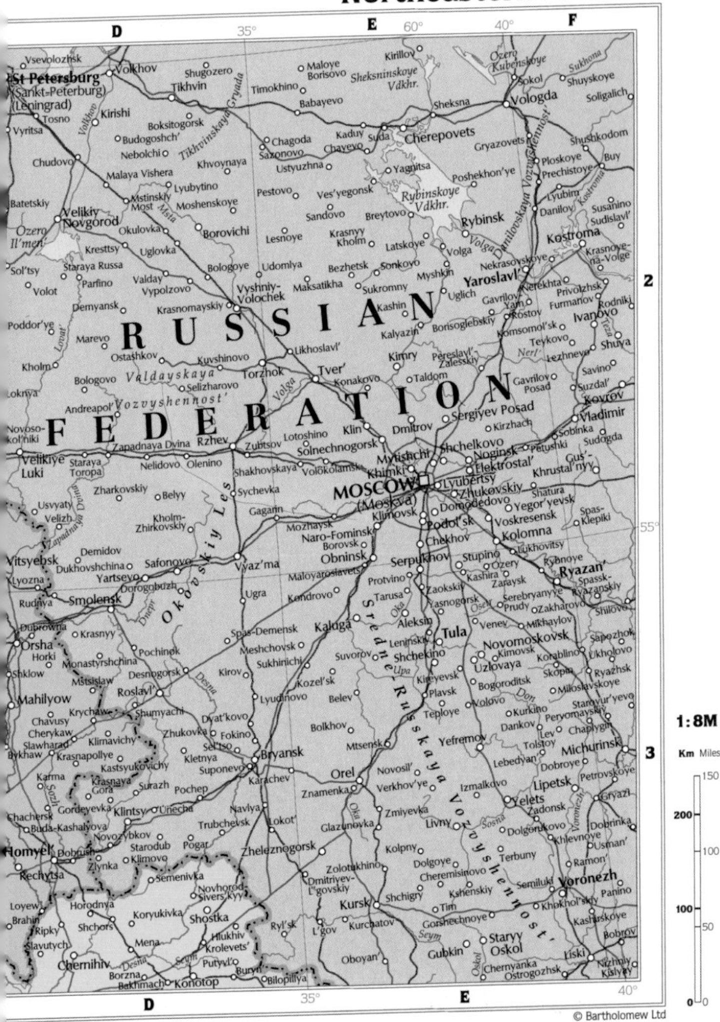
D
E
F
35°
60°
40°
St Petersburg
(Sankt-Peterburg)
(Leningrad)
Vsevolozhsk
Volkhov
Shugozero
Tikhvin
Tikhvinskaya Gryada
Timokhino
Maloye Borisovo
Sheksninskoye Vdkhr.
Kirillov
Ozero Kubenskoye
Sukhona
Sokol
Shuyskoye
Soligalich
Vologda
Sheksna
Babayevo
Tosno
Kirishi
Volkhov
Vyritsa
Boksitogorsk
Budogoshch'
Nebolchi
Chagoda
Sazonovo
Kaduy
Chayevo
Suda
Cherepovets
Gryazovets
Shushkodom
Ploskoye
Buy
Prechistoye
Chudovo
Khvoynaya
Ustyuzhna
Yagnitsa
Poshekhon'ye
Malaya Vishera
Lyubytino
Mstinskiy Most
Moshenskoye
Pestovo
Ves'yegonsk
Rybinskoye Vdkhr.
Lyubim
Kostroma
Danilov
Susanino
Sudislavl'
Batetskiy
Velikiy Novgorod
Msta
Sandovo
Breytovo
Rybinsk
Danilovskaya Vozvyshennost'
Ozero Il'men'
Okulovka
Borovichi
Lesnoye
Krasnyy Kholm
Latskoye
Volga
Kostroma
Krasnoye-na-Volge
Kresttsy
Uglovka
Sol'tsy
Staraya Russa
Bologoye
Udomlya
Bezhetsk
Sonkovo
Myshkin
Nekrasovskoye
Parfino
Valday
Yaroslavl'
Volot
Vypolzovo
Vyshniy-Volochek
Maksatikha
Sukromny
Uglich
Nerekhta
Privolzhsk
Gavrilov-Yam
Furmanov
Rodniki
Demyansk
Krasnomayskiy
Kashin
Rostov
Ivanovo
RUSSIAN
Poddor'ye
Lovat'
Marevo
Kalyazin
Borisoglebskiy
Komsomol'sk
Teza
Shuya
Teykovo
Ostashkov
Likhoslavl'
Kimry
Pereslavl'-Zalesskiy
Nerl'
Lezhnevo
Kuvshinovo
Kholm
Torzhok
Tver'
Savino
Bologovo
Valdayskaya
Konakovo
Taldom
Gavrilov Posad
Suzdal'
Selizharovo
Kovrov
Lokhnya
Vozvyshennost'
Andreapol'
FEDERATION
Sergiyev Posad
Vladimir
Kirzhach
Novosokol'niki
Lotoshino
Klin
Dmitrov
Sobinka
Zapadnaya Dvina
Rzhev
Zubtsov
Solnechnogorsk
Shchelkovo
Petushki
Sudogda
Noginsk
Velikiye Luki
Staraya Toropa
Nelidovo
Olenino
Mytishchi
Elektrostal'
Gus'-Khrustal'nyy
Shakhovskaya
Volokolamsk
Khimki
Lyubertsy
Zharkovskiy
Belyy
MOSCOW
(Moskva)
Zhukovskiy
Shatura
Sychevka
Usvyaty
Domodedovo
Yegor'yevsk
Velizh
Gagarin
Klimovsk
Podol'sk
Voskresensk
Spas-Klepiki
Kholm-Zhirkovskiy
Mozhaysk
Naro-Fominsk
Chekhov
Kolomna
Zapadnaya Dvina
Borovsk
Lukhovitsy
Demidov
Obninsk
Serpukhov
Stupino
Rybnoye
Vitsyebsk
Dukhovshchina
Safonovo
Vyaz'ma
Ozery
Ryazan'
Yartsevo
Maloyaroslavets
Protvino
Kashira
Spassk-Ryazanskiy
Lyozna
Dorogobuzh
Zaraysk
Rudnya
Smolensk
Ugra
Kondrovo
Tarusa
Zaokskiy
Serebryanyye Prudy
Yasnogorsk
Osetr
Zakharovo
Shilovo
Oka
Dnepr
Okovskiy Les
Aleksin
Dubrowna
Krasnyy
Spas-Demensk
Kaluga
Venev
Mikhaylov
Tula
Sapozhok
Orsha
Meshchovsk
Leninskiy
Novomoskovsk
Horki
Pochinok
Suvorov
Shchekino
Kimovsk
Korablino
Ukholovo
Monastyrshchina
Sukhinichi
Upa
Uzlovaya
Shklow
Desnogorsk
Kirov
Skopin
Ryazhsk
Kozel'sk
Kireyevsk
Bogoroditsk
Msitsislaw
Desna
Miloslavskoye
Roslavl'
Plavsk
Mahilyow
Lyudinovo
Belev
Volovo
Don
Starovur'yevo
Krychaw
Shumyachi
Dyat'kovo
Kurkino
Chavusy
Teploye
Pervomayskiy
Cherykaw
Zhukovka
Fokino
Bolkhov
Dankov
Lev
Chaplygin
Slawharad
Klimavichy
Sel'tso
Yefremov
Tolstoy
Bykhaw
Krasnapollye
Kletnya
Bryansk
Mtsensk
Lebedyan'
Michurinsk
Kastsyukovichy
Suponevo
Dobroye
Karma
Krasnaya Gora
Surazh
Karachev
Orel
Novosil'
Lipetsk
Petrovskoye
Pochep
Znamenka
Verkhov'ye
Izmalkovo
Gryazi
Sozh
Gordeyevka
Klintsy
Unecha
Navlya
Yelets
Zadonsk
Chachersk
Buda-Kashalyova
Oka
Zmiyevka
Voronezh
Trubchevsk
Lokot'
Glazunovka
Livny
Sosna
Dolgorukovo
Dobrinka
Novozybkov
Khlevnoye
Usman'
Starodub
Pogar
Homyel'
Dobrush
Klimovo
Zheleznogorsk
Kolpny
Terbuny
Zlynka
Dolgoye
Ramon'
Rechytsa
Zolotukhino
Cheremisinovo
Semenivka
Dmitriyev-L'govskiy
Semiluki
Voronezh
Novhorod-Sivers'kyy
Kshenskiy
Panino
Loyew
Horodnya
Kursk
Shchigry
Khokhol'skiy
Koryukivka
Shostka
Tim
Brahin
Ripky
Shchors
Ryl'sk
L'gov
Kurchatov
Gorshechnoye
Kashirskoye
Hlukhiv
Seym
Staryy Oskol
Bobrov
Slavutych
Mena
Krolevets'
Chernihiv
Desna
Seym
Oboyan'
Gubkin
Oskol
Liski
Putyvl'
Chernyanka
Nizhniy Kislyay
Ostrogozhsk
Buryn'
Borzna
Bakhmach
Konotop
Bilopillya
55°
2
3
1:8M
Km
Miles
150
200
100
100
50
0
0
© Bartholomew Ltd

Conic Equidistant Projection

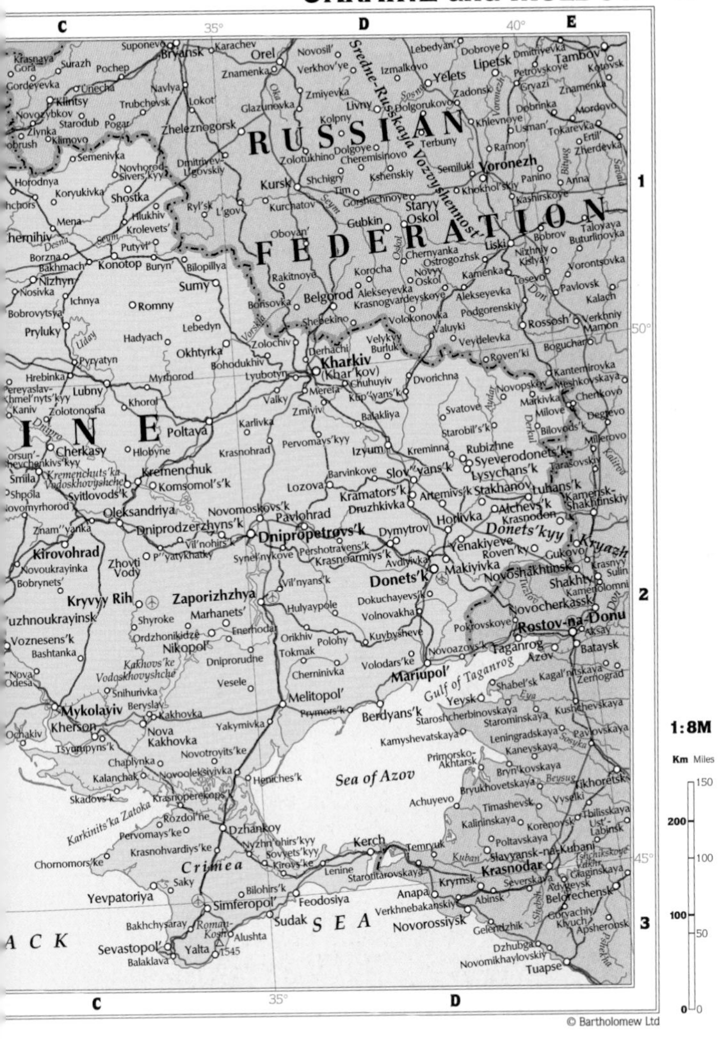
C
D
E
35°
40°
RUSSIAN FEDERATION
Sredne-Russkaya Vozvyshennost'
Bryansk
Suponevo
Karachev
Orel
Novosil'
Lebedyan'
Dobroye
Dmitriyevka
Tambov
Krasnaya Gora
Surazh
Pochep
Znamenka
Verkhov'ye
Izmalkovo
Yelets
Lipetsk
Petrovskoye
Kotovsk
Gordeyevka
Unecha
Navlya
Zmiyevka
Zadonsk
Gryazi
Znamenka
Klintsy
Trubchevsk
Lokot'
Glazunovka
Livny
Dolgorukovo
Dobrinka
Mordovo
Novozybkov
Starodub
Pogar
Zheleznogorsk
Kolpny
Khlevnoye
Usman'
Tokarevka
Zlynka
Klimovo
Dolgoye
Ertil'
Zherdevka
Semenivka
Zolotukhino
Cheremisinovo
Terbuny
Ramon'
Novhorod-Sivers'kyy
Dmitriyev-L'govskiy
Semiluki
Voronezh
Panino
Anna
Horodnya
Koryukivka
Shostka
Kursk
Shchigry
Kshenskiy
Khokhol'skiy
Kashirskoye
Tim
Mena
Hlukhiv
Ryl'sk
L'gov
Kurchatov
Gorshechnoye
Staryy Oskol
Gubkin
Krolevets'
Oboyan'
Bobrov
Taloyaya
Buturlinovka
Putyvl'
Liski
Nizhniy Kislyay
Borzna
Bakhmach
Konotop
Buryn'
Bilopillya
Chernyanka
Ostrogozhsk
Vorontsovka
Nizhyn
Nosivka
Rakitnoye
Korocha
Novyy Oskol
Kamenka
Losevo
Pavlovsk
Sumy
Belgorod
Alekseyevka
Krasnogvardeyskoye
Alekseyevka
Ichnya
Romny
Borisovka
Kalach
Bobrovytsya
Shebekino
Volokonovka
Podgorenskiy
Pryluky
Lebedyn
Valuyki
Rossosh'
Verkhniy Mamon
Hadyach
Okhtyrka
Zolochiv
Velykyy Burluk
Veydelevka
Boguchar
Derhachi
Kharkiv (Khar'kov)
Roven'ki
Pyryatyn
Bohodukhiv
Lyubotyn
Kantemirovka
Hrebinka
Myrhorod
Chuhuyiv
Dvorichna
Meshkovskaya
Pereyaslav-Khmel'nyts'kyy
Lubny
Khorol
Merefa
Kup''yans'k
Novopskov
Markivka
Chertkovo
Kaniv
Zolotonosha
Valky
Zmiyiv
Balakliya
Svatove
Milove
Degtevo
Karlivka
Starobil's'k
Bilovods'k
Poltava
Millerovo
Cherkasy
Hlobyne
Krasnohrad
Pervomays'kyy
Izyum
Kreminna
Rubizhne
Syeverodonets'k
Lysychans'k
Tarasovskiy
Kremenchuk
Barvinkove
Slov''yans'k
Smila
Kremenchuts'ka Vodoskhovyshche
Komsomol's'k
Lozova
Kramators'k
Artemivs'k
Stakhanov
Luhans'k
Kamensk-Shakhtinskiy
Svitlovods'k
Novomoskovs'k
Druzhkivka
Alchevs'k
Oleksandriya
Pavlohrad
Horlivka
Krasnodon
Znam''yanka
Dniprodzerzhyns'k
Dnipropetrovs'k
Dymytrov
Donets'kyy Kryazh
Vil'nohirs'k
Yenakiyeve
Kirovohrad
Zhovti Vody
P''yatykhatky
Synel'nykove
Pershotravens'k
Krasnoarmiys'k
Roven'ky
Gukovo
Novoukrayinka
Avdiyivka
Makiyivka
Krasnyy Sulin
Bobrynets'
Donets'k
Novoshakhtinsk
Shakhty
Vil'nyans'k
Kryvyy Rih
Zaporizhzhya
Dokuchayevs'k
Kamenolomni
Marhanets'
Hulyaypole
Volnovakha
Novocherkassk
Yuzhnoukrayins'k
Shyroke
Enerhodar
Rostov-na-Donu
Pokrovskoye
Aksay
Voznesens'k
Ordzhonikidze
Orikhiv
Polohy
Kuybysheve
Nikopol'
Tokmak
Novoazovs'k
Taganrog
Bataysk
Bashtanka
Dniprorudne
Volodars'ke
Azov
Kakhovs'ke Vodoskhovyshche
Chernihivka
Mariupol'
Gulf of Taganrog
Kagal'nitskaya
Nova Odesa
Zernograd
Vesele
Shabel'sk
Snihurivka
Yeysk
Beryslav
Melitopol'
Mykolayiv
Kakhovka
Prymors'k
Berdyans'k
Staroshcherbinovskaya
Kushchevskaya
Starominskaya
Ochakiv
Kherson
Yakymivka
Nova Kakhovka
Leningradskaya
Pavlovskaya
Kamyshevatskaya
Tsyurupyns'k
Novotroyits'ke
Kanevskaya
Primorsko-Akhtarsk
Chaplynka
Bryn'kovskaya
Sea of Azov
Kalanchak
Novooleksiyivka
Heniches'k
Tikhoretsk
Bryukhovetskaya
Skadovs'k
Krasnoperekops'k
Vyselki
Achuyevo
Timashevsk
Tbilisskaya
Karkinits'ka Zatoka
Rozdol'ne
Kalininskaya
Korenovsk
Ust'-Labinsk
Pervomays'ke
Dzhankoy
Nyzhn'ohirs'kyy
Poltavskaya
Krasnohvardiys'ke
Sovyets'kyy
Kerch
Temryuk
Slavyansk-na-Kubani
Chornomors'ke
Kirovs'ke
Kuban'
Crimea
Lenine
Starotitarovskaya
Krasnodar
Saky
Krymsk
Severskaya
Bilohirs'k
Anapa
Adygeysk
Belorechensk
Yevpatoriya
Simferopol'
Feodosiya
Verkhnebakanskiy
Abinsk
Bakhchysaray
Roman-Kosh
Sudak
SEA
Novorossiysk
Gelendzhik
Apsheronsk
Alushta
Sevastopol'
Yalta
1545
Dzhubga
Balaklava
Novomikhaylovskiy
Tuapse
ACK
INE
1
2
3
50°
45°
1:8M
Km Miles
200
100
0
150
100
50
0

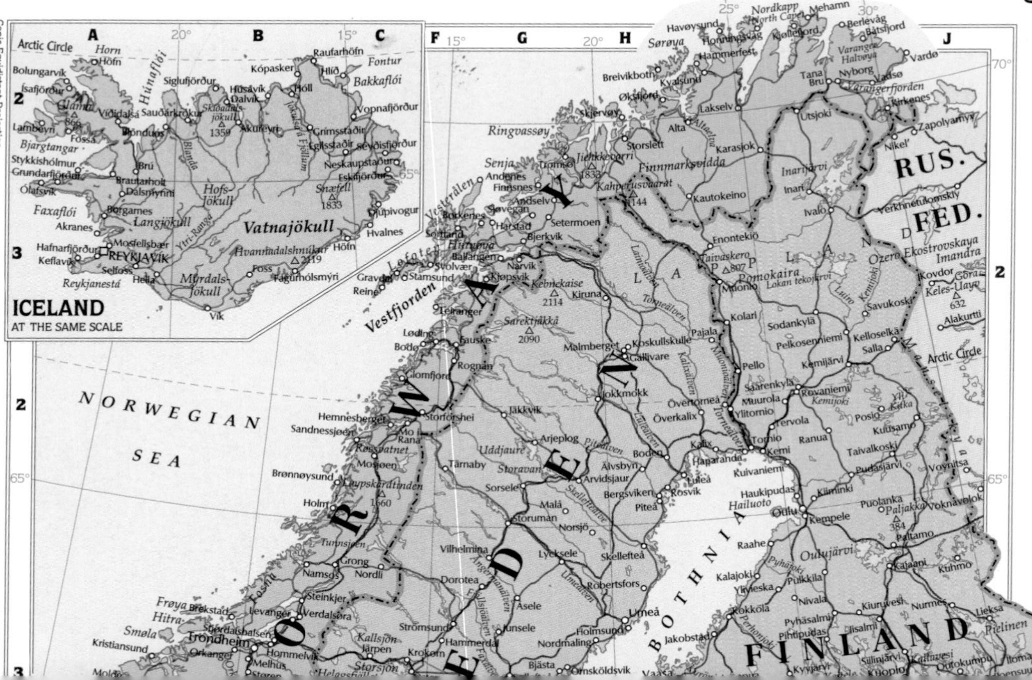
NORWAY
SWEDEN
FINLAND
RUS. FED.
LAPLAND
NORWEGIAN SEA
BOTHNIA
ICELAND
AT THE SAME SCALE
Arctic Circle
Vatnajökull
Hofsjökull
Langjökull
Mýrdalsjökull
REYKJAVÍK
Akureyri
Húsavík
Keflavík
Hafnarfjörður
Akranes
Borgarnes
Selfoss
Hella
Vík
Höfn
Seyðisfjörður
Egilsstaðir
Neskaupstaður
Eskifjörður
Vopnafjörður
Raufarhöfn
Siglufjörður
Sauðárkrókur
Blönduós
Ísafjörður
Bolungarvík
Stykkishólmur
Ólafsvík
Faxaflói
Húnaflói
Bakkaflói
Reykjanestá
Nordkapp
North Cape
Hammerfest
Alta
Tromsø
Narvik
Kiruna
Gällivare
Jokkmokk
Luleå
Piteå
Skellefteå
Umeå
Örnsköldsvik
Oulu
Kemi
Tornio
Haparanda
Rovaniemi
Kajaani
Bodø
Mo i Rana
Mosjøen
Namsos
Steinkjer
Trondheim
Kristiansund
Vestfjorden
Lofoten
Vesterålen
Senja
Ringvassøy
Kebnekaise
2114
Sarektjåkkå
2090
Tana Bru
Vardø
Kirkenes
Inarijärvi
Ivalo
Sodankylä
Kemijärvi
Kuusamo
Torneälven
Luleälven
Skellefteälven
Umeälven
Ångermanälven
Vilhelmina
Storuman
Lycksele
Arvidsjaur
Storsjön
Smøla
Hitra
Frøya
Conic Equidistant Projection

ESTONIA
LATVIA
LITHUANIA
DENMARK
HELSINKI
(Helsingfors)
STOCKHOLM
OSLO
RĪGA
TALLINN
COPENHAGEN
(København)
Gulf of Finland
Gulf of Riga
GULF O
BALTIC SEA
Gotland
(Sweden)
Öland
Saaremaa
Hiiumaa
Åland
Bornholm
(Denmark)
Zealand
Skagerrak
Kattegat
Courland Lagoon
Hanöbukten
Vänern
Vättern
Bergen
Stavanger
Kristiansand
Gothenburg
(Göteborg)
Malmö
Turku
Tampere
Tartu
Daugavpils
Klaipėda
Šiauliai
Liepāja
1:10M
Km Miles
200
100
100
0 0

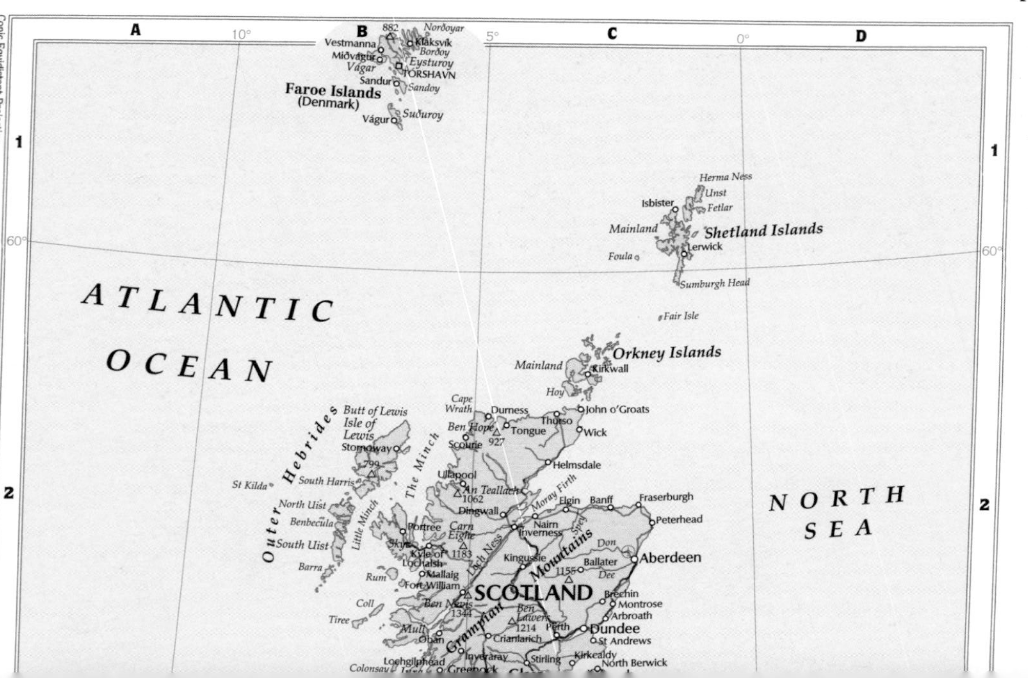

Conic Equidistant Projection

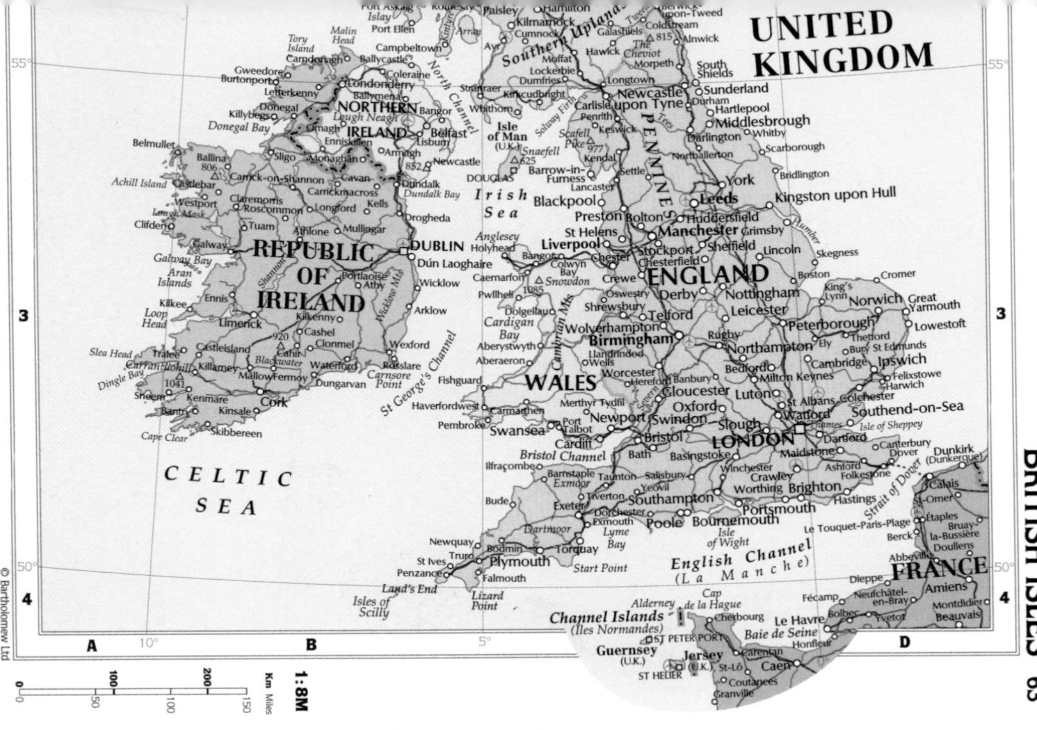
UNITED KINGDOM
REPUBLIC OF IRELAND
NORTHERN IRELAND
ENGLAND
WALES
FRANCE
PENNINES
CELTIC SEA
Irish Sea
English Channel
(La Manche)
Bristol Channel
St George's Channel
North Channel
Strait of Dover
Channel Islands
(Îles Normandes)
LONDON
DUBLIN
Birmingham
Manchester
Liverpool
Leeds
Sheffield
Newcastle upon Tyne
Kingston upon Hull
Middlesbrough
Sunderland
York
Blackpool
Preston
Nottingham
Leicester
Derby
Peterborough
Northampton
Norwich
Ipswich
Southend-on-Sea
Gloucester
Oxford
Swindon
Newport
Cardiff
Swansea
Bristol
Southampton
Portsmouth
Bournemouth
Poole
Brighton
Plymouth
Torquay
Belfast
Londonderry
Cork
Limerick
Galway
Isle of Man
(U.K.)
DOUGLAS
Guernsey
(U.K.)
Jersey
(U.K.)
Le Havre
Caen
Amiens
Calais
Dunkirk
(Dunkerque)
1:8M
Km Miles
0 100 200
0 50 100 150
© Bartholomew Ltd

A
B
C
D
1
2
3
6°
4°
2°
58°
56°
60°
1:4M
Km Miles
Orkney Islands
Shetland Islands
Outer Hebrides
Isle of Lewis
Skye
Mull
Jura
Islay
Arran
Kintyre
The Minch
Little Minch
North Channel
SCOTLAND
GRAMPIAN MOUNTAINS
SOUTHERN UPLANDS
NORTH SEA
NORTHERN IRELAND
ENGLAND
Monadhliath Mountains
Cairngorm Mountains
Cheviot Hills
Sidlaw Hills
Firth of Forth
Firth of Clyde
Firth of Lorn
Firth of Tay
Moray Firth
Solway Firth
Pentland Firth
Glasgow
Edinburgh
Aberdeen
Dundee
Inverness
Perth
Stirling
Oban
Fort William
Ben Nevis
1344
Kirkwall
Stromness
Lerwick
Stornoway
Ullapool
Thurso
Wick
Cape Wrath
Butt of Lewis
Conic Equidistant Projection

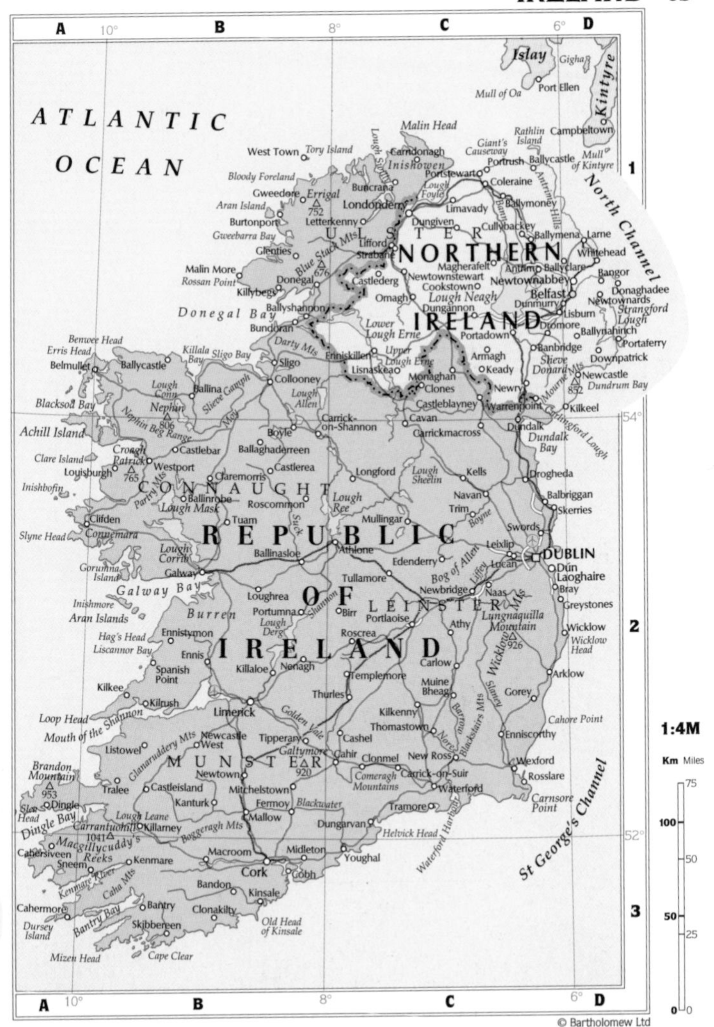
A
B
C
D
10°
8°
6°
1
2
3
54°
52°
ATLANTIC
OCEAN
Islay
Gigha
Kintyre
Port Ellen
Mull of Oa
Rathlin Island
Campbeltown
Mull of Kintyre
Malin Head
Giant's Causeway
Ballycastle
Portrush
Portstewart
Coleraine
North Channel
West Town
Tory Island
Carndonagh
Inishowen
Lough Swilly
Bloody Foreland
Buncrana
Lough Foyle
Bann
Antrim Hills
Gweedore
Errigal
752
Aran Island
Londonderry
Limavady
Ballymoney
Burtonport
Letterkenny
Dungiven
Cullybackey
Gweebarra Bay
Ballymena
Larne
U L S T E R
Lifford
Strabane
Glenties
Blue Stack Mts
676
NORTHERN
Whitehead
Malin More
Rossan Point
Killybegs
Donegal
Castlederg
Magherafelt
Antrim
Ballyclare
Newtownstewart
Newtownabbey
Bangor
Cookstown
Omagh
Lough Neagh
Belfast
Donaghadee
Newtownards
Dunmurry
Dungannon
Strangford Lough
Ballyshannon
Lisburn
Donegal Bay
Bundoran
Lower Lough Erne
IRELAND
Dromore
Ballynahinch
Portadown
Portaferry
Benwee Head
Erris Head
Killala Bay
Sligo Bay
Darty Mts
Enniskillen
Upper Lough Erne
Armagh
Banbridge
Belmullet
Ballycastle
Sligo
Lisnaskea
Keady
Slieve Donard
Mourne Mts
Downpatrick
Monaghan
Newcastle
Lough Conn
Ballina
Slieve Gamph
Collooney
Clones
Newry
852
Dundrum Bay
Lough Allen
Blacksod Bay
Nephin
806
Castleblayney
Warrenpoint
Kilkeel
Nephin Beg Range
Moy
Carrick-on-Shannon
Cavan
Achill Island
Boyle
Carrickmacross
Dundalk
Carlingford Lough
Croagh Patrick
Castlebar
Dundalk Bay
Clare Island
765
Westport
Ballaghaderreen
Louisburgh
Partry Mts
Claremorris
Castlerea
Longford
Lough Sheelin
Kells
Drogheda
Inishbofin
C O N N A U G H T
Ballinrobe
Lough Mask
Roscommon
Lough Ree
Navan
Balbriggan
Trim
Skerries
Clifden
Tuam
Suck
Mullingar
Boyne
Connemara
Slyne Head
R E P U B L I C
Swords
Lough Corrib
Ballinasloe
Athlone
Leixlip
Lucan
DUBLIN
Edenderry
Gorumna Island
Galway
Bog of Allen
Liffey
Dún Laoghaire
Tullamore
Galway Bay
Newbridge
Naas
Bray
Inishmore
Loughrea
O F
Greystones
Aran Islands
Burren
Portumna
Shannon
Birr
L E I N S T E R
Lungnaquilla Mountain
Lough Derg
Portlaoise
926
Ennistymon
Athy
Wicklow
Hag's Head
Roscrea
Wicklow Mts
Wicklow Head
Liscannor Bay
I R E L A N D
Ennis
Carlow
Spanish Point
Killaloe
Nenagh
Templemore
Arklow
Slaney
Muine Bheag
Kilkee
Thurles
Gorey
Kilrush
Barrow
Blackstairs Mts
Limerick
Loop Head
Golden Vale
Kilkenny
Cahore Point
Mouth of the Shannon
Thomastown
Nore
Newcastle West
Tipperary
Cashel
Enniscorthy
Glanaruddery Mts
Listowel
Galtymore
Cahir
Clonmel
New Ross
Wexford
M U N S T E R
920
Brandon Mountain
Newtown
Carrick-on-Suir
Rosslare
Comeragh Mountains
953
Tralee
Castleisland
Mitchelstown
Waterford
St George's Channel
Kanturk
Carnsore Point
Slea Head
Dingle
Fermoy
Blackwater
Tramore
Dingle Bay
Lough Leane
Mallow
Carrantuohill
Killarney
Boggeragh Mts
Dungarvan
Waterford Harbour
1041
Helvick Head
Macgillycuddy's Reeks
Cahersiveen
Macroom
Midleton
Youghal
Sneem
Kenmare
Cork
Kenmare River
Cobh
Caha Mts
Bandon
Kinsale
Cahermore
Bantry
Clonakilty
Bantry Bay
Dursey Island
Skibbereen
Old Head of Kinsale
Mizen Head
Cape Clear
1:4M
Km
Miles
100
50
0
75
50
25
0

Conic Equidistant Projection

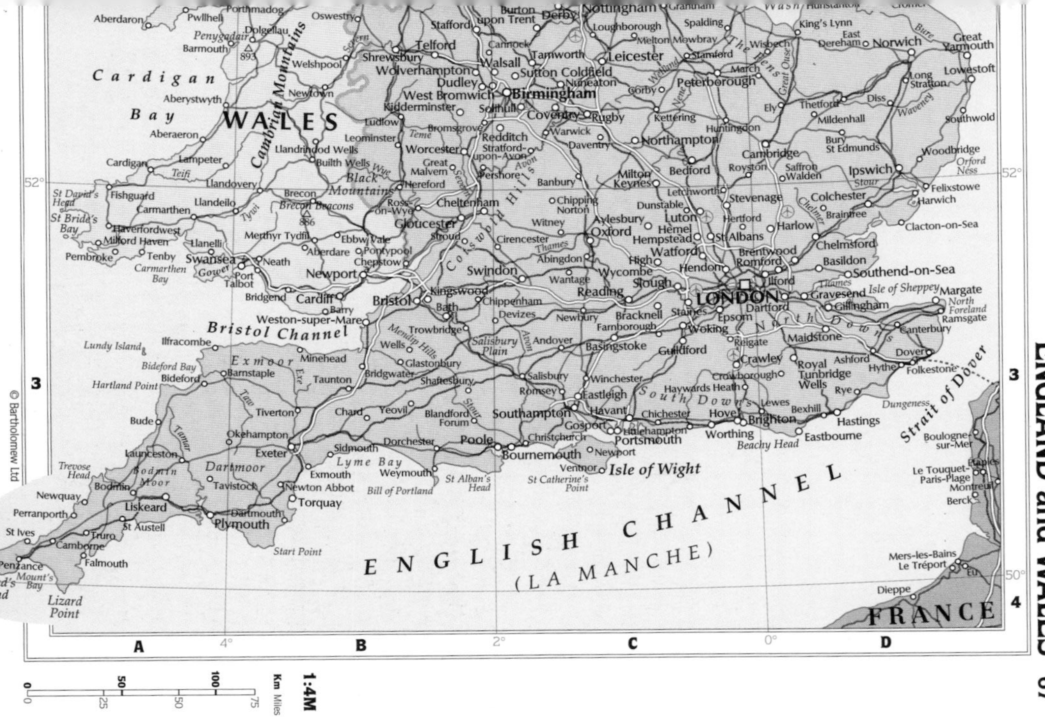

ENGLISH CHANNEL
(LA MANCHE)
WALES
FRANCE
Cardigan Bay
Bristol Channel
Strait of Dover
Cambrian Mountains
Black Mountains
Brecon Beacons
Exmoor
Dartmoor
Bodmin Moor
Cotswold Hills
North Downs
South Downs
The Fens
Salisbury Plain
Mendip Hills
Isle of Wight
Isle of Sheppey
Lundy Island
Lyme Bay
Carmarthen Bay
St Bride's Bay
Bideford Bay
Mount's Bay
Land's End
Lizard Point
Start Point
Hartland Point
Trevose Head
St David's Head
Bill of Portland
St Alban's Head
St Catherine's Point
Beachy Head
Dungeness
North Foreland
Orford Ness
LONDON
Birmingham
Bristol
Cardiff
Swansea
Newport
Plymouth
Exeter
Southampton
Portsmouth
Bournemouth
Brighton
Norwich
Ipswich
Cambridge
Oxford
Reading
Leicester
Nottingham
Coventry
Peterborough
Northampton
Gloucester
Cheltenham
Worcester
Hereford
Shrewsbury
Telford
Wolverhampton
Dudley
West Bromwich
Walsall
Sutton Coldfield
Stafford
Burton upon Trent
Derby
Loughborough
Canterbury
Dover
Folkestone
Margate
Ramsgate
Maidstone
Southend-on-Sea
Colchester
Chelmsford
Great Yarmouth
Lowestoft
Swindon
Bath
Salisbury
Winchester
Taunton
Torquay
Penzance
Truro
Falmouth
Newquay
Boulogne-sur-Mer
Dieppe
Le Touquet-Paris-Plage
Berck
Montreuil
Eu
Le Tréport
Mers-les-Bains
Severn
Thames
Avon
Wye
Teme
Tamar
Taw
Exe
Stour
Nene
Great Ouse
Welland
Waveney
Bure
Chelmer
Teifi
Tywi
886
893
52°
50°
2°
4°
0°
A
B
C
D
3
4
1:4M
Km Miles
0
50
100
25
50
75

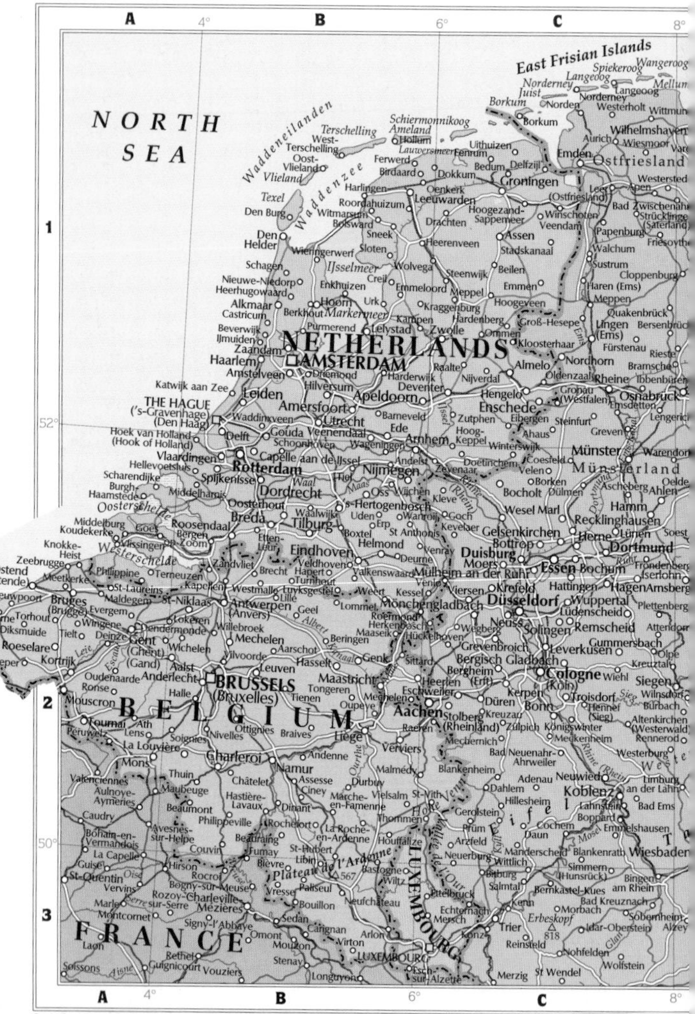

Conic Equidistant Projection

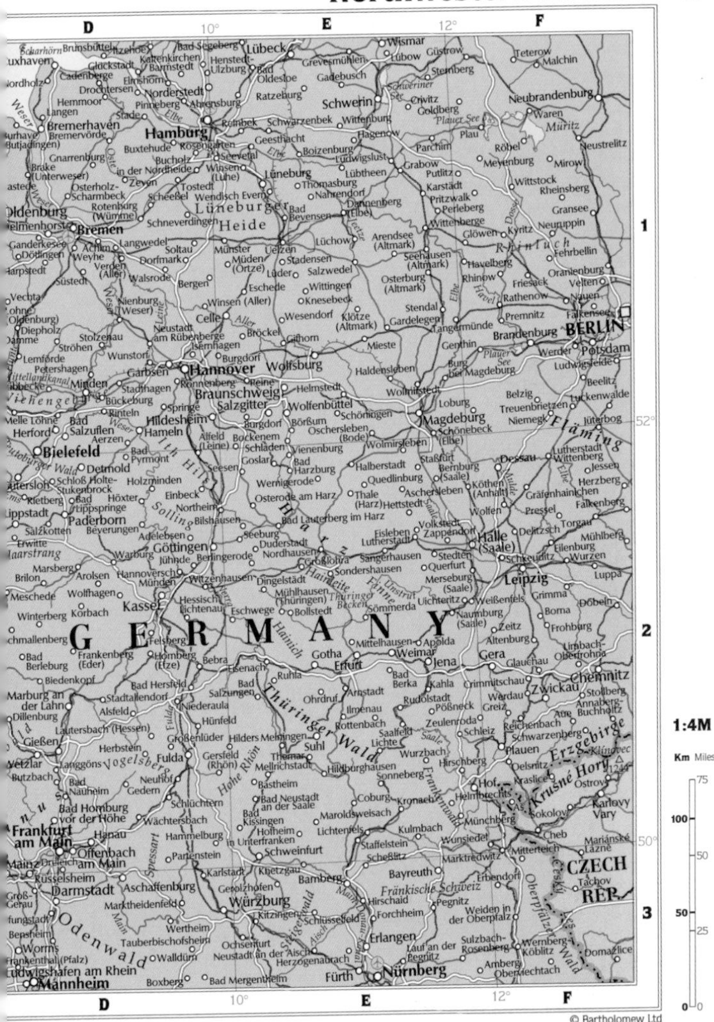
D
E
F
10°
12°
52°
50°
1
2
3
GERMANY
CZECH REP.
Lübeck
Wismar
Güstrow
Teterow
Malchin
Hamburg
Bremerhaven
Schwerin
Neubrandenburg
Waren
Müritz
Neustrelitz
Lüneburg
Lüneburger Heide
Oldenburg
Bremen
Uelzen
Salzwedel
Wittenberge
Neuruppin
Stendal
Celle
Hannover
Wolfsburg
Braunschweig
Salzgitter
Hildesheim
Magdeburg
Brandenburg
BERLIN
Potsdam
Minden
Hameln
Bielefeld
Detmold
Paderborn
Herford
Göttingen
Halberstadt
Quedlinburg
Dessau
Wittenberg
Halle (Saale)
Leipzig
Kassel
Eisenach
Gotha
Erfurt
Weimar
Jena
Gera
Zwickau
Chemnitz
Plauen
Fulda
Marburg an der Lahn
Gießen
Frankfurt am Main
Offenbach am Main
Darmstadt
Aschaffenburg
Würzburg
Schweinfurt
Bamberg
Bayreuth
Erlangen
Fürth
Nürnberg
Mannheim
Ludwigshafen am Rhein
Worms
Coburg
Hof
Thüringer Wald
Harz
Erzgebirge
Krušné Hory
Odenwald
Vogelsberg
Fläming
Karlovy Vary
Cheb
1:4M
Km
Miles
100
75
50
25
0

Conic Equidistant Projection

POLAND
SLOVAKIA
HUNGARY
ROMANIA
CROATIA
YUGOSLAVIA
BELARUS
UKRAINE
RUS. FED.
LITH.
REPUBLIC
WARSAW (Warszawa)
BUDAPEST
VIENNA (Wien)
BRATISLAVA
ZAGREB
Gulf of Gdańsk
Carpathian Mts
Danube (Dunaj)
Balaton
Wisła
Odra
Gdańsk
Gdynia
Kraków
Łódź
Wrocław
Poznań
Lublin
Katowice
Białystok
Košice
Debrecen
Szeged
Pécs
Győr
Miskolc
Brest
Lviv (L'vov)
Uzhhorod
Cluj-Napoca
Timişoara
Oradea
Arad
1:8M
Km Miles
D
E
20°
25°
50°
1
2

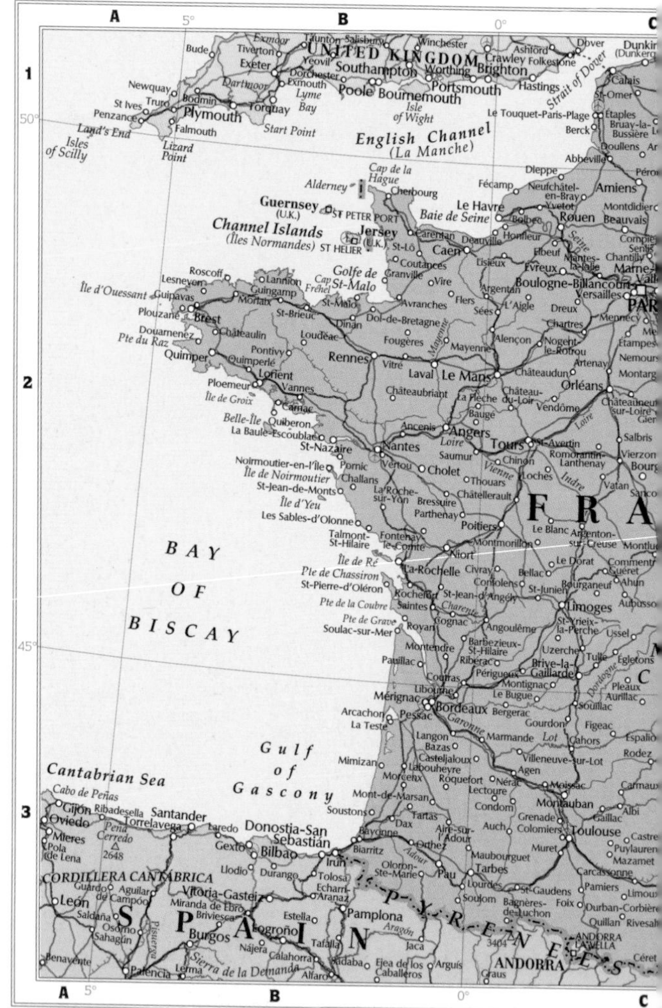

Conic Equidistant Projection

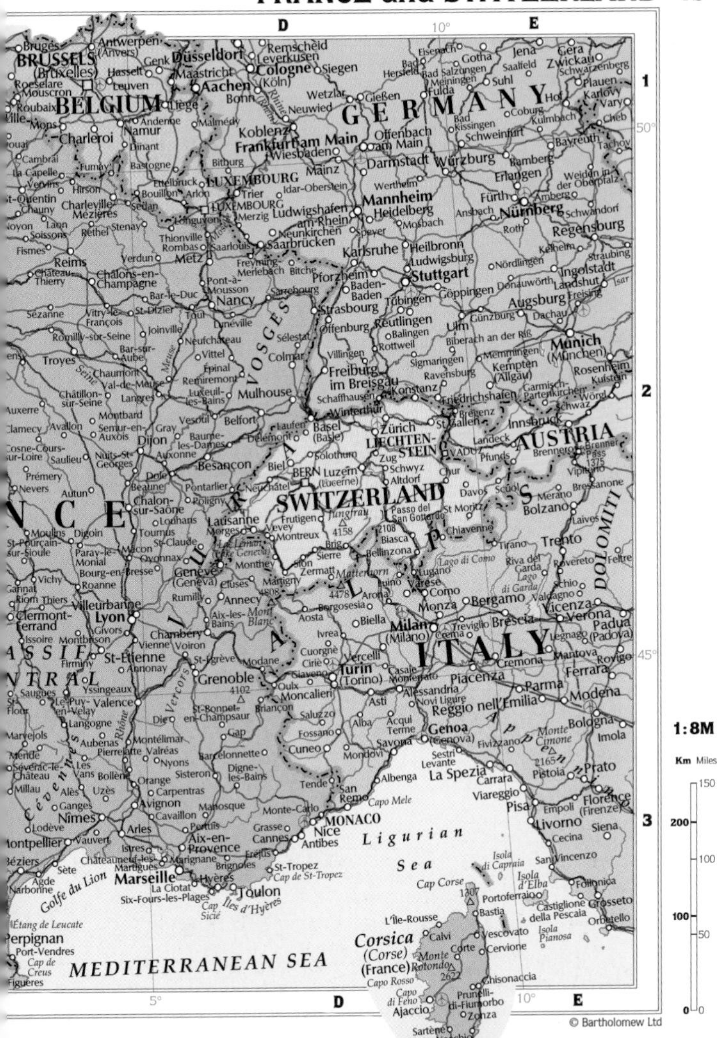
D
E
10°
1
2
3
50°
45°
5°
BELGIUM
BRUSSELS (Bruxelles)
Antwerpen (Anvers)
Brugge
Roeselare
Mouscron
Roubaix
Lille
Mons
Charleroi
Namur
Liège
Leuven
Hasselt
Genk
Maastricht
Aachen
Düsseldorf
Leverkusen
Remscheid
Cologne (Köln)
Siegen
Bonn
Rhine
Wetzlar
Gießen
Neuwied
Koblenz
Frankfurt am Main
Wiesbaden
Mainz
Offenbach am Main
Darmstadt
GERMANY
Eisenach
Gotha
Jena
Gera
Zwickau
Schwarzenberg
Plauen
Karlovy Vary
Cheb
Tachov
Hof
Coburg
Kulmbach
Bayreuth
Bad Hersfeld
Bad Salzungen
Meiningen
Suhl
Saalfeld
Fulda
Bad Kissingen
Schweinfurt
Würzburg
Bamberg
Erlangen
Weiden in der Oberpfalz
Amberg
Fürth
Nürnberg
Schwandorf
Regensburg
Ansbach
Roth
Kelheim
Straubing
Ingolstadt
Landshut
Isar
Nördlingen
Donauwörth
Freising
Augsburg
Dachau
Munich (München)
Rosenheim
Kufstein
Wörgl
Schwaz
Innsbruck
Garmisch-Partenkirchen
Kempten (Allgäu)
Memmingen
Ulm
Günzburg
Biberach an der Riß
Göppingen
Stuttgart
Ludwigsburg
Heilbronn
Mosbach
Heidelberg
Mannheim
Wertheim
Ludwigshafen am Rhein
Speyer
Karlsruhe
Pforzheim
Baden-Baden
Tübingen
Reutlingen
Balingen
Rottweil
Sigmaringen
Ravensburg
Friedrichshafen
Konstanz
Villingen
Freiburg im Breisgau
Offenburg
Strasbourg
Sélestat
Colmar
Mulhouse
VOSGES
Bitburg
Idar-Oberstein
Trier
LUXEMBOURG
Ettelbruck
Arlon
Bouillon
Sedan
Merzig
Neunkirchen
Saarbrücken
Saarlouis
Thionville
Rombas
Metz
Longuyon
Mosel
Freyming-Merlebach
Bitche
Sarrebourg
Pont-à-Mousson
Nancy
Toul
Lunéville
Verdun
Stenay
Andenne
Malmédy
Dinant
Bastogne
Fumay
Cambrai
La Capelle
Vervins
Hirson
St-Quentin
Chauny
Noyon
Laon
Soissons
Fismes
Charleville-Mézières
Rethel
Reims
Château-Thierry
Châlons-en-Champagne
Bar-le-Duc
Vitry-le-François
St-Dizier
Sézanne
Romilly-sur-Seine
Joinville
Neufchâteau
Vittel
Épinal
Remiremont
Meuse
Troyes
Bar-sur-Aube
Chaumont
Val-de-Meuse
Seine
Langres
Luxeuil-les-Bains
Châtillon-sur-Seine
Auxerre
Clamecy
Avallon
Montbard
Semur-en-Auxois
Dijon
Gray
Vesoul
Belfort
Baume-les-Dames
Besançon
Auxonne
Nuits-St-Georges
Cosne-Cours-sur-Loire
Saulieu
Prémery
Nevers
Autun
Dole
Beaune
Pontarlier
Poligny
Chalon-sur-Saône
Louhans
Tournus
Mâcon
St-Claude
Oyonnax
Bourg-en-Bresse
Moulins
Digoin
St-Pourçain-sur-Sioule
Paray-le-Monial
Vichy
Roanne
Gannat
Riom
Thiers
Clermont-Ferrand
Villeurbanne
Lyon
Givors
Issoire
Montbrison
St-Étienne
Firminy
Annonay
Vienne
Voiron
Chambéry
St-Egrève
Grenoble
Vercors
Yssingeaux
Le-Puy-en-Velay
Valence
Saugues
St-Flour
Langogne
Die
St-Bonnet-en-Champsaur
Briançon
Gap
Marvejols
Mende
Aubenas
Montélimar
Pierrelatte
Valréas
Nyons
Sévérac-le-Château
Les Vans
Bollène
Millau
Alès
Uzès
Orange
Sisteron
Carpentras
Avignon
Cavaillon
Manosque
Ganges
Nîmes
Lodève
Arles
Montpellier
Vauvert
Béziers
Sète
Agde
Narbonne
Istres
Châteauneuf-les-Martigues
Marignane
Martigues
Marseille
La Ciotat
Six-Fours-les-Plages
Toulon
Hyères
Brignoles
Pertuis
Aix-en-Provence
Digne-les-Bains
Barcelonnette
Grasse
Cannes
Antibes
Nice
Fréjus
St-Tropez
Cap de St-Tropez
Îles d'Hyères
Cap Sicié
Golfe du Lion
Étang de Leucate
Perpignan
Port-Vendres
Cap de Creus
Figueres
MEDITERRANEAN SEA
Cévennes
Rhône
MASSIF CENTRAL
FRANCE
JURA
ALPS
Laufen
Basel (Basle)
Delémont
Solothurn
Biel
BERN
Neuchâtel
Luzern (Lucerne)
Zug
Schwyz
Altdorf
SWITZERLAND
Zürich
Winterthur
Schaffhausen
St Gallen
Bregenz
LIECHTENSTEIN
VADUZ
Chur
Davos
Scuol
Landeck
Pfunds
AUSTRIA
Brennero
Brenner Pass 1375
Vipiteno
Bressanone
Merano
Bolzano
Laives
DOLOMITI
St Moritz
Passo del San Gottardo
2108
Biasca
Bellinzona
Chiavenna
Frutigen
Jungfrau
4158
Lausanne
Morges
Vevey
Montreux
Lac Léman (Lake Geneva)
Monthey
Brig
Sion
Sierre
Zermatt
Matterhorn
Martigny
4808
4478
Genève (Geneva)
Cluses
Rumilly
Annecy
Mont Blanc
Aix-les-Bains
Modane
Locarno
Lugano
Varese
Como
Lago di Como
Tirano
Riva del Garda
Lago di Garda
Trento
Rovereto
Feltre
Schio
Valdagno
Vicenza
Verona
Padua (Padova)
Legnago
Rovigo
Mantova
Ferrara
Bergamo
Brescia
Monza
Treviglio
Crema
Milan (Milano)
ITALY
Cremona
Arona
Borgosesia
Biella
Aosta
Ivrea
Cuorgnè
Ciriè
Vercelli
Casale Monferrato
Turin (Torino)
Avigliana
Oulx
Moncalieri
4102
Asti
Alessandria
Novi Ligure
Piacenza
Parma
Reggio nell'Emilia
Modena
Bologna
Imola
Saluzzo
Alba
Acqui Terme
Fossano
Cuneo
Mondovì
Savona
Genoa (Genova)
Sestri Levante
Monte Cimone
2165
Fivizzano
Apennines
Pistoia
Prato
Florence (Firenze)
Empoli
Tende
San Remo
Albenga
Capo Mele
MONACO
Monte-Carlo
La Spezia
Carrara
Viareggio
Pisa
Livorno
Cecina
Siena
Ligurian Sea
Isola di Capraia
Isola d'Elba
San Vincenzo
Follonica
Portoferraio
Castiglione della Pescaia
Grosseto
Orbetello
Isola Pianosa
Cap Corse
1307
Bastia
L'Île-Rousse
Calvi
Vescovato
Cervione
Corte
Corsica (Corse) (France)
Monte Rotondo
2622
Ghisonaccia
Capo Rosso
Capo di Feno
Ajaccio
Prunelli-di-Fiumorbo
Zonza
Sartène
Porto-Vecchio
Bonifacio
1:8M
Km
Miles
200
100
0
150
100
50
0

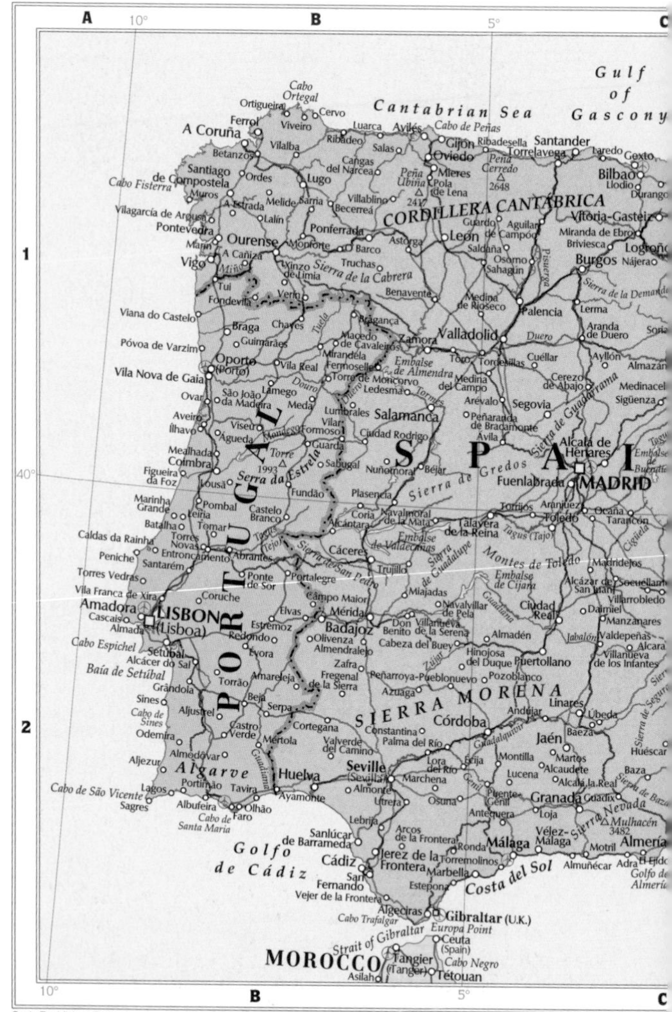

Conic Equidistant Projection

D
E
0°
5°
1
2
40°
FRANCE
PYRENEES
ANDORRA
ANDORRA LA VELLA
3404
Golfe du Lion
Costa Brava
Costa Blanca
MEDITERRANEAN SEA
BALEARIC ISLANDS
(ISLAS BALEARES)
(Spain)
ALGERIA
ALGIERS
(Alger)
Langon
Bazas
Casteljaloux
Labouheyre
Mimizan
Morcenx
Mont-de-Marsan
Soustons
Tartas
Dax
Bayonne
Biarritz
Irún
Orthez
Aire-sur-l'Adour
Auch
Adour
Marmande
Agen
Nérac
Roquefort
Lectoure
Condom
Villeneuve-sur-Lot
Lot
Cahors
Moissac
Montauban
Gaillac
Albi
Carmaux
Rodez
Sévérac-le-Château
Millau
Florac
Cévennes
Alès
Ganges
Uzès
Bollène
Orange
Carpentras
Digne-les-Bains
Sisteron
Avignon
Cavaillon
Manosque
Nîmes
Arles
Pertuis
Aix-en-Provence
Istres
Marignane
Brignoles
Châteauneuf-les-Martigues
Marseille
La Ciotat
Toulon
Hyères
Six-Fours-les-Plages
Cap Sicié
Vauvert
Montpellier
Lodève
Sète
Agde
Béziers
Colomiers
Toulouse
Castres
Puylaurens
Mazamet
Muret
Carcassonne
Maubourguet
Tarbes
Pau
Oloron-Ste-Marie
Lourdes
St-Gaudens
Soulom
Bagnères-de-Luchon
Pamiers
Foix
Quillan
Limoux
Narbonne
Durban-Corbières
Étang de Leucate
Rivesaltes
Perpignan
Port-Vendres
Céret
Cap de Creus
Figueres
Donostia-San Sebastián
Echarri-Aranaz
Estella
Pamplona
Aragón
Tafalla
Jaca
Calahorra
Ejea de los Caballeros
Arguis
Sádaba
Huesca
Graus
Tremp
Berga
Ripoll
Olot
Banyoles
Cap de Begur
Torroella de Montgrí
Torelló
Vic
Girona
Palamós
Blanes
Alfaro
Tudela
Alto del Moncayo
2316
Tarazona
Alagón
Barbastro
Monzón
Binéfar
Tàrrega
Manresa
Sabadell
Igualada
Martorell
Mataró
El Prat de Llobregat
Barcelona
Zaragoza
Fraga
Lleida
Quinto
Calatayud
Cariñena
Escatrón
Caspe
Ebro
Valls
Reus
Tarragona
Vilanova i la Geltrú
Medinaceli
Daroca
Alcañiz
Molina de Aragón
Calamocha
Gandesa
Golf de Sant Jordi
Tortosa
Guadalope
Monreal del Campo
Tordesilos
Amposta
Perales del Alfambra
Morella
Peñarroya
2019
Vinaròs
Teruel
Sarrión
Alcora
Torreblanca
Cuenca
Serranía de Cuenca
Santa Cruz de Moya
Castelló de la Plana
Burriana
Vall de Uxó
Embalse de Alarcón
Cabriel
Turia
Utiel
Llíria
Sagunto
Requena
Valencia
Golfo de Valencia
Motilla del Palancar
Minglanilla
Torrent
La Roda
Júcar
Sueca
Cullera
Algemesí
Carcaixent
Gandía
Oliva
Xátiva
Albacete
Almansa
Alcoy
Cabo de la Nao
Altea
Alcaraz
Hellín
Villena
Benidorm
Segura
Elda
Alicante
Elche
Caravaca de la Cruz
Cieza
Orihuela
Murcia
El Moral
Torrevieja
Alhama de Murcia
Lorca
Cabo de Palos
Cartagena
Huércal-Overa
Águilas
Vera
Cabo de Gata
Menorca
(Minorca)
Pta Nati
Cap de Formentor
Ciutadella de Menorca
Mahón
Mallorca
(Majorca)
Alcúdia
Cap des Freu
Calvià
Sa Dragonera
Palma de Mallorca
Manacor
Felanitx
Cap de ses Salines
Cabrera
Ibiza
(Eivissa)
San Juan Bautista
San Antonio Abad
Eivissa
San Francisco Javier
Formentera
Dellys
Aïn Taya
Tizi Ouzou
Bejaïa
Boumerdes
Koléa
Larba
Blida
Bouira
Bougaa
Ténès
Djebel Bissa
1157
Gouraya
Tipasa
Miliana
Médéa
Bordj Bou Arréridj
Oued Farès
Sidi Ali
Aïn Defla
Khemis Miliana
Berrouaghia
Sour el Ghozlane
Oued Chélif
Ech Chélif
Sidi Aïssa
M'Sila
Mostaganem
Ksar el Boukhari
Gap Carbon
Oran
Arzew
Aïn Tédélès
Relizane
Bordj Bounaama
Oued Tlélat
Sig
Zemmora
Mohammadia
Tissemsilt
1:8M
Km
Miles
150
200
100
100
50
0
0

Conic Equidistant Projection

C
D
20°
1
2
3
45°
40°
HUNGARY
Pécs
Baja
Mohács
Subotica
Senta
Kikinda
Sombor
VOJVODINA
Vrbas
Zrenjanin
Novi Sad
Vršac
Osijek
Vukovar
Đakovo
Bačka Palanka
Ruma
Inđija
Pančevo
BELGRADE
(Beograd)
Smederevo
Požarevac
Mladenovac
Smederevska Palanka
Velika Plana
Crni Vrh
1135
Obrenovac
Lazarevac
Valjevo
Aranđelovac
Gornji Milanovac
Kragujevac
Bor
Zaječar
Povlen
1346
Užice
Požega
Čačak
Paraćin
SERBIA
(SRBIJA)
Trstenik
Kruševac
Aleksinac
Knjaževac
Priboj
YUGOSLAVIA
Niš
Midzhur
2169
Prokuplje
Sjenica
Novi Pazar
Leskovac
Vlasotince
Surdulica
Podujevo
Vučitrn
Kosovska Mitrovica
Priština
Gnjilane
Vranje
Bujanovac
KOSOVO
Peć
Đakovica
Orahovac
Uroševac
Prizren
Preševo
Kumanovo
Kukës
2394
Tetovo
SKOPJE
Veles
Gostivar
Peshkopi
MACEDONIA
(F.Y.R.O.M.)
Štip
Radoviš
Strumica
Kičevo
Debar
Prilep
Struga
Ohrid
Bitola
Gevgelija
Timișoara
Lipova
Deva
Mureș
Hunedoara
Orăștie
Sebeș
Lugoj
Carpații Meridionali
Petroșani
Lupeni
2519
Caransebeș
Reșița
ROMANIA
Târgu Jiu
Vârful Șvinecea Mare
1224
Bela Crkva
Motru
Strehaia
Drobeta - Turnu Severin
Danube
Băilești
Calafat
Poiana Mare
Montana
Berkovitsa
Novi Iskur
SOFIA
(Sofiya)
BUL.
Kyustendil
Blagoevgrad
Kočani
Sveti Nikole
Kilkis
Maribor
Murska Sobota
Ptuj
Velenje
Varaždin
Drava
Čakovec
Nagyatád
Komló
Koprivnica
Szigetvár
Križevci
Barcs
ZAGREB
Suhopolje
Virovitica
Grubišno Polje
Slatina
Samobor
CROATIA
Karlovac
Kutjevo
Požega
984
Ogulin
Nova Gradiška
Bosanska Dubica
Slavonski Brod
Senj
Bosanski Novi
Prijedor
Bosanska Gradiška
Županja
Bosanska Krupa
Banja Luka
Derventa
Doboj
Brčko
Bijeljina
Bihać
Sanski Most
Kotor Varoš
Tešanj
Tuzla
Loznica
Zvornik
Lukavac
Gospić
1758
Ključ
BOSNIA-
HERZEGOVINA
Gračac
Jajce
1943
Zadar
1650
Knin
Glamoč
Bugojno
Gornji Vakuf
Visoko
SARAJEVO
Hrasnica
Livno
Tomislavgrad
Drniš
Šibenik
Sinj
Konjic
Goražde
Foča
Pljevlja
Biograd na Moru
Dalmacija
Posušje
2228
Jablanica
Trogir
Split
Mostar
Nevesinje
Durmitor
2522
Bijelo Polje
Brač
Makarska
Čitluk
Metković
Stolac
MONTE-
NEGRO
(CRNA GORA)
Berane
Vis
Hvar
Ploče
Čapljina
Komiža
Korčula
Trebinje
Nikšić
Lastovo
Dubrovnik
Podgorica
Maja Jezercë
2694
Cetinje
SEA
Bar
Shkodër
Lezhë
Laç
Burrel
Krujë
Durrës
TIRANA
(Tiranë)
Elbasan
Lushnjë
Kuçovë
Fier
Berat
Pogradec
Korçë
ALBANIA
Vlorë
Tepelenë
Ersekë
Gjirokastër
Çorovodë
Grámmos
2503
Kastoria
Florina
Edessa
Thessaloniki
Alexandreia
Veroia
Kalamaria
Ptolemaïda
Kozani
Grevena
Katerini
Olympos
2911
Elassona
Tyrnavos
Larisa
Delvinë
Sarandë
Ioannina
Kalampaka
Kerkyra
Corfu
(Kerkyra)
Igoumenitsa
GREECE
Karditsa
Almyros
Parga
Arta
Filippiada
Kikellos
Domokos
Lamia
Preveza
Lefkada
Karpenisi
Agrinio
Astakos
Parnassos
Amfissa
Delphi
2457
Nafpaktos
Mesolongi
Kefallonia
(Cephalonia)
Korinthiakos Kolpos
Patra
Aigio
Kato Achaia
Kyllini
2376
Argostoli
Lechaina
Ionian Islands
(Ionoi Nisoi)
Amaliada
Olympia
Tripoli
Zakynthos
Zante
(Zakynthos)
Pyrgos
Zacharo
Megalopoli
Kyparissia
Kalamata
Sparti
Isole Tremiti
Termoli
San Severo
Monte Calvo
1055
Vieste
Monte Sant'Angelo
Manfredonia
Lucera
Troia
Foggia
Barletta
Molfetta
Cerignola
Andria
Bari
Bitonto
Monopoli
Ofanto
Le Murge
Melfi
Fasano
Avigliano
Altamura
Brindisi
Eboli
Battipaglia
Potenza
Matera
Francavilla Fontana
Mesagne
Ginosa
Agropoli
Sala Consilina
Montesano sulla Marcellana
Pisticci
Taranto
Lecce
Copertino
Policoro
Gallipoli
Casarano
Sapri
Capo Palinuro
Lauria
Monte Pollino
2248
Golfo di Taranto
Tricase
Capo Sta Maria di Leuca
Strait of Otranto
Scalea
Castrovillari
Rossano
Cetraro
Paola
Monte Botte Donato
1928
Punta Alice
Cirò Marina
Cosenza
San Giovanni in Fiore
Crotone
Amantea
La Sila
Capo Colonna
Isola Stromboli
Catanzaro
Isola di Capo Rizzuto
Parghelia
Vibo Valentia
Golfo di Squillace
Soverato
IONIAN
SEA
Lipari
Isola Vulcano
Rosarno
Palmi
Caulonia
Messina
Montalto
Marina di Gioiosa Ionica
1955
Bianco
Reggio di Calabria
Capo Spartivento
Nebrodi
Monte Etna
3323
Stretta di Messina
Giarre
Acireale
Catania
Lentini
Capo Sta Croce
Augusta
Syracuse
Avola
Pachino
1:8M
Km
Miles
200
100
0
150
100
50
0

Conic Equidistant Projection

GREECE
TURKEY
AEGEAN SEA
IONIAN SEA
MEDITERRANEAN SEA
(F.Y.R.O.M.)
TIRANA
(Tiranë)
Thessaloniki
ATHENS
(Athina)
İstanbul
İzmir
Bursa
Corfu
(Kerkyra)
Ionian Islands
(Ionoi Nisoi)
Kefallonia
(Cephalonia)
Zante
(Zakynthos)
Evvoia
(Euboea)
Lesbos
(Lesvos)
Chios
Cyclades
(Kyklades)
Dodecanese
(Dodekanisos)
Mirtoö Pelagos
Kryiko Pelagos
CRETE
(KRITI)
Rhodes
(Rodos)
Karpathos
(Scarpanto)
Thermaikos Kolpos
Marmara Denizi
Saros Körfezi
Strait of Otranto
1:8M
Km Miles

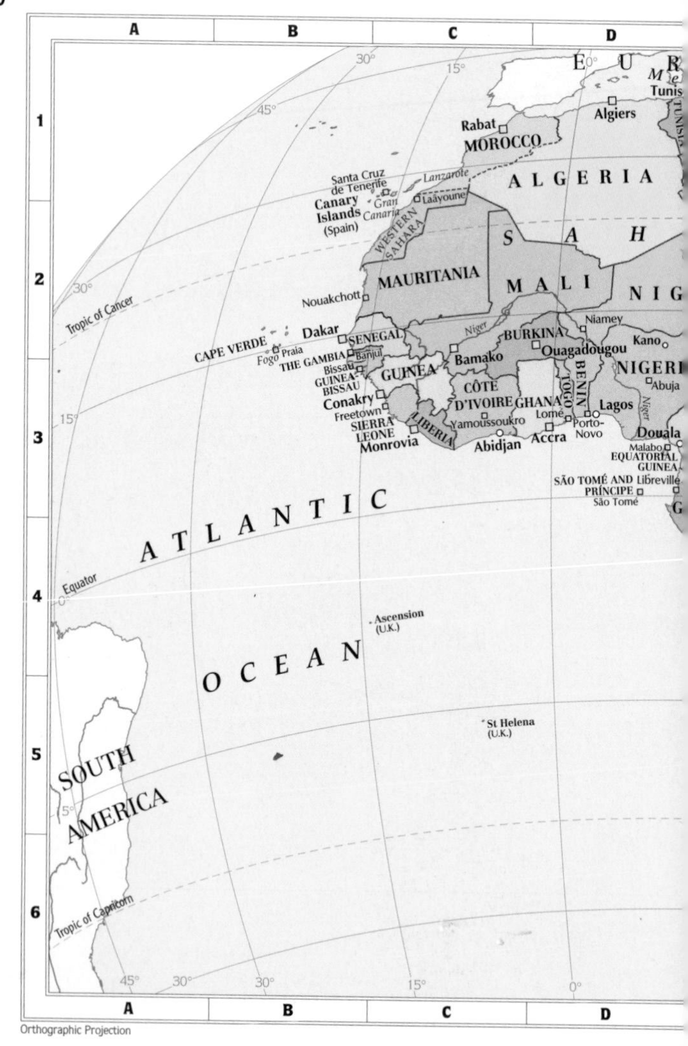

Orthographic Projection

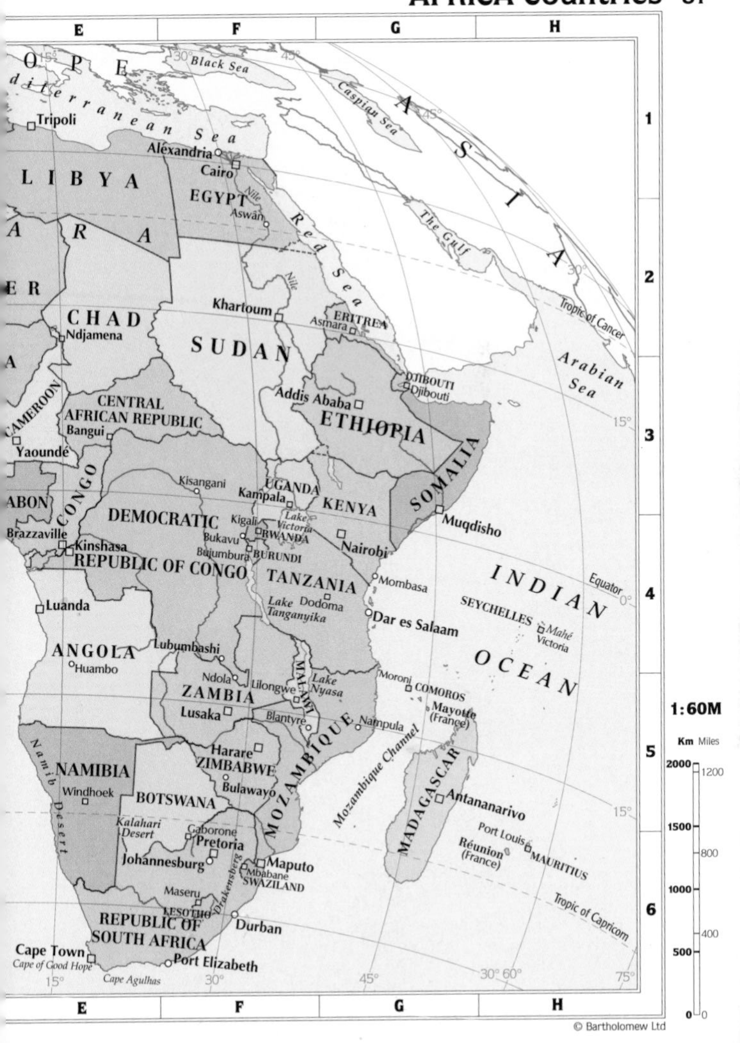
E
F
G
H
1
2
3
4
5
6
Black Sea
Caspian Sea
ASIA
Mediterranean Sea
Tripoli
Alexandria
Cairo
LIBYA
EGYPT
Nile
Aswân
Red Sea
The Gulf
Tropic of Cancer
Khartoum
CHAD
Ndjamena
SUDAN
ERITREA
Asmara
DJIBOUTI
Djibouti
Arabian Sea
CAMEROON
CENTRAL AFRICAN REPUBLIC
Bangui
Yaoundé
Addis Ababa
ETHIOPIA
SOMALIA
CONGO
Kisangani
UGANDA
Kampala
KENYA
Muqdisho
DEMOCRATIC REPUBLIC OF CONGO
Kigali
Lake Victoria
RWANDA
Bukavu
Bujumbura
BURUNDI
Nairobi
Brazzaville
Kinshasa
TANZANIA
Mombasa
INDIAN OCEAN
Equator
Luanda
Lake Tanganyika
Dodoma
Dar es Salaam
SEYCHELLES
Mahé
Victoria
ANGOLA
Huambo
Lubumbashi
Ndola
Lilongwe
MALAWI
Lake Nyasa
Moroni
COMOROS
ZAMBIA
Lusaka
Blantyre
Nampula
Mayotte (France)
Harare
ZIMBABWE
Bulawayo
MOZAMBIQUE
Mozambique Channel
MADAGASCAR
Antananarivo
NAMIBIA
Namib Desert
Windhoek
BOTSWANA
Kalahari Desert
Gaborone
Pretoria
Johannesburg
Maputo
Mbabane
SWAZILAND
Port Louis
Réunion (France)
MAURITIUS
Maseru
LESOTHO
Drakensberg
Tropic of Capricorn
REPUBLIC OF SOUTH AFRICA
Durban
Cape Town
Cape of Good Hope
Port Elizabeth
Cape Agulhas
1:60M
Km Miles
2000
1500
1000
500
0
1200
800
400
0

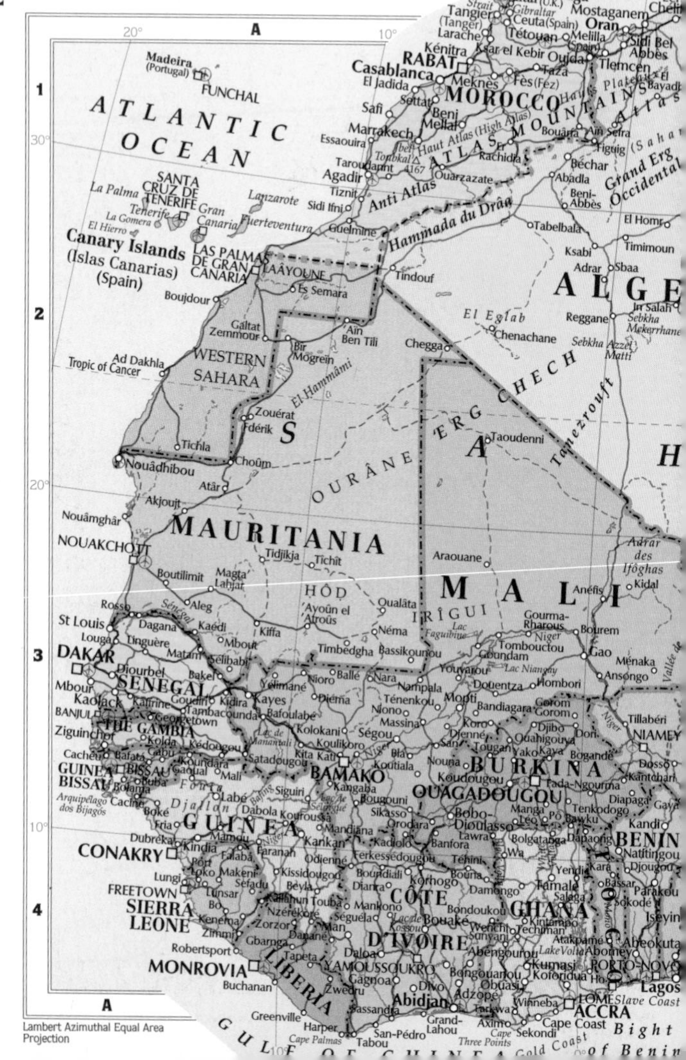

Lambert Azimuthal Equal Area Projection

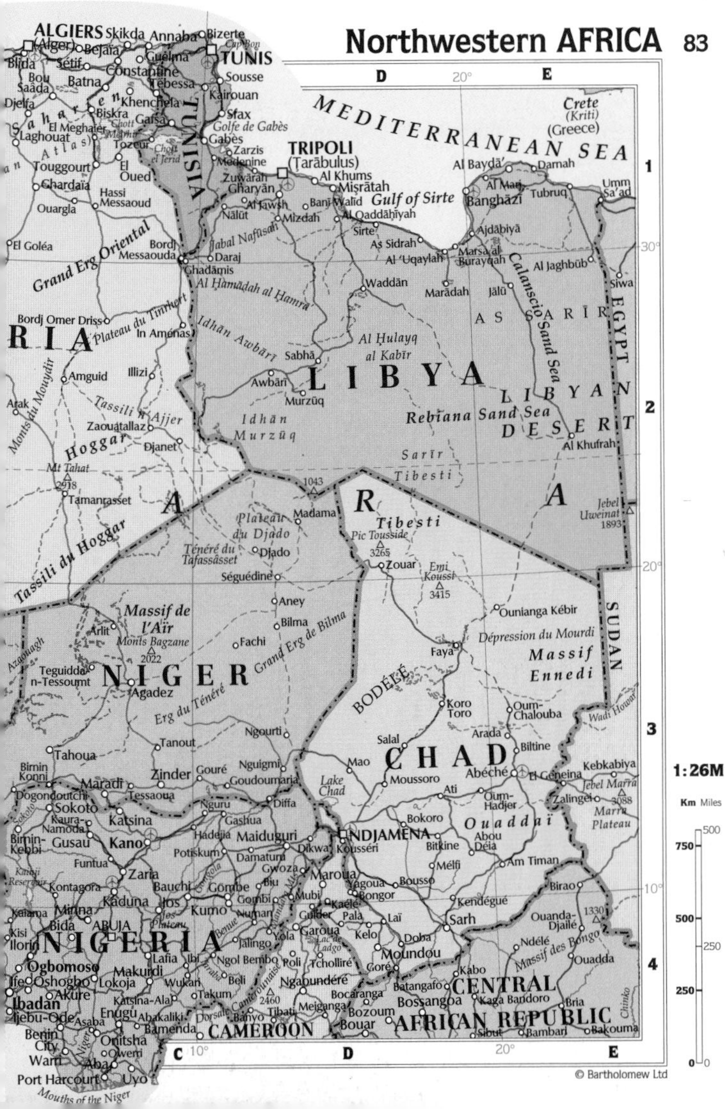

D 20° E

MEDITERRANEAN SEA

ALGIERS (Alger)
Skikda
Annaba
Bizerte
Cap Bon
TUNIS
Blida
Bejaïa
Sétif
Guelma
Constantine
Sousse
Bou Saâda
Batna
Tébessa
Kairouan
Djelfa
Khenchela
Biskra
Gafsa
Sfax
Sahara
El Meghaier
Chott Melrhir
Golfe de Gabès
Laghouat
Atlas
Tozeur
Chott el Jerid
Gabès
Zarzis
TUNISIA
Medenine
TRIPOLI (Ţarābulus)
Touggourt
El Oued
Zuwārah
Al Khums
Ghardaïa
Gharyān
Mişrātah
Hassi Messaoud
Al Jawsh
Banī Walīd
Gulf of Sirte
Ouargla
Nālūt
Mizdah
Al Qaddāhīyah
Sirte
Crete (Kriti) (Greece)
Al Baydā'
Darnah
Al Marj
Banghāzī
Tubruq
Umm Sa'ad
Ajdābiyā
El Goléa
Grand Erg Oriental
Bordj Messaouda
Jabal Nafūsah
Daraj
As Sidrah
Al 'Uqaylah
Marsá al Burayqah
Al Jaghbūb
Siwa
30°
Ghadāmis
Al Hamādah al Hamrā'
Waddān
Marādah
Jālū
Calanscio Sand Sea
AS SARĪR
EGYPT
Bordj Omer Driss
Plateau du Tinrhert
In Amenas
Idhān Awbārī
Al Hulayq al Kabīr
Sabhā
LIBYA
Amguid
Illizi
Awbārī
Murzūq
LIBYAN DESERT
Monts du Mouydir
Arak
Tassili n'Ajjer
Zaouatallaz
Idhān Murzūq
Rebiana Sand Sea
Al Khufrah
Hoggar
Djanet
Sarīr Tibesti
Mt Tahat 2918
1043
Tamanrasset
Madama
Jebel Uweinat 1893
Plateau du Djado
Tibesti
Pic Toussidé 3265
Ténéré du Tafassâsset
Djado
Zouar
Emi Koussi 3415
20°
Tassili du Hoggar
Séguédine
Aney
Ounianga Kébir
Massif de l'Aïr
Bilma
SUDAN
Arlit
Monts Bagzane 2022
Fachi
Grand Erg de Bilma
Dépression du Mourdi
Faya
Massif Ennedi
Azaouagh
Teguidda-n-Tessoumt
NIGER
Agadez
BODÉLÉ
Erg du Ténéré
Koro Toro
Oum-Chalouba
Wadi Howar
Ngourti
Arada
Tanout
Salal
Biltine
Tahoua
CHAD
Birnin Konni
Nguigmi
Mao
Abéché
El Geneina
Kebkabiya
Zinder
Gouré
Goudoumaria
Lake Chad
Moussoro
Jebel Marra 3088
Maradi
Ati
Oum-Hadjer
Zalingei
Dogondoutchi
Tessaoua
Diffa
Marra Plateau
Sokoto
Nguru
Ouaddaï
Kaura-Namoda
Katsina
Gashua
Bokoro
Birnin-Kebbi
Gusau
Kano
Hadejia
Maiduguri
NDJAMENA
Kousséri
Bitkine
Abou Déia
Potiskum
Dikwa
Damaturu
Funtua
Gongola
Gwoza
Maroua
Mélfi
Am Timan
Kainji Reservoir
Zaria
Kontagora
Bauchi
Gombe
Biu
Yagoua
Bousso
Birao
10°
Kaduna
Jos
Gombi
Mubi
Mandara Mts
Bongor
Kendégué
Kaiama
Minna
Kumo
Numan
Kaélé
Jos Plateau
Guider
Pala
Laï
Sarh
1330
Kisi
Bida
ABUJA
Benue
Garoua
Ouanda-Djailé
Ilorin
NIGERIA
Jalingo
Yola
Lac de Lagdo
Kelo
Doba
Ndélé
Massif des Bongo
Ogbomoso
Lafia
Ibi
Ngol Bembo
Roli
Tchollire
Moundou
Ouadda
Makurdi
Taraba
Goré
Kabo
Ife
Oshogbo
Lokoja
Wukari
Beli
Ngaoundéré
Batangafo
CENTRAL AFRICAN REPUBLIC
Ibadan
Akure
Katsina-Ala
Takum
Cameroun
2460
Bocaranga
Bossangoa
Kaga Bandoro
Bria
Ijebu-Ode
Enugu
Abakaliki
Dorsale Camerounaise
Banyo
Tibati
Meiganga
Bozoum
Chinko
Asaba
Bamenda
Benin City
Niger
Onitsha
CAMEROON
Bouar
Sibut
Bambari
Bakouma
Warri
Owerri
Aba
Port Harcourt
Uyo
Mouths of the Niger

C 10° D 20° E

1 2 3 4

1:26M

Km Miles
750 500
500
250
250
0 0

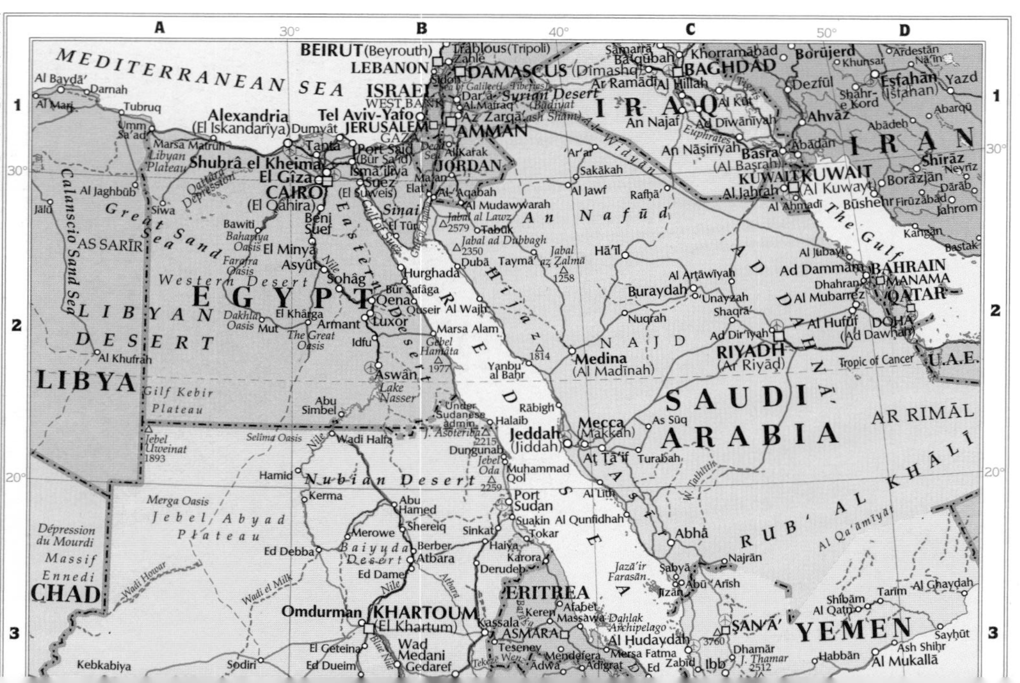
MEDITERRANEAN SEA
EGYPT
LIBYA
SAUDI ARABIA
IRAQ
IRAN
JORDAN
LEBANON
ISRAEL
KUWAIT
BAHRAIN
QATAR
U.A.E.
YEMEN
ERITREA
CHAD
CAIRO
(El Qahira)
BAGHDAD
DAMASCUS (Dimashq)
AMMAN
BEIRUT (Beyrouth)
JERUSALEM
RIYADH
(Ar Riyāḍ)
KHARTOUM
(El Khartum)
ASMARA
SAN'A'
DOHA
(Ad Dawḥah)
MANAMA
Alexandria
(El Iskandarîya)
Tel Aviv-Yafo
Port Said
(Bûr Sa'îd)
Suez
(El Suweis)
Medina
(Al Madīnah)
Mecca
(Makkah)
Jeddah
(Jiddah)
Port Sudan
Omdurman
Aswân
Luxor
Asyût
Basra
(Al Baṣrah)
Eşfahān
(Isfahan)
Shīrāz
Yazd
The Gulf
Tropic of Cancer
AR RIMĀL
RUB' AL KHĀLĪ
An Nafūd
AD DAHNĀ'
NAJD
Syrian Desert
Eastern Desert
Western Desert
LIBYAN DESERT
Great Sand Sea
Nubian Desert
Calanscio Sand Sea
Gulf of Suez
Lake Nasser
Nile
Blue Nile
Euphrates
Tigris
A
B
C
D
1
2
3
30°
40°
50°
20°

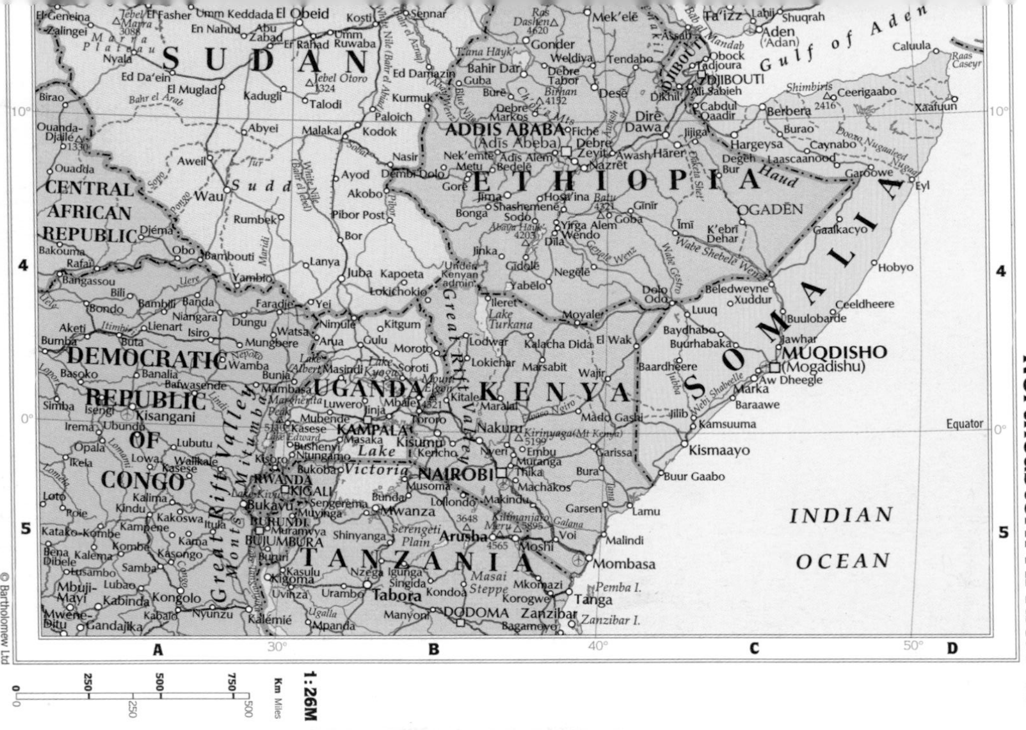
SUDAN
ETHIOPIA
SOMALIA
KENYA
UGANDA
TANZANIA
RWANDA
BURUNDI
DJIBOUTI
CENTRAL AFRICAN REPUBLIC
DEMOCRATIC REPUBLIC OF CONGO
INDIAN OCEAN
Gulf of Aden
ADDIS ABABA (Ādīs Ābeba)
MUQDISHO (Mogadishu)
NAIROBI
KAMPALA
KIGALI
BUJUMBURA
DODOMA
Lake Victoria
Lake Turkana
Great Rift Valley
Equator
OGADĒN
Sudd
Mombasa
Zanzibar
Kismaayo
Hargeysa
Berbera
Aden ('Adan)
Kisangani
1:26M
Km Miles
0
250
500
750
0
250
500

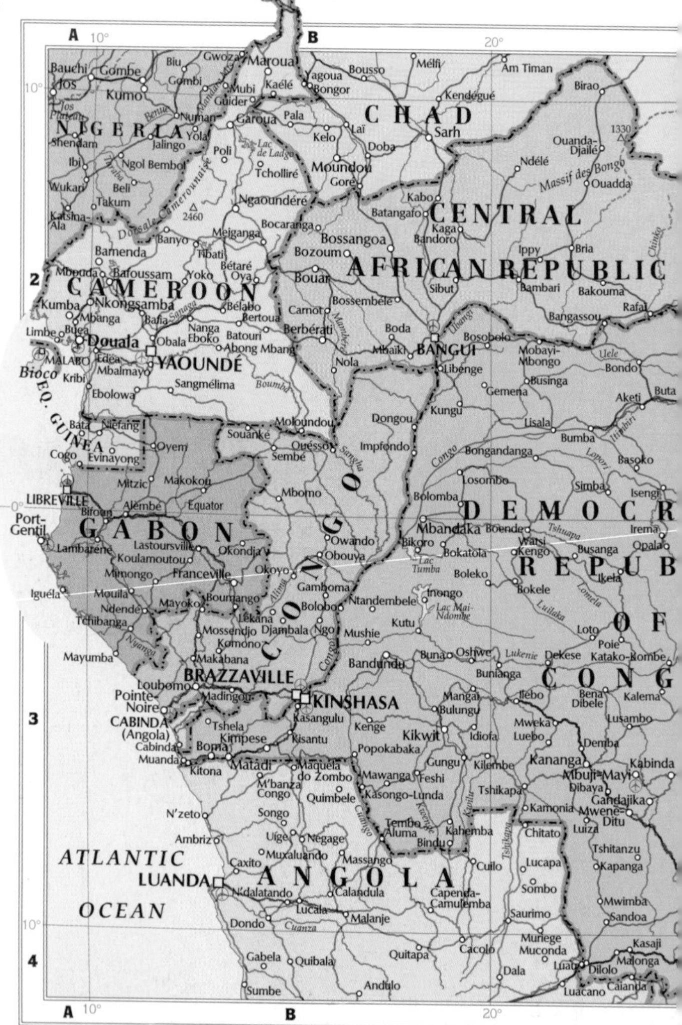

Lambert Azimuthal Equal Area Projection

SUDAN
ETHIOPIA
ADDIS ABABA
(Adis Abeba)
UGANDA
KENYA
TANZANIA
ZAMBIA
MALAWI
MOZAMBIQUE
RWANDA
BURUNDI
KAMPALA
NAIROBI
KIGALI
BUJUMBURA
DODOMA
Lake Victoria
Lake Turkana
Lake Tanganyika
Great Rift Valley
Khartoum
Juba
Wau
Malakal
Kisumu
Nakuru
Mombasa
Arusha
Moshi
Mwanza
Tabora
Dar es Salaam
Zanzibar
Mbeya
Kasama
Lubumbashi
Ndola
Kisangani
Bukavu
Goma
Kigoma
1:20M
Km Miles
© Bartholomew Ltd

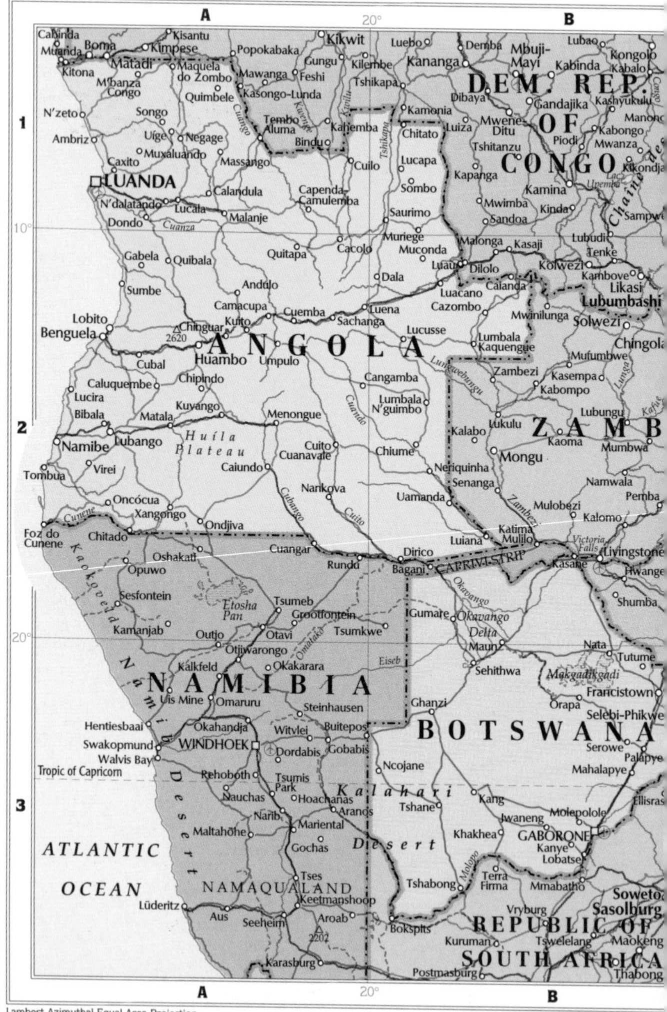

Lambert Azimuthal Equal Area Projection

TANZANIA
MALAWI
MOZAMBIQUE
ZIMBABWE
MADAGASCAR
COMOROS
SWAZILAND
DODOMA
Dar es Salaam
LUSAKA
HARARE
LILONGWE
MAPUTO
MBABANE
PRETORIA
MORONI
ANTANANARIVO
INDIAN OCEAN
Mayotte (France)
Tropic of Capricorn
1:20M
Km Miles
© Bartholomew Ltd

REPUBLIC OF SOUTH AFRICA

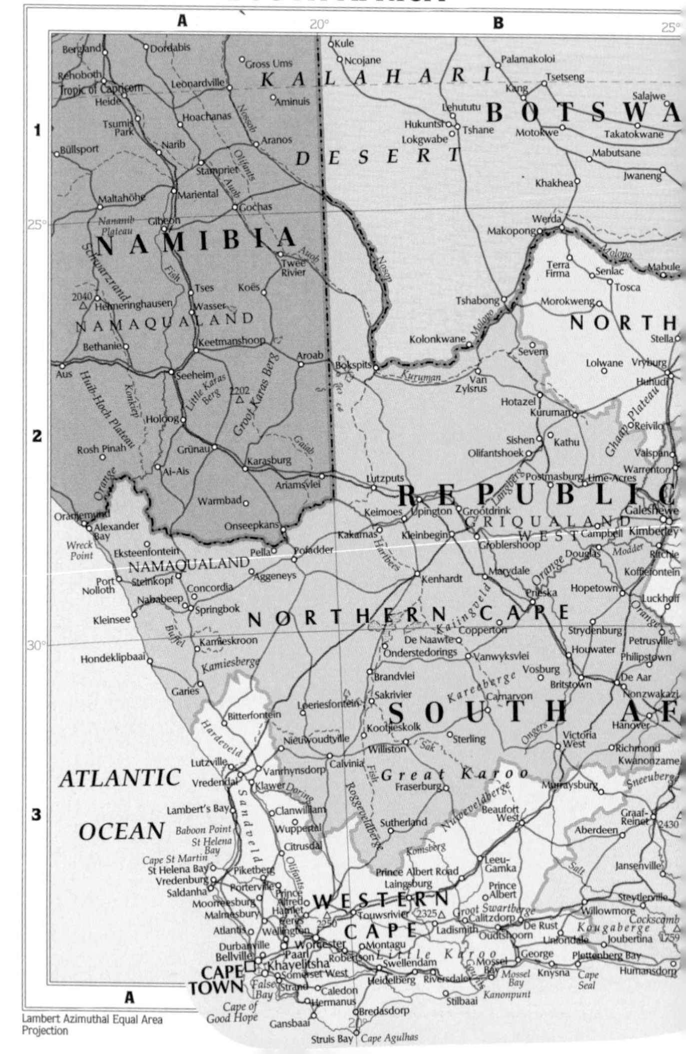

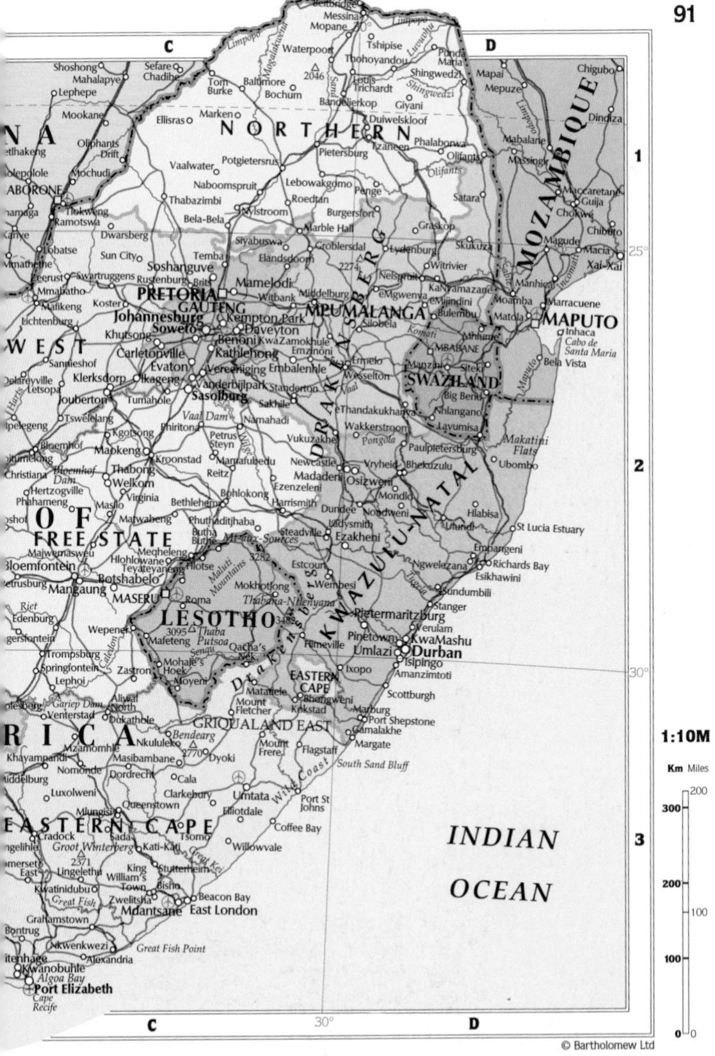
NORTHERN
MPUMALANGA
GAUTENG
PRETORIA
Johannesburg
Soweto
WEST
FREE STATE
LESOTHO
MASERU
SWAZILAND
MBABANE
KWAZULU-NATAL
MOZAMBIQUE
MAPUTO
EASTERN CAPE
GRIQUALAND EAST
Durban
Pietermaritzburg
East London
Port Elizabeth
Bloemfontein
INDIAN
OCEAN
DRAKENSBERG
1:10M
Km Miles

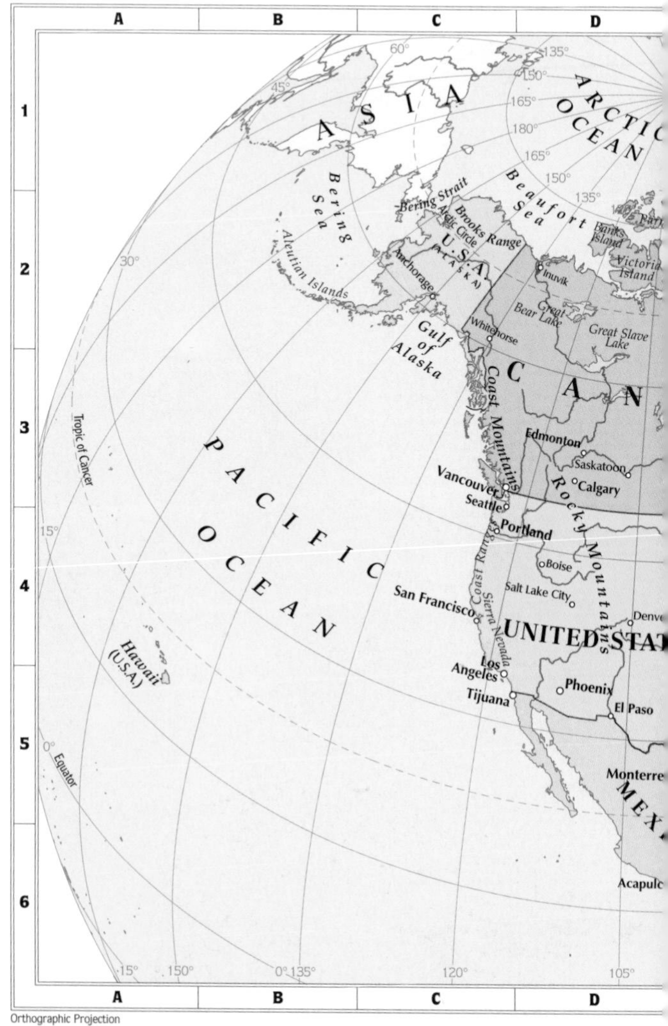

Orthographic Projection

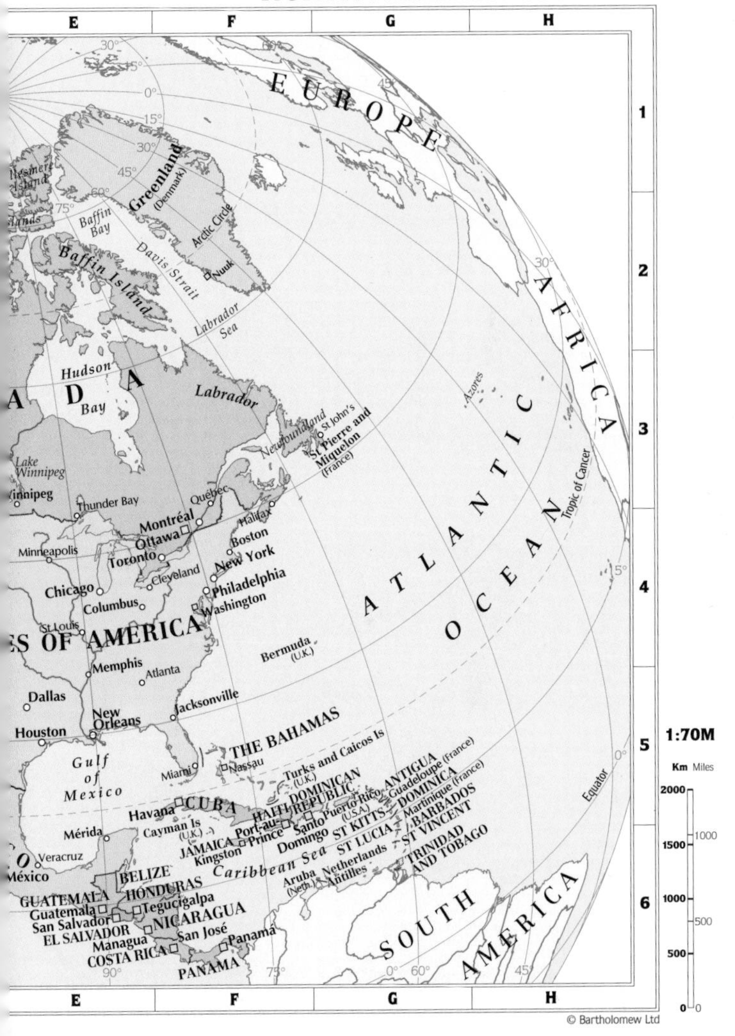
E
F
G
H
1
2
3
4
5
6
EUROPE
AFRICA
Greenland
(Denmark)
Baffin Bay
Baffin Island
Davis Strait
Arctic Circle
Nuuk
Labrador Sea
Hudson Bay
Labrador
Lake Winnipeg
Newfoundland
St John's
St Pierre and Miquelon
(France)
Azores
Tropic of Cancer
Thunder Bay
Québec
Montréal
Ottawa
Halifax
Boston
Minneapolis
Toronto
New York
Cleveland
Chicago
Philadelphia
Columbus
Washington
St Louis
ATLANTIC
OCEAN
Bermuda
(U.K.)
Memphis
Atlanta
Dallas
Jacksonville
New Orleans
Houston
THE BAHAMAS
Nassau
Gulf of Mexico
Miami
Turks and Caicos Is
(U.K.)
ANTIGUA
Guadeloupe (France)
DOMINICAN REPUBLIC
HAITI
CUBA
Havana
Puerto Rico
(U.S.A.)
DOMINICA
Martinique (France)
Equator
BARBADOS
Port-au-Prince
Santo Domingo
ST KITTS
ST LUCIA
ST VINCENT
Mérida
Cayman Is
(U.K.)
JAMAICA
Kingston
Caribbean Sea
Netherlands Antilles
TRINIDAD AND TOBAGO
Veracruz
México
BELIZE
Aruba
(Neth.)
GUATEMALA
HONDURAS
Guatemala
Tegucigalpa
SOUTH AMERICA
San Salvador
NICARAGUA
EL SALVADOR
Managua
San José
Panamá
COSTA RICA
PANAMA
1:70M
Km Miles
2000
1500
1000
500
0
1000
500
0

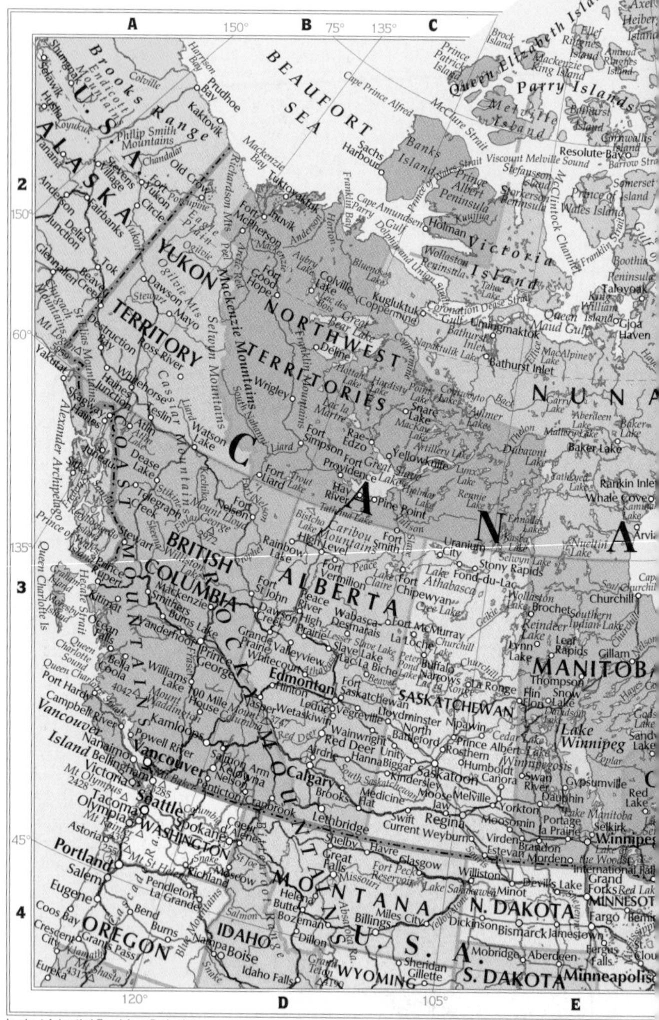
BEAUFORT SEA
YUKON TERRITORY
NORTHWEST TERRITORIES
BRITISH COLUMBIA
ALBERTA
SASKATCHEWAN
MANITOBA
CANADA
ALASKA
U.S.A.
WASHINGTON
OREGON
IDAHO
MONTANA
WYOMING
N. DAKOTA
S. DAKOTA
MINNESOT
Brooks Range
ROCKY MOUNTAINS
COAST MOUNTAINS
Mackenzie Mountains
Victoria Island
Banks Island
Queen Elizabeth Islands
Parry Islands
Melville Island
Great Bear Lake
Great Slave Lake
Lake Athabasca
Lake Winnipeg
Edmonton
Calgary
Vancouver
Seattle
Portland
Regina
Saskatoon
Winnipeg
Yellowknife
Whitehorse
Fairbanks
Juneau
Victoria
Spokane
Boise
Helena
Bismarck
Minneapolis
Churchill
Inuvik
Lambert Azimuthal Equal Area Projection

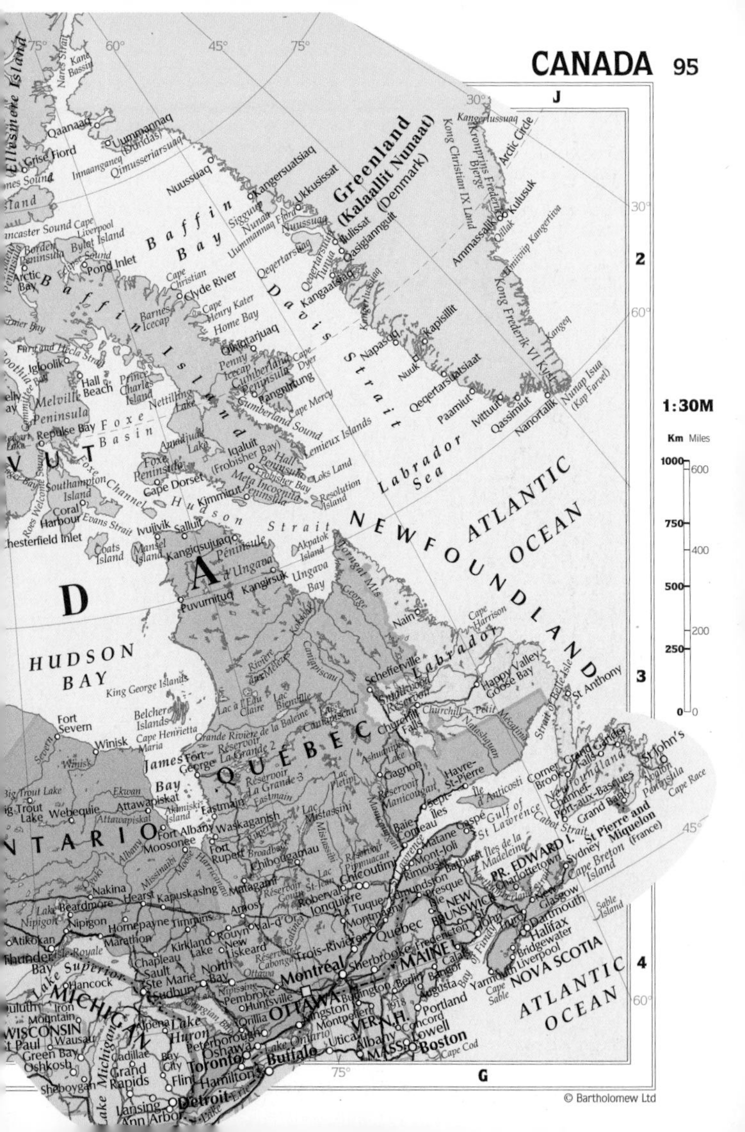
Baffin Bay
Baffin Island
Greenland (Kalaallit Nunaat) (Denmark)
Davis Strait
Labrador Sea
Hudson Strait
HUDSON BAY
James Bay
QUÉBEC
NEWFOUNDLAND
ATLANTIC OCEAN
NOVA SCOTIA
NEW BRUNSWICK
PR. EDWARD I.
St Pierre and Miquelon (France)
Gulf of St Lawrence
MAINE
MICHIGAN
WISCONSIN
ONTARIO
Lake Superior
Lake Huron
Lake Ontario
Montréal
OTTAWA
Québec
Toronto
Buffalo
Detroit
Boston
Halifax
St John's
Nuuk
Iqaluit
Arctic Circle
1:30M
Km Miles
1000 600
750 400
500
250 200
0 0
J
G
2
3
4

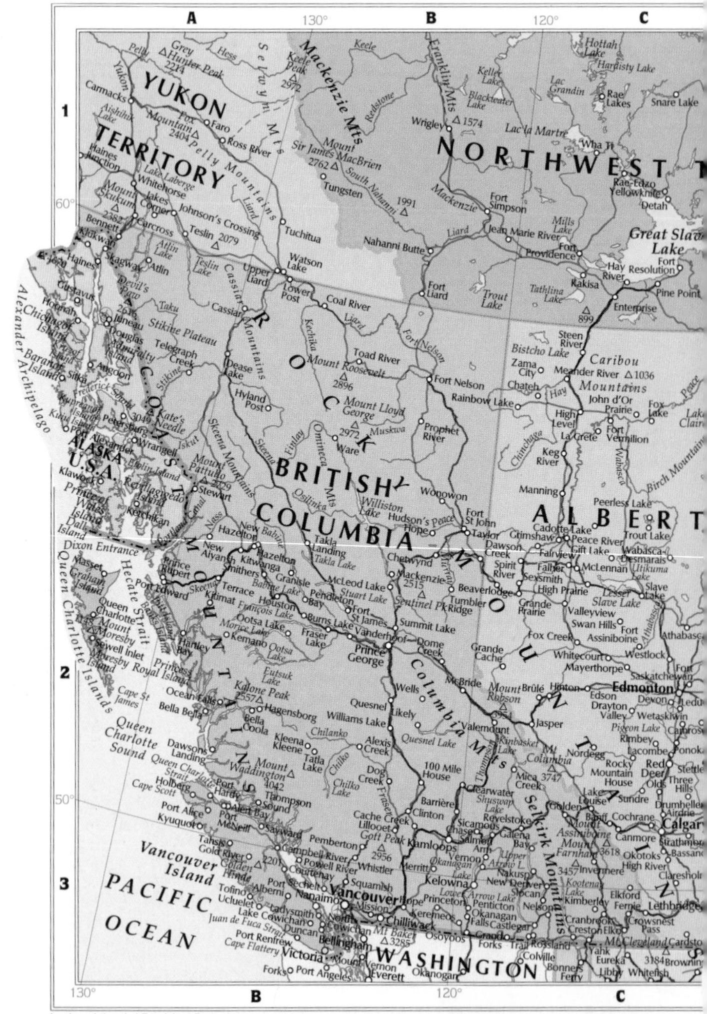

Lambert Azimuthal Equal Area Projection

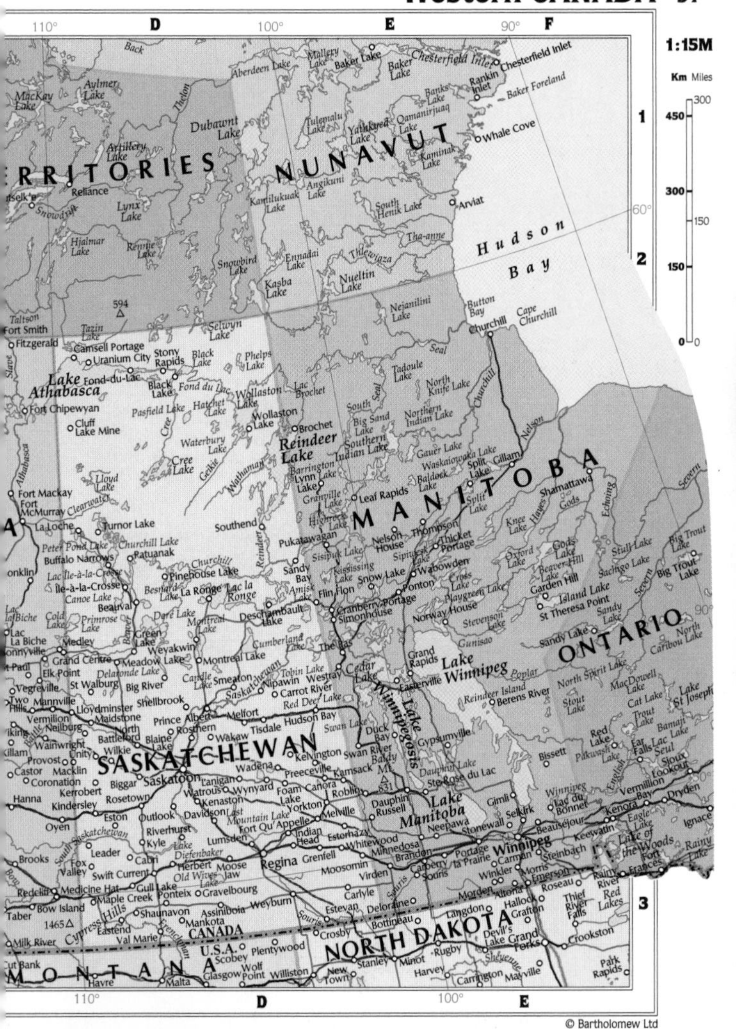
1:15M
Km Miles
NUNAVUT
MANITOBA
SASKATCHEWAN
ONTARIO
NORTH DAKOTA
MONTANA
Hudson Bay
Lake Athabasca
Reindeer Lake
Lake Winnipeg
Lake Manitoba
Lake Winnipegosis
Dubawnt Lake
Churchill
Thompson
Flin Flon
The Pas
Winnipeg
Regina
Saskatoon
Prince Albert
Brandon
Medicine Hat
Moose Jaw
Swift Current
Rankin Inlet
Arviat
Whale Cove
Baker Lake
Chesterfield Inlet
CANADA
U.S.A.
© Bartholomew Ltd

Lambert Azimuthal Equal Area Projection

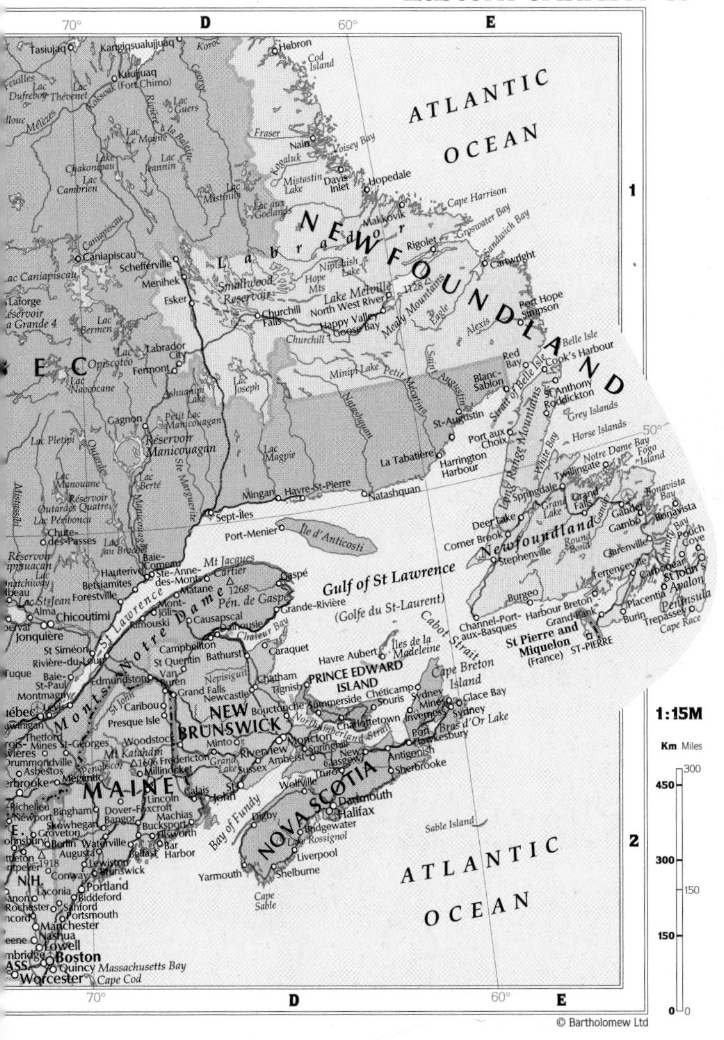
70°
D
60°
E
1
2
ATLANTIC
OCEAN
NEWFOUNDLAND
Labrador
Hebron
Cod Island
Tasiujaq
Kangiqsualujjuaq
Koroc
Kuujjuaq
Koksoak (Fort Chimo)
George
Lac Dufreboy
Lac Thévenet
Mélèzes
Rivière à la Baleine
Lac Guers
Lac Le Moyne
Lac Jeannin
Lake Chakonipau
Lac Cambrien
Lac Mistinibi
Fraser
Nain
Voisey Bay
Kogaluk
Mistastin Lake
Davis Inlet
Hopedale
Lac aux Goélands
Cape Harrison
Makkovik
Groswater Bay
Sandwich Bay
Rigolet
Cartwright
Caniapiscau
Lac Caniapiscau
Schefferville
Menihek
Esker
Smallwood Reservoir
Nipishish Lake
Hope Mts
Lake Melville
North West River
1128
Mealy Mountains
Churchill Falls
Happy Valley
Goose Bay
Churchill
Eagle
Alexis
Port Hope Simpson
Réservoir La Grande 4
Lac Bermen
Labrador City
Fermont
Lac Opiscotéo
Lac Nacocane
Ashuanipi Lake
Minipi Lake
Petit Mécatina
Saint Augustin
Natashquan
Lac Joseph
Red Bay
Belle Isle
Cook's Harbour
Blanc-Sablon
Strait of Belle Isle
St Anthony
Roddickton
Grey Islands
St-Augustin
Port aux Choix
Horse Islands
50°
Gagnon
Petit Lac Manicouagan
Réservoir Manicouagan
Lac Pletipi
Outardes
Ste Marguerite
Lac Magpie
La Tabatière
Harrington Harbour
Long Range Mountains
White Bay
Notre Dame Bay
Fogo Island
Twillingate
Lac Manouane
Réservoir Outardes Quatre
Lac Berté
Lac Péribonca
Mistassibi
Chute-des-Passes
Manicouagan
Mingan
Havre-St-Pierre
Natashquan
Sept-Îles
Springdale
Grand Lake
Grand Falls
Gander
Bonavista Bay
Bonavista
Deer Lake
Gambo
Port-Menier
Île d'Anticosti
Corner Brook
Newfoundland
Round Pond
Trinity Bay
Pouch Cove
Clarenville
Stephenville
Baie-Comeau
Hauterive
Ste-Anne-des-Monts
Mt Jacques Cartier
1268
Gaspé
Gulf of St Lawrence
(Golfe du St-Laurent)
Terrenceville
Carbonear
St John's
Avalon Peninsula
Betsiamites
Forestville
Matane
Pén. de Gaspé
Grande-Rivière
Burgeo
Placentia
Lac St-Jean
Alma
Chicoutimi
Jonquière
St Lawrence
Mont-Joli
Rimouski
Causapscal
Dalhousie
Chaleur Bay
Monts Notre Dame
Cabot Strait
Channel-Port-aux-Basques
Harbour Breton
Grand Bank
Burin
Trepassey
Cape Race
St Pierre and Miquelon (France)
ST-PIERRE
St Siméon
Rivière-du-Loup
Campbellton
Bathurst
Caraquet
Îles de la Madeleine
Havre Aubert
St Quentin
Edmundston
Van Buren
Nepisiguit
Chatham
Tignish
PRINCE EDWARD ISLAND
Cape Breton Island
Baie-St-Paul
Montmagny
Grand Falls
Newcastle
Chéticamp
Sydney Mines
Glace Bay
Sydney
Caribou
Presque Isle
NEW BRUNSWICK
Bouctouche
Summerside
Souris
Inverness
Charlottetown
Northumberland Strait
Bras d'Or Lake
Port Hawkesbury
Thetford Mines
St-Georges
Woodstock
Minto
Moncton
Riverview
Springhill
Amherst
New Glasgow
Antigonish
Sherbrooke
Drummondville
Asbestos
Mt Katahdin
1606
Fredericton
Grand Lake
Sussex
Truro
Millinocket
Mégantic
Penobscot
Wolfville
NOVA SCOTIA
MAINE
St John
Calais
Lincoln
Dartmouth
Halifax
Bingham
Dover-Foxcroft
Machias
Bangor
Bucksport
Ellsworth
Skowhegan
Digby
Bridgewater
Lake Rossignol
Bay of Fundy
Sable Island
Groveton
Berlin
Waterville
Augusta
Bar Harbor
Belfast
Lewiston
Brunswick
1918
Liverpool
Shelburne
Yarmouth
Conway
N.H.
Portland
Laconia
Biddeford
Rochester
Sanford
Portsmouth
Manchester
Nashua
Lowell
Boston
Quincy
Massachusetts Bay
Worcester
Cape Cod
Cape Sable
1:15M
Km Miles
450
300
150
0

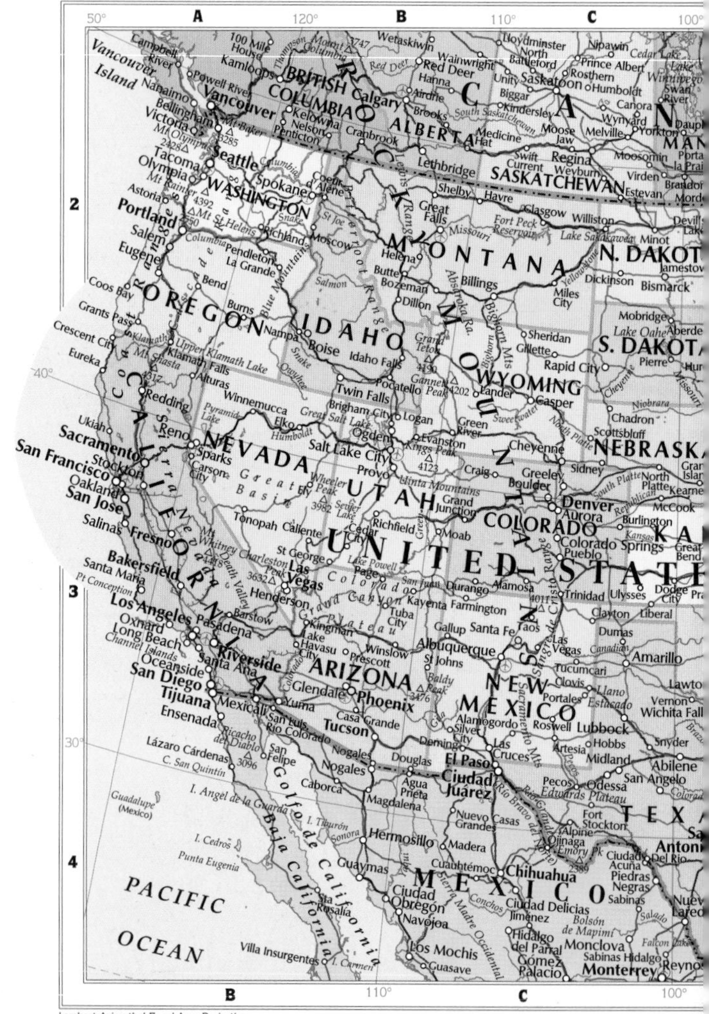

Lambert Azimuthal Equal Area Projection

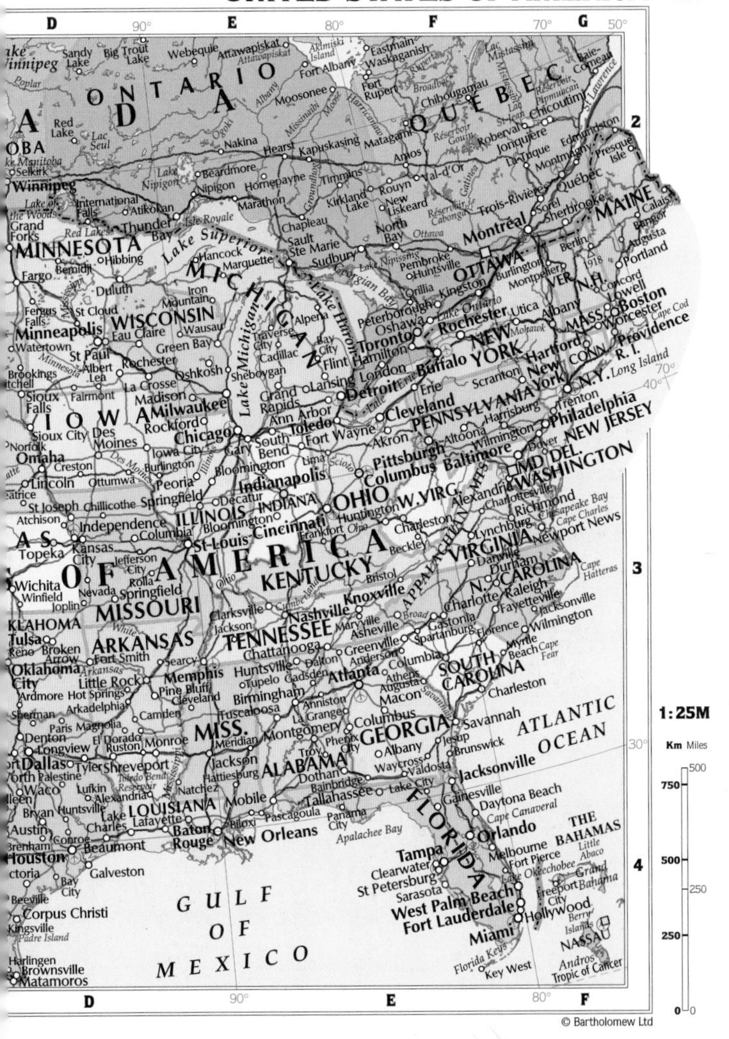
ONTARIO
QUÉBEC
MINNESOTA
MICHIGAN
WISCONSIN
IOWA
ILLINOIS
INDIANA
OHIO
PENNSYLVANIA
NEW YORK
NEW JERSEY
MAINE
VER.
N.H.
MASS.
CONN.
R.I.
MD.
DEL.
WASHINGTON
W. VIRG.
VIRGINIA
KENTUCKY
TENNESSEE
MISSOURI
ARKANSAS
MISS.
ALABAMA
GEORGIA
N. CAROLINA
SOUTH CAROLINA
FLORIDA
LOUISIANA
OF AMERICA
APPALACHIAN MTS
ATLANTIC OCEAN
GULF OF MEXICO
THE BAHAMAS
Chicago
Detroit
Toronto
Montreal
Ottawa
Boston
Philadelphia
Pittsburgh
Cleveland
Indianapolis
Cincinnati
St Louis
Memphis
Nashville
Atlanta
New Orleans
Houston
Dallas
Jacksonville
Orlando
Tampa
Miami
Minneapolis
Milwaukee
Lake Superior
Lake Michigan
Lake Huron
Lake Ontario
Lake Erie
1:25M
Km Miles
750
500
250
0
© Bartholomew Ltd

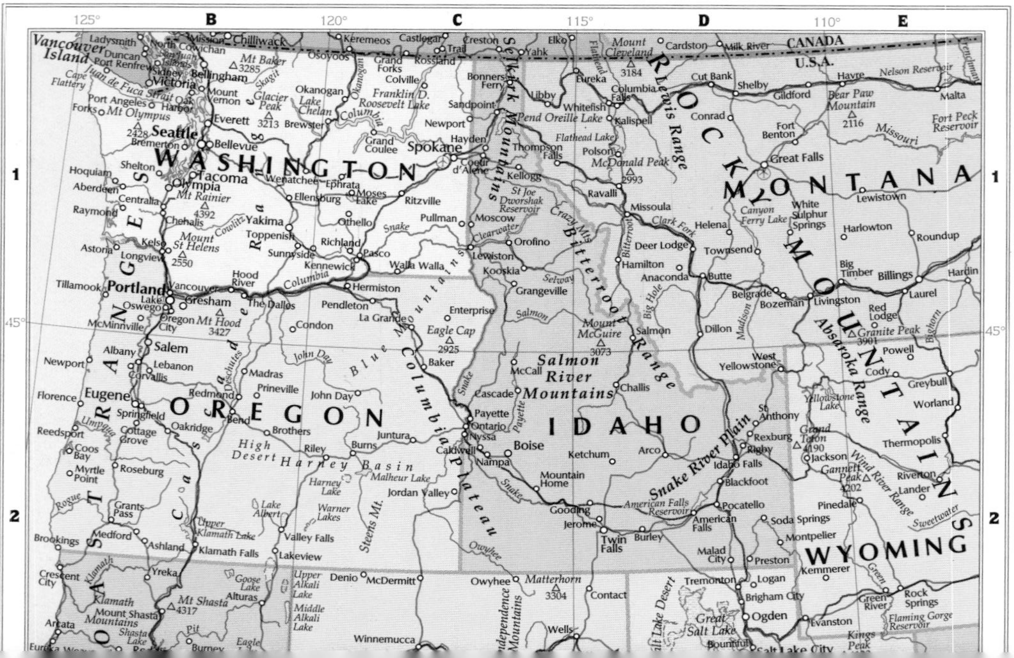

Lambert Azimuthal Equal Area Projection

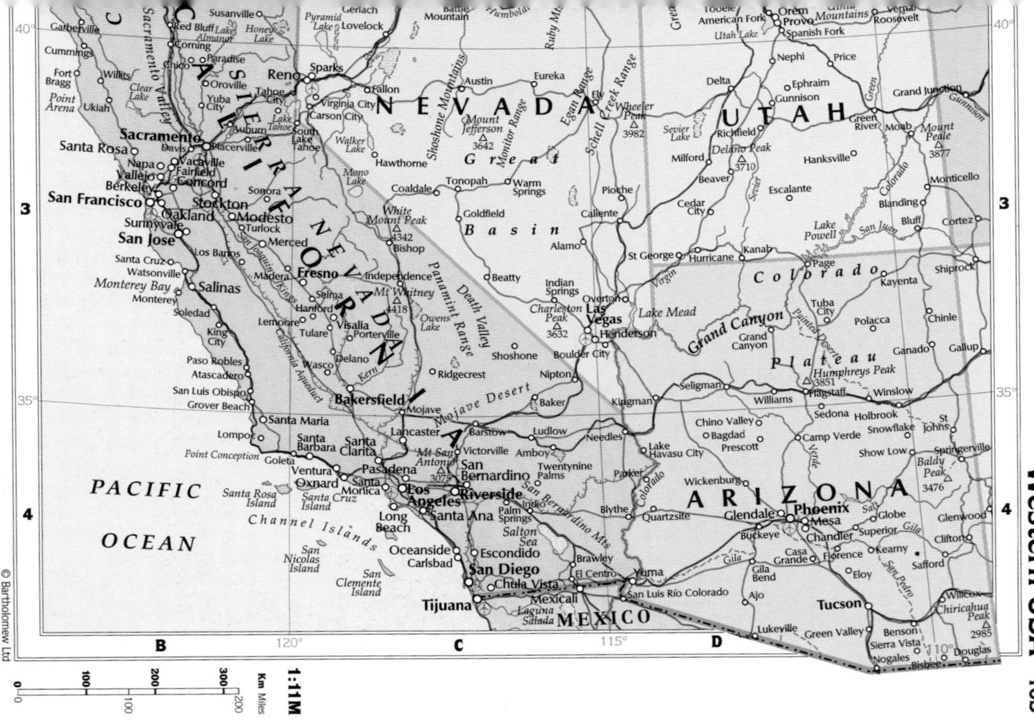
NEVADA
UTAH
ARIZONA
CALIFORNIA
SIERRA NEVADA
MEXICO
PACIFIC OCEAN
Great Basin
Colorado Plateau
Channel Islands
Sacramento Valley
Mojave Desert
Death Valley
Panamint Range
Shoshone Mountains
Monitor Range
Egan Range
Schell Creek Range
San Bernardino Mts
Salton Sea
Lake Mead
Lake Powell
Lake Tahoe
Mono Lake
Walker Lake
Pyramid Lake
Honey Lake
Lake Almanor
Clear Lake
Utah Lake
Sevier Lake
Owens Lake
Laguna Salada
California Aqueduct
Colorado
Green
San Juan
Gila
Verde
Sevier
San Pedro
Kern
Kings
San Joaquin
Virgin
Salt
Gunnison
Grand Canyon
Painted Desert
Humphreys Peak 3851
Mt Whitney 4418
White Mount Peak 4342
Mount Jefferson 3642
Charleston Peak 3632
Wheeler Peak 3982
Delano Peak 3710
Mount Peale 3877
Baldy Peak 3476
Chiricahua Peak 2985
Mt San Antonio 3075
Monterey Bay
Point Conception
Point Arena
Santa Rosa Island
Santa Cruz Island
San Nicolas Island
San Clemente Island
Garberville
Cummings
Fort Bragg
Willits
Ukiah
Red Bluff
Corning
Chico
Paradise
Oroville
Yuba City
Susanville
Sacramento
Santa Rosa
Davis
Napa
Vallejo
Berkeley
Vacaville
Fairfield
Concord
San Francisco
Oakland
Sunnyvale
San Jose
Stockton
Modesto
Turlock
Merced
Los Banos
Santa Cruz
Watsonville
Salinas
Monterey
Soledad
King City
Paso Robles
Atascadero
San Luis Obispo
Grover Beach
Santa Maria
Lompoc
Santa Barbara
Goleta
Ventura
Oxnard
Santa Clarita
Santa Monica
Pasadena
Los Angeles
Long Beach
Santa Ana
Oceanside
Carlsbad
Escondido
San Diego
Chula Vista
Tijuana
Mexicali
El Centro
Brawley
Indio
Palm Springs
Riverside
San Bernardino
Victorville
Barstow
Lancaster
Mojave
Bakersfield
Delano
Wasco
Ridgecrest
Tulare
Visalia
Porterville
Hanford
Lemoore
Selma
Fresno
Madera
Sonora
Auburn
Placerville
Tahoe City
South Lake Tahoe
Reno
Sparks
Virginia City
Carson City
Fallon
Lovelock
Gerlach
Battle Mountain
Austin
Eureka
Ely
Hawthorne
Coaldale
Tonopah
Warm Springs
Goldfield
Bishop
Independence
Beatty
Shoshone
Baker
Nipton
Indian Springs
Las Vegas
Henderson
Boulder City
Overton
Alamo
Caliente
Pioche
Needles
Ludlow
Amboy
Twentynine Palms
Blythe
Parker
Quartzsite
Yuma
San Luis Río Colorado
Lake Havasu City
Kingman
Seligman
Williams
Flagstaff
Winslow
Holbrook
Sedona
Camp Verde
Chino Valley
Bagdad
Prescott
Wickenburg
Glendale
Phoenix
Mesa
Chandler
Buckeye
Gila Bend
Ajo
Casa Grande
Florence
Eloy
Tucson
Lukeville
Green Valley
Nogales
Sierra Vista
Benson
Bisbee
Douglas
Willcox
Safford
Kearny
Superior
Globe
Clifton
Glenwood
Show Low
Snowflake
St Johns
Springerville
Ganado
Gallup
Chinle
Polacca
Tuba City
Kayenta
Shiprock
Page
Grand Canyon
Kanab
Hurricane
St George
Cedar City
Beaver
Milford
Richfield
Delta
Gunnison
Ephraim
Nephi
Price
Spanish Fork
Provo
Orem
American Fork
Tooele
Vernal
Roosevelt
Green River
Grand Junction
Moab
Hanksville
Escalante
Monticello
Blanding
Bluff
Cortez
40°
35°
120°
115°
110°
3
4
B
C
D
1:11M
Km Miles
0
100
200
300
0
100
200

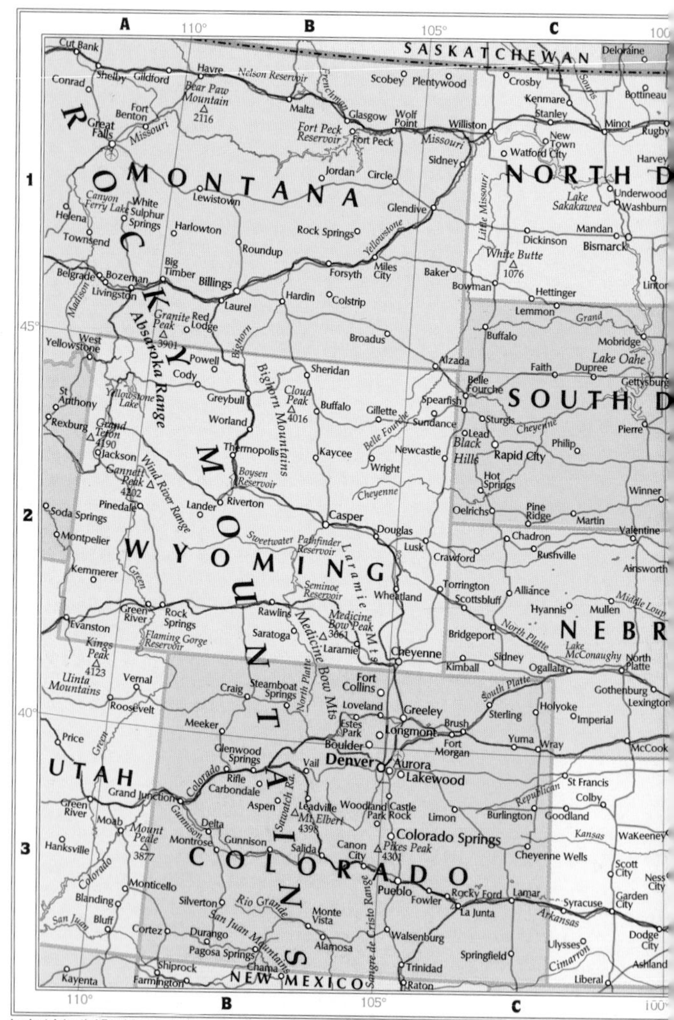
SASKATCHEWAN
MONTANA
NORTH D
SOUTH D
WYOMING
NEBR
UTAH
COLORADO
NEW MEXICO
ROCKY MOUNTAINS
Cut Bank
Shelby
Conrad
Havre
Great Falls
Helena
Townsend
Lewistown
Harlowton
Roundup
Billings
Bozeman
Livingston
Belgrade
Big Timber
Laurel
Hardin
Colstrip
Forsyth
Miles City
Glendive
Sidney
Williston
Circle
Jordan
Glasgow
Malta
Wolf Point
Scobey
Plentywood
Fort Peck
Fort Peck Reservoir
Missouri
Yellowstone
Bear Paw Mountain
2116
Minot
Bismarck
Mandan
Dickinson
Rugby
Bottineau
Crosby
Kenmare
Stanley
New Town
Watford City
Lake Sakakawea
Underwood
Washburn
White Butte
1076
Hettinger
Lemmon
Bowman
Baker
Buffalo
Mobridge
Lake Oahe
Faith
Dupree
Gettysburg
Pierre
Philip
Rapid City
Black Hills
Spearfish
Belle Fourche
Sturgis
Lead
Hot Springs
Winner
Pine Ridge
Martin
Valentine
Oelrichs
Chadron
Rushville
Crawford
Ainsworth
Alliance
Hyannis
Mullen
Scottsbluff
Torrington
Bridgeport
Sidney
Kimball
Ogallala
North Platte
Lake McConaughy
Gothenburg
Lexington
McCook
Imperial
Holyoke
Wray
Yuma
Sterling
Brush
Fort Morgan
Greeley
Longmont
Boulder
Denver
Aurora
Lakewood
Fort Collins
Loveland
Estes Park
Cheyenne
Laramie
Wheatland
Casper
Douglas
Lusk
Gillette
Sundance
Newcastle
Wright
Sheridan
Buffalo
Kaycee
Cloud Peak
4016
Bighorn Mountains
Thermopolis
Worland
Greybull
Cody
Powell
Riverton
Lander
Rawlins
Saratoga
Rock Springs
Green River
Evanston
Kemmerer
Pinedale
Jackson
Grand Teton
4190
Gannett Peak
4202
Wind River Range
Absaroka Range
Granite Peak
3901
Red Lodge
West Yellowstone
Yellowstone Lake
Rexburg
St Anthony
Soda Springs
Montpelier
Alzada
Broadus
Medicine Bow Peak
3661
Medicine Bow Mts
Laramie Mts
Seminoe Reservoir
Pathfinder Reservoir
Boysen Reservoir
Flaming Gorge Reservoir
Kings Peak
4123
Uinta Mountains
Vernal
Roosevelt
Price
Craig
Steamboat Springs
Meeker
Glenwood Springs
Rifle
Carbondale
Vail
Aspen
Leadville
Mt Elbert
4398
Grand Junction
Green River
Moab
Mount Peale
3877
Delta
Montrose
Gunnison
Salida
Canon City
Pikes Peak
4301
Colorado Springs
Woodland Park
Castle Rock
Limon
Burlington
Goodland
Colby
St Francis
WaKeeney
Cheyenne Wells
Scott City
Ness City
Garden City
Syracuse
Dodge City
Lamar
Rocky Ford
La Junta
Pueblo
Fowler
Walsenburg
Trinidad
Raton
Springfield
Ulysses
Liberal
Ashland
Alamosa
Monte Vista
Silverton
Durango
Pagosa Springs
Chama
Cortez
Bluff
Blanding
Monticello
Hanksville
Kayenta
Shiprock
Farmington
San Juan Mountains
Sangre de Cristo Range
Rio Grande
Arkansas
Cimarron
Republican
Kansas
South Platte
North Platte
Colorado
Gunnison
Sawatch Ra.
Green
Sweetwater
Cheyenne
Belle Fourche
Bighorn
Madison
Little Missouri
Grand
Frenchman
Souris
Nelson Reservoir
Canyon Ferry Lake
White Sulphur Springs
Fort Benton
Gildford
Deloraine
Harvey
Linton
A
B
C
1
2
3
110°
105°
45°
40°

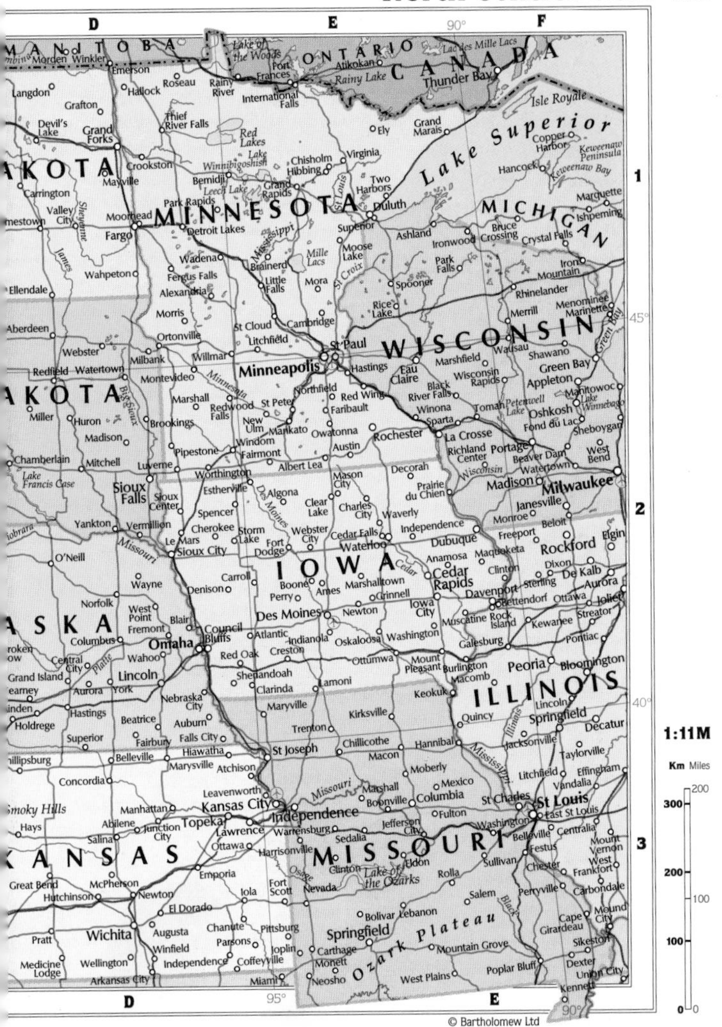

MINNESOTA
WISCONSIN
IOWA
MISSOURI
ILLINOIS
MICHIGAN
KANSAS
CANADA
ONTARIO
MANITOBA
Lake Superior
Lake of the Woods
Minneapolis
St Paul
Duluth
Milwaukee
Madison
Des Moines
Omaha
Kansas City
St Louis
Sioux Falls
Fargo
Thunder Bay
Rochester
Cedar Rapids
Springfield
Ozark Plateau
Lake of the Ozarks
Mississippi
Missouri
1:11M
Km Miles

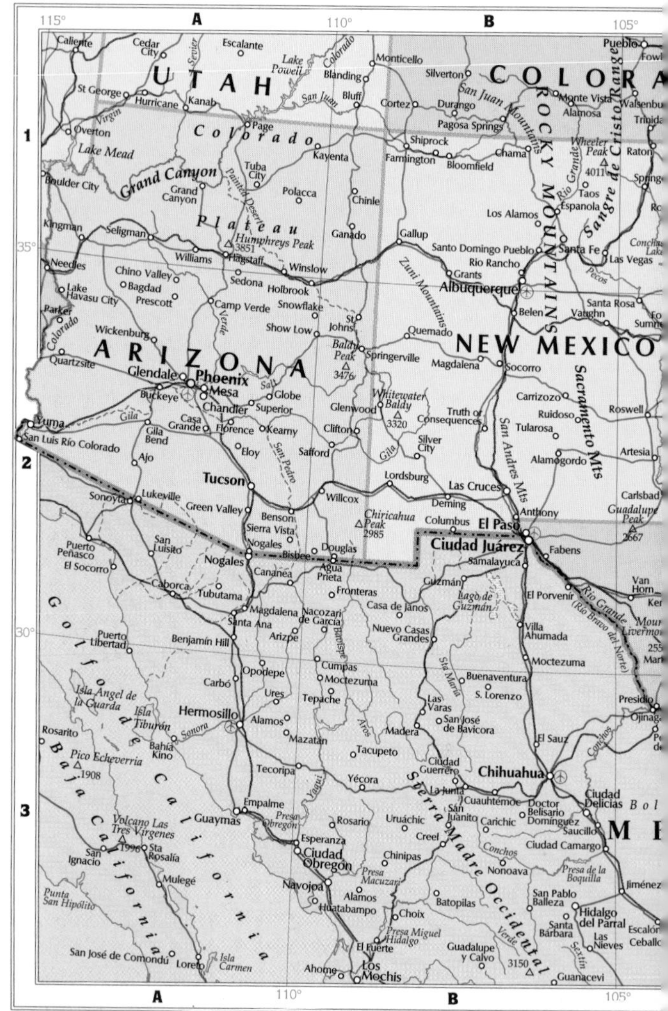
UTAH
COLORADO
ARIZONA
NEW MEXICO
ROCKY MOUNTAINS
Grand Canyon
Colorado Plateau
Sierra Madre Occidental
Golfo de California
Baja California
Sangre de Cristo Range
San Juan Mountains
Sacramento Mts
San Andres Mts
Zuni Mountains
Lake Mead
Lake Powell
Painted Desert
Humphreys Peak 3851
Baldy Peak 3476
Whitewater Baldy 3320
Chiricahua Peak 2985
Wheeler Peak 4011
Guadalupe Peak 2667
Pico Echeverria 1908
Volcano Las Tres Vírgenes 1996
Phoenix
Tucson
Albuquerque
El Paso
Ciudad Juárez
Chihuahua
Hermosillo
Santa Fe
Los Mochis
Ciudad Obregón
Guaymas
Nogales
Flagstaff
Yuma
Lambert Azimuthal Equal Area Projection

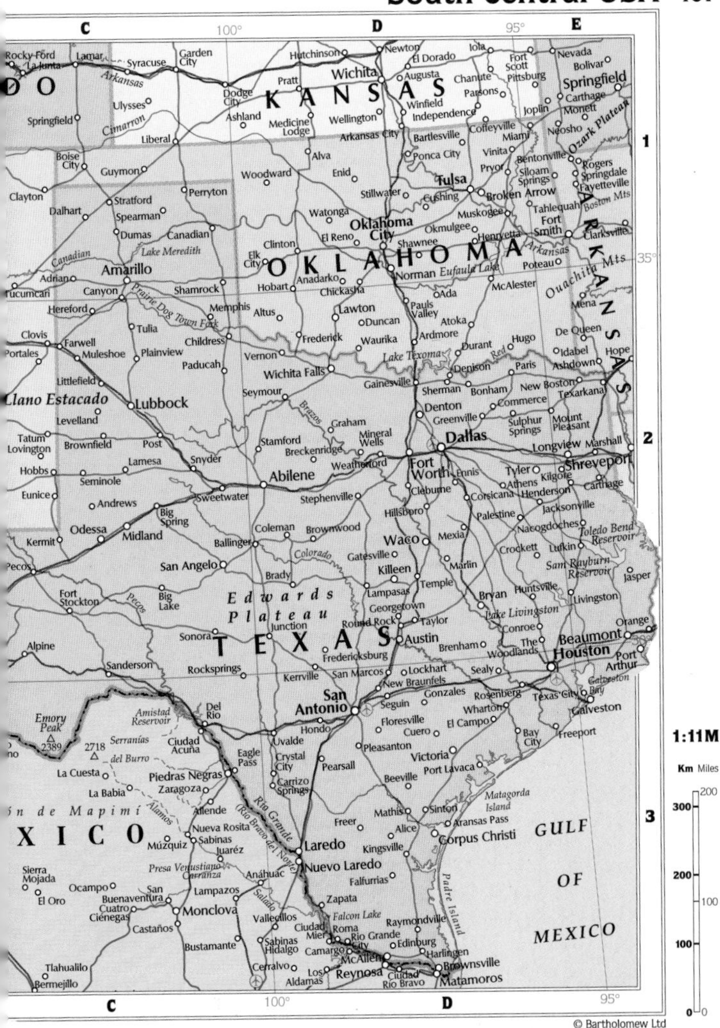
KANSAS
OKLAHOMA
TEXAS
ARKANSAS
MEXICO
GULF OF MEXICO
Wichita
Tulsa
Oklahoma City
Amarillo
Lubbock
Dallas
Fort Worth
Abilene
Shreveport
Springfield
Austin
San Antonio
Houston
Beaumont
Corpus Christi
Laredo
Nuevo Laredo
Brownsville
Matamoros
Llano Estacado
Edwards Plateau
Ozark Plateau
Ouachita Mts
Boston Mts
Rio Grande (Rio Bravo del Norte)
Padre Island
Matagorda Island
Lake Texoma
Lake Livingston
Amistad Reservoir
Emory Peak 2389
Serranías del Burro
1:11M
Km Miles
© Bartholomew Ltd

Lambert Azimuthal Equal Area Projection

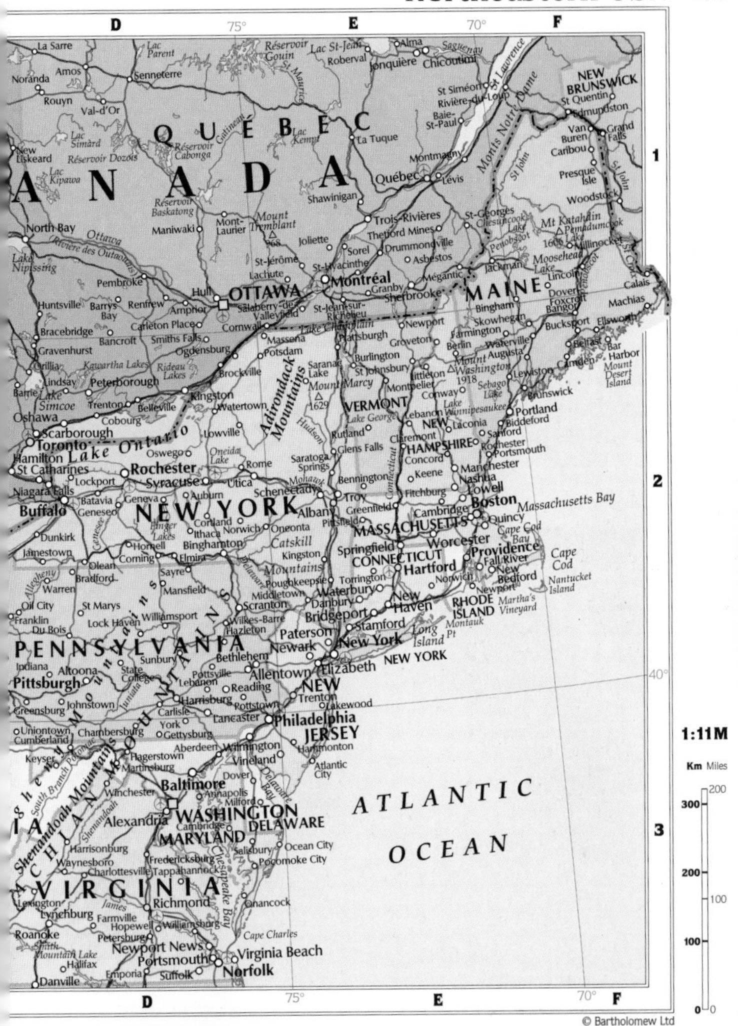

D
E
F
75°
70°
1
2
3
40°
QUÉBEC
CANADA
MAINE
VERMONT
NEW HAMPSHIRE
MASSACHUSETTS
CONNECTICUT
RHODE ISLAND
NEW YORK
PENNSYLVANIA
NEW JERSEY
DELAWARE
MARYLAND
VIRGINIA
NEW BRUNSWICK
ATLANTIC OCEAN
Lake Ontario
Adirondack Mountains
Catskill Mountains
Montréal
OTTAWA
Québec
Toronto
Boston
New York
Philadelphia
Baltimore
WASHINGTON
Pittsburgh
Buffalo
Rochester
Norfolk
Richmond
1:11M
Km Miles
300
200
100
0
200
100
0

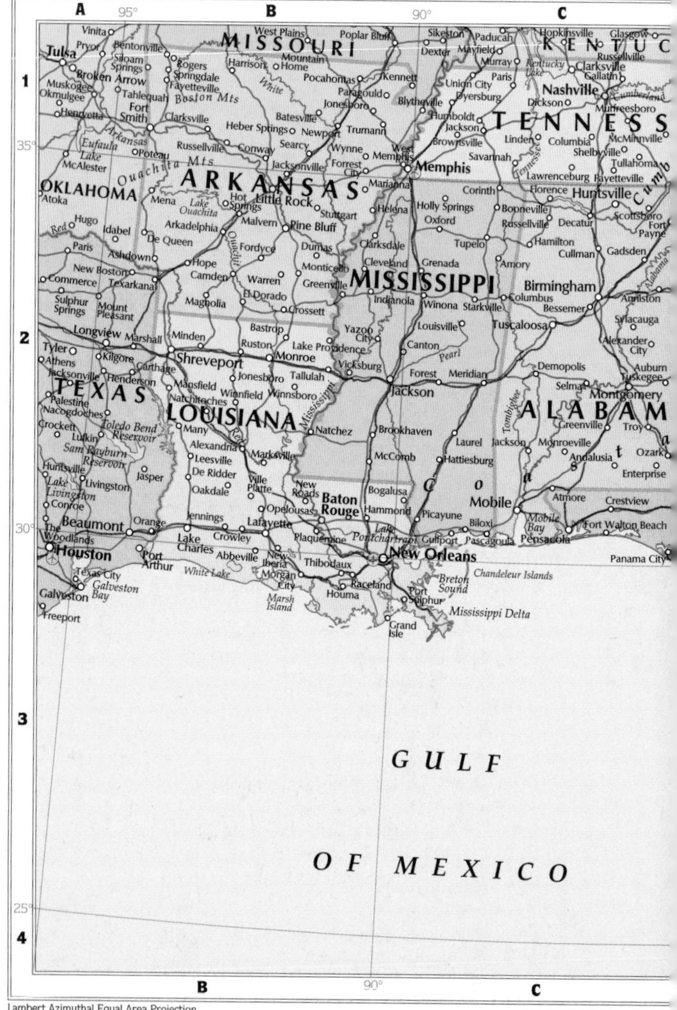

Lambert Azimuthal Equal Area Projection

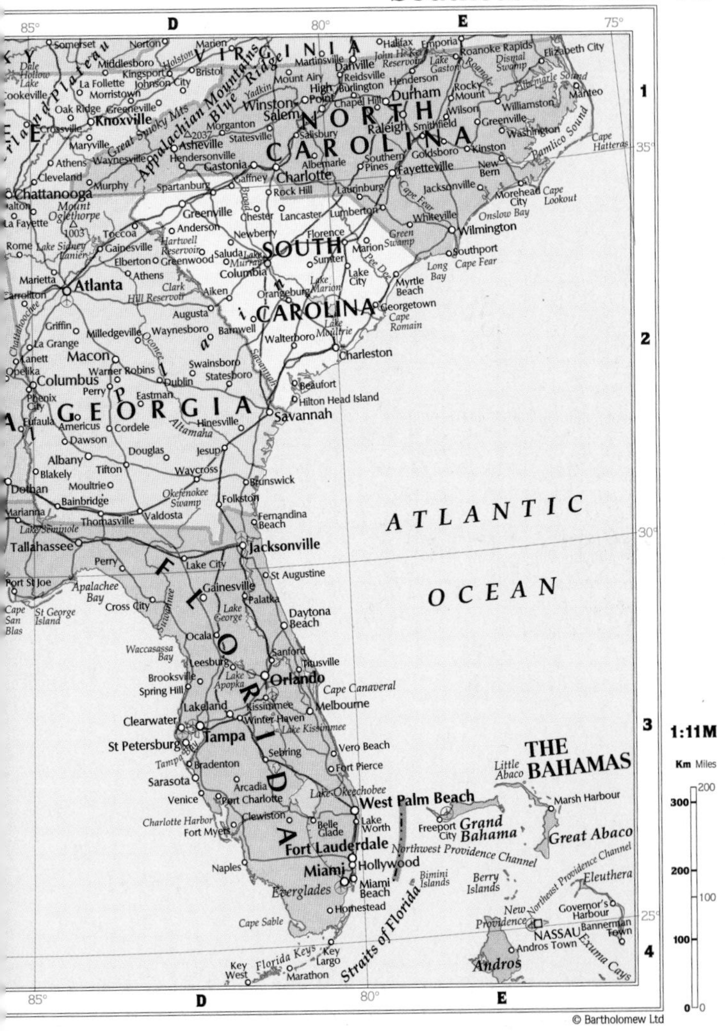
ATLANTIC
OCEAN
NORTH CAROLINA
SOUTH CAROLINA
GEORGIA
FLORIDA
VIRGINIA
THE BAHAMAS
Atlanta
Charlotte
Jacksonville
Savannah
Charleston
Orlando
Tampa
Miami
West Palm Beach
Fort Lauderdale
Knoxville
Chattanooga
Durham
Raleigh
Wilmington
Columbia
Augusta
Macon
Columbus
Tallahassee
Gainesville
St Petersburg
NASSAU
Straits of Florida
Florida Keys
Key West
Everglades
Cape Canaveral
Lake Okeechobee
Great Abaco
Grand Bahama
Andros
Eleuthera
Northwest Providence Channel
1:11M
Km Miles
© Bartholomew Ltd

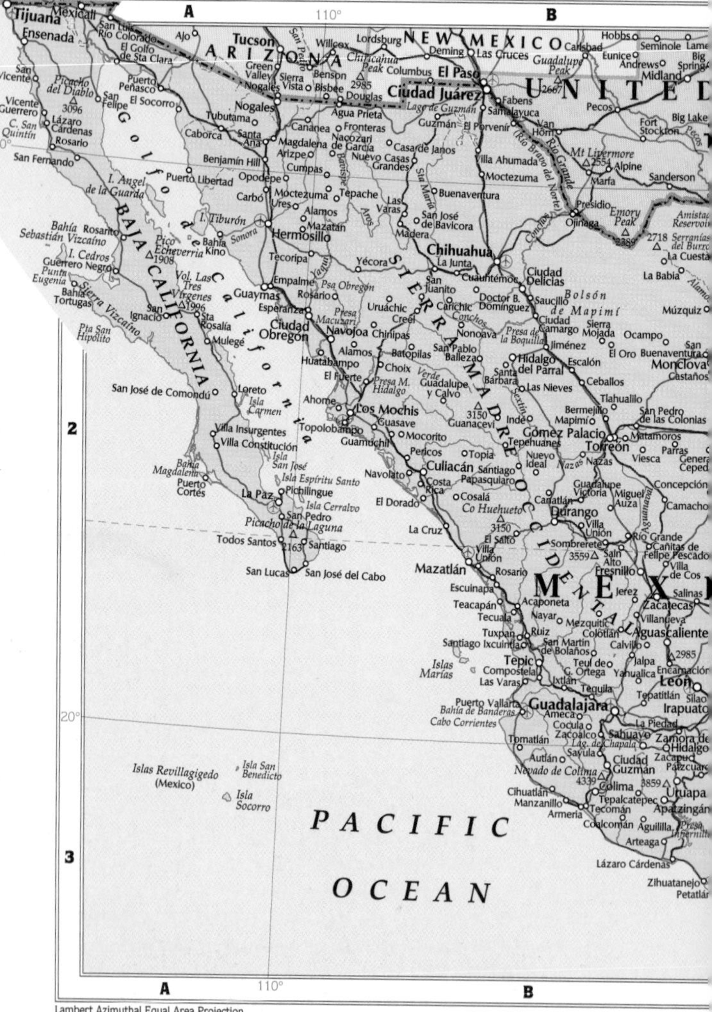
A
110°
B
Tijuana
Mexicali
Ensenada
San Luis Río Colorado
El Golfo de Sta Clara
Ajo
Tucson
ARIZONA
NEW MEXICO
Willcox
Lordsburg
Deming
Las Cruces
Carlsbad
Hobbs
Seminole
Eunice
Andrews
Big Spring
Midland
UNITED
San Vicente
Picacho del Diablo
3096
Puerto Peñasco
San Felipe
El Socorro
Green Valley
Sierra Vista
Nogales
Bisbee
Benson
Chiricahua Peak
2985
Douglas
Columbus
El Paso
Ciudad Juárez
Guadalupe Peak
2667
Fabens
Samalayuca
Pecos
Fort Stockton
Big Lake
Vicente Guerrero
Lázaro Cárdenas
C. San Quintín
Rosario
30°
Tubutama
Caborca
Santa Ana
Agua Prieta
Cananea
Fronteras
Nacozari de García
Magdalena
Arizpe
Casa de Janos
Nuevo Casas Grandes
Lago de Guzmán
Guzmán
El Porvenír
Van Horn
Río Bravo del Norte
Río Grande
Mt Livermore
2554
Alpine
Marfa
Sanderson
San Fernando
Golfo de California
Benjamín Hill
Cumpas
Opodepe
Puerto Libertad
I. Angel de la Guarda
Carbó
Moctezuma
Ures
Tepache
Álamos
Mazatán
Las Varas
Villa Ahumada
Moctezuma
Buenaventura
San José de Bavícora
Madera
Presidio
Ojinaga
Emory Peak
2389
Amistad Reservoir
2718
Serranías del Burro
La Cuesta
Bahía Sebastián Vizcaíno
Rosarito
I. Tiburón
Sonora
Hermosillo
Bahía Kino
Pico Echeverría
1908
BAJA CALIFORNIA
I. Cedros
Guerrero Negro
Punta Eugenia
Bahía Tortugas
Sierra Vizcaíno
Tecoripa
Yaqui
Yécora
Chihuahua
La Junta
Cuauhtémoc
SIERRA MADRE OCCIDENTAL
Ciudad Delicias
La Babia
Vol. Las Tres Vírgenes
1996
San Ignacio
Sta Rosalía
Guaymas
Empalme
Psa Obregón
Rosario
Esperanza
Presa Macuzari
Uruáchic
Creel
San Juanito
Carichic
Doctor B. Domínguez
Saucillo
Bolsón de Mapimí
Múzquiz
Pta San Hipólito
Mulegé
Ciudad Obregón
Navojoa
Chínipas
Conchos
Nonoava
Presa de la Boquilla
Ciudad Camargo
Sierra Mojada
Ocampo
Álamos
Batopilas
San Pablo Balleza
Jiménez
El Oro
Buenaventura
Monclova
Castaños
Huatabampo
El Fuerte
Choix
Presa M. Hidalgo
Guadalupe y Calvo
Santa Bárbara
Hidalgo del Parral
Escalón
Ceballos
Las Nieves
San José de Comondú
Loreto
Isla Carmen
Ahome
Los Mochis
3150
Guanaceví
Indé
Tlahualilo
Bermejillo
Mapimí
San Pedro de las Colonias
2
Villa Insurgentes
Villa Constitución
Topolobampo
Guasave
Mocorito
Guamúchil
Gómez Palacio
Torreón
Matamoros
Tepehuanes
Parras
Viesca
Isla San José
Pericos
Topia
Nuevo Ideal
Nazas
Bahía Magdalena
Puerto Cortés
Isla Espíritu Santo
Navolato
Culiacán
Santiago Papasquiaro
La Paz
Pichilingue
Isla Cerralvo
Costa Rica
Cosalá
Co Huehueto
Guadalupe Victoria
Canatlán
Miguel Auza
Concepción
Camacho
El Dorado
San Pedro
Picacho de la Laguna
Durango
La Cruz
3150
El Salto
Villa Unión
Río Grande
Todos Santos
2163
Santiago
Villa Unión
Sombrerete
Cañitas de Felipe Pescador
3559
Sain Alto
San Lucas
San José del Cabo
Mazatlán
Rosario
Fresnillo
Villa de Cos
Escuinapa
MEX
Jerez
Salinas
Teacapán
Acaponeta
Zacatecas
Tecuala
Nayar
Mezquitic
Villanueva
Tuxpan
Ruiz
Colotlán
Aguascalientes
Santiago Ixcuintla
San Martín de Bolaños
Calvillo
2985
Tepic
Islas Marías
Compostela
Teul de G. Ortega
Jalpa
Yahualica
Encarnación
Las Varas
Ixtlán
León
Tequila
Tepatitlán
Silao
Puerto Vallarta
Bahía de Banderas
Guadalajara
Irapuato
Ameca
Cabo Corrientes
20°
Cocula
La Piedad
Tomatlán
Zacoalco
Sahuayo
Zamora de Hidalgo
Lag. de Chapala
Sayula
Autlán
Ciudad Guzmán
Zacapu
Pátzcuaro
Nevado de Colima
4339
3859
Islas Revillagigedo (Mexico)
Isla San Benedicto
Isla Socorro
Cihuatlán
Manzanillo
Colima
Tepalcatepec
Uruapan
Armería
Tecomán
Apatzingán
PACIFIC OCEAN
Coalcomán
Aguililla
Presa Infiernillo
3
Arteaga
Lázaro Cárdenas
Zihuatanejo
Petatlán
A
110°
B

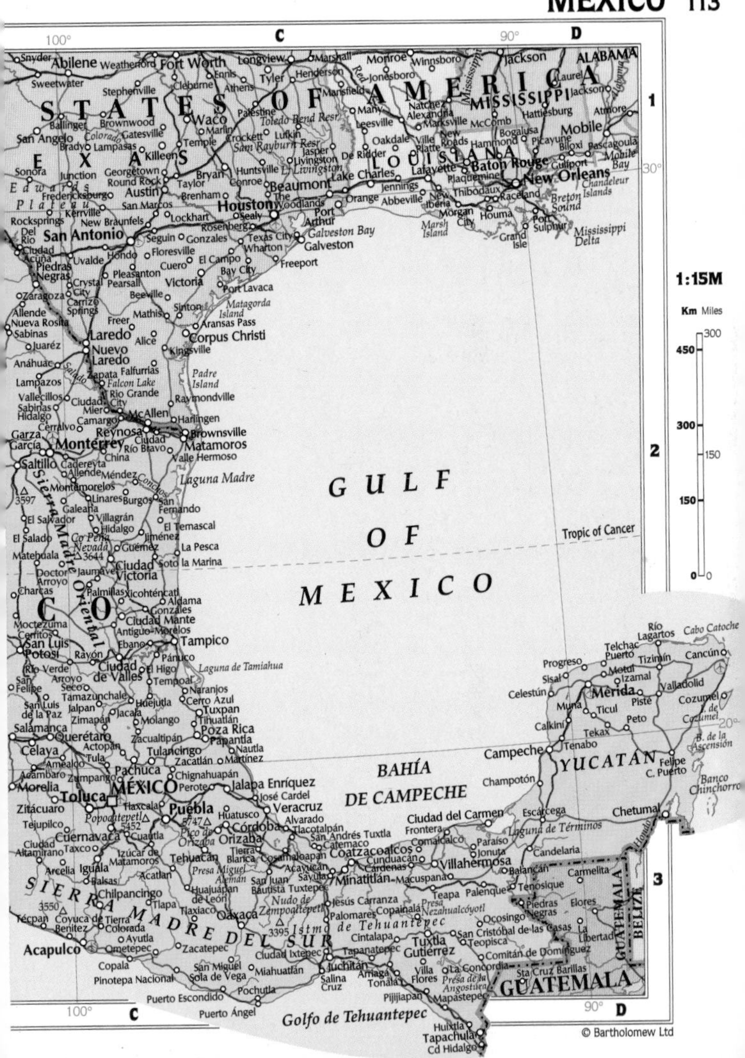

1:15M
Km Miles
STATES OF AMERICA
TEXAS
LOUISIANA
MISSISSIPPI
ALABAMA
MEXICO
GULF OF MEXICO
BAHÍA DE CAMPECHE
YUCATÁN
GUATEMALA
BELIZE
SIERRA MADRE DEL SUR
Sierra Madre Oriental
Tropic of Cancer
Abilene
Fort Worth
Waco
Austin
San Antonio
Houston
Beaumont
Galveston
Corpus Christi
Laredo
Nuevo Laredo
Brownsville
Matamoros
Reynosa
McAllen
Monterrey
Saltillo
Ciudad Victoria
Tampico
Ciudad de Valles
San Luis Potosí
Querétaro
Celaya
Pachuca
Poza Rica
Tuxpan
Jalapa Enríquez
Veracruz
Puebla
MÉXICO
Toluca
Morelia
Cuernavaca
Orizaba
Córdoba
Tehuacán
Oaxaca
Acapulco
Chilpancingo
Coatzacoalcos
Minatitlán
Villahermosa
Ciudad del Carmen
Campeche
Mérida
Progreso
Cancún
Cozumel
Chetumal
Tuxtla Gutiérrez
San Cristóbal de las Casas
Juchitán
Tapachula
Golfo de Tehuantepec
Istmo de Tehuantepec
Laguna Madre
Laguna de Términos
Cabo Catoche
Banco Chinchorro
Mobile
Baton Rouge
New Orleans
Lake Charles
Lafayette
Jackson
Mississippi Delta
Galveston Bay
Padre Island
Matagorda Island

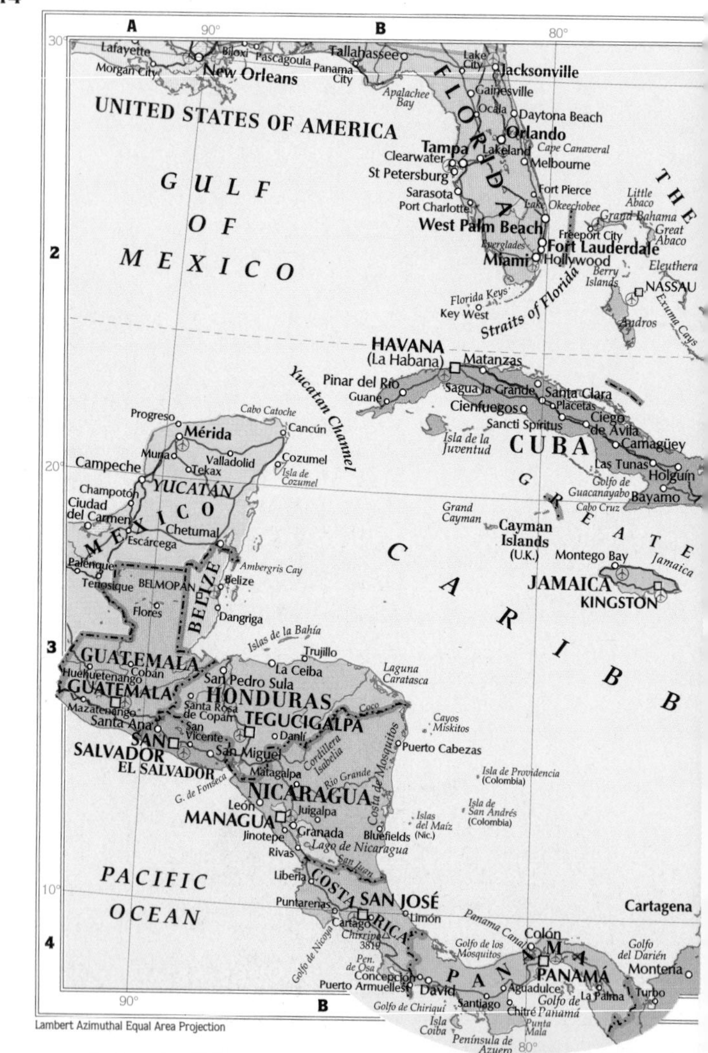

Lambert Azimuthal Equal Area Projection

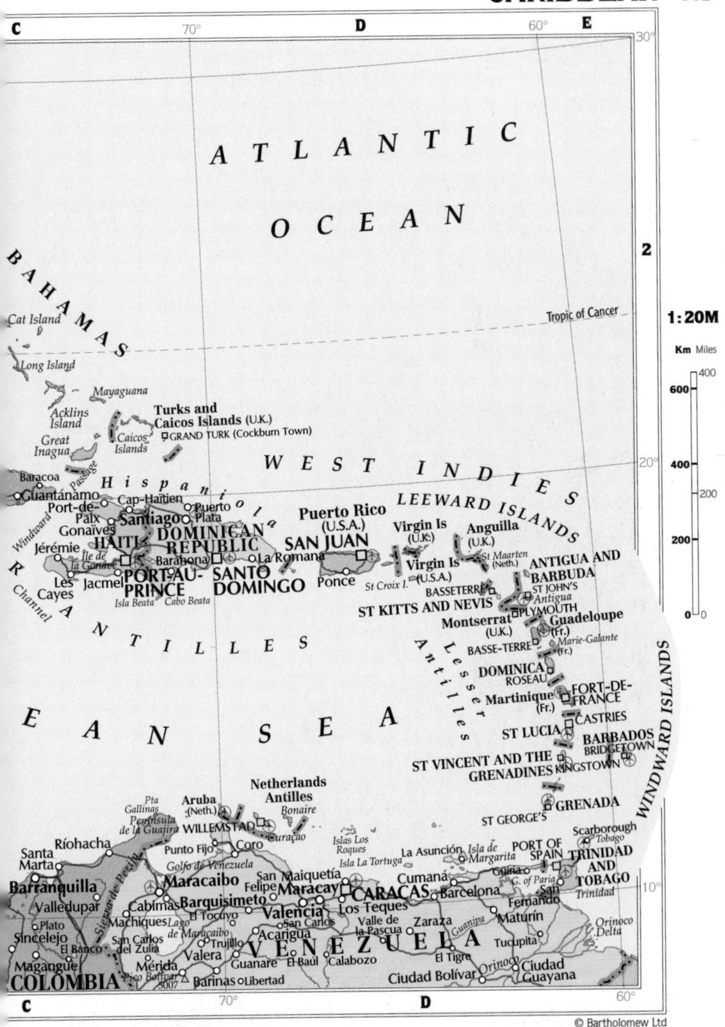
ATLANTIC OCEAN
BAHAMAS
Cat Island
Long Island
Mayaguana
Acklins Island
Great Inagua
Turks and Caicos Islands (U.K.)
Caicos Islands
GRAND TURK (Cockburn Town)
WEST INDIES
LEEWARD ISLANDS
Baracoa
Guantánamo
Windward Passage
Hispaniola
Cap-Haïtien
Port-de-Paix
Gonaïves
Santiago
Puerto Plata
HAITI
DOMINICAN REPUBLIC
Jérémie
Île de la Gonâve
Les Cayes
Jacmel
PORT-AU-PRINCE
Barahona
SANTO DOMINGO
La Romana
Isla Beata
Cabo Beata
Puerto Rico (U.S.A.)
SAN JUAN
Ponce
Virgin Is (U.K.)
Virgin Is (U.S.A.)
St Croix I.
Anguilla (U.K.)
St Maarten (Neth.)
ANTIGUA AND BARBUDA
ST JOHN'S
Antigua
BASSETERRE
ST KITTS AND NEVIS
PLYMOUTH
Montserrat (U.K.)
Guadeloupe (Fr.)
Marie-Galante (Fr.)
BASSE-TERRE
Lesser Antilles
DOMINICA
ROSEAU
Martinique (Fr.)
FORT-DE-FRANCE
CASTRIES
ST LUCIA
BARBADOS
BRIDGETOWN
ST VINCENT AND THE GRENADINES
KINGSTOWN
GRENADA
ST GEORGE'S
WINDWARD ISLANDS
Tropic of Cancer
ANTILLES
EAN SEA
Windward Channel
Netherlands Antilles
Aruba (Neth.)
Bonaire
WILLEMSTAD
Curaçao
Pta Gallinas
Península de la Guajira
Islas Los Roques
Isla La Tortuga
La Asunción
Isla de Margarita
PORT OF SPAIN
TRINIDAD AND TOBAGO
Scarborough
Tobago
Trinidad
Güiria
G. of Paria
San Fernando
Ríohacha
Santa Marta
Barranquilla
Valledupar
Plato
Sincelejo
El Banco
Magangué
COLOMBIA
Sierra de Perijá
Punto Fijo
Coro
Golfo de Venezuela
Maracaibo
Cabimas
Machiques
Lago de Maracaibo
San Carlos del Zulia
Mérida
Pico Bolívar 5007
Barinas
Libertad
Barquisimeto
El Tocuyo
Trujillo
Valera
Guanare
San Felipe
Maiquetía
Maracay
CARACAS
Los Teques
Valencia
San Carlos
Acarigua
El Baúl
Calabozo
VENEZUELA
Valle de la Pascua
Cumaná
Barcelona
Zaraza
Guanipa
Maturín
El Tigre
Ciudad Bolívar
Orinoco
Ciudad Guayana
Tucupita
Orinoco Delta
C
D
E
2
70°
60°
30°
20°
10°
1:20M
Km Miles
600
400
200
0
400
200
0

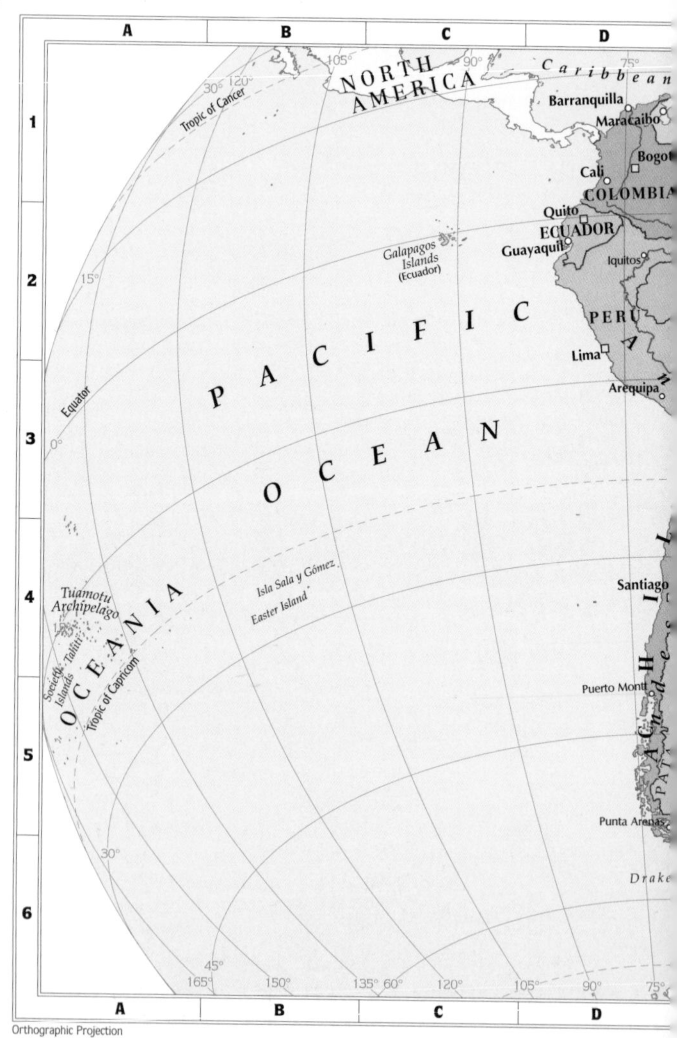

Orthographic Projection

Caracas
Valencia
VENEZUELA
GUYANA
Georgetown
Paramaribo
SUR.
Cayenne
French Guiana
Manaus
Belém
Fortaleza
BRAZIL
Recife
La Paz
BOLIVIA
Santa Cruz
Sucre
Brasília
Salvador
PARAGUAY
Asunción
São Paulo
Curitiba
Rio de Janeiro
Córdoba
Porto Alegre
Buenos Aires
URUGUAY
Montevideo
ARGENTINA
Río de la Plata
Mar del Plata
Bahía Blanca
Falkland Islands
Stanley
Isla Grande de Tierra del Fuego
Cape Horn
South Georgia
South Georgia and South Sandwich Islands (U.K.)
South Sandwich Islands
South Shetland Islands
South Orkney Islands
Antarctic Peninsula
Antarctic Circle
ATLANTIC OCEAN
AFRICA
Tropic of Cancer
Equator
Ascension
St Helena
Tropic of Capricorn
Tristan da Cunha
Cape of Good Hope
1:70M
Km Miles
2000
1500
1000
500
0
1000
500
0

Lambert Azimuthal Equal Area Projection

ATLANTIC OCEAN
1:25M
Km Miles
SURINAME
French Guiana
RAZIL
GEORGETOWN
PARAMARIBO
CAYENNE
BRASÍLIA
Belém
São Luís
Fortaleza
Natal
Recife
Maceió
Aracaju
Salvador
Teresina
Goiânia
Cuiabá
Macapá
Santarém
Mouths of the Amazon
Ilha de Marajó
Serra Tumucumaque
Chapada Diamantina
Serra Geral de Goiás
Serra do Espinhaço
Equator

Lambert Azimuthal Equal Area Projection

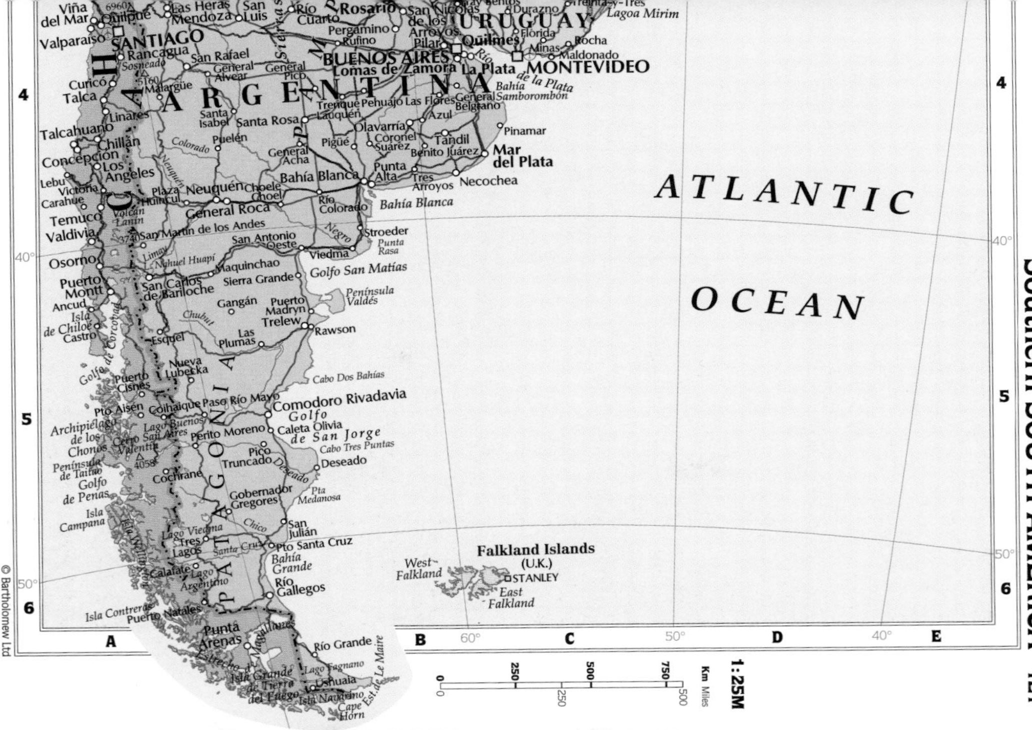
ATLANTIC
OCEAN
ARGENTINA
URUGUAY
PATAGONIA
Viña del Mar
Quilpué
Valparaíso
SANTIAGO
Rancagua
Sosneado
Curicó
Talca
Linares
Talcahuano
Chillán
Concepción
Los Angeles
Lebu
Victoria
Carahue
Temuco
Valdivia
Osorno
Puerto Montt
Ancud
Isla de Chiloé
Castro
Golfo de Corcovado
Puerto Cisnes
Pto Aisén
Coihaique
Archipiélago de los Chonos
Cerro San Valentín
Península de Taitao
Golfo de Penas
Isla Campana
Isla Wellington
Isla Contreras
Puerto Natales
Punta Arenas
Estrecho de Magallanes
Isla Grande de Tierra del Fuego
Isla Navarino
Cape Horn
Est. de Le Maire
Ushuaia
Lago Fagnano
Río Grande
Las Heras
Mendoza
San Luis
Río Cuarto
Rosario
San Nicolás de los Arroyos
Durazno
Lagoa Mirim
Pergamino
Rufino
Pilar
Quilmes
Florida
Minas
Rocha
Maldonado
San Rafael
General Alvear
General Pico
BUENOS AIRES
Lomas de Zamora
La Plata
MONTEVIDEO
Río de la Plata
Bahía Samborombón
Malargüe
Trenque Lauquén
Pehuajó
Las Flores
General Belgrano
Santa Isabel
Santa Rosa
Azul
Olavarría
Pinamar
Puelén
Colorado
Pigüé
Coronel Suárez
Tandil
Benito Juárez
Mar del Plata
General Acha
Punta Alta
Tres Arroyos
Necochea
Neuquén
Plaza Huincul
Choele Choel
Bahía Blanca
Volcán Lanín
General Roca
Río Colorado
San Martín de los Andes
Negro
Stroeder
Punta Rasa
San Antonio Oeste
Viedma
Nahuel Huapí
Maquinchao
Golfo San Matías
San Carlos de Bariloche
Sierra Grande
Península Valdés
Gangán
Puerto Madryn
Chubut
Trelew
Rawson
Esquel
Las Plumas
Nueva Lubecka
Cabo Dos Bahías
Paso Río Mayo
Comodoro Rivadavia
Golfo de San Jorge
Lago Buenos Aires
Perito Moreno
Caleta Olivia
Cabo Tres Puntas
Pico Truncado
Deseado
Cochrane
Gobernador Gregores
Pta Medanosa
Chico
San Julián
Lago Viedma
Tres Lagos
Santa Cruz
Pto Santa Cruz
Bahía Grande
Calafate
Lago Argentino
Río Gallegos
Falkland Islands
(U.K.)
STANLEY
West Falkland
East Falkland
40°
50°
60°
A
B
C
D
E
4
5
6
1:25M
Km
Miles
0
250
500
750

Lambert Azimuthal Equal Area Projection

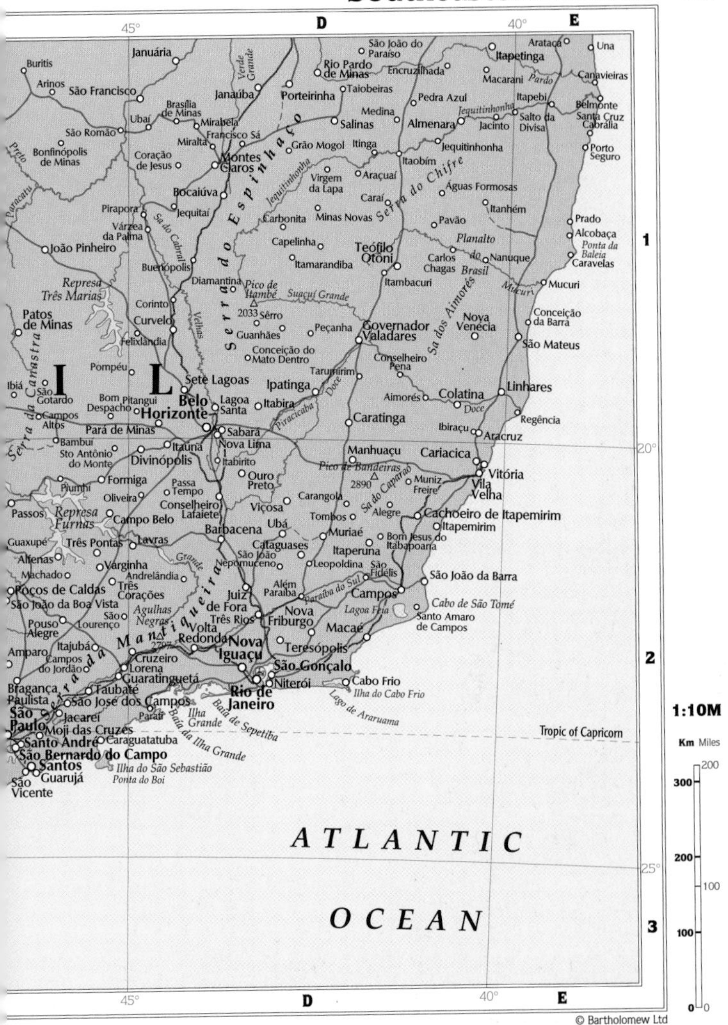
D
E
45°
40°
20°
25°
1
2
3
Januária
Buritis
Arinos
São Francisco
Janaúba
Verde Grande
São João do Paraíso
Rio Pardo de Minas
Encruzilhada
Itapetinga
Arataca
Una
Macarani
Pardo
Canavieiras
Porteirinha
Taiobeiras
Pedra Azul
Itapebi
Belmonte
Santa Cruz Cabrália
Brasília de Minas
Ubaí
Mirabela
Medina
Salinas
Almenara
Jequitinhonha
Salto da Divisa
Jacinto
São Romão
Miralta
Francisco Sá
Grão Mogol
Itinga
Jequitinhonha
Porto Seguro
Preto
Bonfinópolis de Minas
Coração de Jesus
Montes Claros
Serra do Espinhaço
Itaobím
Araçuaí
Virgem da Lapa
Serra do Chifre
Águas Formosas
Paracatu
Bocaiúva
Caraí
Pirapora
Jequitaí
Carbonita
Minas Novas
Itanhém
Pavão
Prado
Várzea da Palma
Sa do Cabral
Capelinha
Planalto
Alcobaça
Ponta da Baleia
Caravelas
João Pinheiro
Teófilo Otoni
Carlos Chagas
Nanuque
Brasil
Buenópolis
Itamarandiba
Diamantina
Pico de Itambé
Itambacuri
Mucuri
Represa Três Marias
Suaçuí Grande
Mucuri
Corinto
2033
Sêrro
Sa dos Aimorés
Patos de Minas
Curvelo
Velhas
Guanhães
Peçanha
Governador Valadares
Nova Venécia
Conceição da Barra
São Mateus
Felixlândia
Conceição do Mato Dentro
Conselheiro Pena
Serra da Canastra
Pompéu
Tarumirim
Doce
Sete Lagoas
Ipatinga
Ibiá
São Gotardo
Linhares
Aimorés
Colatina
Bom Despacho
Pitangui
Belo Horizonte
Lagoa Santa
Itabira
Piracicaba
Caratinga
Doce
Regência
Campos Altos
Pará de Minas
Sabará
Ibiraçu
Aracruz
Bambuí
Itaúna
Nova Lima
Manhuaçu
Cariacica
Sto Antônio do Monte
Divinópolis
Itabirito
Pico de Bandeiras
Vitória
Formiga
Ouro Preto
2890
Muniz Freire
Vila Velha
Piumhi
Passa Tempo
Sa do Caparaó
Oliveira
Carangola
Conselheiro Lafaiete
Viçosa
Alegre
Passos
Represa Furnas
Campo Belo
Tombos
Cachoeiro de Itapemirim
Barbacena
Ubá
Itapemirim
Muriaé
Bom Jesus do Itabapoana
Guaxupé
Três Pontas
Lavras
Cataguases
Itaperuna
São João Nepomuceno
Alfenas
Grande
Leopoldina
São Fidélis
Varginha
Machado
Andrelândia
São João da Barra
Poços de Caldas
Três Corações
Juiz de Fora
Além Paraíba
Paraíba do Sul
Campos
Cabo de São Tomé
São João da Boa Vista
Agulhas Negras
São Lourenço
Três Rios
Nova Friburgo
Lagoa Feia
Santo Amaro de Campos
Pouso Alegre
Serra da Mantiqueira
2787
Volta Redonda
Macaé
Itajubá
Nova Iguaçu
Teresópolis
Amparo
Campos do Jordão
Cruzeiro
Lorena
São Gonçalo
Guaratinguetá
Cabo Frio
Bragança Paulista
Taubaté
Rio de Janeiro
Niterói
Ilha do Cabo Frio
São José dos Campos
Lago de Araruama
São Paulo
Jacareí
Paratí
Ilha Grande
Baía de Sepetiba
Moji das Cruzes
Tropic of Capricorn
Santo André
Caraguatatuba
Baía da Ilha Grande
São Bernardo do Campo
Santos
Ilha do São Sebastião
São Vicente
Guarujá
Ponta do Boi
ATLANTIC
OCEAN
1:10M
Km Miles
300
200
100
0
200
100
0

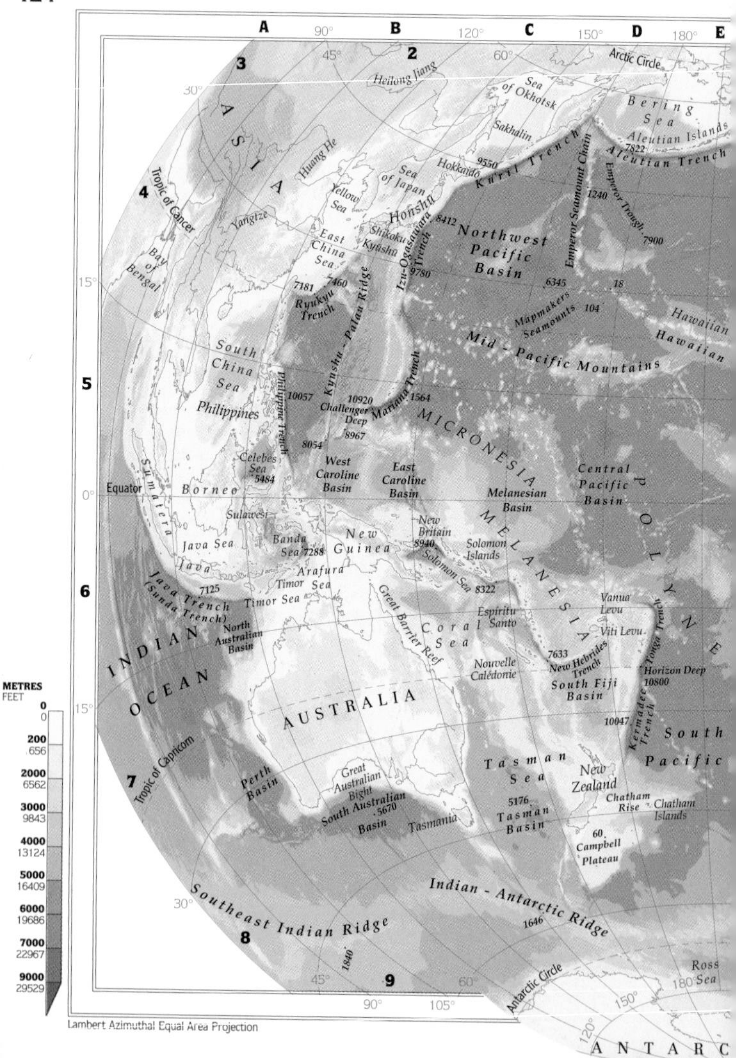

Lambert Azimuthal Equal Area Projection

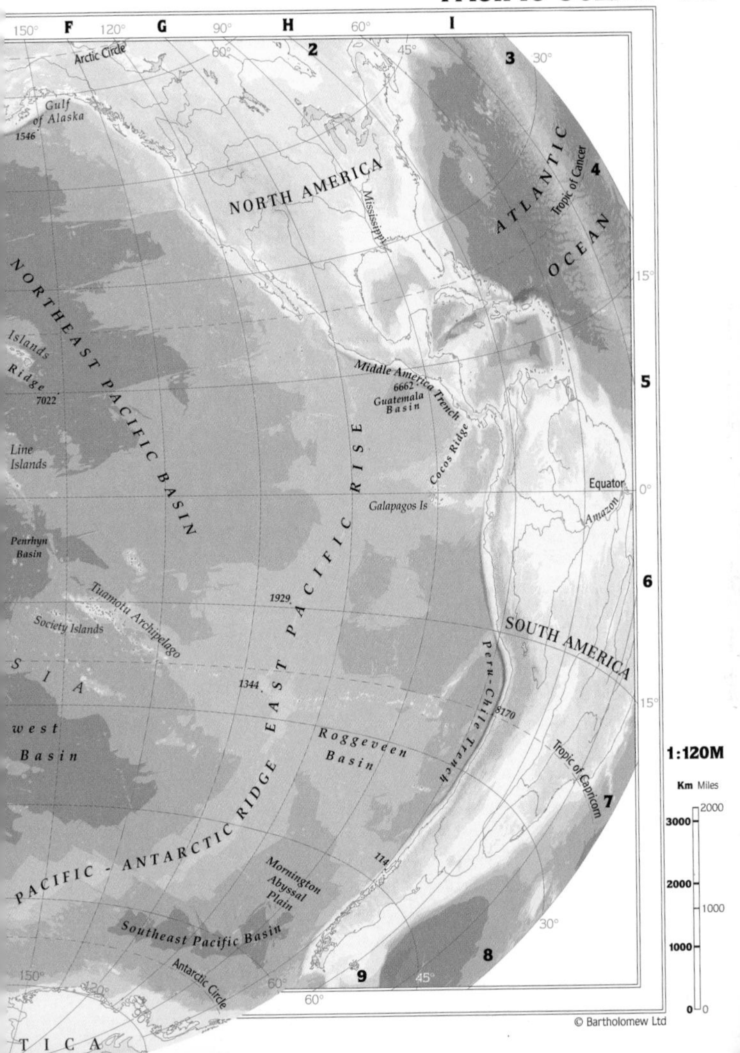
Arctic Circle
Gulf of Alaska
1546
NORTH AMERICA
Mississippi
ATLANTIC OCEAN
Tropic of Cancer
NORTHEAST PACIFIC BASIN
Islands Ridge
7022
Line Islands
Middle America Trench
6662
Guatemala Basin
Cocos Ridge
EAST PACIFIC RISE
Galapagos Is
Equator
Amazon
Penrhyn Basin
Tuamotu Archipelago
Society Islands
1929
SOUTH AMERICA
Peru-Chile Trench
8170
1344
Roggeveen Basin
Tropic of Capricorn
PACIFIC - ANTARCTIC RIDGE
Mornington Abyssal Plain
Southeast Pacific Basin
Antarctic Circle
1:120M
Km Miles
© Bartholomew Ltd

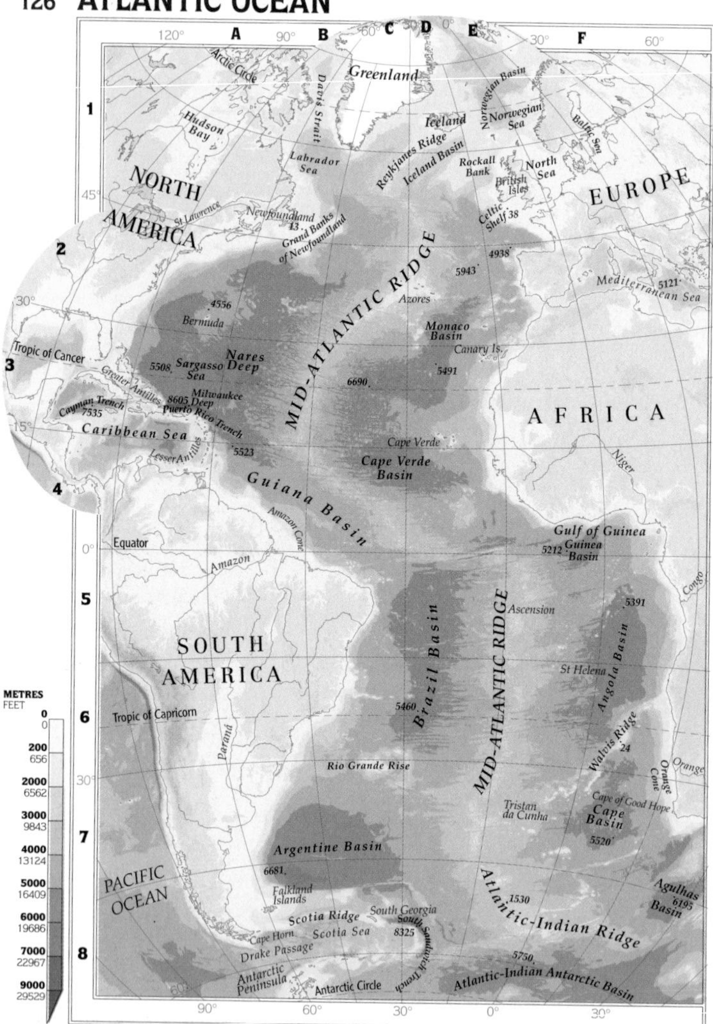
Greenland
Davis Strait
Hudson Bay
Labrador Sea
Iceland
Norwegian Basin
Norwegian Sea
Reykjanes Ridge
Iceland Basin
Rockall Bank
North Sea
British Isles
Baltic Sea
NORTH AMERICA
EUROPE
St Lawrence
Newfoundland
Grand Banks of Newfoundland
Celtic Shelf
MID-ATLANTIC RIDGE
Azores
Mediterranean Sea
Bermuda
Monaco Basin
Canary Is.
Tropic of Cancer
Sargasso Sea
Nares Deep
Greater Antilles
Milwaukee Deep
Puerto Rico Trench
Cayman Trench
Caribbean Sea
Lesser Antilles
AFRICA
Cape Verde
Cape Verde Basin
Niger
Guiana Basin
Amazon Cone
Gulf of Guinea
Guinea Basin
Equator
Amazon
Congo
Ascension
Brazil Basin
SOUTH AMERICA
St Helena
Angola Basin
Tropic of Capricorn
Paraná
Walvis Ridge
Orange
Orange Cone
Rio Grande Rise
Tristan da Cunha
Cape of Good Hope
Cape Basin
Argentine Basin
PACIFIC OCEAN
Falkland Islands
Scotia Ridge
South Georgia
Scotia Sea
Cape Horn
Drake Passage
South Sandwich Trench
Atlantic-Indian Ridge
Agulhas Basin
Antarctic Peninsula
Antarctic Circle
Atlantic-Indian Antarctic Basin
Arctic Circle
METRES
FEET
Lambert Azimuthal Equal Area Projection

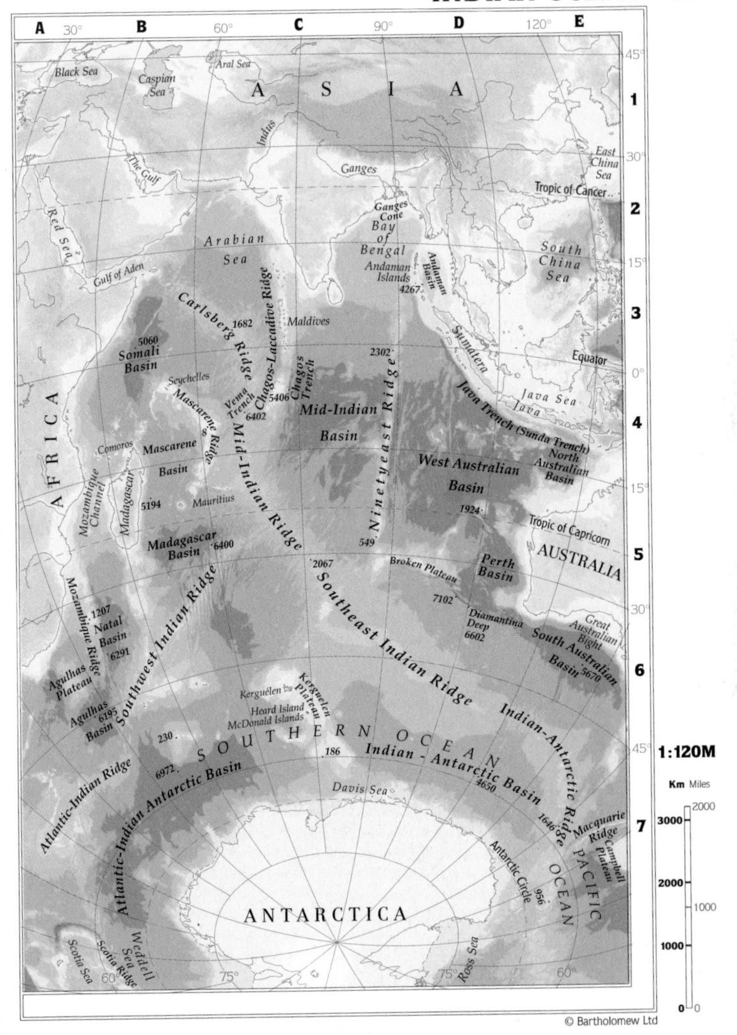
A
B
C
D
E
30°
60°
90°
120°
45°
30°
15°
0°
1
2
3
4
5
6
7
Black Sea
Caspian Sea
Aral Sea
A S I A
Indus
The Gulf
Ganges
East China Sea
Tropic of Cancer
Red Sea
Ganges Cone
Bay of Bengal
Arabian Sea
South China Sea
Gulf of Aden
Andaman Islands
Andaman Basin
4267
Carlsberg Ridge
Chagos-Laccadive Ridge
1682
Maldives
Sumatera
5060
Somali Basin
2302
Equator
Seychelles
Chagos Trench
Ninetyeast Ridge
Java Trench (Sunda Trench)
Java Sea
Java
Mascarene Ridge
Vema Trench
5406
6402
Mid-Indian Basin
AFRICA
Comoros
Mascarene Basin
Mid-Indian Ridge
North Australian Basin
West Australian Basin
Mozambique Channel
Madagascar
5194
Mauritius
1924
Tropic of Capricorn
Madagascar Basin
6400
549
AUSTRALIA
2067
Broken Plateau
Perth Basin
Mozambique Ridge
1207
Natal Basin
Southwest Indian Ridge
Southeast Indian Ridge
7102
Diamantina Deep
6602
Great Australian Bight
South Australian Basin
5670
6291
Agulhas Plateau
Agulhas Basin
6195
Kerguélen
Kerguelen Plateau
Heard Island
McDonald Islands
Indian-Antarctic Ridge
230
SOUTHERN OCEAN
186
Indian - Antarctic Basin
6972
Atlantic-Indian Ridge
Atlantic-Indian Antarctic Basin
Davis Sea
4650
1646
Macquarie Ridge
Campbell Plateau
PACIFIC OCEAN
Antarctic Circle
956
ANTARCTICA
Scotia Sea
Scotia Ridge
Weddell Sea
Ross Sea
60°
75°
75°
60°
1:120M
Km Miles
3000
2000
1000
0
2000
1000
0

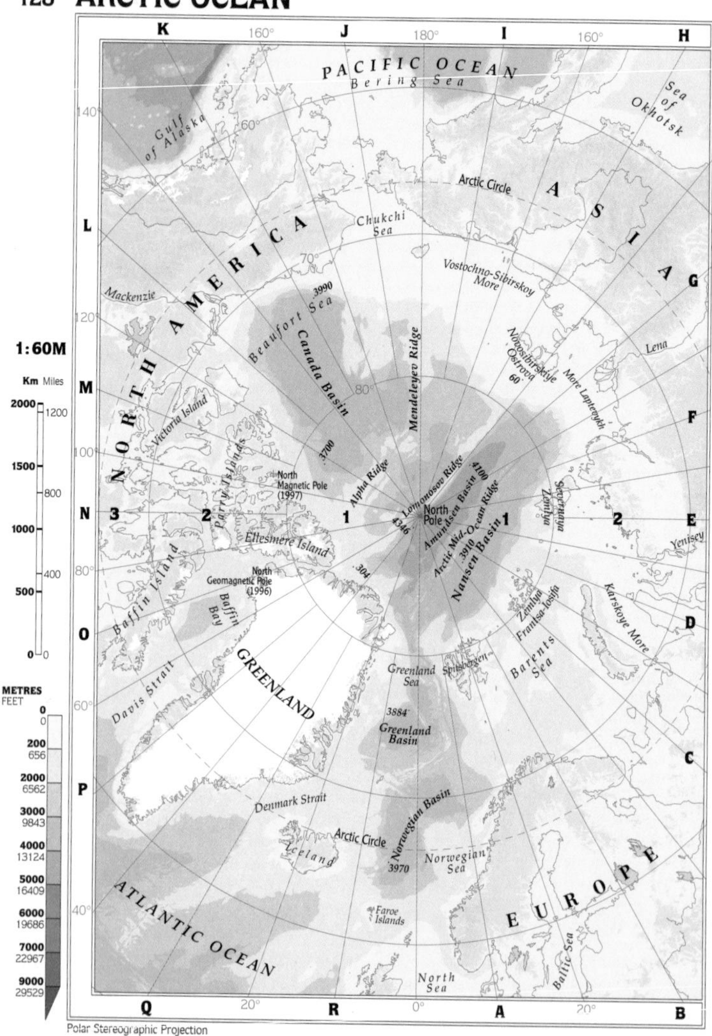

Polar Stereographic Projection

NATIONAL STATISTICS

GEOGRAPHICAL INFORMATION

COUNTRY	AREA sq km	sq mls	POPULATION	CAPITAL
AFGHANISTAN	652 225	251 825	21 354 000	Kābul
ALBANIA	28 748	11 100	3 119 000	Tirana (Tiranë)
ALGERIA	2 381 741	919 595	30 081 000	Algiers (Alger)
American Samoa	197	76	63 000	Fagatogo
ANDORRA	465	180	72 000	Andorra la Vella
ANGOLA	1 246 700	481 354	12 092 000	Luanda
Anguilla (U.K.)	155	60	8 000	The Valley
ANTIGUA AND BARBUDA	442	171	67 000	St John's
ARGENTINA	2 766 889	1 068 302	36 123 000	Buenos Aires
ARMENIA	29 800	11 506	3 536 000	Yerevan (Erevan)
Aruba (Netherlands)	193	75	94 000	Oranjestad
AUSTRALIA	7 682 395	2 966 189	18 520 000	Canberra
AUSTRIA	83 855	32 377	8 140 000	Vienna (Wien)
AZERBAIJAN	86 600	33 436	7 669 000	Baku (Bakı)
Azores (Portugal)	2 300	888	243 600	Ponta Delgada
THE BAHAMAS	13 939	5 382	296 000	Nassau
BAHRAIN	691	267	595 000	Manama
BANGLADESH	143 998	55 598	124 774 000	Dhaka (Dacca)
BARBADOS	430	166	268 000	Bridgetown
BELARUS	207 600	80 155	10 315 000	Minsk
BELGIUM	30 520	11 784	10 141 000	Brussels (Bruxelles)
BELIZE	22 965	8 867	230 000	Belmopan
BENIN	112 620	43 483	5 781 000	Porto-Novo
Bermuda (U.K.)	54	21	64 000	Hamilton
BHUTAN	46 620	18 000	2 004 000	Thimphu
BOLIVIA	1 098 581	424 164	7 957 000	La Paz/Sucre
BOSNIA-HERZEGOVINA	51 130	19 741	3 675 000	Sarajevo
BOTSWANA	581 370	224 468	1 570 000	Gaborone
BRAZIL	8 547 379	3 300 161	165 851 000	Brasília

LANGUAGES	RELIGIONS	CURRENCY
Dari, Pushtu, Uzbek	Muslim	Afghani
Albanian	Muslim, Orthodox, Roman Catholic	Lek
Arabic, French, Berber	Muslim	Dinar
Samoan, English	Protestant, Roman Catholic	US dollar
Catalan, French, Spanish	Roman Catholic	Franc, Peseta
Portuguese, local languages	Roman Catholic, trad. beliefs, Protestant	Kwanza
English	Protestant, Roman Catholic	E. Carib. dollar
English, Creole	Protestant, Roman Catholic	E. Carib. dollar
Spanish, Amerindian languages	Roman Catholic	Peso
Armenian, Azeri, Russian	Orthodox, Roman Catholic, Muslim	Dram
Dutch, Papiamento	Roman Catholic, Protestant	Florin
English, Italian, Greek	Protestant, Roman Catholic, Aboriginal beliefs	Dollar
German	Roman Catholic	Schilling, Euro
Azeri, Armenian, Russian	Muslim	Manat
Portuguese	Roman Catholic, Protestant	Port. escudo
English, Creole, French Creole	Protestant, Roman Catholic	Dollar
Arabic, English	Muslim, Christian	Dinar
Bengali, Bihari, Hindi, English, local languages	Muslim, Hindu	Taka
English, Creole	Protestant, Roman Catholic	Dollar
Belorussian, Russian, Ukrainian	Orthodox, Roman Catholic	Rouble
Flemish, Walloon, German, Italian	Roman Catholic, Protestant	Franc, Euro
English, Creole, Spanish, Mayan	Roman Catholic, Protestant, Hindu	Dollar
French, local languages	Trad. beliefs, Roman Catholic	CFA franc
English	Protestant, Roman Catholic	Dollar
Dzongkha, Nepali	Buddhist, Hindu, Muslim	Ngultrum, Indian rupee
Spanish, Quechua, Aymara	Roman Catholic	Boliviano
Bosnian, Serbian, Croatian	Muslim, Orthodox, Roman Catholic, Protestant	Marka
English, Setswana, local languages	Trad. beliefs, Protestant, Roman Catholic	Pula
Portuguese, Amerindian languages	Roman Catholic	Real

COUNTRY	AREA sq km	AREA sq mls	POPULATION	CAPITAL
BRUNEI	5 765	2 226	315 000	Bandar Seri Begawan
BULGARIA	110 994	42 855	8 336 000	Sofia (Sofiya)
BURKINA	274 200	105 869	11 305 000	Ouagadougou
BURUNDI	27 835	10 747	6 457 000	Bujumbura
CAMBODIA	181 000	69 884	10 716 000	Phnum Pénh
CAMEROON	475 442	183 569	14 305 000	Yaoundé
CANADA	9 970 610	3 849 674	30 563 000	Ottawa
Canary Islands (Spain)	7 447	2 875	1 606 522	Santa Cruz de Tenerife
CAPE VERDE	4 033	1 557	408 000	Praia
Cayman Islands (U.K.)	259	100	36 000	George Town
CENTRAL AFRICAN REPUBLIC	622 436	240 324	3 485 000	Bangui
CHAD	1 284 000	495 755	7 270 000	Ndjamena
CHILE	756 945	292 258	14 824 000	Santiago
CHINA	9 584 492	3 700 593	1 262 817 000	Beijing (Peking)
Christmas Island (Austr.)	135	52	2 195	The Settlement
Cocos Islands (Austr.)	14	5	637	Home Island
COLOMBIA	1 141 748	440 831	40 803 000	Bogotá
COMOROS	1 862	719	658 000	Moroni
CONGO	342 000	132 047	2 785 000	Brazzaville
CONGO, DEMOCRATIC REPUBLIC OF	2 345 410	905 568	49 139 000	Kinshasa
Cook Islands (N.Z.)	293	113	19 000	Avarua
COSTA RICA	51 100	19 730	3 841 000	San José
CÔTE D'IVOIRE	322 463	124 504	14 292 000	Yamoussoukro
CROATIA	56 538	21 829	4 481 000	Zagreb
CUBA	110 860	42 803	11 116 000	Havana (La Habana)
CYPRUS	9 251	3 572	771 000	Nicosia (Lefkosia)
CZECH REPUBLIC	78 864	30 450	10 282 000	Prague (Praha)
DENMARK	43 075	16 631	5 270 000	Copenhagen
DJIBOUTI	23 200	8 958	623 000	Djibouti

LANGUAGES	RELIGIONS	CURRENCY
Malay, English, Chinese	Muslim, Buddhist, Christian	Dollar (Ringgit)
Bulgarian	Orthodox, Muslim	Lev
French, Voltaic languages	Trad. beliefs, Muslim, Roman Catholic	CFA franc
Kirundi, French	Roman Catholic, Protestant	Franc
Khmer, Vietnamese	Buddhist, Muslim	Riel
French, English, local languages	Trad. beliefs, Roman Catholic, Muslim, Protestant	CFA franc
English, French, Amerindian languages, Inuktitut	Roman Catholic, Protestant	Dollar
Spanish	Roman Catholic	Peseta
Portuguese, Portuguese Creole	Roman Catholic	Escudo
English	Protestant, Roman Catholic	Dollar
French, Sango, local languages	Protestant, Roman Catholic, Trad. beliefs	CFA franc
Arabic, French, local languages	Muslim, Trad. beliefs, Roman Catholic	CFA franc
Spanish, Amerindian languages	Roman Catholic	Peso
Chinese, regional languages	Confucian, Taoist, Buddhist	Yuan
English	Buddhist, Muslim, Protestant, Roman Catholic	Austr. dollar
English	Muslim, Christian	Austr. dollar
Spanish, Amerindian languages	Roman Catholic	Peso
Comorian, French, Arabic	Muslim	Franc
French, Kongo, Monokutuba, local languages	Roman Catholic, Protestant	CFA franc
French, Lingala, Swahili, Kongo, local languages	Roman Catholic, Protestant	Franc
English, Maori	Protestant, Roman Catholic	Dollar
Spanish	Roman Catholic, Protestant	Colón
French, Akan, local languages	Trad. beliefs, Muslim, Roman Catholic	CFA franc
Croatian, Serbian	Roman Catholic, Orthodox, Muslim	Kuna
Spanish	Roman Catholic, Protestant	Peso
Greek, Turkish, English	Greek Orthodox, Muslim	Pound
Czech, Moravian, Slovak	Roman Catholic, Protestant	Koruna
Danish	Protestant, Roman Catholic	Krone
Somali, French, Arabic, Issa, Afar	Muslim	Franc

COUNTRY	AREA sq km	sq mls	POPULATION	CAPITAL
DOMINICA	750	290	71 000	Roseau
DOMINICAN REPUBLIC	48 442	18 704	8 232 000	Santo Domingo
ECUADOR	272 045	105 037	12 175 000	Quito
EGYPT	1 000 250	386 199	65 978 000	Cairo (El Qâhira)
EL SALVADOR	21 041	8 124	6 032 000	San Salvador
EQUATORIAL GUINEA	28 051	10 831	431 000	Malabo
ERITREA	117 400	45 328	3 577 000	Asmara
ESTONIA	45 200	17 452	1 429 000	Tallinn
ETHIOPIA	1 133 880	437 794	59 649 000	Addis Ababa (Ādīs Ābeba)
Falkland Islands (U.K.)	12 170	4 699	2 000	Stanley
Faroe Islands (Denmark)	1 399	540	43 000	Tórshavn (Thorshavn)
FIJI	18 330	7 077	796 000	Suva
FINLAND	338 145	130 559	5 154 000	Helsinki (Helsingfors)
FRANCE	543 965	210 026	58 683 000	Paris
French Guiana	90 000	34 749	167 000	Cayenne
French Polynesia	3 265	1 261	227 000	Papeete
GABON	267 667	103 347	1 167 000	Libreville
THE GAMBIA	11 295	4 361	1 229 000	Banjul
GAZA	363	140	1 036 000	Gaza
GEORGIA	69 700	26 911	5 059 000	T'bilisi
GERMANY	357 028	137 849	82 133 000	Berlin
GHANA	238 537	92 100	19 162 000	Accra
Gibraltar (U.K.)	0.7	3	25 000	Gibraltar
GREECE	131 957	50 949	10 600 000	Athens (Athina)
Greenland (Denmark)	2 175 600	840 004	56 000	Nuuk (Godthåb)
GRENADA	378	146	93 000	St George's
Guadeloupe (France)	1 780	687	443 000	Basse-Terre
Guam (U.S.A.)	541	209	161 000	Agana
GUATEMALA	108 890	42 043	10 801 000	Guatemala
Guernsey (U.K.)	78	30	64 555	St Peter Port
GUINEA	245 857	94 926	7 337 000	Conakry

LANGUAGES	RELIGIONS	CURRENCY
English, French Creole	Roman Catholic, Protestant	E. Carib. dollar
Spanish, French Creole	Roman Catholic, Protestant	Peso
Spanish, Amerindian languages	Roman Catholic	Sucre
Arabic, French	Muslim, Coptic Christian	Pound
Spanish	Roman Catholic, Protestant	Colón
Spanish, Fang	Roman Catholic	CFA franc
Tigrinya, Tigre	Muslim, Coptic Christian	Nakfa
Estonian, Russian	Protestant, Orthodox	Kroon
Amharic, Oromo, local languages	Ethiopian Orthodox, Muslim, Trad. beliefs	Birr
English	Protestant, Roman Catholic	Pound
Danish, Faroese	Protestant	Danish krone
English, Fijian, Hindi	Christian, Hindu, Muslim	Dollar
Finnish, Swedish	Protestant, Orthodox	Markka, Euro
French, Arabic	Roman Catholic, Protestant, Muslim	Franc, Euro
French, Creole	Roman Catholic, Protestant	French franc
French, Polynesian languages	Protestant, Roman Catholic	Pacific franc
French, Fang, local languages	Roman Catholic, Protestant	CFA franc
English, Malinke, Fulani	Muslim	Dalasi
Arabic	Muslim	
Georgian, Russian, Armenian	Orthodox, Muslim	Lari
German, Turkish	Protestant, Roman Catholic	Mark, Euro
English, Hausa, Akan, local languages	Protestant, Roman Catholic, Muslim, Trad. beliefs	Cedi
English, Spanish	Roman Catholic, Protestant	Pound
Greek, Macedonian	Greek Orthodox	Drachma, Euro
Greenlandic, Danish	Protestant	Danish krone
English, Creole	Roman Catholic, Protestant	E. Carib. dollar
French, French Creole	Roman Catholic	French franc
Chamorro, English	Roman Catholic	US dollar
Spanish, Mayan languages	Roman Catholic, Protestant	Quetzal
English, French	Protestant, Roman Catholic	Pound
French, Fulani, local languages	Muslim	Franc

COUNTRY	AREA sq km	sq mls	POPULATION	CAPITAL
GUINEA-BISSAU	36 125	13 948	1 161 000	Bissau
GUYANA	214 969	83 000	850 000	Georgetown
HAITI	27 750	10 714	7 952 000	Port-au-Prince
HONDURAS	112 088	43 277	6 147 000	Tegucigalpa
HUNGARY	93 030	35 919	10 116 000	Budapest
ICELAND	102 820	39 699	276 000	Reykjavík
INDIA	3 065 027	1 183 414	982 223 000	New Delhi
INDONESIA	1 919 445	741 102	206 338 000	Jakarta
IRAN	1 648 000	636 296	65 758 000	Tehrān
IRAQ	438 317	169 235	21 800 000	Baghdād
IRELAND, REPUBLIC OF	70 282	27 136	3 681 000	Dublin
Isle of Man (U.K.)	572	221	77 000	Douglas
ISRAEL	20 770	8 019	5 984 000	Jerusalem
ITALY	301 245	116 311	57 369 000	Rome (Roma)
JAMAICA	10 991	4 244	2 538 000	Kingston
JAPAN	377 727	145 841	126 281 000	Tōkyō
Jersey (U.K.)	116	45	89 136	St Helier
JORDAN	89 206	34 443	6 304 000	Ammān
KAZAKHSTAN	2 717 300	1 049 155	16 319 000	Astana (Akmola)
KENYA	582 646	224 961	29 008 000	Nairobi
KIRIBATI	717	277	81 000	Bairiki
KUWAIT	17 818	6 880	1 811 000	Kuwait (Al Kuwayt)
KYRGYZSTAN	198 500	76 641	4 643 000	Bishkek (Frunze)
LAOS	236 800	91 429	5 163 000	Vientiane (Viangchan)
LATVIA	63 700	24 595	2 424 000	Rīga
LEBANON	10 452	4 036	3 191 000	Beirut (Beyrouth)
LESOTHO	30 355	11 720	2 062 000	Maseru
LIBERIA	111 369	43 000	2 666 000	Monrovia
LIBYA	1 759 540	679 362	5 339 000	Tripoli (Ţarābulus)
LIECHTENSTEIN	160	62	32 000	Vaduz
LITHUANIA	65 200	25 174	3 694 000	Vilnius

LANGUAGES	RELIGIONS	CURRENCY
Portuguese, Creole, local languages	Trad. beliefs, Muslim	Peso
English, Creole, Amerindian languages	Protestant, Hindu, Roman Catholic, Muslim	Dollar
French, Creole	Roman Catholic, Protestant	Gourde
Spanish, Amerindian languages	Roman Catholic, Protestant	Lempira
Hungarian	Roman Catholic, Protestant	Forint
Icelandic	Protestant, Roman Catholic	Krôna
Hindi, English, regional languages	Hindu, Muslim, Sikh, Christian	Rupee
Bahasa Indonesian, Dutch, local languages	Muslim, Protestant, Roman Catholic	Rupiah
Farsi, Azeri, Kurdish	Muslim, Baha'i	Rial
Arabic, Kurdish, Turkmen	Muslim	Dinar
English, Irish	Roman Catholic, Protestant	Punt, Euro
English	Protestant, Roman Catholic	Pound
Hebrew, Arabic, Yiddish, English	Jewish, Muslim, Christian	Shekel
Italian	Roman Catholic	Lira, Euro
English, Creole	Protestant, Roman Catholic	Dollar
Japanese	Shintoist, Buddhist	Yen
English, French	Protestant, Roman Catholic	Pound
Arabic	Muslim	Dinar
Kazakh, Russian	Muslim, Orthodox, Protestant	Tenge
Swahili, English, local languages	Roman Catholic, Protestant, Trad. beliefs	Shilling
Kiribati, English	Roman Catholic, Protestant	Austr. dollar
Arabic	Muslim, Christian	Dinar
Kirghiz, Russian, Uzbek	Muslim, Orthodox	Som
Lao, local languages	Buddhist, Trad. beliefs	Kip
Latvian, Russian	Protestant, Roman Catholic, Orthodox	Lat
Arabic, French, Armenian	Muslim, Protestant, Roman Catholic	Pound
Sesotho, English, Zulu	Roman Catholic, Protestant	Loti
English, Creole, local languages	Muslim, Christian	Dollar
Arabic, Berber	Muslim	Dinar
German	Roman Catholic, Protestant	Swiss franc
Lithuanian, Russian, Polish	Roman Catholic, Protestant, Orthodox	Litas

COUNTRY	AREA sq km	sq mls	POPULATION	CAPITAL
LUXEMBOURG	2 586	998	422 000	Luxembourg
MACEDONIA (F.Y.R.O.M.)	25 713	9 928	1 999 000	Skopje
MADAGASCAR	587 041	226 658	15 057 000	Antananarivo
Madeira (Portugal)	779	301	259 000	Funchal
MALAWI	118 484	45 747	10 346 000	Lilongwe
MALAYSIA	332 965	128 559	21 410 000	Kuala Lumpur
MALDIVES	298	115	271 000	Male
MALI	1 240 140	478 821	10 694 000	Bamako
MALTA	316	122	384 000	Valletta
MARSHALL ISLANDS	181	70	60 000	Dalap-Uliga-Darrit
Martinique (France)	1 079	417	389 000	Fort-de-France
MAURITANIA	1 030 700	397 955	2 529 000	Nouakchott
MAURITIUS	2 040	788	1 141 000	Port Louis
Mayotte (France)	373	144	144 944	Dzaoudzi
MEXICO	1 972 545	761 604	95 831 000	México (Mexico City)
MICRONESIA, FEDERATED STATES OF	701	271	114 000	Palikir
MOLDOVA	33 700	13 012	4 378 000	Chișinău (Kishinev)
MONACO	2	1	33 000	Monaco-Ville
MONGOLIA	1 565 000	604 250	2 579 000	Ulaanbaatar (Ulan Bator)
Montserrat (U.K.)	100	39	11 000	Plymouth
MOROCCO	446 550	172 414	27 377 000	Rabat
MOZAMBIQUE	799 380	308 642	18 880 000	Maputo
MYANMAR (BURMA)	676 577	261 228	44 497 000	Yangôn (Rangoon)
NAMIBIA	824 292	318 261	1 660 000	Windhoek
NAURU	21	8	11 000	Yaren
NEPAL	147 181	56 827	22 847 000	Kathmandu
NETHERLANDS	41 526	16 033	15 678 000	Amsterdam/The Hague
Netherlands Antilles	800	309	213 000	Willemstad

LANGUAGES	RELIGIONS	CURRENCY
Luxembourgian, German, French	Roman Catholic, Protestant	Franc, Euro
Macedonian, Albanian	Orthodox, Muslim, Roman Catholic	Denar
Malagasy, French	Trad. beliefs, Roman Catholic, Protestant	Franc
Portuguese	Roman Catholic, Protestant	Port. escudo
English, local languages	Protestant, Roman Catholic, Muslim Trad. beliefs	Kwacha
Malay, English, Chinese, Tamil, local languages	Muslim, Buddhist, Roman Catholic, Christian, Trad. beliefs	Dollar (Ringgit)
Maldivian	Muslim	Rufiyaa
French, local languages	Muslim, Trad. beliefs	CFA franc
Maltese, English	Roman Catholic	Lira
Marshallese, English	Protestant, Roman Catholic	US dollar
French, French Creole	Roman Catholic	French franc
Arabic, French, local languages	Muslim	Ouguiya
English	Hindu, Roman Catholic, Muslim	Rupee
Mahorian (Swahili), French	Muslim, Roman Catholic	French franc
Spanish, Amerindian languages	Roman Catholic	Peso
English, Trukese, Pohnpeian, local languages	Protestant, Roman Catholic	US dollar
Romanian, Russian, Ukrainian	Moldovan Orthodox	Leu
French, Monegasque, Italian	Roman Catholic	French franc
Mongolian, Kazakh, local languages	Buddhist, Muslim, Trad. beliefs	Tugrik
English	Protestant, Roman Catholic	E. Carib. dollar
Arabic, Berber, French, Spanish	Muslim	Dirham
Portuguese, Makua, Tsonga, local languages	Trad. beliefs, Roman Catholic, Muslim	Metical
Burmese, Shan, Karen, local languages	Buddhist, Muslim, Protestant	Kyat
English, Afrikaans, Ovambo, local languages	Protestant, Roman Catholic	Dollar
Nauruan, Kiribati, English	Protestant, Roman Catholic	Austr. dollar
Nepali, English, local languages,	Hindu, Buddhist	Rupee
Dutch	Roman Catholic, Protestant	Guilder, Euro
Dutch, Papiamento	Roman Catholic, Protestant	Guilder

COUNTRY	AREA sq km	sq mls	POPULATION	CAPITAL
New Caledonia (France)	19 058	7 358	206 000	Nouméa
NEW ZEALAND	270 534	104 454	3 796 000	Wellington
NICARAGUA	130 000	50 193	4 807 000	Managua
NIGER	1 267 000	489 191	10 078 000	Niamey
NIGERIA	923 768	356 669	106 409 000	Abuja
Niue (N.Z.)	258	100	2 000	Alofi
Norfolk Island (Austr.)	35	14	2 000	Kingston
Northern Mariana Islands (U.S.A.)	477	184	70 000	Saipan
NORTH KOREA	120 538	46 540	23 348 000	P'yŏngyang
NORWAY	323 878	125 050	4 419 000	Oslo
OMAN	309 500	119 499	2 382 000	Muscat (Masqaṭ)
PAKISTAN	803 940	310 403	148 166 000	Islamabad
PALAU	497	192	19 000	Koror
PANAMA	77 082	29 762	2 767 000	Panamá (Panama City)
PAPUA NEW GUINEA	462 840	178 704	4 600 000	Port Moresby
PARAGUAY	406 752	157 048	5 222 000	Asunción
PERU	1 285 216	496 225	24 797 000	Lima
PHILIPPINES	300 000	115 831	72 944 000	Manila
Pitcairn Islands (U.K.)	45	17	46	Adamstown
POLAND	312 683	120 728	38 718 000	Warsaw (Warszawa)
PORTUGAL	88 940	34 340	9 869 000	Lisbon (Lisboa)
Puerto Rico (U.S.A.)	9 104	3 515	3 810 000	San Juan
QATAR	11 437	4 416	579 000	Doha (Ad Dawḩah)
Réunion (France)	2 551	985	682 000	St-Denis
ROMANIA	237 500	91 699	22 474 000	Bucharest (Bucureşti)
RUSSIAN FEDERATION	17 075 400	6 592 849	147 434 000	Moscow (Moskva
RWANDA	26 338	10 169	6 604 000	Kigali
St Helena and Dependencies (U.K.)	121	47	5 644	Jamestown

LANGUAGES	RELIGIONS	CURRENCY
French, local languages	Roman Catholic, Protestant	Pacific franc
English, Maori	Protestant, Roman Catholic	Dollar
Spanish, Amerindian languages	Roman Catholic, Protestant	Côrdoba
French, Hausa, local languages	Muslim, Trad. beliefs	CFA franc
English, Creole, local languages	Muslim, Protestant, Roman Catholic, Trad. beliefs	Naira
English, Polynesian	Protestant, Roman Catholic	NZ dollar
English	Protestant, Roman Catholic	Austr. dollar
Engllish, Chamorro, Tagalog	Roman Catholic, Protestant	US dollar
Korean	Trad. beliefs, Chondoist, Buddhist, Confucian, Taoist	Won
Norwegian	Protestant, Roman Catholic	Krone
Arabic, Baluchi, Farsi	Muslim	Rial
Urdu, Punjabi, Sindhi, Pushtu, English	Muslim, Christian, Hindu	Rupee
Palauan, English	Roman Catholic, Protestant	US dollar
Spanish, English Creole, Amerindian languages	Roman Catholic	Balboa
English, Papuan languages	Protestant, Roman Catholic, Trad. beliefs	Kina
Spanish, Guarani	Roman Catholic	Guaraní
Spanish, Qechua, Aymara	Roman Catholic	Sol
English, Filipino, local languages	Roman Catholic, Agilipayan, Muslim	Peso
English	Protestant	Dollar
Polish, German	Roman Catholic, Orthodox	Złoty
Portuguese	Roman Catholic, Protestant	Escudo, Euro
Spanish, English	Roman Catholic, Protestant	US dollar
Arabic, Indian languages	Muslim, Christian	Riyal
French	Roman Catholic	French franc
Romanian, Hungarian	Orthodox	Leu
Russian, Tatar, local languages	Orthodox, Muslim, other Christian, Jewish	Rouble
Kinyarwanda, French, English	Roman Catholic, Trad. beliefs	Franc
English	Protestant, Roman Catholic	Pound

COUNTRY	AREA sq km	sq mls	POPULATION	CAPITAL
ST KITTS AND NEVIS	261	101	39 000	Basseterre
ST LUCIA	616	238	150 000	Castries
St Pierre and Miquelon (France)	242	93	7 000	St-Pierre
ST VINCENT AND THE GRENADINES	389	150	112 000	Kingstown
SAMOA	2 831	1 093	174 000	Apia
SAN MARINO	61	24	26 000	San Marino
SÃO TOMÉ AND PRÍNCIPE	964	372	141 000	São Tomé
SAUDI ARABIA	2 200 000	849 425	20 181 000	Riyadh (Ar Riyāḑ)
SENEGAL	196 720	75 954	9 003 000	Dakar
SEYCHELLES	455	176	76 000	Victoria
SIERRA LEONE	71 740	27 699	4 568 000	Freetown
SINGAPORE	639	247	3 476 000	Singapore
SLOVAKIA	49 035	18 933	5 377 000	Bratislava
SLOVENIA	20 251	7 819	1 993 000	Ljubljana
SOLOMON ISLANDS	28 370	10 954	417 000	Honiara
SOMALIA	637 657	246 201	9 237 000	Muqdisho (Mogadishu)
SOUTH AFRICA, REPUBLIC OF	1 219 090	470 693	39 357 000	Pretoria/Cape Town
SOUTH KOREA	99 274	38 330	46 109 000	Seoul (Sŏul)
SPAIN	504 782	194 897	39 628 000	Madrid
SRI LANKA	65 610	25 332	18 455 000	Colombo
SUDAN	2 505 813	967 500	28 292 000	Khartoum
SURINAME	163 820	63 251	414 000	Paramaribo
Svalbard (Norway)	61 229	23 641	2 591	Longyearbyen
SWAZILAND	17 364	6 704	952 000	Mbabane
SWEDEN	449 964	173 732	8 875 000	Stockholm
SWITZERLAND	41 293	15 943	7 299 000	Bern (Berne)
SYRIA	185 180	71 498	15 333 000	Damascus (Dimashq)
TAIWAN	36 179	13 969	21 908 135	T'aipei
TAJIKISTAN	143 100	55 251	6 015 000	Dushanbe

LANGUAGES	RELIGIONS	CURRENCY
English, Creole	Protestant, Roman Catholic	E. Carib. dollar
English, French Creole	Roman Catholic, Protestant	E. Carib. dollar
French	Roman Catholic	French franc
English, Creole	Protestant, Roman Catholic	E. Carib. dollar
Samoan, English	Protestant, Roman Catholic	Tala
Italian	Roman Catholic	Italian lira
Portuguese	Roman Catholic	Dobra
Arabic	Muslim	Riyal
French, Wolof, local languages	Muslim	CFA franc
Seychellois, English	Roman Catholic, Protestant	Rupee
English, Creole	Trad. beliefs, Muslim	Leone
Chinese, English, Malay	Buddhist, Taoist, Muslim, Christian	Dollar
Slovak, Hungarian, Czech	Roman Catholic, Protestant	Koruna
Slovene	Roman Catholic, Protestant	Tôlar
English, Solomon Islands Pidgin, local languages	Protestant, Roman Catholic	Dollar
Somali, Arabic	Muslim	Shilling
Afrikaans, English, local languages	Protestant, Roman Catholic	Rand
Korean	Buddhist, Protestant, Roman Catholic	Won
Spanish, Catalan, Galician, Basque	Roman Catholic	Peseta, Euro
Sinhalese, Tamil, English	Buddhist, Hindu, Muslim	Rupee
Arabic, local languages	Muslim, Trad. beliefs	Dinar
Dutch, Surinamese, English	Hindu, Roman Catholic, Protestant, Muslim	Guilder
Norwegian	Protestant, Roman Catholic	Krone
Swazi, English	Protestant, Roman Catholic, Trad. beliefs	Emalangeni
Swedish	Protestant, Roman Catholic	Krona
German, French, Italian	Roman Catholic, Protestant	Franc
Arabic, Kurdish	Muslim, Christian	Pound
Chinese, local languages	Buddhist, Taoist, Confucian	Dollar
Tajik, Uzbek, Russian	Muslim	Rouble

COUNTRY	AREA sq km	sq mls	POPULATION	CAPITAL
TANZANIA	945 087	364 900	32 102 000	Dodoma
THAILAND	513 115	198 115	60 300 000	Bangkok (Krung Thep)
TOGO	56 785	21 925	4 397 000	Lomé
Tokelau (N.Z.)	10	4	1 000	none
TONGA	748	289	98 000	Nuku'alofa
TRINIDAD AND TOBAGO	5 130	1 981	1 283 000	Port of Spain
TUNISIA	164 150	63 379	9 335 000	Tunis
TURKEY	779 452	300 948	64 479 000	Ankara
TURKMENISTAN	488 100	188 456	4 309 000	Ashgabat (Ashkhabad)
Turks and Caicos Islands (U.K.)	430	166	16 000	Grand Turk (Cockburn Town)
TUVALU	25	10	11 000	Vaiaku
UGANDA	241 038	93 065	20 554 000	Kampala
UKRAINE	603 700	233 090	50 861 000	Kiev (Kyiv)
UNITED ARAB EMIRATES	83 600	32 278	2 377 453	Abu Dhabi (Abū Ẓabī)
UNITED KINGDOM	244 082	94 241	58 649 000	London
UNITED STATES OF AMERICA	9 809 378	3 787 422	274 028 000	Washington
URUGUAY	176 215	68 037	3 289 000	Montevideo
UZBEKISTAN	447 400	172 742	23 574 000	Tashkent
VANUATU	12 190	4 707	182 000	Port Vila
VATICAN CITY	0.5	0.2	480	Vatican City
VENEZUELA	912 050	352 144	23 242 000	Caracas
VIETNAM	329 565	127 246	77 562 000	Ha Nôi (Hanoi)
Virgin Islands (U.K.)	153	59	20 000	Road Town
Virgin Islands (U.S.A.)	352	136	94 000	Charlotte Amalie
Wallis and Futuna Islands (France)	274	106	14 000	Mata-Utu
WESTERN SAHARA	266 000	102 703	275 000	Laâyoune
YEMEN	527 968	203 850	16 887 000	Şan'ā'
YUGOSLAVIA	102 173	39 449	10 635 000	Belgrade (Beograd)
ZAMBIA	752 614	290 586	8 781 000	Lusaka
ZIMBABWE	390 759	150 873	11 377 000	Harare

LANGUAGES	RELIGIONS	CURRENCY
Swahili, English, local languages	Christian, Muslin, Trad. beliefs	Shilling
Thai, Lao, Chinese, Malay	Buddhist, Muslim	Baht
French, local languages	Trad. beliefs, Roman Catholic, Muslim	CFA franc
English, Tokelauan	Protestant, Roman Catholic	NZ dollar
Tongan, English	Protestant, Roman Catholic, Mormon	Pa'anga
English, Creole, Hindi	Roman Catholic, Hindu, Protestant	Dollar
Arabic, French	Muslim	Dinar
Turkish, Kurdish	Muslim	Lira
Turkmen, Russian	Muslim	Manat
English	Protestant	US dollar
Tuvaluan, English	Protestant	Dollar
English, Swahili, local languages	Roman Catholic, Protestant	Shilling
Ukrainian, Russian	Orthodox, Roman Catholic	Hryvnia
Arabic, Hindu, Urdu, Farsi	Muslim, Christian	Dirham
English	Protestant, Roman Catholic, Muslim	Pound
English, Spanish, Amerindian languages	Protestant, Roman Catholic	Dollar
Spanish	Roman Catholic, Protestant	Peso
Uzbek, Russian, Tajik	Muslim, Orthodox	Som
English, Creole	Protestant, Roman Catholic	Vatu
Italian	Roman Catholic	Italian lira
Spanish, Amerindian languages	Roman Catholic	Bolívar
Vietnamese, Thai, local languages	Buddhist, Roman Catholic	Dong
English	Protestant. Roman Catholic	US dollar
English, Spanish	Protestant, Roman Catholic	US dollar
French, Polynesian	Roman Catholic	Pacific franc
Arabic	Muslim	Dirham
Arabic	Muslim	Dinar, Rial
Serbian, Albanian	Serbian and Montenegrin Orthodox, Muslim	Dinar
English, local languages	Christian, Trad. beliefs	Kwacha
English, Shona, Ndebele	Protestant, Roman Catholic, Trad. beliefs	Dollar

ISLANDS

Great Britain
218 476 sq km
84 354 sq miles

Spitsbergen
37 814 sq km
14 600 sq miles

Iceland
102 820 sq km
39 699 sq miles

Novaya Zemlya
90 650 sq km
35 000 sq miles

Ireland
83 045 sq km
32 064 sq miles

Sardegna (Sardinia)
24 090 sq km
9 301 sq miles

Sicilia (Sicily)
25 426 sq km
9 817 sq miles

MOUNTAINS

LAKES

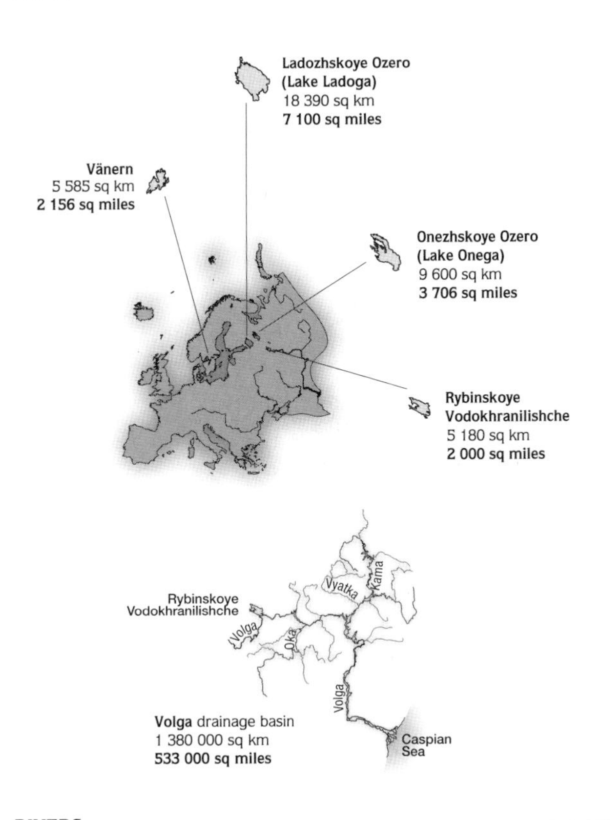

RIVERS

Volga 3 688 km **2 291 miles**

Danube 2 850 km **1 770 miles**

Dnieper 2 285 km **1 419 miles**

Kama 2 028 km **1 260 miles**

Don 1 931 km **1 199 mile**

Pechora 1 802 km **1 119 miles**

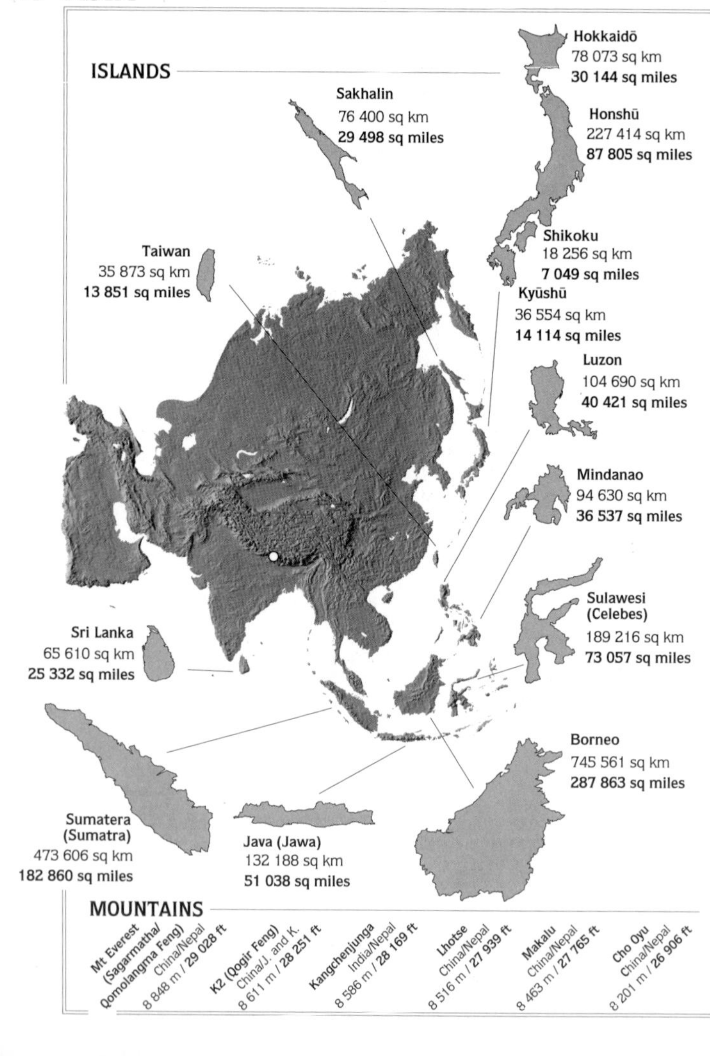
ISLANDS
Hokkaidō
78 073 sq km
30 144 sq miles
Sakhalin
76 400 sq km
29 498 sq miles
Honshū
227 414 sq km
87 805 sq miles
Taiwan
35 873 sq km
13 851 sq miles
Shikoku
18 256 sq km
7 049 sq miles
Kyūshū
36 554 sq km
14 114 sq miles
Luzon
104 690 sq km
40 421 sq miles
Mindanao
94 630 sq km
36 537 sq miles
Sulawesi
(Celebes)
189 216 sq km
73 057 sq miles
Sri Lanka
65 610 sq km
25 332 sq miles
Borneo
745 561 sq km
287 863 sq miles
Sumatera
(Sumatra)
473 606 sq km
182 860 sq miles
Java (Jawa)
132 188 sq km
51 038 sq miles
MOUNTAINS
Mt Everest (Sagarmatha/ Qomolangma Feng) China/Nepal 8 848 m / 29 028 ft
K2 (Qogir Feng) China/J. and K. 8 611 m / 28 251 ft
Kangchenjunga India/Nepal 8 586 m / 28 169 ft
Lhotse China/Nepal 8 516 m / 27 939 ft
Makalu China/Nepal 8 463 m / 27 765 ft
Cho Oyu China/Nepal 8 201 m / 26 906 ft

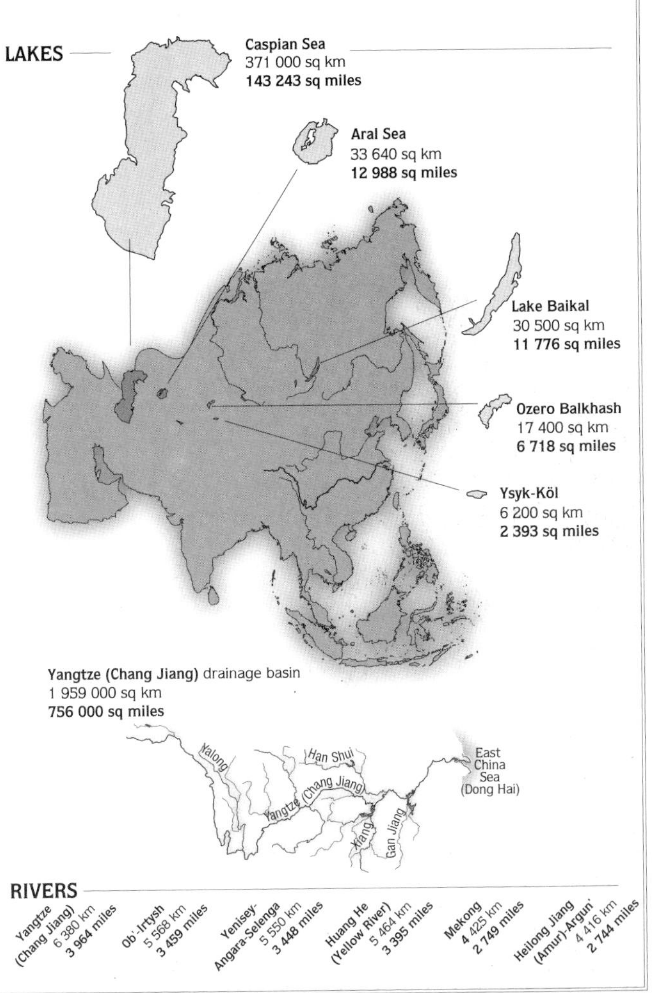
LAKES
Caspian Sea
371 000 sq km
143 243 sq miles
Aral Sea
33 640 sq km
12 988 sq miles
Lake Baikal
30 500 sq km
11 776 sq miles
Ozero Balkhash
17 400 sq km
6 718 sq miles
Ysyk-Köl
6 200 sq km
2 393 sq miles
Yangtze (Chang Jiang) drainage basin
1 959 000 sq km
756 000 sq miles
Yalong
Han Shui
Yangtze (Chang Jiang)
East China Sea (Dong Hai)
Xiang
Gan Jiang
RIVERS
Yangtze (Chang Jiang) 6 380 km 3 964 miles
Ob'-Irtysh 5 568 km 3 459 miles
Yenisey-Angara-Selenga 5 550 km 3 448 miles
Huang He (Yellow River) 5 464 km 3 395 miles
Mekong 4 425 km 2 749 miles
Heilong Jiang (Amur)-Argun' 4 416 km 2 744 miles

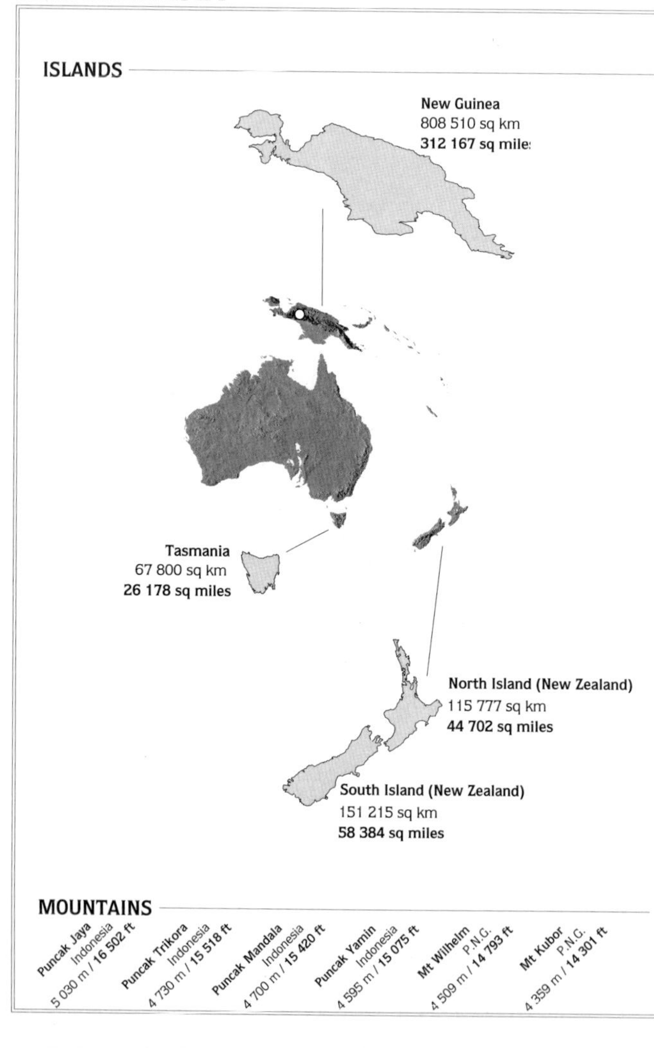
ISLANDS
New Guinea
808 510 sq km
312 167 sq mile
Tasmania
67 800 sq km
26 178 sq miles
North Island (New Zealand)
115 777 sq km
44 702 sq miles
South Island (New Zealand)
151 215 sq km
58 384 sq miles
MOUNTAINS
Puncak Jaya
Indonesia
5 030 m / 16 502 ft
Puncak Trikora
Indonesia
4 730 m / 15 518 ft
Puncak Mandala
Indonesia
4 700 m / 15 420 ft
Puncak Yamin
Indonesia
4 595 m / 15 075 ft
Mt Wilhelm
P.N.G.
4 509 m / 14 793 ft
Mt Kubor
P.N.G.
4 359 m / 14 301 ft

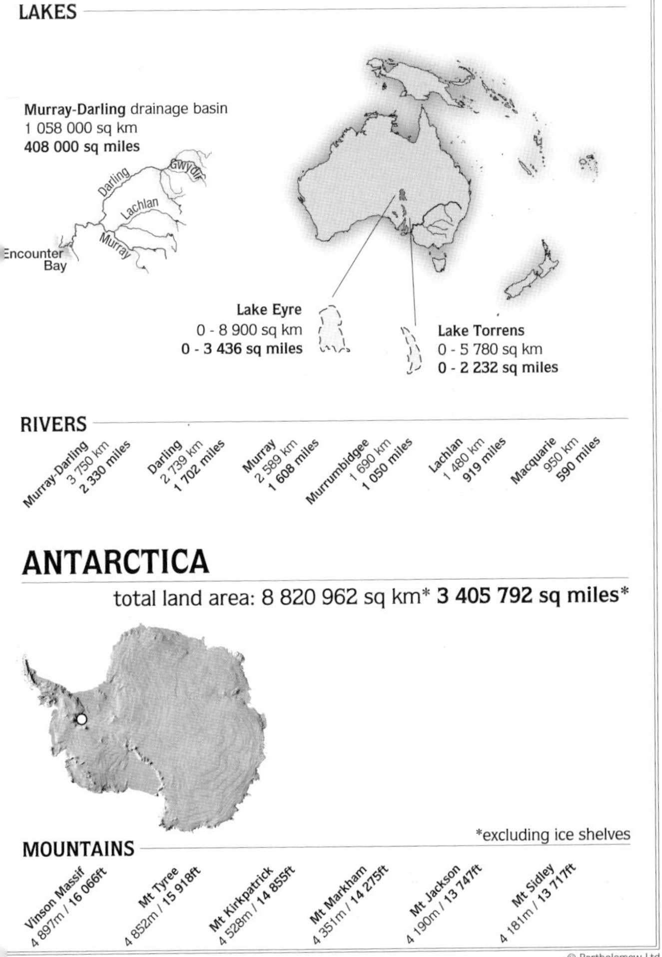
LAKES
Murray-Darling drainage basin
1 058 000 sq km
408 000 sq miles
Darling
Gwydir
Lachlan
Murray
Encounter Bay
Lake Eyre
0 - 8 900 sq km
0 - 3 436 sq miles
Lake Torrens
0 - 5 780 sq km
0 - 2 232 sq miles
RIVERS
Murray-Darling 3 750 km 2 330 miles
Darling 2 739 km 1 702 miles
Murray 2 589 km 1 608 miles
Murrumbidgee 1 690 km 1 050 miles
Lachlan 1 480 km 919 miles
Macquarie 950 km 590 miles
ANTARCTICA
total land area: 8 820 962 sq km* 3 405 792 sq miles*
*excluding ice shelves
MOUNTAINS
Vinson Massif 4 897m / 16 066ft
Mt Tyree 4 852m / 15 918ft
Mt Kirkpatrick 4 528m / 14 855ft
Mt Markham 4 351m / 14 275ft
Mt Jackson 4 190m / 13 747ft
Mt Sidley 4 181m / 13 717ft

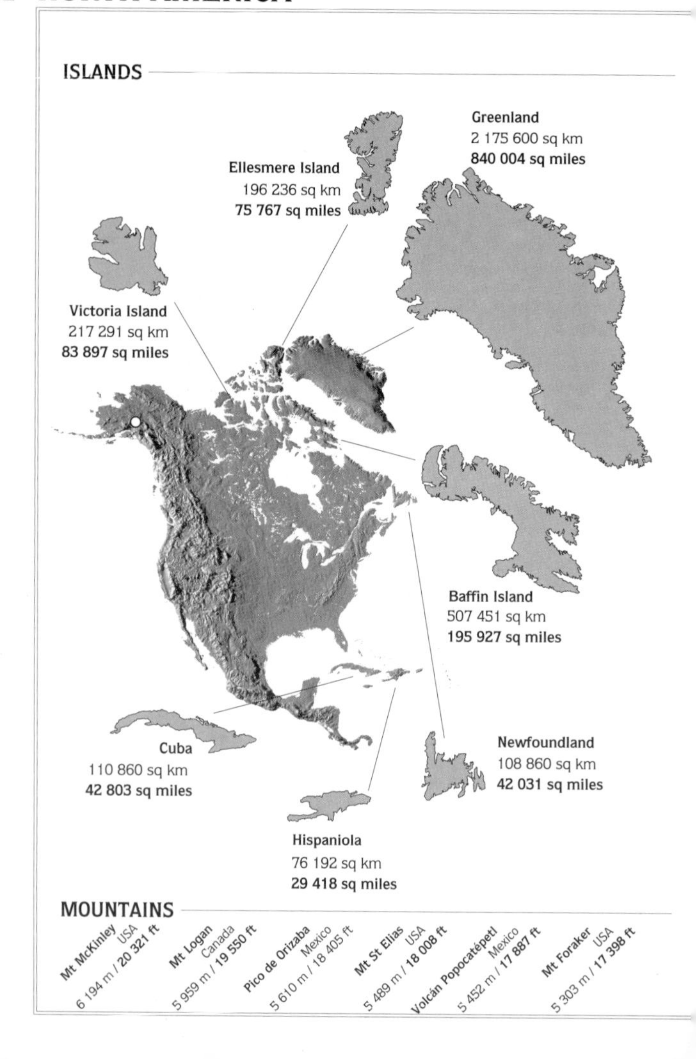
ISLANDS
Greenland
2 175 600 sq km
840 004 sq miles
Ellesmere Island
196 236 sq km
75 767 sq miles
Victoria Island
217 291 sq km
83 897 sq miles
Baffin Island
507 451 sq km
195 927 sq miles
Cuba
110 860 sq km
42 803 sq miles
Newfoundland
108 860 sq km
42 031 sq miles
Hispaniola
76 192 sq km
29 418 sq miles
MOUNTAINS
Mt McKinley USA 6 194 m / 20 321 ft
Mt Logan Canada 5 959 m / 19 550 ft
Pico de Orizaba Mexico 5 610 m / 18 405 ft
Mt St Elias USA 5 489 m / 18 008 ft
Volcán Popocatépetl Mexico 5 452 m / 17 887 ft
Mt Foraker USA 5 303 m / 17 398 ft

LAKES

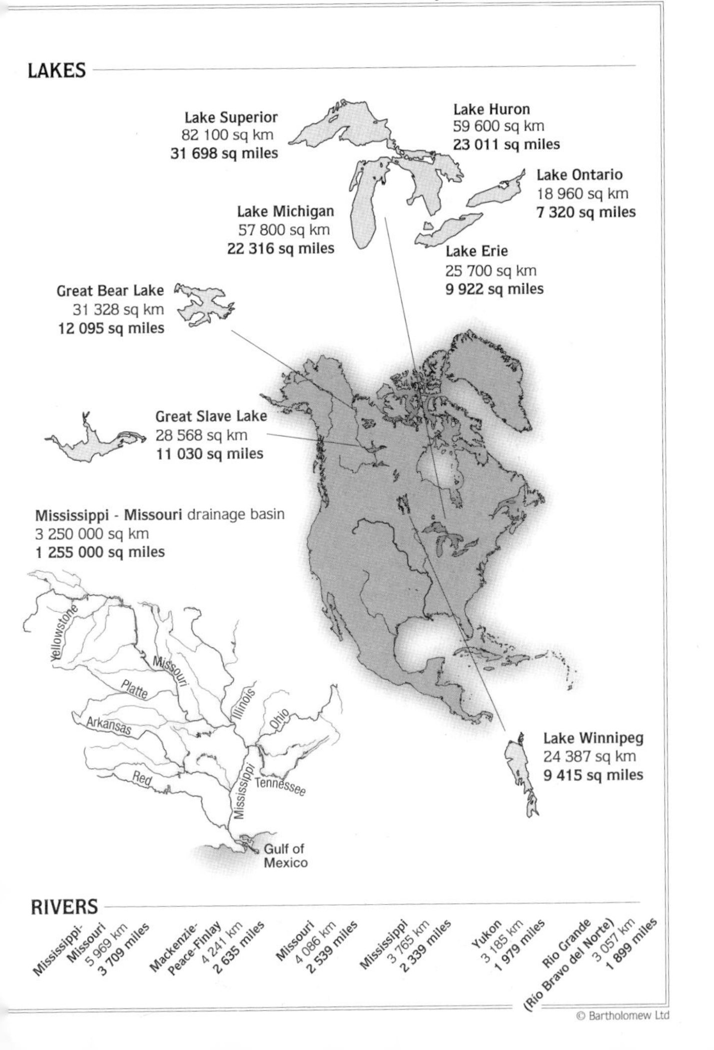

RIVERS

Mississippi-Missouri	5 969 km	**3 709 miles**
Mackenzie-Peace-Finlay	4 241 km	**2 635 miles**
Missouri	4 086 km	**2 539 miles**
Mississippi	3 765 km	**2 339 miles**
Yukon	3 185 km	**1 979 miles**
Rio Grande (Rio Bravo del Norte)	3 057 km	**1 899 miles**

ISLANDS

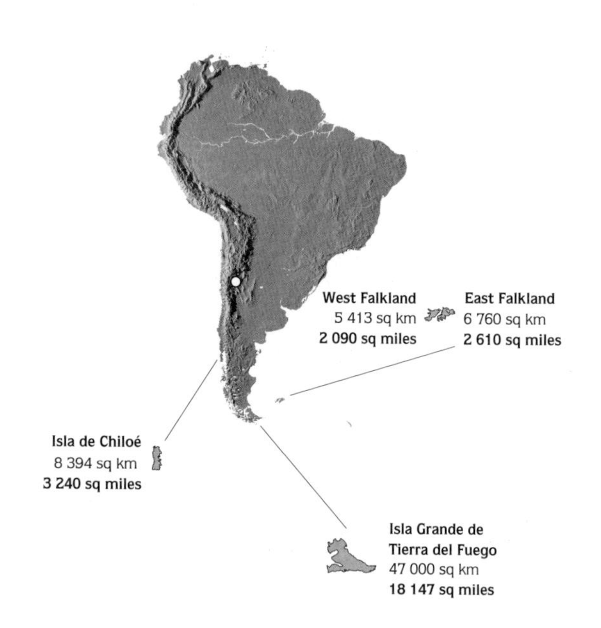

MOUNTAINS

- **Cerro Aconcagua** — Argentina — 6 960 m / **22 834 ft**
- **Nevado Ojos del Salado** — Argentina/Chile — 6 908 m / **22 664 ft**
- **Cerro Bonete** — Argentina — 6 872 m / **22 546 ft**
- **Cerro Pissis** — Argentina — 6 858 m / **22 500 ft**
- **Cerro Tupungato** — Argentina/Chile — 6 800 m / **22 309 ft**
- **Cerro Mercedario** — Argentina — 6 770 m / **22 211 ft**

LAKES

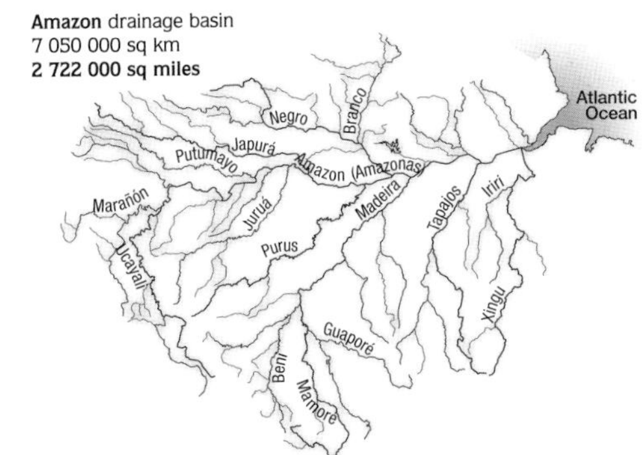

RIVERS

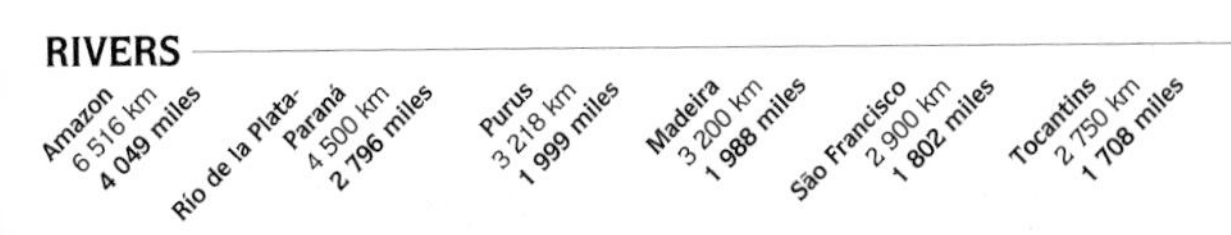

ISLANDS

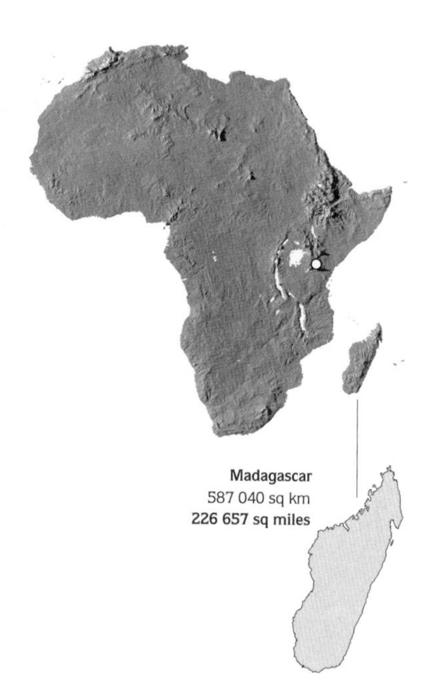

MOUNTAINS

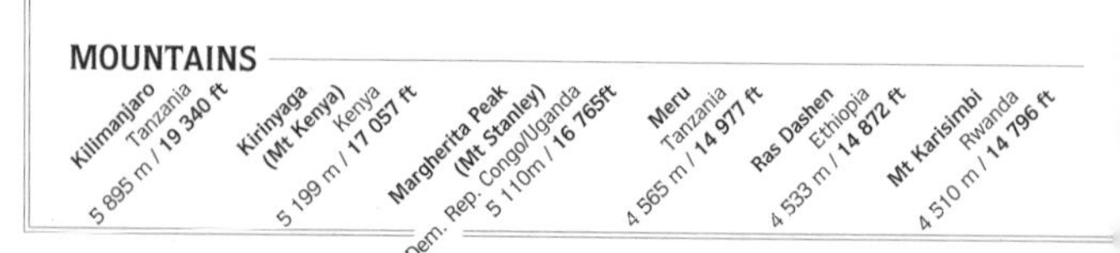

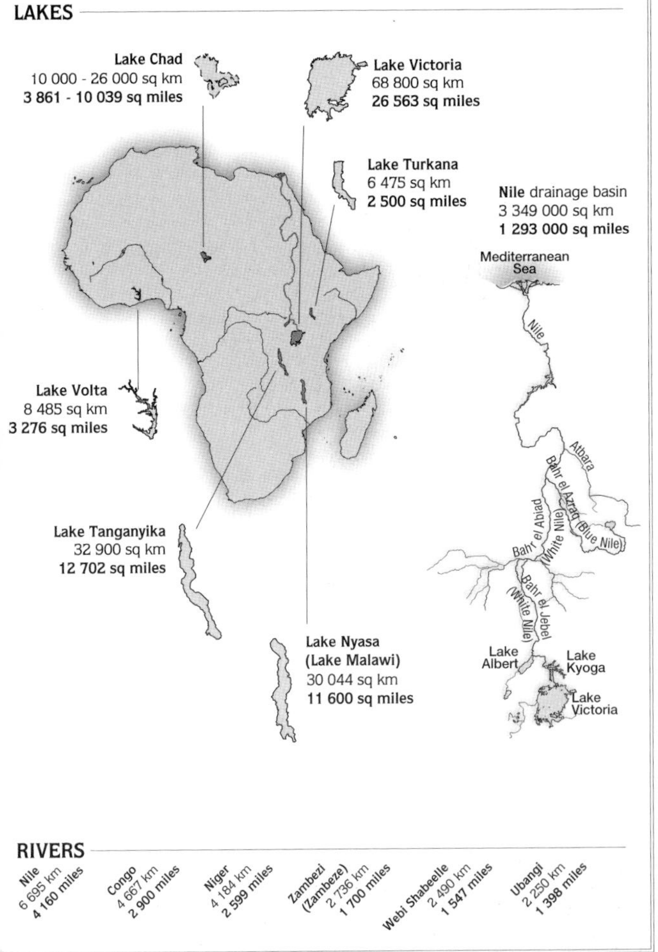
LAKES
Lake Chad
10 000 - 26 000 sq km
3 861 - 10 039 sq miles
Lake Victoria
68 800 sq km
26 563 sq miles
Lake Turkana
6 475 sq km
2 500 sq miles
Nile drainage basin
3 349 000 sq km
1 293 000 sq miles
Mediterranean Sea
Nile
Atbara
Bahr el Azraq (Blue Nile)
Bahr el Abiad (White Nile)
Bahr el Jebel (White Nile)
Lake Albert
Lake Kyoga
Lake Victoria
Lake Volta
8 485 sq km
3 276 sq miles
Lake Tanganyika
32 900 sq km
12 702 sq miles
Lake Nyasa (Lake Malawi)
30 044 sq km
11 600 sq miles
RIVERS
Nile 6 695 km 4 160 miles
Congo 4 667 km 2 900 miles
Niger 4 184 km 2 599 miles
Zambezi (Zambeze) 2 736 km 1 700 miles
Webi Shabeelle 2 490 km 1 547 miles
Ubangi 2 250 km 1 398 miles

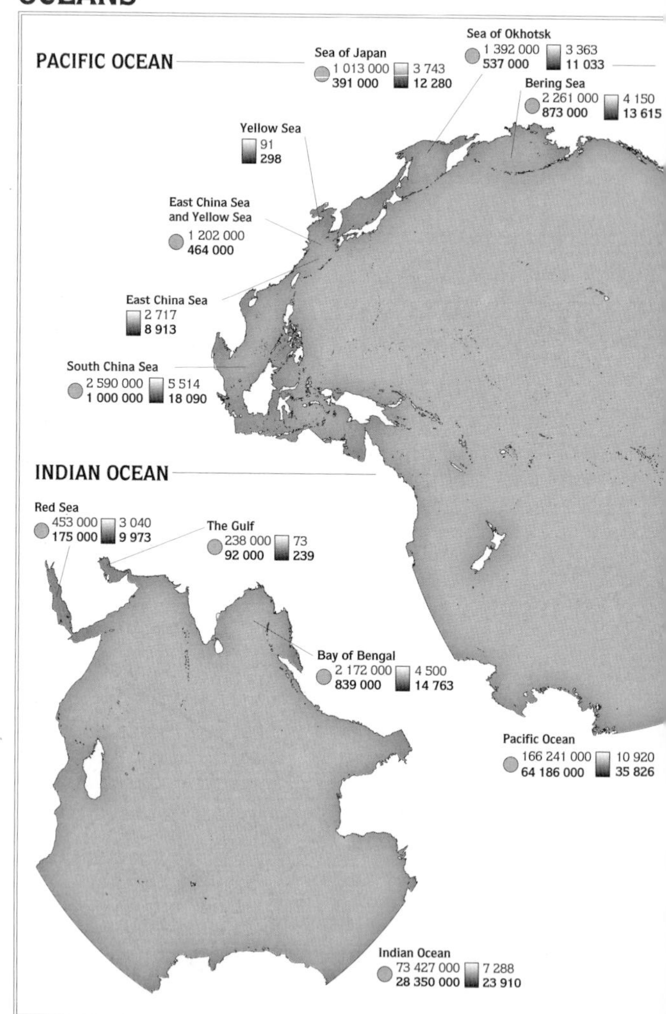
PACIFIC OCEAN
Sea of Japan
1 013 000
391 000
3 743
12 280
Sea of Okhotsk
1 392 000
537 000
3 363
11 033
Bering Sea
2 261 000
873 000
4 150
13 615
Yellow Sea
91
298
East China Sea and Yellow Sea
1 202 000
464 000
East China Sea
2 717
8 913
South China Sea
2 590 000
1 000 000
5 514
18 090
INDIAN OCEAN
Red Sea
453 000
175 000
3 040
9 973
The Gulf
238 000
92 000
73
239
Bay of Bengal
2 172 000
839 000
4 500
14 763
Pacific Ocean
166 241 000
64 186 000
10 920
35 826
Indian Ocean
73 427 000
28 350 000
7 288
23 910

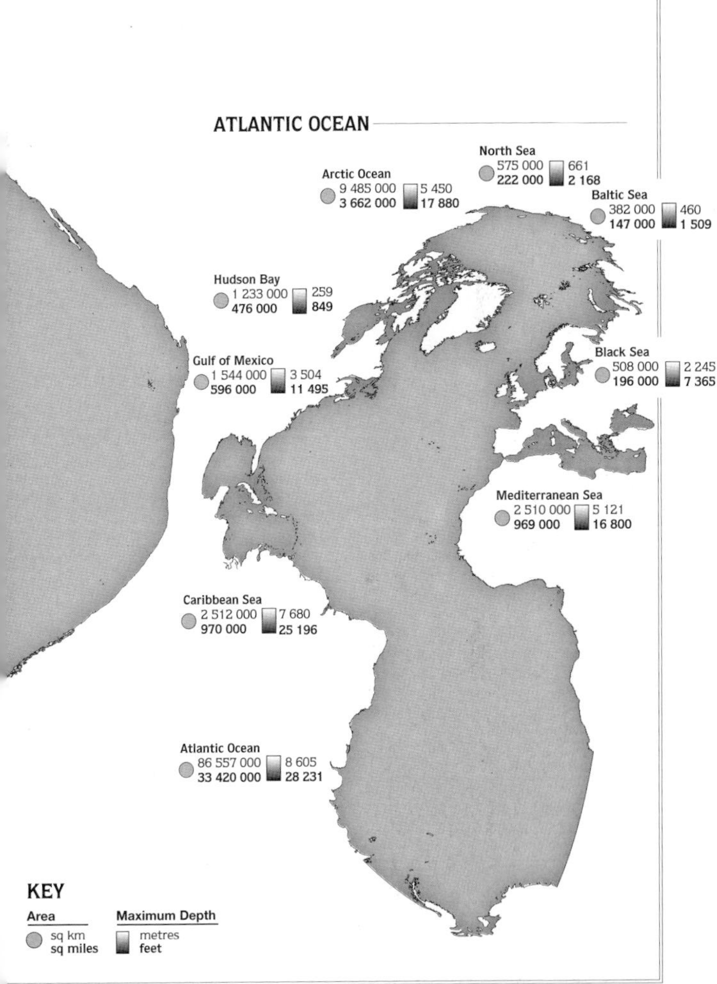
ATLANTIC OCEAN
North Sea
575 000
222 000
661
2 168
Arctic Ocean
9 485 000
3 662 000
5 450
17 880
Baltic Sea
382 000
147 000
460
1 509
Hudson Bay
1 233 000
476 000
259
849
Gulf of Mexico
1 544 000
596 000
3 504
11 495
Black Sea
508 000
196 000
2 245
7 365
Mediterranean Sea
2 510 000
969 000
5 121
16 800
Caribbean Sea
2 512 000
970 000
7 680
25 196
Atlantic Ocean
86 557 000
33 420 000
8 605
28 231
KEY
Area
sq km
sq miles
Maximum Depth
metres
feet

INTRODUCTION TO THE INDEX

The index includes all names shown on the maps in the Atlas of the World. Names are referenced by page number and by a grid reference. The grid reference correlates to the alphanumeric values which appear within each map frame. Each entry also includes the country or geographical area in which the feature is located. Entries relating to names appearing on insets are indicated by a small box symbol: □, followed by a grid reference if the inset has its own alphanumeric values.

Name forms are as they appear on the maps, with additional alternative names or name forms included as cross-references which refer the user to the entry for the map form of the name. Names beginning with Mc or Mac are alphabetized exactly as they appear. The terms Saint, Sainte, etc, are abbreviated to St, Ste, etc, but alphabetized as if in the full form.

Names of physical features beginning with generic, geographical terms are permuted – the descriptive term is placed after the main part of the name. For example, Lake Superior is indexed as Superior, Lake; Mount Everest as Everest, Mount. This policy is applied to all languages.

Entries, other than those for towns and cities, include a descriptor indicating the type of geographical feature. Descriptors are not included where the type of feature is implicit in the name itself.

Administrative divisions are included to differentiate entries of the same name and feature type within the one country. In such cases, duplicate names are alphabetized in order of administrative division. Additional qualifiers are also included for names within selected geographical areas.

INDEX ABBREVIATIONS

admin. div. administrative division
Afgh. Afghanistan
Alg. Algeria
Arg. Argentina
Austr. Australia
aut. reg. autonomous region
aut. rep. autonomous republic
Azer. Azerbaijan
Bangl. Bangladesh
Bol. Bolivia
Bos.-Herz. Bosnia Herzegovina
Bulg. Bulgaria
Can. Canada
C.A.R. Central African Republic
Col. Colombia
Czech Rep. Czech Republic
Dem. Rep. Congo Democratic Republic of Congo
depr. depression
des. desert
Dom. Rep. Dominican Republic
esc. escarpment
est. estuary
Eth. Ethiopia
Fin. Finland
for. forest
g. gulf
Ger. Germany
Guat. Guatemala
hd headland
Hond. Honduras
imp. l. impermanent lake
Indon. Indonesia
isth. isthmus
Kazakh. Kazakhstan
Kyrg. Kyrgyzstan
lag. lagoon
Lith. Lithuania
Lux. Luxembourg
Madag. Madagascar
Maur. Mauritania
Mex. Mexico
Moz. Mozambique
mun. municipality
N. North
Neth. Netherlands
Nic. Nicaragua
N.Z. New Zealand
Pak. Pakistan
Para. Paraguay
Phil. Philippines
plat. plateau
P.N.G. Papua New Guinea
Pol. Poland
Port. Portugal
prov. province
reg. region
Rep. Republic
Rus. Fed. Russian Federation
S. South
Switz. Switzerland
Tajik. Tajikistan
Tanz. Tanzania
terr. territory
Thai. Thailand
Trin. and Tob. Trinidad and Tobago
Turkm. Turkmenistan
U.A.E. United Arab Emirates
U.K. United Kingdom
Ukr. Ukraine
Uru. Uruguay
U.S.A. United States of America
Uzbek. Uzbekistan
val. valley
Venez. Venezuela
Yugo. Yugoslavia

A

68 C2 **Aachen** Ger.
68 B2 **Aalst** Belgium
68 B2 **Aarschot** Belgium
36 C2 **Aba** China
83 C4 **Aba** Nigeria
49 C2 **Ābādān** Iran
49 D2 **Ābādeh** Iran
82 B1 **Abadla** Alg.
83 C4 **Abakaliki** Nigeria
51 H3 **Abakan** Rus. Fed.
118 A3 **Abancay** Peru
49 D2 **Abarqū** Iran
34 D2 **Abashiri** Japan
85 B4 **Ābaya Hāyk'** *l.* Eth.
Ābay Wenz *r.* Eth./Sudan *see* **Blue Nile**
50 G3 **Abaza** Rus. Fed.
76 A2 **Abbasanta** *Sardinia* Italy
72 C1 **Abbeville** France
110 B3 **Abbeville** U.S.A.
42 B1 **Abbottabad** Pak.
23 O2 **Abbott Ice Shelf** Antarctica
83 E3 **Abéché** Chad
82 B4 **Abengourou** Côte d'Ivoire
61 E4 **Åbenrå** Denmark
82 C4 **Abeokuta** Nigeria
67 A2 **Aberaeron** U.K.
64 C2 **Aberchirder** U.K.
67 B3 **Aberdare** U.K.
67 A2 **Aberdaron** U.K.
90 B3 **Aberdeen** S. Africa
64 C2 **Aberdeen** U.K.
109 D3 **Aberdeen** *MD* U.S.A.
105 D1 **Aberdeen** *SD* U.S.A.
102 B1 **Aberdeen** *WA* U.S.A.
97 E1 **Aberdeen Lake** Can.
102 B2 **Abert, Lake** U.S.A.
67 A2 **Aberystwyth** U.K.
54 F2 **Abez'** Rus. Fed.
46 B3 **Abhā** Saudi Arabia
Abiad, Bahr el *r.* Sudan/Uganda *see* **White Nile**
82 B4 **Abidjan** Côte d'Ivoire
105 D3 **Abilene** *KS* U.S.A.
107 D2 **Abilene** *TX* U.S.A.
67 C3 **Abingdon** U.K.
59 D3 **Abinsk** Rus. Fed.
98 B2 **Abitibi, Lake** Can.
Åbo Fin. *see* **Turku**
42 B1 **Abohar** India
82 C4 **Abomey** Benin
28 A1 **Abongabong, Gunung** *mt.* Indon.
86 B2 **Abong Mbang** Cameroon
32 A2 **Aborlan** Phil.
83 D3 **Abou Déia** Chad
74 B2 **Abrantes** Port.
120 B3 **Abra Pampa** Arg.
104 A2 **Absaroka Range** *mts* U.S.A.
49 C1 **Abşeron Yarımadası** *pen.* Azer.
46 B3 **Abū 'Arīsh** Saudi Arabia
47 C2 **Abu Dhabi** U.A.E.
84 B3 **Abu Hamed** Sudan
83 C4 **Abuja** Nigeria
49 C2 **Abū Kamāl** Syria
120 B1 **Abunã** *r.* Bol.
118 B2 **Abunã** Brazil
42 B2 **Abu Road** India
84 B2 **Abu Simbel** Egypt
85 A3 **Abu Zabad** Sudan
Abū Zabī U.A.E. *see* **Abu Dhabi**
85 A4 **Abyei** Sudan
113 B2 **Acambaro** Mex.
74 B1 **A Cañiza** Spain
112 B2 **Acaponeta** Mex.
113 C3 **Acapulco** Mex.
119 D2 **Acará** Brazil
118 B1 **Acarigua** Venez.
113 C3 **Acatlan** Mex.
113 C3 **Acayucán** Mex.
82 B4 **Accra** Ghana
66 B2 **Accrington** U.K.
42 B2 **Achalpur** India
65 A2 **Achill Island** Rep. of Ireland
69 D1 **Achim** Ger.
64 B2 **Achnasheen** U.K.
59 D2 **Achuyevo** Rus. Fed.
79 C3 **Acıpayam** Turkey
77 C3 **Acireale** *Sicily* Italy
115 C2 **Acklins Island** Bahamas
121 B4 **Aconcagua, Cerro** *mt.* Arg.
74 B1 **A Coruña** Spain
76 A2 **Acqui Terme** Italy
71 D2 **Ács** Hungary
113 C2 **Actopán** Mex.
107 D2 **Ada** U.S.A.
47 C2 **Adam** Oman
'Adan Yemen *see* **Aden**
48 B2 **Adana** Turkey
Adapazari Turkey *see* **Sakarya**
76 A1 **Adda** *r.* Italy
46 B2 **Ad Dafīnah** Saudi Arabia
46 B2 **Ad Dahnā'** *des.* Saudi Arabia
46 B2 **Ad Dahnā'** *des.* Saudi Arabia
82 A2 **Ad Dakhla** Western Sahara
47 C2 **Ad Dammām** Saudi Arabia
46 A2 **Ad Dār al Ḥamrā'** Saudi Arabia
46 B3 **Ad Darb** Saudi Arabia
46 B2 **Ad Dawādimī** Saudi Arabia
Ad Dawḥah Qatar *see* **Doha**
46 B2 **Ad Dilam** Saudi Arabia
84 C2 **Ad Dir'īyah** Saudi Arabia
85 B4 **Addis Ababa** Eth.
49 C2 **Ad Dīwānīyah** Iraq
20 A2 **Adelaide** Austr.
18 C1 **Adelaide River** Austr.
69 D2 **Adelebsen** Ger.
23 J2 **Adélie Land** Antarctica
46 B3 **Aden** Yemen
85 C3 **Aden, Gulf of** Somalia/Yemen
68 C2 **Adenau** Ger.
47 C2 **Adh Dhayd** U.A.E.
27 C3 **Adi** *i.* Indon.
46 A3 **Ādī Ārk'ay** Eth.
84 B3 **Ādīgrat** Eth.
43 B3 **Adilabad** India
109 E2 **Adirondack Mountains** U.S.A.
Ādīs Ābeba Eth. *see* **Addis Ababa**
85 B4 **Ādīs Alem** Eth.
78 C1 **Adjud** Romania
18 B1 **Admiralty Gulf** Austr.
96 A2 **Admiralty Island** U.S.A.
72 B3 **Adour** *r.* France
74 C2 **Adra** Spain
82 B2 **Adrar** Alg.
108 C2 **Adrian** *MI* U.S.A.
107 C1 **Adrian** *TX* U.S.A.
76 B2 **Adriatic Sea** Europe
84 B3 **Ādwa** Eth.
51 K2 **Adycha** *r.* Rus. Fed.
59 D3 **Adygeysk** Rus. Fed.
82 B4 **Adzopé** Côte d'Ivoire
79 B3 **Aegean Sea** Greece/Turkey
69 D1 **Aerzen** Ger.
74 B1 **A Estrada** Spain
84 B3 **Afabet** Eritrea
44 C3 **Afghanistan** *country* Asia
46 B2 **'Afīf** Saudi Arabia
48 B2 **Afyon** Turkey
83 C3 **Agadez** Niger
82 B1 **Agadir** Morocco
45 D2 **Agadyr'** Kazakh.
42 B2 **Agar** India
43 D2 **Agartala** India
49 C2 **Ağdam** Azer.
73 C3 **Agde** France
72 C3 **Agen** France
90 A2 **Aggeneys** S. Africa
79 C3 **Agia Vervara** Greece
79 B3 **Agios Dimitrios** Greece
79 C3 **Agios Efstratios** *i.* Greece
79 C3 **Agios Nikolaos** Greece
78 B1 **Agnita** Romania
43 B2 **Agra** India
49 C2 **Ağrı** Turkey
76 B3 **Agrigento** *Sicily* Italy
79 B3 **Agrinio** Greece
77 B2 **Agropoli** Italy
122 B2 **Água Clara** Brazil
114 B4 **Aguadulce** Panama
112 B2 **Aguanaval** *r.* Mex.
112 B1 **Agua Prieta** Mex.
112 B2 **Aguascalientes** Mex.
123 D1 **Águas Formosas** Brazil
74 B1 **Águeda** Port.
74 C1 **Aguilar de Campóo** Spain
75 C2 **Águilas** Spain
112 B3 **Aguililla** Mex.
90 B3 **Agulhas, Cape** S. Africa
123 D2 **Agulhas Negras** *mt.* Brazil
79 C2 **Ağva** Turkey
49 C2 **Ahar** Iran
68 C1 **Ahaus** Ger.
49 C2 **Ahlat** Turkey
68 C2 **Ahlen** Ger.
42 B2 **Ahmadabad** India
41 B3 **Ahmadnagar** India
42 B2 **Ahmadpur East** Pak.
42 B1 **Ahmadpur Sial** Pak.
112 B2 **Ahome** Mex.
47 C2 **Ahram** Iran
69 E1 **Ahrensburg** Ger.
72 C2 **Ahun** France
49 C2 **Ahvāz** Iran
90 A2 **Ai-Ais** Namibia
48 B2 **Aigialousa** Cyprus
79 B3 **Aigio** Greece
111 D2 **Aiken** U.S.A.
123 D1 **Aimorés** Brazil
123 D1 **Aimorés, Serra dos** *hills* Brazil
82 B2 **'Aîn Ben Tili** Maur.
75 D2 **Aïn Defla** Alg.
82 B1 **Aïn Sefra** Alg.
104 D2 **Ainsworth** U.S.A.
Aintab Turkey *see* **Gaziantep**
75 D2 **Aïn Taya** Alg.
75 D2 **Aïn Tédélès** Alg.
28 A1 **Airbangis** Indon.
96 C2 **Airdrie** Can.
72 B3 **Aire-sur-l'Adour** France
69 E3 **Aisch** *r.* Ger.
96 A1 **Aishihik Lake** Can.
68 A3 **Aisne** *r.* France
27 D3 **Aitape** P.N.G.
78 B1 **Aiud** Romania
73 D3 **Aix-en-Provence** France
73 D2 **Aix-les-Bains** France
30 A1 **Aizawl** India
56 C2 **Aizkraukle** Latvia
35 C3 **Aizu-wakamatsu** Japan
73 D3 **Ajaccio** *Corsica* France
83 E1 **Ajdābiyā** Libya
42 B2 **Ajmer** India
106 A2 **Ajo** U.S.A.
55 E3 **Akbulak** Rus. Fed.
48 B2 **Akçakale** Turkey
48 B2 **Akdağmadeni** Turkey
56 A2 **Åkersberga** Sweden

86 C2 **Aketi** Dem. Rep. Congo
55 D4 **Akhalk'alak'i** Georgia
47 C2 **Akhḍar, Jabal** *mts* Oman
79 C3 **Akhisar** Turkey
55 D4 **Akhtubinsk** Rus. Fed.
98 B1 **Akimiski Island** Can.
34 D3 **Akita** Japan
82 A3 **Akjoujt** Maur.
45 D1 **Akkol'** Kazakh.
56 B2 **Akmenrags** *pt* Latvia
Akmola Kazakh. *see* **Astana**
85 B4 **Akobo** Sudan
42 B2 **Akola** India
46 A3 **Akordat** Eritrea
95 G2 **Akpatok Island** Can.
60 □A3 **Akranes** Iceland
79 B2 **Akrathos, Akra** *pt* Greece
108 C2 **Akron** U.S.A.
43 B1 **Aksai Chin** *terr.* Asia
48 B2 **Aksaray** Turkey
44 B1 **Aksay** Kazakh.
59 D2 **Aksay** Rus. Fed.
48 B2 **Akşehir** Turkey
44 C2 **Akshiganak** Kazakh.
45 E2 **Aksu** China
46 A3 **Äksum** Eth.
44 B2 **Aktau** Kazakh.
45 D2 **Aktogay** Kazakh.
56 C3 **Aktsyabrski** Belarus
44 B1 **Aktyubinsk** Kazakh.
83 C4 **Akure** Nigeria
60 □B2 **Akureyri** Iceland
Akyab Myanmar *see* **Sittwe**
110 C2 **Alabama** *r.* U.S.A.
110 C2 **Alabama** *state* U.S.A.
79 C3 **Alaçatı** Turkey
49 C1 **Alagir** Rus. Fed.
119 E3 **Alagoinhas** Brazil
75 C1 **Alagón** Spain
47 B2 **Al Aḥmadī** Kuwait
45 E2 **Alakol', Ozero** *salt l.* Kazakh.
60 J2 **Alakurtti** Rus. Fed.
46 B3 **Al 'Alayyah** Saudi Arabia
49 C2 **Al 'Amādīyah** Iraq
103 C3 **Alamo** U.S.A.
106 B2 **Alamogordo** U.S.A.
112 A2 **Alamos** *Sonora* Mex.
112 B2 **Alamos** *Sonora* Mex.
112 B2 **Alamos** *r.* Mex.
104 B3 **Alamosa** U.S.A.
61 G3 **Åland** *is* Fin.
48 B2 **Alanya** Turkey
48 B3 **Al 'Aqabah** Jordan
46 B2 **Al 'Aqīq** Saudi Arabia
75 C2 **Alarcón, Embalse de** *resr* Spain
46 B2 **Al Arṭāwīyah** Saudi Arabia
29 C2 **Alas** Indon.
79 C3 **Alaşehir** Turkey
96 A2 **Alaska** *state* U.S.A.
92 C2 **Alaska, Gulf of** U.S.A.
49 C2 **Älät** Azer.
55 D3 **Alatyr'** Rus. Fed.
118 A2 **Alausí** Ecuador
61 H3 **Alavus** Fin.
20 B2 **Alawoona** Austr.
76 A2 **Alba** Italy
75 C2 **Albacete** Spain
78 B1 **Alba Iulia** Romania
77 C2 **Albania** *country* Europe
18 A3 **Albany** Austr.
98 B1 **Albany** *r.* Can.
111 D2 **Albany** *GA* U.S.A.
109 E2 **Albany** *NY* U.S.A.
102 B2 **Albany** *OR* U.S.A.
Al Başrah Iraq *see* **Basra**
19 D1 **Albatross Bay** Austr.
83 E1 **Al Bayḍā'** Libya
46 B3 **Al Bayḍā'** Yemen
111 D1 **Albemarle** U.S.A.
111 E1 **Albemarle Sound** *sea chan.* U.S.A.
76 A2 **Albenga** Italy
19 C2 **Alberga** *watercourse* Austr.
87 D2 **Albert, Lake** Dem. Rep.Congo/Uganda
96 C2 **Alberta** *prov.* Can.
68 B2 **Albert Kanaal** *canal* Belgium
105 E2 **Albert Lea** U.S.A.
72 C3 **Albi** France
46 A2 **Al Bi'r** Saudi Arabia
46 B3 **Al Birk** Saudi Arabia
46 B2 **Al Biyāḍh** *reg.* Saudi Arabia
61 E4 **Ålborg** Denmark
49 C2 **Alborz, Reshteh-ye** *mts* Iran
74 B2 **Albufeira** Port.
106 B1 **Albuquerque** U.S.A.
47 C2 **Al Buraymī** Oman
21 C3 **Albury** Austr.
74 B2 **Alcácer do Sal** Port.
74 C1 **Alcalá de Henares** Spain
74 C2 **Alcalá la Real** Spain
76 B3 **Alcamo** *Sicily* Italy
75 C1 **Alcañiz** Spain
74 B2 **Alcántara** Spain
74 C2 **Alcaraz** Spain
74 C2 **Alcaraz, Sierra de** *mts* Spain
74 C2 **Alcaudete** Spain
74 C2 **Alcázar de San Juan** Spain
59 D2 **Alchevs'k** Ukr.
123 E1 **Alcobaça** Brazil
75 C1 **Alcora** Spain
75 C2 **Alcoy** Spain
75 D2 **Alcúdia** Spain
113 C2 **Aldama** Mex.
51 J3 **Aldan** Rus. Fed.
51 J2 **Aldan** *r.* Rus. Fed.
63 C4 **Alderney** *i.* Channel Is
82 A3 **Aleg** Maur.
123 D2 **Alegre** Brazil
120 C3 **Alegrete** Brazil
51 K3 **Aleksandrovsk-Sakhalinskiy** Rus. Fed.
59 D1 **Alekseyevka** *Belgorodskaya Oblast'* Rus. Fed.
59 D1 **Alekseyevka** *Belgorodskaya Oblast'* Rus. Fed.
57 E3 **Aleksin** Rus. Fed.
77 D2 **Aleksinac** Yugo.
86 B3 **Alèmbé** Gabon
123 D2 **Além Paraíba** Brazil
61 F3 **Ålen** Norway
72 C2 **Alençon** France
48 B2 **Aleppo** Syria
118 A3 **Alerta** Peru
96 B2 **Alert Bay** Can.
73 C3 **Alès** France
78 B1 **Aleşd** Romania
76 A2 **Alessandria** Italy
61 E3 **Ålesund** Norway
92 B2 **Aleutian Islands** U.S.A.
51 L3 **Alevina, Mys** *c.* Rus. Fed.
96 A2 **Alexander Archipelago** *is* U.S.A.
90 A2 **Alexander Bay** S. Africa
110 C2 **Alexander City** U.S.A.
23 O2 **Alexander Island** Antarctica
21 C3 **Alexandra** Austr.
22 A3 **Alexandra** N.Z.
79 B2 **Alexandreia** Greece
Alexandretta Turkey *see* **İskenderun**
84 A1 **Alexandria** Egypt
78 C2 **Alexandria** Romania
91 C3 **Alexandria** S. Africa
110 B2 **Alexandria** *LA* U.S.A.
105 D1 **Alexandria** *MN* U.S.A.
109 D3 **Alexandria** *VA* U.S.A.
20 A3 **Alexandrina, Lake** Austr.
79 C2 **Alexandroupoli** Greece
99 E1 **Alexis** *r.* Can.
96 B2 **Alexis Creek** Can.
45 E1 **Aleysk** Rus. Fed.
75 C1 **Alfaro** Spain
49 C3 **Al Fāw** Iraq
69 D2 **Alfeld (Leine)** Ger.
123 C2 **Alfenas** Brazil
Al Furāt *r.* Iraq/Syria *see* **Euphrates**
74 B2 **Algeciras** Spain
75 C2 **Algemesí** Spain
46 A3 **Algena** Eritrea
Alger Alg. *see* **Algiers**
82 C1 **Algeria** *country* Africa
47 C3 **Al Ghayḍah** Yemen
76 A2 **Alghero** *Sardinia* Italy
47 B2 **Al Ghwaybiyah** Saudi Arabia
83 C1 **Algiers** Alg.
91 C3 **Algoa Bay** S. Africa
105 E2 **Algona** U.S.A.
49 C2 **Al Ḥadīthah** Iraq
47 C2 **Al Ḥajar al Gharbī** *mts* Oman
83 D2 **Al Ḥamādah al Ḥamrā'** *plat.* Libya
75 C2 **Alhama de Murcia** Spain
46 B2 **Al Ḥanākīyah** Saudi Arabia
49 C2 **Al Ḥasakah** Syria
49 C2 **Al Ḥayy** Iraq
46 B3 **Al Ḥazm al-Jawf** Yemen
47 C3 **Al Ḥibāk** *des.* Saudi Arabia
49 C2 **Al Ḥillah** Iraq
46 B2 **Al Ḥillah** Saudi Arabia
47 B2 **Al Ḥinnāh** Saudi Arabia
46 B3 **Al Ḥudaydah** Yemen
47 B2 **Al Hufūf** Saudi Arabia
83 D2 **Al Ḥulayq al Kabīr** *hills* Libya
47 C2 **'Alīābād** Iran
79 C3 **Aliağa** Turkey
79 B2 **Aliakmonas** *r.* Greece
75 C2 **Alicante** Spain
107 D3 **Alice** U.S.A.
77 C3 **Alice, Punta** *pt* Italy
19 C2 **Alice Springs** Austr.
43 B2 **Aligarh** India
49 C2 **Alīgūdarz** Iran
37 E1 **Alihe** China
86 B3 **Alima** *r.* Congo
79 C3 **Aliova** *r.* Turkey
85 C3 **Ali Sabieh** Djibouti
91 C3 **Aliwal North** S. Africa
83 E2 **Al Jaghbūb** Libya
46 B2 **Al Jahrah** Kuwait
47 C2 **Al Jamalīyah** Qatar
46 A2 **Al Jawf** Saudi Arabia
83 D1 **Al Jawsh** Libya
74 B2 **Aljezur** Port.
47 B2 **Al Jubayl** Saudi Arabia
46 B2 **Al Junaynah** Saudi Arabia
74 B2 **Aljustrel** Port.
46 B2 **Al Kahfah** Saudi Arabia
47 C2 **Al Kāmil** Oman
48 B2 **Al Karak** Jordan
47 C2 **Al Khābūrah** Oman
46 B2 **Al Khamāsīn** Saudi Arabia
47 C2 **Al Khaşab** Oman
46 B3 **Al Khawkhah** Yemen
47 C2 **Al Khawr** Qatar
83 E2 **Al Khufrah** Libya
83 D1 **Al Khums** Libya
47 B2 **Al Khunn** Saudi Arabia
47 C2 **Al Kir'ānah** Qatar
68 B1 **Alkmaar** Neth.
49 C2 **Al Kūt** Iraq
Al Kuwayt Kuwait *see* **Kuwait**
48 B2 **Al Lādhiqīyah** Syria
43 C2 **Allahabad** India
51 K2 **Allakh-Yun'** Rus. Fed.

30 A2 **Allanmyo** Myanmar
109 D2 **Allegheny** *r.* U.S.A.
108 C3 **Allegheny Mountains** U.S.A.
65 B1 **Allen, Lough** *l.* Rep. of Ireland
113 B2 **Allende** *Coahuila* Mex.
113 B2 **Allende** *Nuevo León* Mex.
109 D2 **Allentown** U.S.A.
41 B4 **Alleppey** India
69 D1 **Aller** *r.* Ger.
104 C2 **Alliance** *NE* U.S.A.
108 C2 **Alliance** *OH* U.S.A.
46 B2 **Al Līth** Saudi Arabia
64 C2 **Alloa** U.K.
99 C2 **Alma** Can.
Alma-Ata Kazakh. *see* **Almaty**
74 B2 **Almada** Port.
74 C2 **Almadén** Spain
Al Madīnah Saudi Arabia *see* **Medina**
84 B1 **Al Mafraq** Jordan
46 B3 **Al Maḥwīt** Yemen
46 B2 **Al Majma'ah** Saudi Arabia
103 B2 **Almanor, Lake** U.S.A.
75 C2 **Almansa** Spain
47 C2 **Al Mariyyah** U.A.E.
83 E1 **Al Marj** Libya
45 D2 **Almaty** Kazakh.
49 C2 **Al Mawṣil** Iraq
74 C1 **Almazán** Spain
119 C2 **Almeirim** Brazil
68 C1 **Almelo** Neth.
123 D1 **Almenara** Brazil
74 B1 **Almendra, Embalse de** *resr* Spain
74 B2 **Almendralejo** Spain
74 C2 **Almería** Spain
74 C2 **Almería, Golfo de** *b.* Spain
55 E3 **Al'met'yevsk** Rus. Fed.
46 B2 **Al Mindak** Saudi Arabia
47 B2 **Al Mish'āb** Saudi Arabia
74 B2 **Almodôvar** Port.
74 B2 **Almonte** Spain
43 B2 **Almora** India
47 B2 **Al Mubarrez** Saudi Arabia
47 C2 **Al Muḍaibī** Oman
48 B3 **Al Mudawwarah** Jordan
47 B3 **Al Mukallā** Yemen
46 B3 **Al Mukhā** Yemen
74 C2 **Almuñécar** Spain
46 A2 **Al Muwayliḥ** Saudi Arabia
79 B3 **Almyros** Greece
64 B2 **Alness** U.K.
66 C1 **Alnwick** U.K.
79 B3 **Alonnisos** *i.* Greece
27 C3 **Alor** *i.* Indon.
27 C3 **Alor, Kepulauan** *is* Indon.
28 B1 **Alor Setar** Malaysia
Alost Belgium *see* **Aalst**
54 C2 **Alozero** Rus. Fed.
108 C1 **Alpena** U.S.A.
107 C2 **Alpine** U.S.A.
52 D5 **Alps** *mts* Europe
47 B3 **Al Qa'āmīyāt** *reg.* Saudi Arabia
83 D1 **Al Qaddāḥīyah** Libya
46 B2 **Al Qā'īyah** Saudi Arabia
49 C2 **Al Qāmishlī** Syria
48 B2 **Al Qaryatayn** Syria
47 B3 **Al Qaṭn** Yemen
48 B2 **Al Qunayṭirah** Syria
46 B3 **Al Qunfidhah** Saudi Arabia
46 B2 **Al Quwayīyah** Saudi Arabia
69 D2 **Alsfeld** Ger.
60 H2 **Alta** Norway
60 H2 **Altaelva** *r.* Norway
36 B1 **Altai Mountains** Asia
111 D2 **Altamaha** *r.* U.S.A.
119 C2 **Altamira** Brazil
77 C2 **Altamura** Italy
36 B1 **Altay** China
36 C1 **Altay** Mongolia
73 D2 **Altdorf** Switz.
75 C2 **Altea** Spain
69 F2 **Altenburg** Ger.
68 C2 **Altenkirchen (Westerwald)** Ger.
79 C3 **Altınoluk** Turkey
79 D3 **Altıntaş** Turkey
120 B2 **Altiplano** *plain* Bol.
122 B1 **Alto Araguaia** Brazil
75 C1 **Alto del Moncayo** *mt.* Spain
122 B1 **Alto Garças** Brazil
89 C2 **Alto Molócuè** Moz.
97 E3 **Altona** Can.
109 D2 **Altoona** U.S.A.
122 B1 **Alto Sucuriú** Brazil
70 C2 **Altötting** Ger.
36 B2 **Altun Shan** *mts* China
102 B2 **Alturas** U.S.A.
107 D2 **Altus** U.S.A.
56 C2 **Alūksne** Latvia
46 A2 **Al 'Ulā** Saudi Arabia
83 D1 **Al 'Uqaylah** Libya
59 C3 **Alushta** Ukr.
107 D1 **Alva** U.S.A.
113 C3 **Alvarado** Mex.
61 F3 **Älvdalen** Sweden
61 F3 **Alvdalen** *val.* Sweden
60 H2 **Älvsbyn** Sweden
46 A2 **Al Wajh** Saudi Arabia
42 B2 **Alwar** India
49 C2 **Al Widyān** *plat.* Iraq/Saudi Arabia
Alxa Youqi China *see* **Ehen Hudag**
Alxa Zuoqi China *see* **Bayan Hot**
19 C1 **Alyangula** Austr.
56 B3 **Alytus** Lith.
104 C1 **Alzada** U.S.A.
68 D3 **Alzey** Ger.
18 C2 **Amadeus, Lake** *salt flat* Austr.
95 G2 **Amadjuak Lake** Can.
74 B2 **Amadora** Port.
46 B2 **Amā'ir** Saudi Arabia
61 F4 **Åmål** Sweden
79 B3 **Amaliada** Greece
27 D3 **Amamapare** Indon.
122 A2 **Amambaí** Brazil
122 B2 **Amambaí** *r.* Brazil
37 E3 **Amami-Ō-shima** *i.* Japan
37 E3 **Amami-shotō** *is* Japan
45 C1 **Amangel'dy** Kazakh.
77 C3 **Amantea** Italy
91 D3 **Amanzimtoti** S. Africa
119 C1 **Amapá** Brazil
74 B2 **Amareleja** Port.
107 C1 **Amarillo** U.S.A.
76 B2 **Amaro, Monte** *mt.* Italy
48 B1 **Amasya** Turkey
118 C1 **Amazon** *r.* S. America
119 D1 **Amazon, Mouths of the** Brazil
Amazonas *r.* S. America *see* **Amazon**
42 B1 **Ambala** India
89 □D3 **Ambalavao** Madag.
89 □D2 **Ambanja** Madag.
118 A2 **Ambato** Ecuador
89 □D2 **Ambato Boeny** Madag.
89 □D3 **Ambato Finandrahana** Madag.
89 □D2 **Ambatolampy** Madag.
89 □D2 **Ambatondrazaka** Madag.
69 E3 **Amberg** Ger.
114 B3 **Ambergris Cay** *i.* Belize
43 C2 **Ambikapur** India
89 □D2 **Ambilobe** Madag.
66 B1 **Ambleside** U.K.
89 □D3 **Amboasary** Madag.
89 □D3 **Ambohimahasoa** Madag.
27 C3 **Ambon** Indon.
27 C3 **Ambon** *i.* Indon.
89 □D3 **Ambositra** Madag.
89 □D3 **Ambovombe** Madag.
103 C4 **Amboy** U.S.A.
88 A1 **Ambriz** Angola
36 C2 **Amdo** China
113 B2 **Amealco** Mex.
112 B2 **Ameca** Mex.
68 B1 **Ameland** *i.* Neth.
102 D2 **American Falls** U.S.A.
102 D2 **American Falls Reservoir** U.S.A.
103 D2 **American Fork** U.S.A.
17 G4 **American Samoa** *terr.* S. Pacific Ocean
111 D2 **Americus** U.S.A.
68 B1 **Amersfoort** Neth.
23 F2 **Amery Ice Shelf** Antarctica
105 E2 **Ames** U.S.A.
79 B3 **Amfissa** Greece
51 J2 **Amga** Rus. Fed.
34 C1 **Amgu** Rus. Fed.
83 C2 **Amguid** Alg.
51 K3 **Amgun'** *r.* Rus. Fed.
99 D2 **Amherst** Can.
72 C2 **Amiens** France
41 B3 **Amindivi Islands** India
90 A1 **Aminuis** Namibia
42 A2 **Amir Chah** Pak.
97 D2 **Amisk Lake** Can.
107 C3 **Amistad Reservoir** Mex./U.S.A.
66 A2 **Amlwch** U.K.
48 B2 **'Ammān** Jordan
95 I2 **Ammassalik** Greenland
46 B3 **Am Nābiyah** Yemen
49 D2 **Amol** Iran
79 C3 **Amorgos** *i.* Greece
110 C2 **Amory** U.S.A.
98 C2 **Amos** Can.
Amoy China *see* **Xiamen**
123 C2 **Amparo** Brazil
75 D1 **Amposta** Spain
43 B2 **Amravati** India
42 B1 **Amritsar** India
68 B1 **Amstelveen** Neth.
68 B1 **Amsterdam** Neth.
71 C2 **Amstetten** Austria
83 E3 **Am Timan** Chad
44 B2 **Amudar'ya** *r.* Asia
94 E1 **Amund Ringnes Island** Can.
94 C2 **Amundsen Gulf** Can.
23 M2 **Amundsen Sea** Antarctica
29 C2 **Amuntai** Indon.
Amur *r.* China/Rus. Fed. *see* **Heilong Jiang**
46 A3 **'Amur, Wadi** *watercourse* Sudan
51 I2 **Anabar** *r.* Rus. Fed.
51 I2 **Anabarskiy Zaliv** *b.* Rus. Fed.
102 D1 **Anaconda** U.S.A.
107 D1 **Anadarko** U.S.A.
48 B1 **Anadolu Dağları** *mts* Turkey
51 M2 **Anadyr'** *r.* Rus. Fed.
49 C2 **'Ānah** Iraq
113 B2 **Anáhuac** Mex.
28 B1 **Anambas, Kepulauan** *is* Indon.
105 E2 **Anamosa** U.S.A.
48 B2 **Anamur** Turkey
35 B4 **Anan** Japan
41 B3 **Anantapur** India
42 B1 **Anantnag** Jammu and Kashmir

58 B2 **Anan'yiv** Ukr.
59 D3 **Anapa** Rus. Fed.
122 C1 **Anápolis** Brazil
120 B3 **Añatuya** Arg.
33 B2 **Anbyon** N. Korea
72 B2 **Ancenis** France
76 B2 **Ancona** Italy
121 A5 **Ancud** Chile
Anda China *see* **Daqing**
61 E3 **Åndalsnes** Norway
110 C2 **Andalusia** U.S.A.
41 D3 **Andaman Islands** India
31 A2 **Andaman Sea** Indian Ocean
89 □D2 **Andapa** Madag.
68 B2 **Andelst** Neth.
60 G2 **Andenes** Norway
68 B2 **Andenne** Belgium
68 B2 **Anderlecht** Belgium
94 C2 **Anderson** *r.* Can.
94 B2 **Anderson** *AK* U.S.A.
108 B2 **Anderson** *IN* U.S.A.
111 D2 **Anderson** *SC* U.S.A.
116 D2 **Andes** *mts* S. America
89 □D2 **Andilamena** Madag.
89 □D2 **Andilanatoby** Madag.
45 D2 **Andizhan** Uzbek.
89 □D2 **Andoany** Madag.
Andong China *see* **Dandong**
33 B2 **Andong** S. Korea
75 D1 **Andorra** *country* Europe
75 D1 **Andorra la Vella** Andorra
67 C3 **Andover** U.K.
122 B2 **Andradina** Brazil
57 D2 **Andreapol'** Rus. Fed.
123 D2 **Andrelândia** Brazil
107 C2 **Andrews** U.S.A.
77 C2 **Andria** Italy
89 □D3 **Androka** Madag.
114 C2 **Andros** *i.* Bahamas
79 B3 **Andros** *i.* Greece
111 E4 **Andros Town** Bahamas
41 B3 **Ândrott** *i.* India
58 B1 **Andrushivka** Ukr.
60 G2 **Andselv** Norway
74 C2 **Andújar** Spain
88 A2 **Andulo** Angola
82 C3 **Anéfis** Mali
83 D3 **Aney** Niger
51 H3 **Angara** *r.* Rus. Fed.
37 C1 **Angarsk** Rus. Fed.
61 G3 **Ånge** Sweden
112 A2 **Angel de la Guarda, Isla** *i.* Mex.
32 B1 **Angeles** Phil.
118 B1 **Angel Falls** *waterfall* Venez.
61 F4 **Ängelholm** Sweden
60 G3 **Ångermanälven** *r.* Sweden
72 B2 **Angers** France
97 E1 **Angikuni Lake** *l.* Can.
66 A2 **Anglesey** *i.* U.K.
89 C2 **Angoche** Moz.
47 C2 **Angohrān** Iran
88 A2 **Angola** *country* Africa
108 C2 **Angola** U.S.A.
96 A2 **Angoon** U.S.A.
72 C2 **Angoulême** France
45 D2 **Angren** Uzbek.
115 D3 **Anguilla** *terr.* West Indies
61 F4 **Anholt** *i.* Denmark
39 B3 **Anhua** China
38 B2 **Anhui** *prov.* China
122 B1 **Anhumas** Brazil
122 C1 **Anicuns** Brazil
34 D1 **Aniva, Mys** *c.* Rus. Fed.
33 B2 **Anjŭ** N. Korea
38 A2 **Ankang** China
48 B2 **Ankara** Turkey
38 B2 **Anlu** China
59 E1 **Anna** Rus. Fed.
83 C1 **Annaba** Alg.
69 F2 **Annaberg-Buchholtz** Ger.
46 B2 **An Nafūd** *des.* Saudi Arabia
49 C2 **An Najaf** Iraq
64 C3 **Annan** U.K.
109 D3 **Annapolis** U.S.A.
43 C2 **Annapurna I** *mt.* Jammu and Kashmir/Nepal
108 C2 **Ann Arbor** U.S.A.
118 C1 **Anna Regina** Guyana
49 C2 **An Nāşirīyah** Iraq
73 D2 **Annecy** France
46 B3 **An Nimāş** Saudi Arabia
30 B1 **Anning** China
110 C2 **Anniston** U.S.A.
73 C2 **Annonay** France
47 B2 **An Nu'ayrīyah** Saudi Arabia
89 □D2 **Anorontany, Tanjona** *hd* Madag.
79 C3 **Ano Viannos** Greece
39 B3 **Anpu** China
38 B2 **Anqing** China
70 C2 **Ansbach** Ger.
38 C1 **Anshan** China
39 A3 **Anshun** China
82 C3 **Ansongo** Mali
118 A3 **Antabamba** Peru
48 B2 **Antakya** Turkey
89 □E2 **Antalaha** Madag.
48 B2 **Antalya** Turkey
48 B2 **Antalya Körfezi** *g.* Turkey
89 □D2 **Antananarivo** Madag.
23 Q3 **Antarctic Peninsula** Antarctica
64 B2 **An Teallach** *mt.* U.K.
74 C2 **Antequera** Spain
106 B2 **Anthony** U.S.A.
82 B2 **Anti Atlas** *mts* Morocco
73 D3 **Antibes** France
99 D2 **Anticosti, Île d'** *i.* Can.
99 D2 **Antigonish** Can.
115 D3 **Antigua** *i.* Antigua
115 D3 **Antigua and Barbuda** *country* West Indies
113 C2 **Antiguo-Morelos** Mex.
79 B3 **Antikythira** *i.* Greece
Antioch Turkey *see* **Antakya**
120 A3 **Antofagasta** Chile
122 C3 **Antonina** Brazil
65 C1 **Antrim** U.K.
65 C1 **Antrim Hills** U.K.
89 □D2 **Antsalova** Madag.
89 □D2 **Antsirabe** Madag.
89 □D2 **Antsirañana** Madag.
89 □D2 **Antsohihy** Madag.
68 B2 **Antwerpen** Belgium
An Uaimh Rep. of Ireland *see* Navan
43 C2 **Anugul** India
42 B2 **Anupgarh** India
41 C4 **Anuradhapura** Sri Lanka
Anvers Belgium *see* Antwerpen
36 C2 **Anxi** China
19 C3 **Anxious Bay** Austr.
38 B2 **Anyang** China
33 B2 **Anyang** S. Korea
76 B2 **Anzio** Italy
34 D2 **Aomori** Japan
Aoraki *mt.* N.Z. *see* Cook, Mount
76 A1 **Aosta** Italy
111 D3 **Apalachee Bay** U.S.A.
118 B2 **Apaporis** *r.* Col.
122 B2 **Aparecida do Tabuado** Brazil
32 B1 **Aparri** Phil.
54 C2 **Apatity** Rus. Fed.
112 B3 **Apatzingán** Mex.
68 B1 **Apeldoorn** Neth.
68 C1 **Apen** Ger.
122 C2 **Apiaí** Brazil
32 B2 **Apo, Mount** *vol.* Phil.
69 E2 **Apolda** Ger.
20 B3 **Apollo Bay** Austr.
111 D3 **Apopka, Lake** U.S.A.
122 B1 **Aporé** Brazil
122 B1 **Aporé** *r.* Brazil
48 B2 **Apostolos Andreas, Cape** Cyprus
101 E3 **Appalachian Mountains** U.S.A.
76 A2 **Appennino** *mts* Italy
21 D2 **Appin** Austr.
108 B2 **Appleton** U.S.A.
76 B2 **Aprilia** Italy
59 D3 **Apsheronsk** Rus. Fed.
122 B2 **Apucarana** Brazil
122 B2 **Apucarana, Serra da** *hills* Brazil
46 A2 **Aqaba, Gulf of** Asia
122 A1 **Aquidauana** *r.* Brazil
43 C2 **Ara** India
85 A4 **Arab, Bahr el** *watercourse* Sudan
24 B4 **Arabian Sea** Indian Ocean
119 E3 **Aracaju** Brazil
119 E2 **Aracati** Brazil
122 B2 **Araçatuba** Brazil
123 D1 **Aracruz** Brazil
123 D1 **Araçuaí** Brazil
78 B1 **Arad** Romania
83 E3 **Arada** Chad
124 C6 **Arafura Sea** Austr./Indon.
122 B1 **Aragarças** Brazil
75 C1 **Aragón** *r.* Spain
119 D2 **Araguaia** *r.* Brazil
119 D2 **Araguaína** Brazil
122 C1 **Araguari** Brazil
35 C3 **Arai** Japan
83 C2 **Arak** Alg.
49 C2 **Arāk** Iran
30 A1 **Arakan Yoma** *mts* Myanmar
49 C1 **Araks** *r.* Armenia/Turkey
44 C2 **Aral Sea** *salt l.* Kazakh./Uzbek.
44 C2 **Aral'sk** Kazakh.
Aral'skoye More *salt l.* Kazakh./Uzbek. *see* **Aral Sea**
74 C1 **Aranda de Duero** Spain
77 D2 **Aranđelovac** Yugo.
65 B1 **Aran Island** Rep. of Ireland
65 B2 **Aran Islands** Rep. of Ireland
74 C1 **Aranjuez** Spain
90 A1 **Aranos** Namibia
107 D3 **Aransas Pass** U.S.A.
35 B4 **Arao** Japan
82 B3 **Araouane** Mali
119 E2 **Arapiraca** Brazil
122 B2 **Arapongas** Brazil
122 C3 **Araquari** Brazil
46 B1 **'Ar'ar** Saudi Arabia
122 C2 **Araraquara** Brazil
119 C2 **Araras** Brazil
122 B3 **Araras, Serra das** *mts* Brazil
20 B3 **Ararat** Austr.
123 D2 **Araruama, Lago de** *lag.* Brazil
123 E1 **Arataca** Brazil
Aratürük China *see* **Yiwu**
118 A1 **Arauca** Col.
122 C1 **Araxá** Brazil
49 C2 **Arbīl** Iraq
64 C2 **Arbroath** U.K.
42 A2 **Arbu Lut, Dasht-e** *des.* Afgh.
72 B3 **Arcachon** France
111 D3 **Arcadia** U.S.A.
102 B2 **Arcata** U.S.A.
113 B3 **Arcelia** Mex.
54 D2 **Archangel** Rus. Fed.
19 D1 **Archer** *r.* Austr.

102 D2 Arco U.S.A.
74 B2 Arcos de la Frontera Spain
95 F2 Arctic Bay Can.
128 Arctic Ocean
94 C2 Arctic Red r. Can.
49 C2 Ardabīl Iran
49 C1 Ardahan Turkey
61 E3 Årdalstangen Norway
49 D2 Ardestān Iran
21 C2 Ardlethan Austr.
107 D2 Ardmore U.S.A.
64 A2 Ardnamurchan, Point of U.K.
20 A2 Ardrossan Austr.
64 B3 Ardrossan U.K.
103 B3 Arena, Point U.S.A.
61 E4 Arendal Norway
69 E1 Arendsee (Altmark) Ger.
118 A3 Arequipa Peru
119 C2 Arere Brazil
74 C1 Arévalo Spain
76 B2 Arezzo Italy
72 B2 Argentan France
121 C3 Argentina country S. America
121 A6 Argentino, Lago l. Arg.
72 C2 Argenton-sur-Creuse France
78 C2 Argeș r. Romania
42 A1 Arghandab r. Afgh.
79 B3 Argolikos Kolpos b. Greece
79 B3 Argos Greece
79 B3 Argostoli Greece
75 C1 Arguís Spain
37 E1 Argun' r. China/Rus. Fed.
18 B1 Argyle, Lake Austr.
61 F4 Århus Denmark
90 A2 Ariamsvlei Namibia
120 A2 Arica Chile
64 A2 Arinagour U.K.
123 C1 Arinos Brazil
118 C3 Aripuanã Brazil
118 B2 Aripuanã r. Brazil
118 B2 Ariquemes Brazil
122 B1 Ariranhá r. Brazil
106 A2 Arizona state U.S.A.
112 A1 Arizpe Mex.
46 B2 'Arjah Saudi Arabia
60 G2 Arjeplog Sweden
110 B2 Arkadelphia U.S.A.
45 C1 Arkalyk Kazakh.
110 B2 Arkansas r. U.S.A.
110 B1 Arkansas state U.S.A.
105 D3 Arkansas City U.S.A.
Arkhangel'sk Rus. Fed. see Archangel
65 C2 Arklow Rep. of Ireland
70 C1 Arkona, Kap c. Ger.
50 G1 Arkticheskogo Instituta, Ostrova is Rus. Fed.
73 C3 Arles France
83 C3 Arlit Niger
68 B3 Arlon Belgium
65 C1 Armagh U.K.
84 B2 Armant Egypt
49 C1 Armavir Rus. Fed.
49 C1 Armenia country Asia
118 A1 Armenia Col.
112 B3 Armeria Mex.
21 D2 Armidale Austr.
98 B1 Armstrong Can.
48 B1 Armyans'k Ukr.
48 B2 Arnaoutis, Cape Cyprus
68 B2 Arnhem Neth.
19 C1 Arnhem, Cape Austr.
19 C1 Arnhem Bay Austr.
19 C1 Arnhem Land reg. Austr.
76 B2 Arno r. Italy
20 A2 Arno Bay Austr.
98 C2 Arnprior Can.
68 D2 Arnsberg Ger.
69 E2 Arnstadt Ger.
90 A2 Aroab Namibia
69 D2 Arolsen Ger.
46 A3 Aroma Sudan
76 A1 Arona Italy
112 B2 Aros r. Mex.
49 C2 Ar Ramādī Iraq
64 B3 Arran i. U.K.
48 B2 Ar Raqqah Syria
72 C1 Arras France
46 B2 Ar-Rass Saudi Arabia
113 C3 Arriagá Mex.
47 C2 Ar Rimāl reg. Saudi Arabia
Ar Riyāḍ Saudi Arabia see Riyadh
113 C2 Arroyo Seco Mex.
47 C2 Ar Rustāq Oman
48 C2 Ar Ruṭbah Iraq
49 D3 Arsenajān Iran
34 B2 Arsen'yev Rus. Fed.
79 B3 Arta Greece
112 B3 Arteaga Mex.
34 B2 Artem Rus. Fed.
59 D2 Artemivs'k Ukr.
72 C2 Artenay France
106 C2 Artesia U.S.A.
19 E2 Arthur Point Austr.
22 B2 Arthur's Pass N.Z.
120 C4 Artigas Uru.
97 D1 Artillery Lake Can.
58 B2 Artsyz Ukr.
45 D3 Artux China
49 C1 Artvin Turkey
27 C3 Aru, Kepulauan is Indon.
87 D2 Arua Uganda
115 D3 Aruba terr. West Indies
87 D3 Arusha Tanz.
37 C1 Arvayheer Mongolia
97 E1 Arviat Can.
60 G2 Arvidsjaur Sweden
61 F4 Arvika Sweden
55 D3 Arzamas Rus. Fed.
75 C2 Arzew Alg.
68 C2 Arzfeld Ger.
69 F2 Aš Czech Rep.
83 C4 Asaba Nigeria
34 D2 Asahi-dake vol. Japan
34 D2 Asahikawa Japan
46 B3 Āsalē l. Eth.
43 C2 Asansol India
99 C2 Asbestos Can.
77 C2 Ascea Italy
120 B2 Ascensión Bol.
80 C4 Ascension i. S. Atlantic Ocean
69 D3 Aschaffenburg Ger.
68 C2 Ascheberg Ger.
69 E2 Aschersleben Ger.
76 B2 Ascoli Piceno Italy
60 G3 Åsele Sweden
78 B2 Asenovgrad Bulg.
18 A2 Ashburton watercourse Austr.
22 B2 Ashburton N.Z.
110 B2 Ashdown U.S.A.
111 D1 Asheville U.S.A.
21 D1 Ashford Austr.
67 D3 Ashford U.K.
44 B3 Ashgabat Turkm.
66 C1 Ashington U.K.
35 B4 Ashizuri-misaki pt Japan
Ashkhabad Turkm. see Ashgabat
104 D3 Ashland KS U.S.A.
108 C3 Ashland KY U.S.A.
108 C2 Ashland OH U.S.A.
102 B2 Ashland OR U.S.A.
108 A1 Ashland WI U.S.A.
56 C3 Ashmyany Belarus
46 B3 Ash Sharawrah Saudi Arabia
49 C2 Ash Sharqāṭ Iraq
49 C2 Ash Shaṭrah Iraq
47 B3 Ash Shiḥr Yemen
47 C2 Ash Ṣhināṣ Oman
46 B2 Ash Shu'bah Saudi Arabia
46 B2 Ash Shumlūl Saudi Arabia
108 C2 Ashtabula U.S.A.
99 D1 Ashuanipi Lake Can.
74 B2 Asilah Morocco
76 A2 Asinara, Golfo dell' b. Sardinia Italy
50 G3 Asino Rus. Fed.
56 C3 Asipovichy Belarus
46 B2 'Asīr reg. Saudi Arabia
61 F4 Askim Norway
36 C1 Askiz Rus. Fed.
84 B3 Asmara Eritrea
61 F4 Åsnen l. Sweden
84 B2 Asoteriba, Jebel mt. Sudan
71 D2 Aspang-Markt Austria
104 B3 Aspen U.S.A.
22 A2 Aspiring, Mount N.Z.
85 C3 Assab Eritrea
46 B2 As Salamiyah Saudi Arabia
49 C2 As Samāwah Iraq
47 C2 Aṣ Ṣanām reg. Saudi Arabia
83 E2 As Sarīr reg. Libya
68 C1 Assen Neth.
68 B2 Assesse Belgium
83 D1 As Sidrah Libya
97 D3 Assiniboia Can.
96 C2 Assiniboine, Mount Can.
122 B2 Assis Brazil
46 B2 Aṣ Ṣubayḥīyah Kuwait
49 C2 As Sulaymānīyah Iraq
46 B2 As Sulaymī Saudi Arabia
46 B2 As Sulayyil Saudi Arabia
46 B2 As Sūq Saudi Arabia
48 B2 As Suwaydā' Syria
47 C2 As Suwayq Oman
79 B3 Astakos Greece
45 D1 Astana Kazakh.
49 C2 Astara Iran
76 A2 Asti Italy
42 B1 Astor Jammu and Kashmir
74 B1 Astorga Spain
102 B1 Astoria U.S.A.
55 D4 Astrakhan' Rus. Fed.
56 C3 Astravyets Belarus
79 C3 Astypalaia i. Greece
120 C3 Asunción Para.
84 B2 Aswân Egypt
84 B2 Asyût Egypt
120 B3 Atacama, Desierto de des. Chile
120 B3 Atacama, Salar de salt flat Chile
82 C4 Atakpamé Togo
79 B3 Atalanti Greece
118 A3 Atalaya Peru
46 B3 'Ataq Yemen
82 A2 Atâr Maur.
103 B3 Atascadero U.S.A.
45 D2 Atasu Kazakh.
84 B3 Atbara Sudan
84 B3 Atbara r. Sudan
45 C1 Atbasar Kazakh.
105 D3 Atchison U.S.A.
76 B2 Aterno r. Italy
76 B2 Atessa Italy
68 A2 Ath Belgium
96 C2 Athabasca Can.
97 C2 Athabasca r. Can.
97 D2 Athabasca, Lake Can.
79 B3 Athens Greece
111 D2 Athens GA U.S.A.
108 C3 Athens OH U.S.A.
111 D1 Athens TN U.S.A.
107 D2 Athens TX U.S.A.

Athina Greece see Athens
65 C2 Athlone Rep. of Ireland
79 B2 Athos mt. Greece
65 C2 Athy Rep. of Ireland
83 D3 Ati Chad
98 A2 Atikokan Can.
55 D3 Atkarsk Rus. Fed.
111 D2 Atlanta U.S.A.
105 D2 Atlantic U.S.A.
109 E3 Atlantic City U.S.A.
126 Atlantic Ocean
90 A3 Atlantis S. Africa
82 B1 Atlas Mountains mts Africa
82 C1 Atlas Saharien mts Alg.
96 A2 Atlin Can.
96 A2 Atlin Lake Can.
110 C2 Atmore U.S.A.
107 D2 Atoka U.S.A.
43 C2 Atrai r. India
46 B2 Aṭ Ṭā'if Saudi Arabia
31 B2 Attapu Laos
79 C3 Attavyros mt. Greece
98 B1 Attawapiskat Can.
98 B1 Attawapiskat r. Can.
98 B1 Attawapiskat Lake Can.
68 C2 Attendorn Ger.
46 B3 At Turbah Yemen
44 B2 Atyrau Kazakh.
73 C3 Aubenas France
94 C2 Aubry Lake Can.
110 C2 Auburn AL U.S.A.
103 B3 Auburn CA U.S.A.
105 D2 Auburn NE U.S.A.
109 D2 Auburn NY U.S.A.
72 C2 Aubusson France
72 C3 Auch France
22 B1 Auckland N.Z.
69 F2 Aue Ger.
70 C2 Augsburg Ger.
77 C3 Augusta Sicily Italy
111 D2 Augusta GA U.S.A.
105 D3 Augusta KS U.S.A.
109 F2 Augusta ME U.S.A.
18 A2 Augustus, Mount Austr.
68 A2 Aulnoye-Aymeries France
90 B2 Auob watercourse Namibia/S. Africa
42 B3 Aurangābād India
68 C1 Aurich Ger.
122 B1 Aurilândia Brazil
72 C3 Aurillac France
104 C3 Aurora CO U.S.A.
108 B2 Aurora IL U.S.A.
105 D2 Aurora NE U.S.A.
90 A2 Aus Namibia
105 E2 Austin MN U.S.A.
103 C3 Austin NV U.S.A.
107 D2 Austin TX U.S.A.
18 B2 Australia country Oceania
21 C3 Australian Capital Territory admin. div. Austr.
70 C2 Austria country Europe
112 B3 Autlán Mex.
73 C2 Autun France
73 C2 Auxerre France
73 D2 Auxonne France
73 C2 Avallon France
99 E2 Avalon Peninsula Can.
122 C2 Avaré Brazil
59 D2 Avdiyivka Ukr.
74 B1 Aveiro Port.
77 B2 Avellino Italy
68 A2 Avesnes-sur-Helpe France
61 G3 Avesta Sweden
76 B2 Avezzano Italy
64 C2 Aviemore U.K.
77 C2 Avigliano Italy
73 C3 Avignon France
74 C1 Ávila Spain
74 B1 Avilés Spain
77 C3 Avola Sicily Italy
67 C3 Avon r. England U.K.
67 B2 Avon r. England U.K.
72 B2 Avranches France
22 B1 Awanui N.Z.
85 C4 Āwash Eth.
85 C3 Āwash r. Eth.
83 D2 Awbārī Libya
85 C4 Aw Dheegle Somalia
64 B2 Awe, Loch l. U.K.
85 A4 Aweil Sudan
94 E1 Axel Heiberg Island Can.
82 B4 Axim Ghana
118 A3 Ayacucho Peru
45 E2 Ayagoz Kazakh.
36 B2 Ayakkum Hu salt l. China
74 B2 Ayamonte Spain
51 K3 Ayan Rus. Fed.
118 A3 Ayaviri Peru
44 A2 Aybas Kazakh.
59 D2 Aydar r. Ukr.
45 C2 Aydarkul', Ozero l. Uzbek.
79 C3 Aydın Turkey
45 C2 Ayeat, Gora hill Kazakh.
18 C2 Ayers Rock hill Austr.
67 C3 Aylesbury U.K.
74 C1 Ayllón Spain
97 D1 Aylmer Lake Can.
85 B4 Ayod Sudan
51 M2 Ayon, Ostrov i. Rus. Fed.
82 B3 'Ayoûn el 'Atroûs Maur.
19 D1 Ayr Austr.
64 B3 Ayr U.K.
66 A1 Ayre, Point of Isle of Man
44 C2 Ayteke Bi Kazakh.
78 C2 Aytos Bulg.
113 C3 Ayutla Mex.
31 B2 Ayutthaya Thai.
79 C3 Ayvacık Turkey
79 C3 Ayvalık Turkey
82 C3 Azaouagh, Vallée de watercourse Mali/Niger
Azbine mts Niger see L'Aïr, Massif de
49 C1 Azerbaijan country Asia
54 D2 Azopol'ye Rus. Fed.
52 A5 Azores aut. reg. N. Atlantic Ocean
59 D2 Azov Rus. Fed.
59 D2 Azov, Sea of Rus. Fed./Ukr.
Azraq, Bahr el r. Eth./Sudan see Blue Nile
74 B2 Azuaga Spain
114 B4 Azuero, Península de pen. Panama
121 C4 Azul Arg.
48 B2 Az Zarqā' Jordan
46 B3 Az Zaydīyah Yemen
82 C2 Azzel Matti, Sebkha salt pan Alg.
46 B2 Az Zilfī Saudi Arabia
46 B3 Az Zuqur i. Yemen

B

85 C4 Baardheere Somalia
45 C3 Bābā, Kūh-e mts Afgh.
78 C2 Babadag Romania
79 C2 Babaeski Turkey
85 C3 Bāb al Mandab str. Africa/Asia
29 C2 Babana Indon.
27 C3 Babar i. Indon.
87 D3 Babati Tanz.
57 E2 Babayevo Rus. Fed.
96 B2 Babine r. Can.
96 B2 Babine Lake Can.
27 C3 Babo Indon.
49 D2 Bābol Iran
90 A3 Baboon Point S. Africa
56 C3 Babruysk Belarus
32 B1 Babuyan i. Phil.
32 B1 Babuyan Islands Phil.
119 D2 Bacabal Brazil
27 C3 Bacan i. Indon.
78 C1 Bacău Romania
20 B3 Bacchus Marsh Austr.
45 D3 Bachu China
97 E1 Back r. Can.
77 C1 Bačka Palanka Yugo.
30 B1 Bac Lac Vietnam
31 B3 Bac Liêu Vietnam
32 B1 Bacolod Phil.
98 C1 Bacqueville, Lac l. Can.
74 B2 Badajoz Spain
30 A1 Badarpur India
69 E2 Bad Berka Ger.
69 D2 Bad Berleburg Ger.
69 E1 Bad Bevensen Ger.
68 C2 Bad Ems Ger.
71 D2 Baden Austria
70 B2 Baden-Baden Ger.
69 E2 Bad Harzburg Ger.
69 D2 Bad Hersfeld Ger.
70 C2 Bad Hofgastein Austria
69 D2 Bad Homburg vor der Höhe Ger.
42 A2 Badin Pak.
48 B2 Bādiyat ash Shām des. Asia
69 E2 Bad Kissingen Ger.
68 C3 Bad Kreuznach Ger.
69 E2 Bad Lauterberg im Harz Ger.
69 D2 Bad Lippspringe Ger.
69 D3 Bad Mergentheim Ger.
69 D2 Bad Nauheim Ger.
68 C2 Bad Neuenahr-Ahrweiler Ger.
69 E2 Bad Neustadt an der Saale Ger.
69 E1 Bad Oldesloe Ger.
69 D2 Bad Pyrmont Ger.
46 A2 Badr Ḥunayn Saudi Arabia
69 D1 Bad Salzuflen Ger.
69 E2 Bad Salzungen Ger.
70 C1 Bad Schwartau Ger.
69 E1 Bad Segeberg Ger.
41 C4 Badulla Sri Lanka
68 D1 Bad Zwischenahn Ger.
74 C2 Baeza Spain
82 A3 Bafatá Guinea-Bissau
128 O2 Baffin Bay sea Can./Greenland
95 G2 Baffin Island Can.
86 B2 Bafia Cameroon
82 A3 Bafing r. Guinea/Mali
82 A3 Bafoulabé Mali
86 B2 Bafoussam Cameroon
44 B3 Bāfq Iran
48 B1 Bafra Turkey
47 C2 Bāft Iran
87 C2 Bafwasende Dem. Rep. Congo
87 D3 Bagamoyo Tanz.
28 B1 Bagan Datuk Malaysia
88 B2 Bagani Namibia
28 B1 Bagansiapiapi Indon.
106 A2 Bagdad U.S.A.
120 C4 Bagé Brazil
49 C2 Baghdād Iraq
45 C3 Baghlān Afgh.
72 C3 Bagnères-de-Luchon France
56 B3 Bagrationovsk Rus. Fed.
Bagrax China see Bohu
32 B1 Baguio Phil.
83 C3 Bagzane, Monts mts Niger

Bahamas *country* West Indies see **The Bahamas**
43 C2 **Baharampur** India
84 A2 **Bahariya Oasis** Egypt
28 B1 **Bahau** Malaysia
40 B2 **Bahawalnagar** Pak.
42 B2 **Bahawalpur** Pak.
121 B4 **Bahía Blanca** Arg.
112 A2 **Bahía Kino** Mex.
120 C3 **Bahía Negra** Para.
112 A2 **Bahía Tortugas** Mex.
85 B3 **Bahir Dar** Eth.
43 C2 **Bahraich** India
47 C2 **Bahrain** *country* Asia
47 D2 **Bāhū Kālāt** Iran
78 B1 **Baia Mare** Romania
37 E1 **Baicheng** China
99 D2 **Baie-Comeau** Can.
Baie-du-Poste Can. see **Mistissini**
99 C2 **Baie-St-Paul** Can.
33 B1 **Baihe** China
37 D1 **Baikal, Lake** Rus. Fed.
78 B2 **Băileşti** Romania
36 C2 **Baima** China
43 C2 **Bainang** China
111 D2 **Bainbridge** U.S.A.
Bairin Youqi China see **Daban**
21 C3 **Bairnsdale** Austr.
33 B1 **Baishan** *Jilin* China
33 B1 **Baishan** *Jilin* China
38 A2 **Baiyin** China
84 B3 **Baiyuda Desert** Sudan
71 D2 **Baja** Hungary
112 A1 **Baja California** *pen.* Mex.
46 B3 **Bājil** Yemen
82 A3 **Bakel** Senegal
103 C3 **Baker** *CA* U.S.A.
104 C1 **Baker** *MT* U.S.A.
102 C2 **Baker** *OR* U.S.A.
102 B1 **Baker, Mount** *vol.* U.S.A.
97 E1 **Baker Foreland** *hd* Can.
97 E1 **Baker Lake** Can.
97 E1 **Baker Lake** *l.* Can.
103 C3 **Bakersfield** U.S.A.
59 C3 **Bakhchysaray** Ukr.
44 B3 **Bakherden** Turkm.
59 C1 **Bakhmach** Ukr.
Bākhtarān Iran see **Kermānshāh**
Bakı Azer. see **Baku**
79 C2 **Bakırköy** Turkey
60 □C2 **Bakkaflói** *b.* Iceland
86 C2 **Bakouma** C.A.R.
49 C1 **Baku** Azer.
32 A2 **Balabac** Phil.
32 A2 **Balabac** *i.* Phil.
32 A2 **Balabac Strait** Malaysia/Phil.
29 C2 **Balaiberkuak** Indon.
20 A2 **Balaklava** Austr.
59 C3 **Balaklava** Ukr.
59 D2 **Balakliya** Ukr.
55 D3 **Balakovo** Rus. Fed.
44 C3 **Bālā Morghāb** Afgh.
113 C3 **Balancán** Mex.
79 C3 **Balan Dağı** *hill* Turkey
32 B1 **Balanga** Phil.
43 C2 **Balāngīr** India
55 D3 **Balashov** Rus. Fed.
71 D2 **Balaton** *l.* Hungary
71 D2 **Balatonboglár** Hungary
118 C2 **Balbina, Represa de** *resr* Brazil
65 C2 **Balbriggan** Rep. of Ireland
20 A2 **Balcanoona** Austr.
78 C2 **Balchik** Bulg.
22 A3 **Balclutha** N.Z.
97 E2 **Baldock Lake** Can.
97 D2 **Baldy Mountain** Can.
106 B2 **Baldy Peak** U.S.A.
Baleares, Islas *is* Spain see **Balearic Islands**
75 D2 **Balearic Islands** *is* Spain
123 E1 **Baleia, Ponta da** *pt* Brazil
43 C2 **Bāleshwar** India
29 C2 **Bali** *i.* Indon.
28 A1 **Balige** Indon.
43 C2 **Baliguda** India
79 C3 **Balıkesir** Turkey
29 C2 **Balikpapan** Indon.
27 D3 **Balimo** P.N.G.
70 B2 **Balingen** Ger.
29 C2 **Bali Sea** Indon.
78 B2 **Balkan Mountains** Bulg./Yugo.
45 D2 **Balkhash** Kazakh.
45 D2 **Balkhash, Ozero** *l.* Kazakh.
18 B3 **Balladonia** Austr.
65 B2 **Ballaghaderreen** Rep. of Ireland
60 G2 **Ballangen** Norway
20 B3 **Ballarat** Austr.
18 B2 **Ballard, Lake** *salt flat* Austr.
64 C2 **Ballater** U.K.
82 B3 **Ballé** Mali
21 D1 **Ballina** Austr.
65 B1 **Ballina** Rep. of Ireland
65 B2 **Ballinasloe** Rep. of Ireland
107 D2 **Ballinger** U.S.A.
65 B2 **Ballinrobe** Rep. of Ireland
65 B1 **Ballycastle** Rep. of Ireland
65 C1 **Ballycastle** U.K.
65 D1 **Ballyclare** U.K.
65 C1 **Ballymena** U.K.
65 C1 **Ballymoney** U.K.
65 D1 **Ballynahinch** U.K.
65 B1 **Ballyshannon** Rep. of Ireland
63 B2 **Ballyvoy** U.K.
19 D2 **Balonne** *r.* Austr.
42 B2 **Balotra** India
43 C2 **Balrampur** India
20 B2 **Balranald** Austr.
78 B2 **Balş** Romania
119 D2 **Balsas** Brazil
113 C3 **Balsas** Mex.
58 B2 **Balta** Ukr.
58 B2 **Bălţi** Moldova
61 G4 **Baltic Sea** *g.* Europe
48 B2 **Baltîm** Egypt
91 C1 **Baltimore** S. Africa
109 D3 **Baltimore** U.S.A.
56 A3 **Baltiysk** Rus. Fed.
30 A1 **Balu** India
56 C2 **Balvi** Latvia
45 D2 **Balykchy** Kyrg.
55 E4 **Balykshi** Kazakh.
47 C2 **Bam** Iran
19 D1 **Bamaga** Austr.
98 A1 **Bamaji Lake** Can.
82 B3 **Bamako** Mali
86 C2 **Bambari** C.A.R.
69 E3 **Bamberg** Ger.
87 C2 **Bambili** Dem. Rep. Congo
87 C2 **Bambouti** C.A.R.
123 C2 **Bambuí** Brazil
86 B2 **Bamenda** Cameroon
42 A1 **Bāmīān** Afgh.
120 B2 **Bañados del Izozog** *swamp* Bol.
87 C2 **Banalia** Dem. Rep. Congo
119 C3 **Bananal, Ilha do** *i.* Brazil
42 B2 **Banas** *r.* India
79 C3 **Banaz** Turkey
30 B2 **Ban Ban** Laos
65 C1 **Banbridge** U.K.
67 C2 **Banbury** U.K.
98 C2 **Bancroft** Can.
87 C2 **Banda** Dem. Rep. Congo
43 C2 **Banda** India
27 C3 **Banda, Kepulauan** *is* Indon.
28 A1 **Banda Aceh** Indon.
47 C2 **Bandar-e 'Abbās** Iran
49 C2 **Bandar-e Anzalī** Iran
47 C2 **Bandar-e Chārak** Iran
49 C2 **Bandar-e Emām Khomeynī** Iran
47 C2 **Bandar-e Lengeh** Iran
47 C2 **Bandar-e Maqām** Iran
29 C1 **Bandar Seri Begawan** Brunei
27 C3 **Banda Sea** Indon.
123 D2 **Bandeiras, Pico de** *mt.* Brazil
91 C1 **Bandelierkop** S. Africa
112 B2 **Banderas, Bahía de** *b.* Mex.
82 B3 **Bandiagara** Mali
79 C2 **Bandırma** Turkey
65 B3 **Bandon** Rep. of Ireland
86 B3 **Bandundu** Dem. Rep. Congo
29 B2 **Bandung** Indon.
96 C2 **Banff** Can.
64 C2 **Banff** U.K.
82 B3 **Banfora** Burkina
32 B2 **Banga** Phil.
41 B3 **Bangalore** India
86 C2 **Bangassou** C.A.R.
27 C3 **Banggai** Indon.
27 C3 **Banggai, Kepulauan** *is* Indon.
29 C1 **Banggi** *i. Sabah* Malaysia
83 E1 **Banghāzī** Libya
28 B2 **Bangka** *i.* Indon.
29 C2 **Bangkalan** Indon.
28 B1 **Bangkinang** Indon.
28 B2 **Bangko** Indon.
31 B2 **Bangkok** Thai.
43 C2 **Bangladesh** *country* Asia
65 D1 **Bangor** *Northern Ireland* U.K.
66 A2 **Bangor** *Wales* U.K.
109 F2 **Bangor** U.S.A.
31 A2 **Bang Saphan Yai** Thai.
32 B1 **Bangued** Phil.
86 B2 **Bangui** C.A.R.
89 B2 **Bangweulu, Lake** Zambia
30 B1 **Ban Houayxay** Laos
83 D1 **Banī Walīd** Libya
48 B2 **Bāniyās** Syria
77 C2 **Banja Luka** Bos.-Herz.
29 C2 **Banjarmasin** Indon.
82 A3 **Banjul** Gambia
96 A2 **Banks Island** *B.C.* Can.
94 C2 **Banks Island** *N.W.T.* Can.
97 E1 **Banks Lake** *l.* Can.
22 B2 **Banks Peninsula** N.Z.
43 C2 **Bankura** India
30 A1 **Banmauk** Myanmar
30 B2 **Ban Mouang** Laos
65 C1 **Bann** *r.* U.K.
30 B2 **Ban Napè** Laos
31 A3 **Ban Na San** Thai.
111 E4 **Bannerman Town** Bahamas
42 B1 **Bannu** Pak.
42 B2 **Banswara** India
31 A3 **Ban Tha Kham** Thai.
30 A2 **Ban Tha Song Yang** Thai.
31 B2 **Ban Tôp** Laos
65 B3 **Bantry** Rep. of Ireland
65 B3 **Bantry Bay** Rep. of Ireland
86 B2 **Banyo** Cameroon
75 D1 **Banyoles** Spain
29 C2 **Banyuwangi** Indon.
Bao'an China see **Shenzhen**
37 D2 **Baochang** China
38 B2 **Baoding** China
38 A2 **Baoji** China
31 B2 **Bao Lôc** Vietnam
34 B1 **Baoqing** China
30 A1 **Baoshan** China

38 B1 **Baotou** China
49 C2 **Ba'qūbah** Iraq
77 C2 **Bar** Yugo.
85 C4 **Baraawe** Somalia
115 C2 **Baracoa** Cuba
21 C2 **Baradine** Austr.
115 C3 **Barahona** Dom. Rep.
84 B3 **Baraka** *watercourse* Eritrea/Sudan
29 C1 **Baram** *r. Sarawak* Malaysia
56 C3 **Baranavichy** Belarus
58 B1 **Baranivka** Ukr.
96 A2 **Baranof Island** U.S.A.
27 C3 **Barat Daya, Kepulauan** *is* Indon.
123 D2 **Barbacena** Brazil
115 E3 **Barbados** *country* West Indies
75 D1 **Barbastro** Spain
72 B2 **Barbezieux-St-Hilaire** France
19 D2 **Barcaldine** Austr.
75 D1 **Barcelona** Spain
118 B1 **Barcelona** Venez.
73 D3 **Barcelonnette** France
118 B2 **Barcelos** Brazil
Barcoo Creek *watercourse* Austr. *see* **Cooper Creek**
71 D2 **Barcs** Hungary
43 C2 **Barddhamān** India
71 E2 **Bardejov** Slovakia
47 C2 **Bardsīr** Iran
43 B2 **Bareilly** India
128 B2 **Barents Sea** Arctic Ocean
46 A3 **Barentu** Eritrea
43 C2 **Barh** India
109 F2 **Bar Harbor** U.S.A.
77 C2 **Bari** Italy
42 B1 **Barikot** Afgh.
118 A1 **Barinas** Venez.
43 C2 **Baripada** India
43 D2 **Barisal** Bangl.
28 B2 **Barisan, Pegunungan** *mts* Indon.
29 C2 **Barito** *r.* Indon.
47 C2 **Barkā** Oman
56 C2 **Barkava** Latvia
19 C1 **Barkly Tableland** *reg.* Austr.
36 C2 **Barkol** China
78 C1 **Bârlad** Romania
73 D2 **Bar-le-Duc** France
18 A2 **Barlee, Lake** *salt flat* Austr.
77 C2 **Barletta** Italy
21 C2 **Barmedman** Austr.
42 B2 **Barmer** India
67 A2 **Barmouth** U.K.
69 D1 **Barmstedt** Ger.
66 C1 **Barnard Castle** U.K.
21 B2 **Barnato** Austr.
50 G3 **Barnaul** Rus. Fed.
95 G2 **Barnes Icecap** Can.
68 B1 **Barneveld** Neth.
66 C2 **Barnsley** U.K.
67 A3 **Barnstaple** U.K.
111 D2 **Barnwell** U.S.A.
Baroda India *see* **Vadodara**
118 B1 **Barquisimeto** Venez.
64 A2 **Barra** *i.* U.K.
21 D2 **Barraba** Austr.
119 D2 **Barra do Corda** Brazil
122 B1 **Barra do Garças** Brazil
118 C2 **Barra do São Manuel** Brazil
118 A3 **Barranca** *Lima* Peru
118 A2 **Barranca** *Loreto* Peru
120 C3 **Barranqueras** Arg.
118 A1 **Barranquilla** Col.
119 D3 **Barreiras** Brazil
122 C2 **Barretos** Brazil
98 C2 **Barrie** Can.
96 B2 **Barrière** Can.
20 B2 **Barrier Range** *hills* Austr.
21 D2 **Barrington, Mount** Austr.
97 D2 **Barrington Lake** Can.
21 C1 **Barringun** Austr.
65 C2 **Barrow** *r.* Rep. of Ireland
19 C2 **Barrow Creek** Austr.
66 B1 **Barrow-in-Furness** U.K.
18 A2 **Barrow Island** Austr.
94 E2 **Barrow Strait** Can.
67 B3 **Barry** U.K.
98 C2 **Barrys Bay** Can.
42 B2 **Barsalpur** India
103 C4 **Barstow** U.S.A.
73 C2 **Bar-sur-Aube** France
48 B1 **Bartın** Turkey
19 D1 **Bartle Frere, Mount** Austr.
107 D1 **Bartlesville** U.S.A.
71 E1 **Bartoszyce** Pol.
29 C2 **Barung** *i.* Indon.
37 D1 **Baruun Urt** Mongolia
59 D2 **Barvinkove** Ukr.
21 C2 **Barwon** *r.* Austr.
56 C3 **Barysaw** Belarus
78 C2 **Basarabi** Romania
73 D2 **Basel** Switz.
59 C2 **Bashtanka** Ukr.
32 B2 **Basilan** *i.* Phil.
67 D3 **Basildon** U.K.
67 C3 **Basingstoke** U.K.
49 C2 **Başkale** Turkey
98 C2 **Baskatong, Réservoir** *resr* Can.
Basle Switz. *see* **Basel**
86 C2 **Basoko** Dem. Rep. Congo
49 C2 **Basra** Iraq
96 C2 **Bassano** Can.
82 C4 **Bassar** Togo
31 A2 **Bassein** Myanmar
115 D3 **Basse-Terre** Guadeloupe
115 D3 **Basseterre** St Kitts and Nevis
82 B3 **Bassikounou** Maur.
19 D3 **Bass Strait** Austr.
47 C2 **Bastak** Iran
69 E2 **Bastheim** Ger.
43 C2 **Basti** India
73 D3 **Bastia** *Corsica* France
68 B2 **Bastogne** Belgium
110 B2 **Bastrop** U.S.A.
Basuo China *see* **Dongfang**
86 A2 **Bata** Equat. Guinea
51 J2 **Batagay** Rus. Fed.
122 B2 **Bataguaçu** Brazil
74 B2 **Batalha** Port.
39 C3 **Batan** *i.* Phil.
86 B2 **Batangafo** C.A.R.
32 B1 **Batangas** Phil.
28 B2 **Batanghari** *r.* Indon.
39 C3 **Batan Islands** Phil.
109 D2 **Batavia** U.S.A.
59 D2 **Bataysk** Rus. Fed.
98 B2 **Batchawana Mountain** Can.
18 C1 **Batchelor** Austr.
31 B2 **Bătdâmbâng** Cambodia
21 D3 **Batemans Bay** Austr.
110 B1 **Batesville** U.S.A.
57 D2 **Batetskiy** Rus. Fed.
67 B3 **Bath** U.K.
42 B1 **Bathinda** India
21 C2 **Bathurst** Austr.
99 D2 **Bathurst** Can.
94 D2 **Bathurst Inlet** Can.
94 D2 **Bathurst Inlet** *inlet* Can.
18 C1 **Bathurst Island** Austr.
94 E1 **Bathurst Island** Can.
46 B1 **Bāṭin, Wādī al** *watercourse* Asia
49 C2 **Batman** Turkey
83 C1 **Batna** Alg.
110 B2 **Baton Rouge** U.S.A.
112 B2 **Batopilas** Mex.
86 B2 **Batouri** Cameroon
122 B1 **Batovi** Brazil
60 I1 **Båtsfjord** Norway
41 C4 **Batticaloa** Sri Lanka
77 B2 **Battipaglia** Italy
97 D2 **Battle** *r.* Can.
108 B2 **Battle Creek** U.S.A.
103 C2 **Battle Mountain** U.S.A.
42 B1 **Battura Glacier** Jammu and Kashmir
85 B4 **Batu** *mt.* Eth.
28 A2 **Batu, Pulau-pulau** *is* Indon.
49 C1 **Bat'umi** Georgia
28 B1 **Batu Pahat** Malaysia
27 C3 **Baubau** Indon.
83 C3 **Bauchi** Nigeria
72 B2 **Baugé** France
73 D2 **Baume-les-Dames** France
122 C2 **Bauru** Brazil
122 B1 **Baús** Brazil
56 B2 **Bauska** Latvia
70 C1 **Bautzen** Ger.
112 B2 **Bavispe** *r.* Mex.
55 E3 **Bavly** Rus. Fed.
30 A1 **Bawdwin** Myanmar
29 C2 **Bawean** *i.* Indon.
84 A2 **Bawiti** Egypt
82 B3 **Bawku** Ghana
114 C2 **Bayamo** Cuba
Bayan Gol China *see* Dengkou
36 C1 **Bayanhongor** Mongolia
38 A2 **Bayan Hot** China
32 B2 **Bayawan** Phil.
48 C1 **Bayburt** Turkey
108 C2 **Bay City** *MI* U.S.A.
107 D3 **Bay City** *TX* U.S.A.
54 F2 **Baydaratskaya Guba** *b.* Rus. Fed.
85 C4 **Baydhabo** Somalia
49 C2 **Bayjī** Iraq
Baykal, Ozero *l.* Rus. Fed. *see* **Baikal, Lake**
51 I3 **Baykal'skiy Khrebet** *mts* Rus. Fed.
44 C2 **Baykonur** Kazakh.
55 E3 **Baymak** Rus. Fed.
32 B1 **Bayombong** Phil.
72 B3 **Bayonne** France
79 C3 **Bayramiç** Turkey
69 E3 **Bayreuth** Ger.
46 B3 **Bayt al Faqīh** Yemen
74 C2 **Baza** Spain
74 C2 **Baza, Sierra de** *mts* Spain
44 A2 **Bazardyuzi, Gora** *mt.* Azer./Rus. Fed.
72 B3 **Bazas** France
42 A2 **Bazdar** Pak.
38 A2 **Bazhong** China
47 D2 **Bazmān** Iran
47 D2 **Bazmān, Kūh-e** *mt.* Iran
67 D3 **Beachy Head** *hd* U.K.
91 C3 **Beacon Bay** S. Africa
18 B1 **Beagle Gulf** Austr.
89 □D2 **Bealanana** Madag.
98 B2 **Beardmore** Can.
50 C2 **Bear Island** *i.* Arctic Ocean
102 E1 **Bear Paw Mountain** U.S.A.
115 C3 **Beata, Cabo** *c.* Dom. Rep.
115 C3 **Beata, Isla** *i.* Dom. Rep.
105 D2 **Beatrice** U.S.A.
103 C3 **Beatty** U.S.A.
21 D1 **Beaudesert** Austr.
20 B3 **Beaufort** Austr.
29 C1 **Beaufort** *Sabah* Malaysia
111 D2 **Beaufort** U.S.A.
128 L2 **Beaufort Sea** Can./U.S.A.
90 B3 **Beaufort West** S. Africa
64 B2 **Beauly** *r.* U.K.

68 B2 **Beaumont** Belgium
22 A3 **Beaumont** N.Z.
107 E2 **Beaumont** U.S.A.
73 C2 **Beaune** France
68 B2 **Beauraing** Belgium
97 E2 **Beauséjour** Can.
72 C2 **Beauvais** France
97 D2 **Beauval** Can.
97 D2 **Beaver** *r.* Can.
103 D3 **Beaver** U.S.A.
94 B2 **Beaver Creek** Can.
108 B2 **Beaver Dam** U.S.A.
97 E2 **Beaver Hill Lake** Can.
108 B1 **Beaver Island** U.S.A.
96 C2 **Beaverlodge** Can.
42 B2 **Beawar** India
122 C2 **Bebedouro** Brazil
69 D2 **Bebra** Ger.
74 B1 **Becerreá** Spain
82 B1 **Béchar** Alg.
108 C3 **Beckley** U.S.A.
85 B4 **Bedelē** Eth.
67 C2 **Bedford** U.K.
108 B3 **Bedford** U.S.A.
68 C1 **Bedum** Neth.
21 D2 **Beecroft Peninsula** Austr.
69 F1 **Beelitz** Ger.
21 D1 **Beenleigh** Austr.
48 B2 **Be'ér Sheva'** Israel
107 D3 **Beeville** U.S.A.
21 C3 **Bega** Austr.
75 D1 **Begur, Cap de** *c.* Spain
49 D2 **Behshahr** Iran
37 E1 **Bei'an** China
39 A3 **Beihai** China
38 B2 **Beijing** China
68 C1 **Beilen** Neth.
64 B3 **Beinn an Oir** *hill* U.K.
64 A2 **Beinn Mhòr** *hill* U.K.
89 C2 **Beira** Moz.
48 B2 **Beirut** Lebanon
91 C1 **Beitbridge** Zimbabwe
74 B2 **Beja** Port.
83 C1 **Bejaïa** Alg.
74 B1 **Béjar** Spain
42 A2 **Beji** *r.* Pak.
44 B2 **Bekdash** Turkm.
71 E2 **Békés** Hungary
71 E2 **Békéscsaba** Hungary
89 □D3 **Bekily** Madag.
42 A2 **Bela** Pak.
91 C1 **Bela-Bela** S. Africa
86 B2 **Bélabo** Cameroon
77 D2 **Bela Crkva** Yugo.
29 C1 **Belaga** *Sarawak* Malaysia
29 C2 **Belangiran** Indon.
56 C3 **Belarus** *country* Europe
Belau *country* N. Pacific Ocean *see* **Palau**
91 D2 **Bela Vista** Moz.
28 A1 **Belawan** Indon.
51 M2 **Belaya** *r.* Rus. Fed.
71 D1 **Bełchatów** Pol.
98 C1 **Belcher Islands** Can.
85 C4 **Beledweyne** Somalia
119 D2 **Belém** Brazil
106 B2 **Belen** U.S.A.
57 E3 **Belev** Rus. Fed.
65 D1 **Belfast** U.K.
109 F2 **Belfast** U.S.A.
73 D2 **Belfort** France
41 B3 **Belgaum** India
68 B2 **Belgium** *country* Europe
59 D1 **Belgorod** Rus. Fed.
102 D1 **Belgrade** U.S.A.
77 D2 **Belgrade** Yugo.
83 D4 **Beli** Nigeria
28 B2 **Belinyu** Indon.
29 B2 **Belitung** *i.* Indon.
114 B3 **Belize** Belize
114 B3 **Belize** *country* Central America
51 K1 **Bel'kovskiy, Ostrov** *i.* Rus. Fed.
96 B2 **Bella Bella** Can.
72 C2 **Bellac** France
96 B2 **Bella Coola** Can.
21 C1 **Bellata** Austr.
104 C2 **Belle Fourche** U.S.A.
104 C2 **Belle Fourche** *r.* U.S.A.
111 D3 **Belle Glade** U.S.A.
72 B2 **Belle-Île** *i.* France
99 E1 **Belle Isle** *i.* Can.
99 E1 **Belle Isle, Strait of** Can.
98 C2 **Belleville** Can.
108 B3 **Belleville** *IL* U.S.A.
105 D3 **Belleville** *KS* U.S.A.
102 B1 **Bellevue** U.S.A.
102 B1 **Bellingham** U.S.A.
23 O2 **Bellingshausen Sea** Antarctica
73 D2 **Bellinzona** Switz.
76 B1 **Belluno** Italy
90 A3 **Bellville** S. Africa
123 E1 **Belmonte** Brazil
114 B3 **Belmopan** Belize
65 B1 **Belmullet** Rep. of Ireland
37 E1 **Belogorsk** Rus. Fed.
89 □D3 **Beloha** Madag.
123 D1 **Belo Horizonte** Brazil
108 B2 **Beloit** U.S.A.
54 C2 **Belomorsk** Rus. Fed.
59 D3 **Belorechensk** Rus. Fed.
55 E3 **Beloretsk** Rus. Fed.
Belorussia *country* Europe *see* **Belarus**
54 F2 **Beloyarskiy** Rus. Fed.
54 C2 **Beloye, Ozero** *l.* Rus. Fed.
Beloye More *sea* Rus. Fed. *see* **White Sea**
54 C2 **Belozersk** Rus. Fed.
45 E2 **Belukha, Gora** *mt.* Kazakh./Rus. Fed.
54 D2 **Belush'ye** Rus. Fed.
57 D2 **Belyy** Rus. Fed.
50 F2 **Belyy, Ostrov** *i.* Rus. Fed.
69 F1 **Belzig** Ger.
105 E1 **Bemidji** U.S.A.
86 C3 **Bena Dibele** Dem. Rep. Congo
21 C3 **Benalla** Austr.
74 B1 **Benavente** Spain
64 A2 **Benbecula** *i.* U.K.
102 B2 **Bend** U.S.A.
91 C3 **Bendearg** *mt.* S. Africa
20 B3 **Bendigo** Austr.
89 C2 **Bene** Moz.
70 C2 **Benešov** Czech Rep.
77 B2 **Benevento** Italy
127 C2 **Bengal, Bay of** *sea* Indian Ocean
38 B2 **Bengbu** China
29 B1 **Bengkayang** Indon.
28 B2 **Bengkulu** Indon.
88 A2 **Benguela** Angola
64 B1 **Ben Hope** *hill* U.K.
120 B2 **Beni** *r.* Bol.
87 C2 **Beni** Dem. Rep. Congo
82 B1 **Beni-Abbès** Alg.
75 C2 **Benidorm** Spain
82 B1 **Beni Mellal** Morocco
82 C4 **Benin** *country* Africa
82 C4 **Benin, Bight of** *g.* Africa
83 C4 **Benin City** Nigeria
84 B2 **Beni Suef** Egypt
121 C4 **Benito Juárez** Arg.
118 B2 **Benjamim Constant** Brazil
112 A1 **Benjamín Hill** Mex.
27 C3 **Benjina** Indon.
64 B2 **Ben Lawers** *mt.* U.K.
64 B2 **Ben Lomond** *hill* U.K.
64 C2 **Ben Macdui** *mt.* U.K.
64 A2 **Ben More** *hill* U.K.
22 B2 **Benmore, Lake** N.Z.
64 B1 **Ben More Assynt** *hill* U.K.
96 A2 **Bennett** Can.
51 K1 **Bennetta, Ostrov** *i.* Rus. Fed.
64 B2 **Ben Nevis** *mt.* U.K.
109 E2 **Bennington** U.S.A.
91 C2 **Benoni** S. Africa
69 D3 **Bensheim** Ger.
106 A2 **Benson** U.S.A.
26 C3 **Benteng** Indon.
108 B2 **Benton Harbor** U.S.A.
110 B1 **Bentonville** U.S.A.
83 C4 **Benue** *r.* Nigeria
65 B1 **Benwee Head** *hd* Rep. of Ireland
64 B2 **Ben Wyvis** *mt.* U.K.
38 C1 **Benxi** China
Beograd Yugo. *see* **Belgrade**
43 C2 **Beohari** India
35 B4 **Beppu** Japan
77 C2 **Berane** Yugo.
77 C2 **Berat** Albania
27 C3 **Berau, Teluk** *b.* Indon.
84 B3 **Berber** Sudan
85 C3 **Berbera** Somalia
86 B2 **Berbérati** C.A.R.
72 C1 **Berck** France
59 D2 **Berdyans'k** Ukr.
58 B2 **Berdychiv** Ukr.
58 A2 **Berehove** Ukr.
27 D3 **Bereina** P.N.G.
46 A2 **Berenice** Egypt
97 E2 **Berens River** Can.
58 A2 **Berezhany** Ukr.
58 C2 **Berezivka** Ukr.
58 B1 **Berezne** Ukr.
54 D2 **Bereznik** Rus. Fed.
54 E3 **Berezniki** Rus. Fed.
54 F2 **Berezovo** Rus. Fed.
75 D1 **Berga** Spain
79 C3 **Bergama** Turkey
76 A1 **Bergamo** Italy
70 C1 **Bergen** *Mecklenburg-Vorpommern* Ger.
69 D1 **Bergen** *Niedersachsen* Ger.
61 E3 **Bergen** Norway
68 B2 **Bergen op Zoom** Neth.
72 C3 **Bergerac** France
68 C2 **Bergheim (Erft)** Ger.
68 C2 **Bergisch Gladbach** Ger.
90 A1 **Bergland** Namibia
60 H2 **Bergsviken** Sweden
51 M3 **Beringa, Ostrov** *i.* Rus. Fed.
68 B2 **Beringen** Belgium
92 B1 **Bering Sea** N. Pacific Ocean
51 N2 **Bering Strait** Rus. Fed./U.S.A.
103 B3 **Berkeley** U.S.A.
68 B1 **Berkhout** Neth.
23 Q2 **Berkner Island** Antarctica
78 B2 **Berkovitsa** Bulg.
60 I1 **Berlevåg** Norway
69 F1 **Berlin** Ger.
109 E2 **Berlin** U.S.A.
69 E2 **Berlingerode** Ger.
21 D3 **Bermagui** Austr.
112 B2 **Bermejíllo** Mex.
120 B3 **Bermejo** Bol.
93 F4 **Bermuda** *terr.* N. Atlantic Ocean
73 D2 **Bern** Switz.
69 E2 **Bernburg (Saale)** Ger.
95 F2 **Bernier Bay** Can.
68 C3 **Bernkastel-Kues** Ger.
89 □D3 **Beroroha** Madag.
20 B2 **Berri** Austr.

75 D2 **Berrouaghia** Alg.
114 C2 **Berry Islands** Bahamas
68 C1 **Bersenbrück** Ger.
58 B2 **Bershad'** Ukr.
99 D1 **Berté, Lac** *l.* Can.
86 B2 **Bertoua** Cameroon
118 B2 **Beruri** Brazil
66 B1 **Berwick-upon-Tweed** U.K.
59 C2 **Beryslav** Ukr.
89 □D2 **Besalampy** Madag.
73 D2 **Besançon** France
97 D2 **Besnard Lake** Can.
110 C2 **Bessemer** U.S.A.
44 B2 **Besshoky, Gora** *hill* Kazakh.
74 B1 **Betanzos** Spain
86 B2 **Bétaré Oya** Cameroon
90 A2 **Bethanie** Namibia
91 C2 **Bethlehem** S. Africa
109 D2 **Bethlehem** U.S.A.
89 □D3 **Betioky** Madag.
45 D2 **Betpak-Dala** *plain* Kazakh.
89 □D3 **Betroka** Madag.
99 D2 **Betsiamites** Can.
89 □D2 **Betsiboka** *r.* Madag.
105 E2 **Bettendorf** U.S.A.
43 C2 **Bettiah** India
43 B2 **Betul** India
43 B2 **Betwa** *r.* India
66 B2 **Betws-y-coed** U.K.
66 C2 **Beverley** U.K.
69 D2 **Beverungen** Ger.
68 B1 **Beverwijk** Neth.
67 D3 **Bexhill** U.K.
79 C2 **Beykoz** Turkey
82 B4 **Beyla** Guinea
44 B2 **Beyneu** Kazakh.
48 B1 **Beypazarı** Turkey
Beyrouth Lebanon *see* **Beirut**
48 B2 **Beyşehir** Turkey
48 B2 **Beyşehir Gölü** *l.* Turkey
59 D2 **Beysug** *r.* Rus. Fed.
56 C2 **Bezhanitsy** Rus. Fed.
57 E2 **Bezhetsk** Rus. Fed.
73 C3 **Béziers** France
43 C2 **Bhadrakh** India
41 B3 **Bhadravati** India
43 C2 **Bhagalpur** India
42 A2 **Bhairi Hol** *mt.* Pak.
42 B1 **Bhakkar** Pak.
30 A1 **Bhamo** Myanmar
43 C3 **Bhanjanagar** India
42 B2 **Bharatpur** India
42 B2 **Bharuch** India
42 B2 **Bhavnagar** India
43 C3 **Bhawanipatna** India
91 D2 **Bhekuzulu** S. Africa
42 B2 **Bhilwara** India
41 B3 **Bhima** *r.* India
42 B2 **Bhiwani** India
91 C3 **Bhongweni** S. Africa
42 B2 **Bhopal** India
43 C2 **Bhubaneshwar** India
42 A2 **Bhuj** India
30 A2 **Bhumiphol Dam** Thai.
42 B2 **Bhusawal** India
43 D2 **Bhutan** *country* Asia
30 B2 **Bia, Phou** *mt.* Laos
27 D3 **Biak** Indon.
27 D3 **Biak** *i.* Indon.
71 E1 **Biała Podlaska** Pol.
71 D1 **Białogard** Pol.
71 E1 **Białystok** Pol.
77 C3 **Bianco** Italy
72 B3 **Biarritz** France
73 D2 **Biasca** Switz.
34 D2 **Bibai** Japan
88 A2 **Bibala** Angola
70 B2 **Biberach an der Riß** Ger.
83 C4 **Bida** Nigeria
109 E2 **Biddeford** U.S.A.
67 A3 **Bideford** U.K.
67 A3 **Bideford Bay** U.K.
69 D2 **Biedenkopf** Ger.
73 D2 **Biel** Switz.
69 D1 **Bielefeld** Ger.
76 A1 **Biella** Italy
71 D2 **Bielsko-Biała** Pol.
31 B2 **Biên Hoa** Vietnam
98 C1 **Bienville, Lac** *l.* Can.
68 B3 **Bièvre** Belgium
86 B3 **Bifoun** Gabon
79 C2 **Biga** Turkey
91 D2 **Big Bend** Swaziland
97 D2 **Biggar** Can.
64 C3 **Biggar** U.K.
102 D1 **Big Hole** *r.* U.S.A.
104 B1 **Bighorn** *r.* U.S.A.
104 B2 **Bighorn Mountains** U.S.A.
107 C2 **Big Lake** U.S.A.
108 B2 **Big Rapids** U.S.A.
97 D2 **Big River** Can.
97 E2 **Big Sand Lake** Can.
105 D2 **Big Sioux** *r.* U.S.A.
107 C2 **Big Spring** U.S.A.
102 E1 **Big Timber** U.S.A.
98 B1 **Big Trout Lake** Can.
98 A1 **Big Trout Lake** *l.* Can.
77 C2 **Bihać** Bos.-Herz.
43 C2 **Bihar Sharif** India
78 B1 **Bihor, Vârful** *mt.* Romania
82 A3 **Bijagós, Arquipélago dos** *is* Guinea-Bissau
49 C2 **Bījār** Iran
77 C2 **Bijeljina** Bos.-Herz.
77 C2 **Bijelo Polje** Yugo.
39 A3 **Bijie** China
42 B2 **Bikaner** India
37 E1 **Bikin** Rus. Fed.
86 B3 **Bikoro** Dem. Rep. Congo
47 C2 **Bilād Banī Bū 'Alī** Oman
43 C2 **Bilaspur** India
58 C2 **Bila Tserkva** Ukr.
31 A2 **Bilauktaung Range** *mts* Myanmar/Thai.
74 C1 **Bilbao** Spain
79 C2 **Bilecik** Turkey
71 E1 **Biłgoraj** Pol.
58 C2 **Bilhorod-Dnistrovs'kyy** Ukr.
87 C2 **Bili** Dem. Rep. Congo
51 M2 **Bilibino** Rus. Fed.
102 E1 **Billings** U.S.A.
67 B3 **Bill of Portland** *hd* U.K.
83 D3 **Bilma** Niger
19 E2 **Biloela** Austr.
59 C2 **Bilohirs'k** Ukr.
58 B1 **Bilohir"ya** Ukr.
59 C1 **Bilopillya** Ukr.
59 D2 **Bilovods'k** Ukr.
110 C2 **Biloxi** U.S.A.
19 C2 **Bilpa Morea Claypan** *salt flat* Austr.
69 E2 **Bilshausen** Ger.
83 E3 **Biltine** Chad
58 C2 **Bilyayivka** Ukr.
111 E3 **Bimini Islands** Bahamas
43 B2 **Bina-Etawa** India
27 C3 **Binaija, Gunung** *mt.* Indon.
86 B3 **Bindu** Dem. Rep. Congo
89 C2 **Bindura** Zimbabwe
75 D1 **Binefar** Spain
21 D1 **Bingara** Austr.
68 C3 **Bingen am Rhein** Ger.
109 F1 **Bingham** U.S.A.
109 D2 **Binghamton** U.S.A.
49 C2 **Bingöl** Turkey
28 A1 **Binjai** Indon.
28 B2 **Bintuhan** Indon.
29 C1 **Bintulu** *Sarawak* Malaysia
38 B2 **Binzhou** China
86 A2 **Bioco** *i.* Equat. Guinea
77 C2 **Biograd na Moru** Croatia
86 C1 **Birao** C.A.R.
43 C2 **Biratnagar** Nepal
96 C2 **Birch Mountains** Can.
68 B1 **Birdaard** Neth.
19 C2 **Birdsville** Austr.
48 B2 **Birecik** Turkey
28 A1 **Bireun** Indon.
43 C2 **Birganj** Nepal
85 B3 **Birhan** *mt.* Eth.
122 B2 **Birigüi** Brazil
44 B3 **Bīrjand** Iran
66 B2 **Birkenhead** U.K.
67 C2 **Birmingham** U.K.
110 C2 **Birmingham** U.S.A.
82 A2 **Bîr Mogreïn** Maur.
83 C3 **Birnin-Kebbi** Nigeria
83 C3 **Birnin Konni** Niger
37 E1 **Birobidzhan** Rus. Fed.
65 C2 **Birr** Rep. of Ireland
64 C1 **Birsay** U.K.
46 A2 **Bi'r Shalatein** Egypt
56 B2 **Biržai** Lith.
43 B2 **Bisalpur** India
106 B2 **Bisbee** U.S.A.
72 A2 **Biscay, Bay of** *sea* France/Spain
70 C2 **Bischofshofen** Austria
98 B2 **Biscotasi Lake** Can.
45 D2 **Bishkek** Kyrg.
91 C3 **Bisho** S. Africa
103 C3 **Bishop** U.S.A.
37 E1 **Bishui** China
118 B1 **Bisinaca** Col.
83 C1 **Biskra** Alg.
104 C1 **Bismarck** U.S.A.
75 D2 **Bissa, Djebel** *mt.* Alg.
82 A3 **Bissau** Guinea-Bissau
97 E2 **Bissett** Can.
96 C2 **Bistcho Lake** Can.
78 B1 **Bistrița** Romania
78 C1 **Bistrița** *r.* Romania
68 C3 **Bitburg** Ger.
73 D2 **Bitche** France
83 D3 **Bitkine** Chad
77 D2 **Bitola** Macedonia
77 C2 **Bitonto** Italy
90 A3 **Bitterfontein** S. Africa
102 D1 **Bitterroot** *r.* U.S.A.
102 C1 **Bitterroot Range** *mts* U.S.A.
59 D1 **Bityug** *r.* Rus. Fed.
83 D3 **Biu** Nigeria
35 C3 **Biwa-ko** *l.* Japan
36 B1 **Biysk** Rus. Fed.
83 C1 **Bizerte** Tunisia
60 □A2 **Bjargtangar** *hd* Iceland
60 G3 **Bjästa** Sweden
60 G2 **Bjerkvik** Norway
Björneborg Fin. *see* **Pori**
Bjørnøya *i.* Arctic Ocean *see* **Bear Island**
82 B3 **Bla** Mali
105 E3 **Black** *r.* U.S.A.
19 D2 **Blackall** Austr.
66 B2 **Blackburn** U.K.
102 D2 **Blackfoot** U.S.A.
104 C2 **Black Hills** U.S.A.
64 B2 **Black Isle** *pen.* U.K.
97 D2 **Black Lake** Can.
97 D2 **Black Lake** *l.* Can.
67 B3 **Black Mountains** U.K.
66 B2 **Blackpool** U.K.
Black River *r.* Vietnam *see* **Đa, Sông**
108 A2 **Black River Falls** U.S.A.
108 C3 **Blacksburg** U.S.A.
48 B1 **Black Sea** Asia/Europe
65 A1 **Blacksod Bay** Rep. of Ireland
65 C2 **Blackstairs Mountains** Rep. of Ireland

82 B4 **Black Volta** *r.* Africa
65 C2 **Blackwater** *r.* Rep. of Ireland
96 B1 **Blackwater Lake** *l.* Can.
18 A3 **Blackwood** *r.* Austr.
44 A2 **Blagodarnyy** Rus. Fed.
78 B2 **Blagoevgrad** Bulg.
37 E1 **Blagoveshchensk** Rus. Fed.
97 D2 **Blaine Lake** Can.
105 D2 **Blair** U.S.A.
64 C2 **Blair Atholl** U.K.
64 C2 **Blairgowrie** U.K.
111 D2 **Blakely** U.S.A.
121 B4 **Blanca, Bahía** *b.* Arg.
20 A1 **Blanche, Lake** *salt flat* Austr.
120 B2 **Blanco** *r.* Bol.
99 E1 **Blanc-Sablon** Can.
60 □A2 **Blanda** *r.* Iceland
67 B3 **Blandford Forum** U.K.
103 E3 **Blanding** U.S.A.
75 D1 **Blanes** Spain
28 A1 **Blangkejeren** Indon.
68 C2 **Blankenheim** Ger.
68 C2 **Blankenrath** Ger.
71 D2 **Blansko** Czech Rep.
89 C2 **Blantyre** Malawi
66 C1 **Blaydon** U.K.
21 C2 **Blayney** Austr.
22 B2 **Blenheim** N.Z.
83 C1 **Blida** Alg.
98 B2 **Blind River** Can.
91 C2 **Bloemfontein** S. Africa
91 C2 **Bloemhof** S. Africa
91 C2 **Bloemhof Dam** S. Africa
60 □A2 **Blönduós** Iceland
65 B1 **Bloody Foreland** *pt* Rep. of Ireland
106 B1 **Bloomfield** U.S.A.
108 B2 **Bloomington** *IL* U.S.A.
108 B3 **Bloomington** *IN* U.S.A.
108 C3 **Bluefield** U.S.A.
114 B3 **Bluefields** Nic.
21 C2 **Blue Mountains** Austr.
102 C1 **Blue Mountains** U.S.A.
84 B3 **Blue Nile** *r.* Eth./Sudan
94 D2 **Bluenose Lake** Can.
108 C3 **Blue Ridge** *mts* U.S.A.
65 B1 **Blue Stack Mountains** Rep. of Ireland
22 A3 **Bluff** N.Z.
103 E3 **Bluff** U.S.A.
120 D3 **Blumenau** Brazil
20 A2 **Blyth** Austr.
66 C1 **Blyth** U.K.
103 D4 **Blythe** U.S.A.
110 C1 **Blytheville** U.S.A.
82 A4 **Bo** Sierra Leone
32 B1 **Boac** Phil.
119 D2 **Boa Esperança, Açude** *resr* Brazil
118 B1 **Boa Vista** Brazil
21 C2 **Bobadah** Austr.
39 B3 **Bobai** China
89 □D2 **Bobaomby, Tanjona** *c.* Madag.
82 B3 **Bobo-Dioulasso** Burkina
57 F3 **Bobrov** Rus. Fed.
59 C1 **Bobrovytsya** Ukr.
59 C2 **Bobrynets'** Ukr.
89 □D3 **Boby** *mt.* Madag.
118 B2 **Boca do Acre** Brazil
123 D1 **Bocaiúva** Brazil
86 B2 **Bocaranga** C.A.R.
71 E2 **Bochnia** Pol.
68 C2 **Bocholt** Ger.
68 C2 **Bochum** Ger.
91 C1 **Bochum** S. Africa
69 E1 **Bockenem** Ger.
86 B2 **Boda** C.A.R.
51 I3 **Bodaybo** Rus. Fed.
83 D3 **Bodélé** *reg.* Chad
60 H2 **Boden** Sweden
67 A3 **Bodmin** U.K.
67 A3 **Bodmin Moor** *moorland* U.K.
60 F2 **Bodø** Norway
79 C3 **Bodrum** Turkey
86 C3 **Boende** Dem. Rep. Congo
110 C2 **Bogalusa** U.S.A.
82 B3 **Bogandé** Burkina
36 B2 **Bogda Shan** *mts* China
21 D1 **Boggabilla** Austr.
65 B2 **Boggeragh Mountains** Rep. of Ireland
68 B3 **Bogny-sur-Meuse** France
65 C2 **Bog of Allen** *reg.* Rep. of Ireland
21 C3 **Bogong, Mount** Austr.
28 B2 **Bogor** Indon.
57 E3 **Bogoroditsk** Rus. Fed.
118 A1 **Bogotá** Col.
50 G3 **Bogotol** Rus. Fed.
51 H3 **Boguchany** Rus. Fed.
59 E2 **Boguchar** Rus. Fed.
38 B2 **Bo Hai** *g.* China
68 A3 **Bohain-en-Vermandois** France
38 B2 **Bohai Wan** *b.* China
91 C2 **Bohlokong** S. Africa
70 C2 **Böhmer Wald** *mts* Ger.
59 D1 **Bohodukhiv** Ukr.
32 B2 **Bohol** *i.* Phil.
32 B2 **Bohol Sea** Phil.
36 B2 **Bohu** China
123 C2 **Boi, Ponta do** *pt* Brazil
122 B1 **Bois** *r.* Brazil
94 C2 **Bois, Lac des** *l.* Can.
102 C2 **Boise** U.S.A.
107 C1 **Boise City** U.S.A.
91 C2 **Boitumelong** S. Africa
69 E1 **Boizenburg** Ger.
44 B3 **Bojnūrd** Iran
86 B3 **Bokatola** Dem. Rep. Congo
82 A3 **Boké** Guinea
86 C3 **Bokele** Dem. Rep. Congo
61 E4 **Boknafjorden** *sea chan.* Norway
83 D3 **Bokoro** Chad
31 A2 **Bokpyin** Myanmar
57 D2 **Boksitogorsk** Rus. Fed.
90 B2 **Bokspits** S. Africa
82 A3 **Bolama** Guinea-Bissau
72 C2 **Bolbec** France
45 E2 **Bole** China
86 B3 **Boleko** Dem. Rep. Congo
82 B3 **Bolgatanga** Ghana
58 B2 **Bolhrad** Ukr.
34 B1 **Boli** China
78 C2 **Bolintin-Vale** Romania
105 E3 **Bolivar** U.S.A.
118 A1 **Bolívar, Pico** *mt.* Venez.
120 B2 **Bolivia** *country* S. America
57 E3 **Bolkhov** Rus. Fed.
73 C3 **Bollène** France
61 G3 **Bollnäs** Sweden
21 C1 **Bollon** Austr.
69 E2 **Bollstedt** Ger.
61 F4 **Bolmen** *l.* Sweden
86 B3 **Bolobo** Dem. Rep. Congo
76 B2 **Bologna** Italy
57 D2 **Bologovo** Rus. Fed.
57 D2 **Bologoye** Rus. Fed.
86 B2 **Bolomba** Dem. Rep. Congo
31 B2 **Bolovens, Phouphieng** *plat.* Laos
51 H1 **Bol'shevik, Ostrov** *i.* Rus. Fed.
54 E2 **Bol'shezemel'skaya Tundra** *lowland* Rus. Fed.
51 L2 **Bol'shoy Aluy** *r.* Rus. Fed.
34 B2 **Bol'shoy Kamen'** Rus. Fed.
Bol'shoy Kavkaz *mts* Asia/Europe *see* **Caucasus**
51 K2 **Bol'shoy Lyakhovskiy, Ostrov** *i.* Rus. Fed.
68 B1 **Bolsward** Neth.
66 B2 **Bolton** U.K.
48 B1 **Bolu** Turkey
60 □A2 **Bolungarvík** Iceland
76 B1 **Bolzano** Italy
86 B3 **Boma** Dem. Rep. Congo
21 D2 **Bomaderry** Austr.
21 C3 **Bombala** Austr.
Bombay India *see* **Mumbai**
123 C1 **Bom Despacho** Brazil
43 D2 **Bomdila** India
119 D3 **Bom Jesus da Lapa** Brazil
123 D2 **Bom Jesus do Itabapoana** Brazil
83 D1 **Bon, Cap** *c.* Tunisia
115 D3 **Bonaire** *i.* Neth. Antilles
18 B1 **Bonaparte Archipelago** *is* Austr.
99 E2 **Bonavista** Can.
99 E2 **Bonavista Bay** Can.
86 C2 **Bondo** Dem. Rep. Congo
82 B4 **Bondoukou** Côte d'Ivoire
26 C3 **Bonerate, Kepulauan** *is* Indon.
123 C1 **Bonfinópolis de Minas** Brazil
85 B4 **Bonga** Eth.
43 D2 **Bongaigaon** India
86 C2 **Bongandanga** Dem. Rep. Congo
86 C2 **Bongo, Massif des** *mts* C.A.R.
83 D3 **Bongor** Chad
82 B4 **Bongouanou** Côte d'Ivoire
31 B2 **Bông Sơn** Vietnam
107 D2 **Bonham** U.S.A.
73 D3 **Bonifacio** *Corsica* France
76 A2 **Bonifacio, Strait of** France/Italy
68 C2 **Bonn** Ger.
102 C1 **Bonners Ferry** U.S.A.
18 A3 **Bonnie Rock** Austr.
97 C2 **Bonnyville** Can.
76 A2 **Bonorva** *Sardinia* Italy
29 C1 **Bontang** Indon.
32 B1 **Bontoc** Phil.
26 B3 **Bontosunggu** Indon.
91 C3 **Bontrug** S. Africa
20 B2 **Booligal** Austr.
21 C1 **Boomi** Austr.
21 D1 **Boonah** Austr.
105 E2 **Boone** U.S.A.
110 C2 **Booneville** U.S.A.
105 E3 **Boonville** U.S.A.
21 C2 **Boorowa** Austr.
94 F2 **Boothia, Gulf of** Can.
94 E2 **Boothia Peninsula** Can.
68 C2 **Boppard** Ger.
85 B4 **Bor** Sudan
48 B2 **Bor** Turkey
77 D2 **Bor** Yugo.
61 F4 **Borås** Sweden
47 C2 **Borāzjān** Iran
118 C2 **Borba** Brazil
72 B3 **Bordeaux** France
95 F2 **Borden Peninsula** Can.
20 B3 **Bordertown** Austr.
75 D2 **Bordj Bou Arréridj** Alg.
75 D2 **Bordj Bounaama** Alg.
83 C1 **Bordj Messaouda** Alg.
83 C2 **Bordj Omer Driss** Alg.
62 B1 **Borðoy** *i.* Faroe Is
60 □A3 **Borgarnes** Iceland
76 A1 **Borgosesia** Italy
55 D3 **Borisoglebsk** Rus. Fed.
57 E2 **Borisoglebskiy** Rus. Fed.
59 D1 **Borisovka** Rus. Fed.
68 C2 **Borken** Ger.

60 G2 **Borkenes** Norway
68 C1 **Borkum** Ger.
68 C1 **Borkum** *i.* Ger.
61 G3 **Borlänge** Sweden
69 F2 **Borna** Ger.
29 C1 **Borneo** *i.* Asia
61 F4 **Bornholm** *i.* Denmark
79 C3 **Bornova** Turkey
58 B1 **Borodyanka** Ukr.
57 D2 **Borovichi** Rus. Fed.
57 E2 **Borovsk** Rus. Fed.
19 C1 **Borroloola** Austr.
78 B1 **Borşa** Romania
58 B2 **Borshchiv** Ukr.
37 D1 **Borshchovochnyy Khrebet** *mts* Rus. Fed.
69 E1 **Börßum** Ger.
Bortala China *see* **Bole**
49 C2 **Borūjerd** Iran
58 A2 **Boryslav** Ukr.
58 C1 **Boryspil'** Ukr.
59 C1 **Borzna** Ukr.
37 D1 **Borzya** Rus. Fed.
77 C1 **Bosanska Dubica** Bos.-Herz.
77 C1 **Bosanska Gradiška** Bos.-Herz.
77 C2 **Bosanska Krupa** Bos.-Herz.
77 C1 **Bosanski Novi** Bos.-Herz.
77 C2 **Bosansko Grahovo** Bos.-Herz.
39 A3 **Bose** China
91 C2 **Boshof** S. Africa
77 C2 **Bosnia-Herzegovina** *country* Europe
86 B2 **Bosobolo** Dem. Rep. Congo
78 C2 **Bosporus** *str.* Turkey
86 B2 **Bossangoa** C.A.R.
86 B2 **Bossembélé** C.A.R.
36 B2 **Bosten Hu** *l.* China
66 C2 **Boston** U.K.
109 E2 **Boston** U.S.A.
110 B1 **Boston Mountains** U.S.A.
21 D2 **Botany Bay** Austr.
78 B2 **Botev** *mt.* Bulg.
61 G3 **Bothnia, Gulf of** Fin./Sweden
78 C1 **Botoşani** Romania
91 C2 **Botshabelo** S. Africa
88 B3 **Botswana** *country* Africa
77 C3 **Botte Donato, Monte** *mt.* Italy
104 C1 **Bottineau** U.S.A.
68 C2 **Bottrop** Ger.
122 C2 **Botucatu** Brazil
82 B4 **Bouaké** Côte d'Ivoire
86 B2 **Bouar** C.A.R.
82 B1 **Bouârfa** Morocco
99 D2 **Bouctouche** Can.
75 E2 **Bougaa** Alg.
82 B3 **Bougouni** Mali
68 B3 **Bouillon** Belgium
75 D2 **Bouira** Alg.
82 A2 **Boujdour** Western Sahara
104 B2 **Boulder** U.S.A.
103 D3 **Boulder City** U.S.A.
19 C2 **Boulia** Austr.
72 C2 **Boulogne-Billancourt** France
67 D3 **Boulogne-sur-Mer** France
86 B3 **Boumango** Gabon
86 B2 **Boumba** *r.* Cameroon
75 D2 **Boumerdes** Alg.
82 B4 **Bouna** Côte d'Ivoire
82 B4 **Boundiali** Côte d'Ivoire
102 D2 **Bountiful** U.S.A.
82 B3 **Bourem** Mali
72 C2 **Bourganeuf** France
73 D2 **Bourg-en-Bresse** France
72 C2 **Bourges** France
21 C2 **Bourke** Austr.
67 C3 **Bournemouth** U.K.
83 C1 **Bou Saâda** Alg.
83 D3 **Bousso** Chad
82 A3 **Boutilimit** Maur.
97 C3 **Bow** *r.* Can.
19 D2 **Bowen** Austr.
97 C3 **Bow Island** Can.
108 B3 **Bowling Green** *KY* U.S.A.
108 C2 **Bowling Green** *OH* U.S.A.
104 C1 **Bowman** U.S.A.
69 D3 **Boxberg** Ger.
68 B2 **Boxtel** Neth.
48 B1 **Boyabat** Turkey
39 B3 **Boyang** China
65 B2 **Boyle** Rep. of Ireland
65 C2 **Boyne** *r.* Rep. of Ireland
104 B2 **Boysen Reservoir** U.S.A.
120 B3 **Boyuibe** Bol.
79 C3 **Bozburun** Turkey
79 C3 **Bozcaada** *i.* Turkey
79 C3 **Bozdağ** *mt.* Turkey
79 C3 **Boz Dağları** *mts* Turkey
79 C3 **Bozdoğan** Turkey
102 D1 **Bozeman** U.S.A.
86 B2 **Bozoum** C.A.R.
79 D3 **Bozüyük** Turkey
77 C2 **Brač** *i.* Croatia
98 C2 **Bracebridge** Can.
99 D2 **Brachet, Lac au** *l.* Can.
67 C3 **Bracknell** U.K.
77 C2 **Bradano** *r.* Italy
111 D3 **Bradenton** U.S.A.
66 C2 **Bradford** U.K.
109 D2 **Bradford** U.S.A.
107 D2 **Brady** U.S.A.
64 C2 **Braemar** U.K.
74 B1 **Braga** Port.
119 D2 **Bragança** Brazil
74 B1 **Bragança** Port.
123 C2 **Bragança Paulista** Brazil
57 D3 **Brahin** Belarus
41 C3 **Brahmapur** India
30 A1 **Brahmaputra** *r.* China/India
78 C1 **Brăila** Romania
105 E1 **Brainerd** U.S.A.
67 D3 **Braintree** U.K.
68 B2 **Braives** Belgium
69 D1 **Brake (Unterweser)** Ger.
68 D1 **Bramsche** Ger.
118 B2 **Branco** *r.* Brazil
69 F1 **Brandenburg** Ger.
97 E3 **Brandon** Can.
65 A2 **Brandon Mountain** Rep. of Ireland
90 B3 **Brandvlei** S. Africa
71 D1 **Braniewo** Pol.
98 B2 **Brantford** Can.
99 D2 **Bras d'Or Lake** Can.
123 D1 **Brasil, Planalto do** *plat.* Brazil
122 C1 **Brasilândia** Brazil
122 C1 **Brasília** Brazil
123 D1 **Brasília de Minas** Brazil
56 C2 **Braslaw** Belarus
78 C1 **Braşov** Romania
71 D2 **Bratislava** Slovakia
51 H3 **Bratsk** Rus. Fed.
70 C2 **Braunau am Inn** Austria
69 E1 **Braunschweig** Ger.
60 □A2 **Brautarholt** Iceland
Bravo del Norte, Río *r.* Mex./U.S.A. *see* **Rio Grande**
103 C4 **Brawley** U.S.A.
65 C2 **Bray** Rep. of Ireland
118 C2 **Brazil** *country* S. America
107 D3 **Brazos** *r.* U.S.A.
86 B3 **Brazzaville** Congo
77 C2 **Brčko** Bos.-Herz.
64 C2 **Brechin** U.K.
68 B2 **Brecht** Belgium
107 D2 **Breckenridge** U.S.A.
71 D2 **Břeclav** Czech Rep.
67 B3 **Brecon** U.K.
67 B3 **Brecon Beacons** *reg.* U.K.
68 B2 **Breda** Neth.
90 B3 **Bredasdorp** S. Africa
70 B2 **Bregenz** Austria
60 H1 **Breivikbotn** Norway
60 E3 **Brekstad** Norway
69 D1 **Bremen** Ger.
69 D1 **Bremerhaven** Ger.
102 B1 **Bremerton** U.S.A.
69 D1 **Bremervörde** Ger.
107 D2 **Brenham** U.S.A.
76 B1 **Brennero** Italy
70 C2 **Brenner Pass** Austria/Italy
67 D3 **Brentwood** U.K.
76 B1 **Brescia** Italy
76 B1 **Bressanone** Italy
64 □ **Bressay** *i.* U.K.
72 B2 **Bressuire** France
56 B3 **Brest** Belarus
72 B2 **Brest** France
110 C3 **Breton Sound** *b.* U.S.A.
119 C2 **Breves** Brazil
21 C1 **Brewarrina** Austr.
102 C1 **Brewster** U.S.A.
57 E2 **Breytovo** Rus. Fed.
77 C1 **Brezovo Polje** *hill* Croatia
86 C2 **Bria** C.A.R.
73 D3 **Briançon** France
58 B2 **Briceni** Moldova
67 B3 **Bridgend** U.K.
109 E2 **Bridgeport** *CT* U.S.A.
104 C2 **Bridgeport** *NE* U.S.A.
115 E3 **Bridgetown** Barbados
99 D2 **Bridgewater** Can.
67 B3 **Bridgwater** U.K.
66 C1 **Bridlington** U.K.
66 C1 **Bridlington Bay** U.K.
73 D2 **Brig** Switz.
102 D2 **Brigham City** U.S.A.
22 B3 **Brighton** N.Z.
67 C3 **Brighton** U.K.
73 D3 **Brignoles** France
69 D2 **Brilon** Ger.
77 C2 **Brindisi** Italy
21 D1 **Brisbane** Austr.
67 B3 **Bristol** U.K.
111 D1 **Bristol** U.S.A.
67 A3 **Bristol Channel** *est.* U.K.
96 B2 **British Columbia** *prov.* Can.
24 C6 **British Indian Ocean Territory** *terr.* Indian Ocean
91 C2 **Brits** S. Africa
90 B3 **Britstown** S. Africa
72 C2 **Brive-la-Gaillarde** France
74 C1 **Briviesca** Spain
71 D2 **Brno** Czech Rep.
111 D2 **Broad** *r.* U.S.A.
98 C1 **Broadback** *r.* Can.
64 A1 **Broad Bay** U.K.
64 C3 **Broad Law** *hill* U.K.
104 B1 **Broadus** U.S.A.
97 D2 **Brochet** Can.
97 D2 **Brochet, Lac** *l.* Can.
69 E1 **Bröckel** Ger.
94 D1 **Brock Island** Can.
98 C2 **Brockville** Can.
95 F2 **Brodeur Peninsula** Can.
64 B3 **Brodick** U.K.
71 D1 **Brodnica** Pol.
58 B1 **Brody** Ukr.
107 D1 **Broken Arrow** U.S.A.
105 D2 **Broken Bow** U.S.A.
20 B2 **Broken Hill** Austr.
67 B2 **Bromsgrove** U.K.
61 E4 **Brønderslev** Denmark
60 F2 **Brønnøysund** Norway
32 A2 **Brooke's Point** Phil.

110 B2 **Brookhaven** U.S.A.
102 B2 **Brookings** *OR* U.S.A.
105 D2 **Brookings** *SD* U.S.A.
97 C2 **Brooks** Can.
94 B2 **Brooks Range** *mts* U.S.A.
111 D3 **Brooksville** U.S.A.
64 B2 **Broom, Loch** *inlet* U.K.
18 B1 **Broome** Austr.
102 B2 **Brothers** U.S.A.
Broughton Island Can. *see* Qikiqtarjuaq
58 C1 **Brovary** Ukr.
107 C2 **Brownfield** U.S.A.
96 C3 **Browning** U.S.A.
110 C1 **Brownsville** *TN* U.S.A.
107 D3 **Brownsville** *TX* U.S.A.
107 D2 **Brownwood** U.S.A.
60 □A2 **Brú** Iceland
72 C1 **Bruay-la-Bussière** France
108 B1 **Bruce Crossing** U.S.A.
71 D2 **Bruck an der Mur** Austria
68 A2 **Bruges** Belgium
Brugge Belgium *see* **Bruges**
96 C2 **Brûlé** Can.
119 D3 **Brumado** Brazil
61 F3 **Brumunddal** Norway
29 C1 **Brunei** *country* Asia
70 C2 **Brunico** Italy
69 D1 **Brunsbüttel** Ger.
111 D2 **Brunswick** *GA* U.S.A.
109 F2 **Brunswick** *ME* U.S.A.
21 D1 **Brunswick Head** Austr.
104 C2 **Brush** U.S.A.
68 B2 **Brussels** Belgium
Bruxelles Belgium *see* Brussels
107 D2 **Bryan** U.S.A.
57 D3 **Bryansk** Rus. Fed.
59 D2 **Bryn'kovskaya** Rus. Fed.
59 D2 **Bryukhovetskaya** Rus. Fed.
71 D1 **Brzeg** Pol.
82 A3 **Buba** Guinea-Bissau
48 B2 **Bucak** Turkey
118 A1 **Bucaramanga** Col.
21 C3 **Buchan** Austr.
82 A4 **Buchanan** Liberia
78 C2 **Bucharest** Romania
69 D1 **Bucholz in der Nordheide** Ger.
78 C1 **Bucin, Pasul** *pass* Romania
69 D1 **Bückeburg** Ger.
106 A2 **Buckeye** U.S.A.
64 C2 **Buckhaven** U.K.
64 C2 **Buckie** U.K.
19 C1 **Buckingham Bay** Austr.
19 D2 **Buckland Tableland** *reg.* Austr.
20 A2 **Buckleboo** Austr.
109 F2 **Bucksport** U.S.A.
București Romania *see* Bucharest
57 D3 **Buda-Kashalyova** Belarus
71 D2 **Budapest** Hungary
43 B2 **Budaun** India
76 A2 **Buddusò** *Sardinia* Italy
67 A3 **Bude** U.K.
55 D4 **Budennovsk** Rus. Fed.
57 D2 **Budogoshch'** Rus. Fed.
76 A2 **Budoni** *Sardinia* Italy
86 A2 **Buea** Cameroon
112 B2 **Buenaventura** Mex.
74 C1 **Buendia, Embalse de** *resr* Spain
123 D1 **Buenópolis** Brazil
121 C4 **Buenos Aires** Arg.
121 A5 **Buenos Aires, Lago** *l.* Arg./Chile
109 D2 **Buffalo** *NY* U.S.A.
104 C1 **Buffalo** *SD* U.S.A.
104 B2 **Buffalo** *WY* U.S.A.
97 D2 **Buffalo Narrows** Can.
90 A2 **Buffels** *watercourse* S. Africa
78 C2 **Buftea** Romania
71 E1 **Bug** *r.* Pol.
29 C2 **Bugel, Tanjung** *pt* Indon.
77 C2 **Bugojno** Bos.-Herz.
32 A2 **Bugsuk** *i.* Phil.
55 E3 **Buguruslan** Rus. Fed.
78 C1 **Buhuși** Romania
67 B2 **Builth Wells** U.K.
37 D1 **Buir Nur** *l.* Mongolia
88 A3 **Buitepos** Namibia
77 D2 **Bujanovac** Yugo.
87 C3 **Bujumbura** Burundi
37 D1 **Bukachacha** Rus. Fed.
87 C3 **Bukavu** Dem. Rep. Congo
44 C3 **Bukhara** Uzbek.
28 B2 **Bukittinggi** Indon.
87 D3 **Bukoba** Tanz.
21 D2 **Bulahdelal** Austr.
89 B3 **Bulawayo** Zimbabwe
79 C3 **Buldan** Turkey
91 D2 **Bulembu** Swaziland
37 C1 **Bulgan** Mongolia
78 C2 **Bulgaria** *country* Europe
22 B2 **Buller** *r.* N.Z.
20 B1 **Bulloo Downs** Austr.
90 A1 **Büllsport** Namibia
26 C3 **Bulukumba** Indon.
86 B3 **Bulungu** Dem. Rep. Congo
86 C2 **Bumba** Dem. Rep. Congo
30 A1 **Bumhkang** Myanmar
86 B3 **Buna** Dem. Rep. Congo
18 A3 **Bunbury** Austr.
65 C1 **Buncrana** Rep. of Ireland
87 D3 **Bunda** Tanz.
19 E2 **Bundaberg** Austr.
21 D2 **Bundarra** Austr.
42 B2 **Bundi** India
65 B1 **Bundoran** Rep. of Ireland
21 C3 **Bungendore** Austr.
35 B4 **Bungo-suidō** *sea chan.* Japan
87 D2 **Bunia** Dem. Rep. Congo
86 C3 **Bunianga** Dem. Rep. Congo
31 B2 **Buôn Mê Thuôt** Vietnam
87 D3 **Bura** Kenya
43 C1 **Burang** China
85 C4 **Burao** Somalia
46 B2 **Buraydah** Saudi Arabia
68 D2 **Burbach** Ger.
48 B2 **Burdur** Turkey
85 B3 **Burē** Eth.
67 D2 **Bure** *r.* U.K.
78 C2 **Burgas** Bulg.
69 E1 **Burg bei Magdeburg** Ger.
69 E1 **Burgdorf** *Niedersachsen* Ger.
69 E1 **Burgdorf** *Niedersachsen* Ger.
99 E2 **Burgeo** Can.
91 D1 **Burgersfort** S. Africa
68 A2 **Burgh-Haamstede** Neth.
113 C2 **Burgos** Mex.
74 C1 **Burgos** Spain
79 C3 **Burhaniye** Turkey
42 B2 **Burhanpur** India
69 D1 **Burhave (Butjadingen)** Ger.
99 E2 **Burin** Can.
119 D2 **Buriti Bravo** Brazil
123 C1 **Buritis** Brazil
19 C1 **Burketown** Austr.
82 B3 **Burkina** *country* Africa
102 D2 **Burley** U.S.A.
104 C3 **Burlington** *CO* U.S.A.
105 E2 **Burlington** *IA* U.S.A.
111 E1 **Burlington** *NC* U.S.A.
109 E2 **Burlington** *VT* U.S.A.
Burma *country* Asia *see* Myanmar
102 B2 **Burney** U.S.A.
19 D4 **Burnie** Austr.
66 B2 **Burnley** U.K.
102 C2 **Burns** U.S.A.
96 B2 **Burns Lake** Can.
45 E2 **Burqin** China
20 A2 **Burra** Austr.
77 D2 **Burrel** Albania
65 B2 **Burren** *reg.* Rep. of Ireland
21 C2 **Burrendong Reservoir** Austr.
21 C2 **Burren Junction** Austr.
75 C2 **Burriana** Spain
21 C2 **Burrinjuck Reservoir** Austr.
112 B2 **Burro, Serranías del** *mts* Mex.
79 C2 **Bursa** Turkey
84 B2 **Bûr Safâga** Egypt
Bûr Sa'îd Egypt *see* Port Said
98 C1 **Burton, Lac** *l.* Can.
65 B1 **Burtonport** Rep. of Ireland
67 C2 **Burton upon Trent** U.K.
27 C3 **Buru** *i.* Indon.
87 D3 **Burundi** *country* Africa
87 C3 **Bururi** Burundi
59 C1 **Buryn'** Ukr.
44 B2 **Burynshyk** Kazakh.
67 D2 **Bury St Edmunds** U.K.
86 C3 **Busanga** Dem. Rep. Congo
47 C2 **Būshehr** Iran
87 D3 **Bushenyi** Uganda
86 C2 **Businga** Dem. Rep. Congo
18 A3 **Busselton** Austr.
107 C3 **Bustamante** Mex.
32 A1 **Busuanga** Phil.
86 C2 **Buta** Dem. Rep. Congo
87 C3 **Butare** Rwanda
91 C2 **Butha Buthe** Lesotho
108 D2 **Butler** U.S.A.
27 C3 **Buton** *i.* Indon.
102 D1 **Butte** U.S.A.
28 B1 **Butterworth** Malaysia
64 A1 **Butt of Lewis** *hd* U.K.
97 E2 **Button Bay** Can.
32 B2 **Butuan** Phil.
59 E1 **Buturlinovka** Rus. Fed.
43 C2 **Butwal** Nepal
69 D2 **Butzbach** Ger.
85 C4 **Buulobarde** Somalia
85 C5 **Buur Gaabo** Somalia
85 C4 **Buurhabaka** Somalia
69 D1 **Buxtehude** Ger.
57 F2 **Buy** Rus. Fed.
55 D4 **Buynaksk** Rus. Fed.
79 C3 **Büyükmenderes** *r.* Turkey
78 C1 **Buzău** Romania
89 C2 **Búzi** Moz.
55 E3 **Buzuluk** Rus. Fed.
78 C2 **Byala** Bulg.
56 C3 **Byalynichy** Belarus
56 D3 **Byarezina** *r.* Belarus
56 B3 **Byaroza** Belarus
71 D1 **Bydgoszcz** Pol.
56 C3 **Byerazino** Belarus
56 C2 **Byeshankovichy** Belarus
57 D3 **Bykhaw** Belarus
95 F2 **Bylot Island** Can.
21 C2 **Byrock** Austr.
21 D1 **Byron Bay** Austr.
51 J2 **Bytantay** *r.* Rus. Fed.
71 D1 **Bytom** Pol.
71 D1 **Bytów** Pol.

C

122 B2 **Caarapó** Brazil
32 B1 **Cabanatuan** Phil.
85 C3 **Cabdul Qaadir** Somalia

74 B2 **Cabeza del Buey** Spain
120 B2 **Cabezas** Bol.
118 A1 **Cabimas** Venez.
88 A1 **Cabinda** Angola
86 B3 **Cabinda** *prov.* Angola
123 D2 **Cabo Frio** Brazil
123 D2 **Cabo Frio, Ilha do** *i.* Brazil
98 C2 **Cabonga, Réservoir** *resr* Can.
19 E2 **Caboolture** Austr.
118 A2 **Cabo Pantoja** Peru
112 A1 **Caborca** Mex.
99 D2 **Cabot Strait** Can.
123 D1 **Cabral, Serra do** *mts* Brazil
75 D2 **Cabrera** *i.* Spain
97 D2 **Cabri** Can.
75 C2 **Cabriel** *r.* Spain
120 C3 **Caçador** Brazil
77 D2 **Čačak** Yugo.
76 A2 **Caccia, Capo** *c. Sardinia* Italy
119 C3 **Cáceres** Brazil
74 B2 **Cáceres** Spain
96 B2 **Cache Creek** Can.
82 A3 **Cacheu** Guinea-Bissau
119 C2 **Cachimbo, Serra do** *hills* Brazil
122 B1 **Cachoeira Alta** Brazil
123 D2 **Cachoeiro de Itapemirim** Brazil
82 A3 **Cacine** Guinea-Bissau
88 A2 **Cacolo** Angola
122 B1 **Caçu** Brazil
71 D2 **Čadca** Slovakia
69 D1 **Cadenberge** Ger.
113 B2 **Cadereyta** Mex.
108 B2 **Cadillac** U.S.A.
74 B2 **Cádiz** Spain
74 B2 **Cádiz, Golfo de** *g.* Spain
96 C2 **Cadotte Lake** Can.
72 B2 **Caen** France
66 A2 **Caernarfon** U.K.
66 A2 **Caernarfon Bay** U.K.
120 B3 **Cafayate** Arg.
32 B2 **Cagayan de Oro** Phil.
76 B2 **Cagli** Italy
76 A3 **Cagliari** *Sardinia* Italy
76 A3 **Cagliari, Golfo di** *b. Sardinia* Italy
65 B3 **Caha Mountains** Rep. of Ireland
65 A3 **Cahermore** Rep. of Ireland
65 A3 **Cahersiveen** Rep. of Ireland
65 C2 **Cahir** Rep. of Ireland
89 C2 **Cahora Bassa, Lago de** *resr* Moz.
65 C2 **Cahore Point** Rep. of Ireland
72 C3 **Cahors** France
58 B2 **Cahul** Moldova
89 C2 **Caia** Moz.
119 C3 **Caiabis, Serra dos** *hills* Brazil
88 B2 **Caianda** Angola
122 B1 **Caiapó, Serra do** *mts* Brazil
122 B1 **Caiapônia** Brazil
115 C2 **Caicos Islands** Turks and Caicos Is
64 C2 **Cairngorm Mountains** U.K.
66 A1 **Cairnryan** U.K.
19 D1 **Cairns** Austr.
84 B1 **Cairo** Egypt
66 C2 **Caistor** U.K.
88 A2 **Caiundo** Angola
118 A2 **Cajamarca** Peru
77 C1 **Čakovec** Croatia
91 C3 **Cala** S. Africa
118 B1 **Calabozo** Venez.
78 B2 **Calafat** Romania
121 A6 **Calafate** Arg.
75 C1 **Calahorra** Spain
72 C1 **Calais** France
109 F1 **Calais** U.S.A.
120 B3 **Calama** Chile
32 A1 **Calamian Group** *is* Phil.
75 C1 **Calamocha** Spain
88 A1 **Calandula** Angola
28 A1 **Calang** Indon.
83 E1 **Calanscio Sand Sea** *des.* Libya
32 B1 **Calapan** Phil.
78 C2 **Călărași** Romania
75 C1 **Calatayud** Spain
32 B1 **Calayan** *i.* Phil.
32 B1 **Calbayog** Phil.
119 E2 **Calcanhar, Ponta do** *pt* Brazil
119 C1 **Calçoene** Brazil
43 C2 **Calcutta** India
74 B2 **Caldas da Rainha** Port.
122 C1 **Caldas Novas** Brazil
120 A3 **Caldera** Chile
102 C2 **Caldwell** U.S.A.
91 C3 **Caledon** *r.* Lesotho/S. Africa
90 A3 **Caledon** S. Africa
121 B5 **Caleta Olivia** Arg.
66 A1 **Calf of Man** *i.* Isle of Man
96 C2 **Calgary** Can.
118 A1 **Cali** Col.
41 B3 **Calicut** India
103 D3 **Caliente** U.S.A.
103 B2 **California** *state* U.S.A.
112 A1 **California, Golfo de** *g.* Mex.
103 B3 **California Aqueduct** *canal* U.S.A.
90 B3 **Calitzdorp** S. Africa
113 C2 **Calkiní** Mex.
20 B1 **Callabonna, Lake** *salt flat* Austr.
64 B2 **Callander** U.K.
118 A3 **Callao** Peru
76 B3 **Caltagirone** *Sicily* Italy
76 B3 **Caltanissetta** *Sicily* Italy
88 A2 **Caluquembe** Angola
85 D3 **Caluula** Somalia
73 D3 **Calvi** *Corsica* France
75 D2 **Calvià** Spain
112 B2 **Calvillo** Mex.
90 A3 **Calvinia** S. Africa
77 C2 **Calvo, Monte** *mt.* Italy
112 B2 **Camacho** Mex.
88 A2 **Camacupa** Angola
114 C2 **Camagüey** Cuba
118 A3 **Camana** Peru
122 B1 **Camapuã** Brazil
113 C2 **Camargo** Mex.
31 B3 **Ca Mau** Vietnam
31 B2 **Cambodia** *country* Asia
67 A3 **Camborne** U.K.
73 C1 **Cambrai** France
67 B2 **Cambrian Mountains** U.K.
22 C1 **Cambridge** N.Z.
67 D2 **Cambridge** U.K.
109 E2 **Cambridge** *MA* U.S.A.
109 D3 **Cambridge** *MD* U.S.A.
105 E1 **Cambridge** *MN* U.S.A.
108 C2 **Cambridge** *OH* U.S.A.
99 D1 **Cambrien, Lac** *l.* Can.
21 D2 **Camden** Austr.
110 B2 **Camden** *AR* U.S.A.
109 F2 **Camden** *ME* U.S.A.
86 B2 **Cameroon** *country* Africa
119 D2 **Cametá** Brazil
32 B1 **Camiguin** *i.* Phil.
120 B3 **Camiri** Bol.
119 D2 **Camocim** Brazil
19 C1 **Camooweal** Austr.
31 A3 **Camorta** *i.* India
121 A5 **Campana, Isla** *i.* Chile
90 B2 **Campbell** S. Africa
22 B2 **Campbell, Cape** N.Z.
96 B2 **Campbell River** Can.
108 B3 **Campbellsville** U.S.A.
99 D2 **Campbellton** Can.
64 B3 **Campbeltown** U.K.
113 C3 **Campeche** Mex.
113 C3 **Campeche, Bahía de** *g.* Mex.
20 B3 **Camperdown** Austr.
78 C1 **Câmpina** Romania
119 E2 **Campina Grande** Brazil
122 C2 **Campinas** Brazil
122 C1 **Campina Verde** Brazil
76 B2 **Campobasso** Italy
123 C2 **Campo Belo** Brazil
122 C1 **Campo Florido** Brazil
120 B3 **Campo Gallo** Arg.
122 B2 **Campo Grande** Brazil
119 D2 **Campo Maior** Brazil
74 B2 **Campo Maior** Port.
122 B2 **Campo Mourão** Brazil
123 D2 **Campos** Brazil
123 C1 **Campos Altos** Brazil
123 C2 **Campos do Jordão** Brazil
78 C1 **Câmpulung** Romania
106 A2 **Camp Verde** U.S.A.
31 B2 **Cam Ranh** Vietnam
96 C2 **Camrose** Can.
97 D2 **Camsell Portage** Can.
79 C2 **Çan** Turkey
94 F3 **Canada** *country* N. America
107 C1 **Canadian** U.S.A.
107 D1 **Canadian** *r.* U.S.A.
79 C2 **Çanakkale** Turkey
112 A1 **Cananea** Mex.
122 C2 **Cananéia** Brazil
Canarias, Islas *terr.* N. Atlantic Ocean *see* Canary Islands
82 A2 **Canary Islands** *terr.* N. Atlantic Ocean
123 C1 **Canastra, Serra da** *mts* Brazil
112 B2 **Canatlán** Mex.
111 D3 **Canaveral, Cape** U.S.A.
123 E1 **Canavieiras** Brazil
21 C3 **Canberra** Austr.
113 D2 **Cancún** Mex.
79 C3 **Çandarlı** Turkey
113 C3 **Candelaria** Mex.
97 D2 **Candle Lake** *l.* Can.
88 A2 **Cangamba** Angola
74 B1 **Cangas del Narcea** Spain
120 C4 **Canguçu** Brazil
38 B2 **Cangzhou** China
99 D1 **Caniapiscau** Can.
99 D1 **Caniapiscau** *r.* Can.
99 C1 **Caniapiscau, Lac** *l.* Can.
76 B3 **Canicattì** *Sicily* Italy
119 E2 **Canindé** Brazil
112 B2 **Cañitas de Felipe Pescador** Mex.
48 B1 **Çankırı** Turkey
96 C2 **Canmore** Can.
64 A2 **Canna** *i.* U.K.
41 B3 **Cannanore** India
73 D3 **Cannes** France
67 B2 **Cannock** U.K.
21 C3 **Cann River** Austr.
120 C3 **Canoas** Brazil
97 D2 **Canoe Lake** *l.* Can.
122 B3 **Canoinhas** Brazil
104 B3 **Canon City** U.S.A.
97 D2 **Canora** Can.
21 C2 **Canowindra** Austr.
74 B1 **Cantabrian Sea** *sea* Spain
74 C1 **Cantábrica, Cordillera** *mts* Spain
67 D3 **Canterbury** U.K.
22 B2 **Canterbury Bight** *b.* N.Z.
22 B2 **Canterbury Plains** N.Z.
31 B2 **Cân Thơ** Vietnam
119 D2 **Canto do Buriti** Brazil

Canton China *see* Guangzhou
110 C2 **Canton** *MS* U.S.A.
108 C2 **Canton** *OH* U.S.A.
107 C1 **Canyon** U.S.A.
102 D1 **Canyon Ferry Lake** U.S.A.
30 B1 **Cao Bằng** Vietnam
122 C2 **Capão Bonito** Brazil
123 D2 **Caparaó, Serra do** *mts* Brazil
19 D4 **Cape Barren Island** Austr.
20 A3 **Cape Borda** Austr.
99 D2 **Cape Breton Island** Can.
82 B4 **Cape Coast** Ghana
109 E2 **Cape Cod Bay** U.S.A.
95 F2 **Cape Dorset** Can.
111 E2 **Cape Fear** *r.* U.S.A.
105 F3 **Cape Girardeau** U.S.A.
123 D1 **Capelinha** Brazil
68 B2 **Capelle aan de IJssel** Neth.
88 A1 **Capenda-Camulemba** Angola
90 A3 **Cape Town** S. Africa
80 B2 **Cape Verde** *country* N. Atlantic Ocean
19 D1 **Cape York Peninsula** Austr.
115 C3 **Cap-Haïtien** Haiti
119 D2 **Capim** *r.* Brazil
77 C2 **Čapljina** Bos.-Herz.
77 B3 **Capo d'Orlando** *Sicily* Italy
76 A2 **Capraia, Isola di** *i.* Italy
76 A2 **Caprara, Punta** *pt Sardinia* Italy
76 B2 **Capri, Isola di** *i.* Italy
88 B2 **Caprivi Strip** *reg.* Namibia
118 B2 **Caquetá** *r.* Col.
78 B2 **Caracal** Romania
118 B1 **Caracas** Venez.
119 D2 **Caracol** Brazil
123 C2 **Caraguatatuba** Brazil
121 A4 **Carahue** Chile
123 D1 **Caraí** Brazil
123 D2 **Carangola** Brazil
78 B1 **Caransebeş** Romania
99 D2 **Caraquet** Can.
114 B3 **Caratasca, Laguna** *lag.* Hond.
123 D1 **Caratinga** Brazil
118 B2 **Carauari** Brazil
75 C2 **Caravaca de la Cruz** Spain
123 E1 **Caravelas** Brazil
97 E3 **Carberry** Can.
112 A2 **Carbó** Mex.
76 A3 **Carbonara, Capo** *c. Sardinia* Italy
104 B3 **Carbondale** *CO* U.S.A.
108 B3 **Carbondale** *IL* U.S.A.
99 E2 **Carbonear** Can.
123 D1 **Carbonita** Brazil
75 C2 **Carcaixent** Spain
72 C3 **Carcassonne** France
96 A1 **Carcross** Can.
113 C3 **Cárdenas** Mex.
67 B3 **Cardiff** U.K.
67 A2 **Cardigan** U.K.
67 A2 **Cardigan Bay** U.K.
96 C3 **Cardston** Can.
78 B1 **Carei** Romania
72 B2 **Carentan** France
18 B2 **Carey, Lake** *salt flat* Austr.
123 D2 **Cariacica** Brazil
114 B3 **Caribbean Sea** N. Atlantic Ocean
109 F1 **Caribou** U.S.A.
98 B1 **Caribou Lake** Can.
96 C2 **Caribou Mountains** Can.
112 B2 **Carichic** Mex.
68 B3 **Carignan** France
21 C2 **Carinda** Austr.
75 C1 **Cariñena** Spain
98 C2 **Carleton Place** Can.
91 C2 **Carletonville** S. Africa
65 C1 **Carlingford Lough** *inlet* Rep. of Ireland/U.K.
66 B1 **Carlisle** U.K.
109 D2 **Carlisle** U.S.A.
123 D1 **Carlos Chagas** Brazil
65 C2 **Carlow** Rep. of Ireland
103 C4 **Carlsbad** *CA* U.S.A.
106 C2 **Carlsbad** *NM* U.S.A.
97 D3 **Carlyle** Can.
96 A1 **Carmacks** Can.
97 E3 **Carman** Can.
67 A3 **Carmarthen** U.K.
67 A3 **Carmarthen Bay** U.K.
72 C3 **Carmaux** France
113 C3 **Carmelita** Guat.
112 A2 **Carmen, Isla** *i.* Mex.
72 B2 **Carnac** France
90 B3 **Carnarvon** S. Africa
65 C1 **Carndonagh** Rep. of Ireland
18 B2 **Carnegie, Lake** *salt flat* Austr.
64 B2 **Carn Eighe** *mt.* U.K.
31 A3 **Car Nicobar** *i.* India
86 B2 **Carnot** C.A.R.
20 A2 **Carnot, Cape** Austr.
65 C2 **Carnsore Point** Rep. of Ireland
119 D2 **Carolina** Brazil
27 D2 **Caroline Islands** N. Pacific Ocean
71 D2 **Carpathian Mountains** Europe
78 B1 **Carpații Meridionali** *mts* Romania
19 C1 **Carpentaria, Gulf of** Austr.
73 D3 **Carpentras** France
65 B3 **Carrantuohill** *mt.* Rep. of Ireland
73 E3 **Carrara** Italy
65 C2 **Carrickmacross** Rep. of Ireland
65 B2 **Carrick-on-Shannon** Rep. of Ireland
65 C2 **Carrick-on-Suir** Rep. of Ireland
105 D1 **Carrington** U.S.A.
107 D3 **Carrizo Springs** U.S.A.
106 B2 **Carrizozo** U.S.A.
105 E2 **Carroll** U.S.A.
111 C2 **Carrollton** U.S.A.
97 D2 **Carrot River** Can.
103 C3 **Carson City** U.S.A.
118 A1 **Cartagena** Col.
75 C2 **Cartagena** Spain
114 B4 **Cartago** Costa Rica
105 E3 **Carthage** *MO* U.S.A.
107 E2 **Carthage** *TX* U.S.A.
99 E1 **Cartwright** Can.
119 E2 **Caruarú** Brazil
82 B1 **Casablanca** Morocco
122 C2 **Casa Branca** Brazil
112 B1 **Casa de Janos** Mex.
106 A2 **Casa Grande** U.S.A.
76 A1 **Casale Monferrato** Italy
77 C2 **Casarano** Italy
102 C2 **Cascade** U.S.A.
102 B2 **Cascade Range** *mts* Can./U.S.A.
74 B2 **Cascais** Port.
122 B2 **Cascavel** Brazil
76 B2 **Caserta** Italy
85 D3 **Caseyr, Raas** *c.* Somalia
65 C2 **Cashel** Rep. of Ireland
21 D1 **Casino** Austr.
75 C1 **Caspe** Spain
104 B2 **Casper** U.S.A.
49 C1 **Caspian Sea** Asia/Europe
122 C2 **Cássia** Brazil
96 B2 **Cassiar** Can.
96 A2 **Cassiar Mountains** Can.
122 B1 **Cassilândia** Brazil
76 B2 **Cassino** Italy
64 B2 **Cassley** *r.* U.K.
119 D2 **Castanhal** Brazil
120 B4 **Castaño** *r.* Arg.
112 B2 **Castaños** Mex.
72 C3 **Casteljaloux** France
75 C2 **Castelló de la Plana** Spain
74 B2 **Castelo Branco** Port.
76 B3 **Castelvetrano** *Sicily* Italy
20 B3 **Casterton** Austr.
76 B2 **Castiglione della Pescaia** Italy
65 B2 **Castlebar** Rep. of Ireland
64 A2 **Castlebay** U.K.
65 C1 **Castleblayney** Rep. of Ireland
65 C1 **Castlederg** U.K.
64 C3 **Castle Douglas** U.K.
96 C3 **Castlegar** Can.
65 B2 **Castleisland** Rep. of Ireland
20 B3 **Castlemaine** Austr.
65 B2 **Castlerea** Rep. of Ireland
21 C2 **Castlereagh** *r.* Austr.
104 C3 **Castle Rock** U.S.A.
97 C2 **Castor** Can.
72 C3 **Castres** France
68 B1 **Castricum** Neth.
115 D3 **Castries** St Lucia
122 C2 **Castro** Brazil
121 A5 **Castro** Chile
74 B2 **Castro Verde** Port.
77 C3 **Castrovillari** Italy
118 A2 **Catacaos** Peru
123 D2 **Cataguases** Brazil
122 C1 **Catalão** Brazil
120 B3 **Catamarca** Arg.
32 B1 **Catanduanes** *i.* Phil.
122 C2 **Catanduva** Brazil
122 B3 **Catanduvas** Brazil
77 C3 **Catania** *Sicily* Italy
77 C3 **Catanzaro** Italy
32 B1 **Catarman** Phil.
32 B1 **Catbalogan** Phil.
113 C3 **Catemaco** Mex.
115 C2 **Cat Island** Bahamas
98 A1 **Cat Lake** *l.* Can.
113 D2 **Catoche, Cabo** *c.* Mex.
109 E2 **Catskill Mountains** U.S.A.
32 B2 **Cauayan** Phil.
118 A1 **Cauca** *r.* Col.
119 E2 **Caucaia** Brazil
49 C1 **Caucasus** *mts* Asia/Europe
68 A2 **Caudry** France
77 C3 **Caulonia** Italy
99 D2 **Causapscal** Can.
73 D3 **Cavaillon** France
119 D3 **Cavalcante** Brazil
65 C2 **Cavan** Rep. of Ireland
122 B3 **Cavernoso, Serra do** *mts* Brazil
119 D2 **Caxias** Brazil
120 C3 **Caxias do Sul** Brazil
88 A1 **Caxito** Angola
119 C1 **Cayenne** Fr. Guiana
114 B3 **Cayman Islands** *terr.* West Indies
85 C4 **Caynabo** Somalia
88 B2 **Cazombo** Angola
112 B2 **Ceballos** Mex.
32 B1 **Cebu** Phil.
32 B1 **Cebu** *i.* Phil.
76 B2 **Cecina** Italy
105 F2 **Cedar** *r.* U.S.A.
103 D3 **Cedar City** U.S.A.
105 F2 **Cedar Falls** U.S.A.
97 D2 **Cedar Lake** *l.* Can.

105 E2 **Cedar Rapids** U.S.A.
112 A2 **Cedros, Isla** *i.* Mex.
19 C3 **Ceduna** Austr.
85 C4 **Ceeldheere** Somalia
85 C3 **Ceerigaabo** Somalia
76 B3 **Cefalù** *Sicily* Italy
113 B2 **Celaya** Mex.
Celebes *i.* Indon. *see* **Sulawesi**
124 B5 **Celebes Sea** Indon./Phil.
113 C2 **Celestún** Mex.
69 E1 **Celle** Ger.
63 B3 **Celtic Sea** Rep. of Ireland/U.K.
27 D3 **Cenderawasih, Teluk** *b.* Indon.
118 A1 **Central, Cordillera** *mts* Col.
118 A3 **Central, Cordillera** *mts* Peru
32 B1 **Central, Cordillera** *mts* Phil.
86 C2 **Central African Republic** *country* Africa
42 A2 **Central Brahui Range** *mts* Pak.
105 D2 **Central City** U.S.A.
108 B3 **Centralia** *IL* U.S.A.
102 B1 **Centralia** *WA* U.S.A.
42 A2 **Central Makran Range** *mts* Pak.
27 D3 **Central Range** *mts* P.N.G.
Central Siberian Plateau Rus. Fed. *see* **Sredne-Sibirskoye Ploskogor'ye**
Cephalonia *i.* Greece *see* **Kefallonia**
120 B3 **Ceres** Arg.
122 C1 **Ceres** Brazil
90 A3 **Ceres** S. Africa
72 C3 **Céret** France
74 C1 **Cerezo de Abajo** Spain
77 C2 **Cerignola** Italy
78 C2 **Cernavodă** Romania
113 C2 **Cerralvo** Mex.
112 B2 **Cerralvo, Isla** *i.* Mex.
113 B2 **Cerritos** Mex.
122 C2 **Cerro Azul** Brazil
113 C2 **Cerro Azul** Mex.
118 A3 **Cerro de Pasco** Peru
73 D3 **Cervione** *Corsica* France
74 B1 **Cervo** Spain
76 B2 **Cesena** Italy
56 C2 **Cēsis** Latvia
70 C2 **České Budějovice** Czech Rep.
69 F3 **Český Les** *mts* Czech Rep./Ger.
79 C3 **Çeşme** Turkey
21 D2 **Cessnock** Austr.
77 C2 **Cetinje** Yugo.
77 C3 **Cetraro** Italy
74 B2 **Ceuta** N. Africa
73 C3 **Cévennes** *mts* France
Ceylon *country* Asia *see* **Sri Lanka**
47 D2 **Chābahār** Iran
43 C1 **Chabyêr Caka** *salt l.* China
118 A2 **Chachapoyas** Peru
57 D3 **Chachersk** Belarus
31 B2 **Chachoengsao** Thai.
83 D3 **Chad** *country* Africa
83 D3 **Chad, Lake** Africa
37 C1 **Chadaasan** Mongolia
36 C1 **Chadan** Rus. Fed.
91 C1 **Chadibe** Botswana
104 C2 **Chadron** U.S.A.
33 B2 **Chaeryŏng** N. Korea
42 A2 **Chagai** Pak.
45 C3 **Chaghcharān** Afgh.
57 E2 **Chagoda** Rus. Fed.
44 B2 **Chagyl** Turkm.
43 C2 **Chāībāsa** India
31 B2 **Chainat** Thai.
31 B2 **Chaiyaphum** Thai.
120 C4 **Chajarí** Arg.
99 D1 **Chakonipau, Lake** Can.
118 A3 **Chala** Peru
42 A1 **Chalap Dalan** *mts* Afgh.
99 D2 **Chaleur Bay** *inlet* Can.
42 B2 **Chalisgaon** India
79 C3 **Chalki** *i.* Greece
79 B3 **Chalkida** Greece
72 B2 **Challans** France
102 D2 **Challis** U.S.A.
73 C2 **Châlons-en-Champagne** France
73 C2 **Chalon-sur-Saône** France
106 B1 **Chama** U.S.A.
89 C2 **Chama** Zambia
42 A1 **Chaman** Pak.
42 B1 **Chamba** India
42 B2 **Chambal** *r.* India
105 D2 **Chamberlain** U.S.A.
109 D3 **Chambersburg** U.S.A.
73 D2 **Chambéry** France
89 C2 **Chambeshi** Zambia
108 B2 **Champaign** U.S.A.
109 E2 **Champlain, Lake** Can./U.S.A.
113 C3 **Champotón** Mex.
120 A3 **Chañaral** Chile
94 B2 **Chandalar** *r.* U.S.A.
110 C3 **Chandeleur Islands** U.S.A.
42 B1 **Chandigarh** India
106 A2 **Chandler** U.S.A.
43 B3 **Chandrapur** India
31 B2 **Chang, Ko** *i.* Thai.
Chang'an China *see* **Rong'an**
89 C3 **Changane** *r.* Moz.
89 C2 **Changara** Moz.
33 B1 **Changbai** China
33 B1 **Changbai Shan** *mts* China/N. Korea
37 E2 **Changchun** China
39 B3 **Changde** China
33 B2 **Ch'angdo** N. Korea
39 C3 **Changhua** Taiwan
33 B3 **Changhŭng** S. Korea
Chang Jiang *r.* China *see* **Yangtze**
33 B1 **Changjin** N. Korea
33 B1 **Changjin-gang** *r.* N. Korea
39 B3 **Changsha** China
33 B2 **Changsŏng** S. Korea
39 B3 **Changting** China
33 B2 **Ch'angwŏn** S. Korea
38 B2 **Changyuan** China
38 B2 **Changzhi** China
38 B2 **Changzhou** China
79 B3 **Chania** Greece
63 C4 **Channel Islands** Europe
103 C4 **Channel Islands** U.S.A.
99 E2 **Channel-Port-aux-Basques** Can.
31 B2 **Chanthaburi** Thai.
72 C2 **Chantilly** France
105 D3 **Chanute** U.S.A.
50 G3 **Chany, Ozero** *salt l.* Rus. Fed.
39 B3 **Chaoyang** China
Chaoyang China *see* **Huinan**
39 B3 **Chaozhou** China
112 B2 **Chapala, Laguna de** *l.* Mex.
44 B1 **Chapayevo** Kazakh.
120 C3 **Chapecó** Brazil
111 E1 **Chapel Hill** U.S.A.
98 B2 **Chapleau** Can.
57 E3 **Chaplygin** Rus. Fed.
59 C2 **Chaplynka** Ukr.
113 B2 **Charcas** Mex.
67 B3 **Chard** U.K.
44 C3 **Chardzhev** Turkm.
72 B2 **Charente** *r.* France
45 C3 **Chārīkār** Afgh.
54 E2 **Charkayuvom** Rus. Fed.
68 B2 **Charleroi** Belgium
109 D3 **Charles, Cape** U.S.A.
105 E2 **Charles City** U.S.A.
108 B3 **Charleston** *IL* U.S.A.
111 E2 **Charleston** *SC* U.S.A.
108 C3 **Charleston** *WV* U.S.A.
103 C3 **Charleston Peak** U.S.A.
19 D2 **Charleville** Austr.
73 C2 **Charleville-Mézières** France
111 D1 **Charlotte** U.S.A.
111 D3 **Charlotte Harbor** *b.* U.S.A.
109 D3 **Charlottesville** U.S.A.
99 D2 **Charlottetown** Can.
20 B3 **Charlton** Austr.
98 C1 **Charlton Island** Can.
19 D2 **Charters Towers** Austr.
72 C2 **Chartres** France
96 C2 **Chase** Can.
56 C3 **Chashniki** Belarus
22 A3 **Chaslands Mistake** *c.* N.Z.
33 B1 **Chasŏng** N. Korea
72 B2 **Chassiron, Pointe de** *pt* France
72 B2 **Châteaubriant** France
72 C2 **Château-du-Loir** France
72 C2 **Châteaudun** France
72 B2 **Châteaulin** France
73 D3 **Châteauneuf-les-Martigues** France
72 C2 **Châteauneuf-sur-Loire** France
73 C2 **Château-Thierry** France
96 C2 **Chateh** Can.
68 B2 **Châtelet** Belgium
72 C2 **Châtellerault** France
99 D2 **Chatham** *N.B.* Can.
108 C2 **Chatham** *Ont.* Can.
73 C2 **Châtillon-sur-Seine** France
111 D2 **Chattahoochee** *r.* U.S.A.
111 C1 **Chattanooga** U.S.A.
31 B2 **Châu Đôc** Vietnam
30 A1 **Chauk** Myanmar
73 D2 **Chaumont** France
73 C2 **Chauny** France
119 D2 **Chaves** Brazil
74 B1 **Chaves** Port.
98 C1 **Chavigny, Lac** *l.* Can.
57 D3 **Chavusy** Belarus
57 E2 **Chayevo** Rus. Fed.
54 E3 **Chaykovskiy** Rus. Fed.
70 C1 **Cheb** Czech Rep.
55 D3 **Cheboksary** Rus. Fed.
108 C1 **Cheboygan** U.S.A.
33 B2 **Chech'ŏn** S. Korea
30 A2 **Cheduba Island** Myanmar
82 B2 **Chegga** Maur.
102 B1 **Chehalis** U.S.A.
33 B3 **Cheju** S. Korea
33 B3 **Cheju-do** *i.* S. Korea
33 B3 **Cheju-haehyŏp** *sea chan.* S. Korea
57 E2 **Chekhov** Rus. Fed.
102 B1 **Chelan, Lake** U.S.A.
49 D2 **Cheleken** Turkm.
75 D2 **Chélif, Oued** *r.* Alg.
44 B2 **Chelkar** Kazakh.
71 E1 **Chełm** Pol.
67 D3 **Chelmer** *r.* U.K.
71 D1 **Chełmno** Pol.
67 D3 **Chelmsford** U.K.
67 B3 **Cheltenham** U.K.
55 F3 **Chelyabinsk** Rus. Fed.
69 F2 **Chemnitz** Ger.
82 B2 **Chenachane** Alg.
38 B1 **Chengde** China
38 A2 **Chengdu** China
39 B4 **Chengmai** China

Chengshou China *see* Yingshan
38 A2 **Chengxian** China
Chengyang China *see* **Juxian**
Chennai India *see* **Madras**
39 B3 **Chenzhou** China
67 B3 **Chepstow** U.K.
72 B2 **Cherbourg** France
57 E3 **Cheremisinovo** Rus. Fed.
37 C1 **Cheremkhovo** Rus. Fed.
57 E2 **Cherepovets** Rus. Fed.
59 C2 **Cherkasy** Ukr.
55 D4 **Cherkessk** Rus. Fed.
59 C1 **Chernihiv** Ukr.
59 D2 **Cherninivka** Ukr.
58 B2 **Chernivtsi** Ukr.
58 B1 **Chernyakhiv** Ukr.
56 B3 **Chernyakhovsk** Rus. Fed.
57 E3 **Chernyanka** Rus. Fed.
51 I2 **Chernyshevskiy** Rus. Fed.
105 D2 **Cherokee** U.S.A.
51 K2 **Cherskogo, Khrebet** *mts* Rus. Fed.
59 E2 **Chertkovo** Rus. Fed.
58 A1 **Chervonohrad** Ukr.
56 C3 **Chervyen'** Belarus
57 D3 **Cherykaw** Belarus
109 D3 **Chesapeake Bay** U.S.A.
54 D2 **Cheshskaya Guba** *b.* Rus. Fed.
66 B2 **Chester** U.K.
108 B3 **Chester** *IL* U.S.A.
111 D2 **Chester** *SC* U.S.A.
66 C2 **Chesterfield** U.K.
97 E1 **Chesterfield Inlet** Can.
97 E1 **Chesterfield Inlet** *inlet* Can.
109 F1 **Chesuncook Lake** U.S.A.
99 D2 **Chéticamp** Can.
113 D3 **Chetumal** Mex.
96 B2 **Chetwynd** Can.
66 B1 **Cheviot Hills** U.K.
104 C2 **Cheyenne** U.S.A.
104 C2 **Cheyenne** *r.* U.S.A.
104 C3 **Cheyenne Wells** U.S.A.
43 C2 **Chhapra** India
43 B2 **Chhatarpur** India
39 C3 **Chiai** Taiwan
30 A2 **Chiang Dao** Thai.
30 A2 **Chiang Mai** Thai.
30 A2 **Chiang Rai** Thai.
76 A1 **Chiavenno** Italy
89 C3 **Chiboma** Moz.
98 C2 **Chibougamau** Can.
91 D1 **Chibuto** Moz.
43 D1 **Chibuzhang Hu** *l.* China
108 B2 **Chicago** U.S.A.
96 A2 **Chichagof Island** U.S.A.
67 C3 **Chichester** U.K.
18 A2 **Chichester Range** *mts* Austr.
107 D1 **Chickasha** U.S.A.
118 A2 **Chiclayo** Peru
121 B5 **Chico** *r.* Arg.
103 B3 **Chico** U.S.A.
99 C2 **Chicoutimi** Can.
76 B2 **Chieti** Italy
37 D2 **Chifeng** China
123 D1 **Chifre, Serra do** *mts* Brazil
45 D2 **Chiganak** Kazakh.
113 C3 **Chignahuapán** Mex.
89 C3 **Chigubo** Moz.
112 B2 **Chihuahua** Mex.
56 C2 **Chikhachevo** Rus. Fed.
35 C3 **Chikuma-gawa** *r.* Japan
96 B2 **Chilanko** *r.* Can.
42 B1 **Chilas** Jammu and Kashmir
107 C2 **Childress** U.S.A.
121 A4 **Chile** *country* S. America
120 B3 **Chilecito** Arg.
43 C3 **Chilika Lake** India
89 B2 **Chililabombwe** Zambia
96 B2 **Chilko** *r.* Can.
96 B2 **Chilko Lake** Can.
121 A4 **Chillán** Chile
105 E3 **Chillicothe** *MO* U.S.A.
108 C3 **Chillicothe** *OH* U.S.A.
96 B3 **Chilliwack** Can.
121 A5 **Chiloé, Isla de** *i.* Chile
113 C3 **Chilpancingo** Mex.
21 C3 **Chiltern** Austr.
39 C3 **Chilung** Taiwan
87 D3 **Chimala** Tanz.
120 B4 **Chimbas** Arg.
55 E4 **Chimbay** Uzbek.
118 A2 **Chimborazo** *mt.* Ecuador
118 A2 **Chimbote** Peru
Chimkent Kazakh. *see* Shymkent
89 C2 **Chimoio** Moz.
45 C3 **Chimtargha, Qullai** *mt.* Tajik.
36 B2 **China** *country* Asia
113 C2 **China** Mex.
118 A3 **Chincha Alta** Peru
96 C2 **Chinchaga** *r.* Can.
113 D3 **Chinchorro, Banco** *sea feature* Mex.
89 C2 **Chinde** Moz.
33 B3 **Chindo** S. Korea
33 B3 **Chin-do** *i.* S. Korea
36 C2 **Chindu** China
30 A1 **Chindwin** *r.* Myanmar
33 B2 **Chinghwa** N. Korea
88 B2 **Chingola** Zambia
88 A2 **Chinguar** Angola
33 B2 **Chinhae** S. Korea
89 C2 **Chinhoyi** Zimbabwe
42 B1 **Chiniot** Pak.
112 B2 **Chinipas** Mex.
33 B2 **Chinju** S. Korea
86 C2 **Chinko** *r.* C.A.R.
106 B1 **Chinle** U.S.A.
39 B3 **Chinmen** Taiwan
35 C3 **Chino** Japan
72 C2 **Chinon** France
106 A2 **Chino Valley** U.S.A.
89 C2 **Chinsali** Zambia
76 B1 **Chioggia** Italy
79 C3 **Chios** Greece
79 C3 **Chios** *i.* Greece
89 C2 **Chipata** Zambia
88 A2 **Chipindo** Angola
89 C3 **Chipinge** Zimbabwe
41 B3 **Chiplun** India
67 B3 **Chippenham** U.K.
67 C3 **Chipping Norton** U.K.
45 C2 **Chirchik** Uzbek.
89 C3 **Chiredzi** Zimbabwe
106 B2 **Chiricahua Peak** U.S.A.
114 B4 **Chiriquí, Golfo de** *b.* Panama
33 B2 **Chiri-san** *mt.* S. Korea
114 B4 **Chirripo** *mt.* Costa Rica
89 B2 **Chirundu** Zimbabwe
105 E1 **Chisholm** U.S.A.
58 B2 **Chişinău** Moldova
55 E3 **Chistopol'** Rus. Fed.
37 D1 **Chita** Rus. Fed.
88 A2 **Chitado** Angola
89 C2 **Chitambo** Zambia
88 B1 **Chitato** Angola
89 C1 **Chitipa** Malawi
41 B3 **Chitradurga** India
42 B1 **Chitral** Pak.
114 B4 **Chitré** Panama
43 D2 **Chittagong** Bangl.
42 B2 **Chittaurgarh** India
89 C2 **Chitungwiza** Zimbabwe
88 B2 **Chiume** Angola
89 C2 **Chivhu** Zimbabwe
121 B4 **Choele Choel** Arg.
112 B2 **Choix** Mex.
71 D1 **Chojnice** Pol.
85 B3 **Ch'ok'ē Mountains** Eth.
51 K2 **Chokurdakh** Rus. Fed.
89 C3 **Chókwé** Moz.
72 B2 **Cholet** France
70 C1 **Chomutov** Czech Rep.
51 I2 **Chona** *r.* Rus. Fed.
33 B2 **Ch'ŏnan** S. Korea
118 A2 **Chone** Ecuador
33 B1 **Ch'ŏngjin** N. Korea
33 B2 **Chŏngju** N. Korea
33 B2 **Chŏngp'yŏng** N. Korea
38 A3 **Chongqing** China
38 A2 **Chongqing** *mun.* China
39 A3 **Chongzuo** China
33 B2 **Chŏnju** S. Korea
122 B3 **Chopimzinho** Brazil
66 B2 **Chorley** U.K.
59 C2 **Chornomors'ke** Ukr.
58 B2 **Chortkiv** Ukr.
33 B2 **Ch'ŏrwŏn** S. Korea
33 B1 **Ch'osan** N. Korea
35 D3 **Chōshi** Japan
71 D1 **Choszczno** Pol.
82 A2 **Choûm** Maur.
37 D1 **Choybalsan** Mongolia
37 D1 **Choyr** Mongolia
22 B2 **Christchurch** N.Z.
67 C3 **Christchurch** U.K.
95 G2 **Christian, Cape** Can.
91 C2 **Christiana** S. Africa
22 A2 **Christina, Mount** N.Z.
16 B4 **Christmas Island** *terr.* Indian Ocean
79 C3 **Chrysi** *i.* Greece
121 B5 **Chubut** *r.* Arg.
58 B1 **Chudniv** Ukr.
57 D2 **Chudovo** Rus. Fed.
Chudskoye Ozero *l.* Estonia/Rus. Fed. *see* **Peipus, Lake**
94 B2 **Chugach Mountains** U.S.A.
35 B4 **Chūgoku-sanchi** *mts* Japan
34 B2 **Chuguyevka** Rus. Fed.
59 D2 **Chuhuyiv** Ukr.
128 J3 **Chukchi Sea** Rus. Fed./U.S.A.
51 N2 **Chukotskiy Poluostrov** *pen.* Rus. Fed.
103 C4 **Chula Vista** U.S.A.
50 G3 **Chulym** Rus. Fed.
120 B3 **Chumbicha** Arg.
51 K3 **Chumikan** Rus. Fed.
31 A2 **Chumphon** Thai.
33 B2 **Ch'unch'ŏn** S. Korea
Chungking China *see* Chongqing
51 H2 **Chunya** *r.* Rus. Fed.
118 A3 **Chuquibamba** Peru
120 B3 **Chuquicamata** Chile
73 D2 **Chur** Switz.
30 A1 **Churachandpur** India
97 E2 **Churchill** Can.
97 E2 **Churchill** *r. Man.* Can.
99 D1 **Churchill** *r. Nfld.* Can.
97 E2 **Churchill, Cape** Can.
99 D1 **Churchill Falls** Can.
97 D2 **Churchill Lake** Can.
42 B2 **Churu** India
99 C2 **Chute-des-Passes** Can.
30 B1 **Chuxiong** China
58 B2 **Ciadâr-Lunga** Moldova
29 B2 **Ciamis** Indon.
28 B2 **Cianjur** Indon.
122 B2 **Cianorte** Brazil
71 E1 **Ciechanów** Pol.
114 C2 **Ciego de Avila** Cuba
114 B2 **Cienfuegos** Cuba
75 C2 **Cieza** Spain

74 C2 **Cigüela** *r.* Spain
48 B2 **Cihanbeyli** Turkey
112 B3 **Cihuatlán** Mex.
74 C2 **Cijara, Embalse de** *resr* Spain
29 B2 **Cilacap** Indon.
107 C1 **Cimarron** *r.* U.S.A.
58 B2 **Cimişlia** Moldova
76 B2 **Cimone, Monte** *mt.* Italy
108 C3 **Cincinnati** U.S.A.
79 C3 **Çine** Turkey
68 B2 **Ciney** Belgium
113 C3 **Cintalapa** Mex.
94 B2 **Circle** *AK* U.S.A.
104 B1 **Circle** *MT* U.S.A.
26 B3 **Cirebon** Indon.
67 C3 **Cirencester** U.K.
76 A1 **Ciriè** Italy
77 C3 **Cirò Marina** Italy
77 C2 **Čitluk** Bos.-Herz.
90 A3 **Citrusdal** S. Africa
113 B2 **Ciudad Acuña** Mex.
113 B3 **Ciudad Altamirano** Mex.
118 B1 **Ciudad Bolívar** Venez.
112 B2 **Ciudad Camargo** Mex.
113 C3 **Ciudad del Carmen** Mex.
112 B2 **Ciudad Delicias** Mex.
113 C2 **Ciudad de Valles** Mex.
118 B1 **Ciudad Guayana** Venez.
106 B3 **Ciudad Guerrero** Mex.
112 B3 **Ciudad Guzmán** Mex.
113 C3 **Ciudad Hidalgo** Mex.
113 C3 **Ciudad Ixtepec** Mex.
112 B1 **Ciudad Juárez** Mex.
113 C2 **Ciudad Mante** Mex.
113 C2 **Ciudad Mier** Mex.
112 B2 **Ciudad Obregón** Mex.
74 C2 **Ciudad Real** Spain
113 C2 **Ciudad Río Bravo** Mex.
74 B1 **Ciudad Rodrigo** Spain
113 C2 **Ciudad Victoria** Mex.
75 D1 **Ciutadella de Menorca** Spain
79 C3 **Cıvan Dağ** *mt.* Turkey
76 B1 **Cividale del Friuli** Italy
76 B2 **Civitanova Marche** Italy
76 B2 **Civitavecchia** Italy
72 C2 **Civray** France
79 C3 **Çivril** Turkey
38 C2 **Cixi** China
67 D3 **Clacton-on-Sea** U.K.
96 C2 **Claire, Lake** Can.
73 C2 **Clamecy** France
90 A3 **Clanwilliam** S. Africa
20 A2 **Clare** Austr.
65 A2 **Clare Island** Rep. of Ireland
109 E2 **Claremont** U.S.A.
65 B2 **Claremorris** Rep. of Ireland
22 B2 **Clarence** N.Z.
99 E2 **Clarenville** Can.
96 C2 **Claresholm** Can.
105 D2 **Clarinda** U.S.A.
91 C3 **Clarkebury** S. Africa
102 C1 **Clark Fork** *r.* U.S.A.
111 D2 **Clark Hill Reservoir** U.S.A.
108 C3 **Clarksburg** U.S.A.
110 B2 **Clarksdale** U.S.A.
110 B1 **Clarksville** *AR* U.S.A.
110 C1 **Clarksville** *TN* U.S.A.
122 B1 **Claro** *r.* Brazil
107 C1 **Clayton** U.S.A.
65 B3 **Clear, Cape** Rep. of Ireland
105 E2 **Clear Lake** U.S.A.
103 B3 **Clear Lake** *l.* U.S.A.
96 C2 **Clearwater** Can.
97 C2 **Clearwater** *r.* Can.
111 D3 **Clearwater** U.S.A.
102 C1 **Clearwater** *r.* U.S.A.
107 D2 **Cleburne** U.S.A.
19 D2 **Clermont** Austr.
73 C2 **Clermont-Ferrand** France
20 A2 **Cleve** Austr.
110 B2 **Cleveland** *MS* U.S.A.
108 C2 **Cleveland** *OH* U.S.A.
111 D1 **Cleveland** *TN* U.S.A.
102 D1 **Cleveland, Mount** U.S.A.
111 D3 **Clewiston** U.S.A.
65 A2 **Clifden** Rep. of Ireland
21 D1 **Clifton** Austr.
106 B2 **Clifton** U.S.A.
96 B2 **Clinton** Can.
105 E2 **Clinton** *IA* U.S.A.
105 E3 **Clinton** *MO* U.S.A.
107 D1 **Clinton** *OK* U.S.A.
64 A2 **Clisham** *hill* U.K.
66 B2 **Clitheroe** U.K.
65 B3 **Clonakilty** Rep. of Ireland
19 D2 **Cloncurry** Austr.
65 C1 **Clones** Rep. of Ireland
65 C2 **Clonmel** Rep. of Ireland
68 D1 **Cloppenburg** Ger.
104 B2 **Cloud Peak** U.S.A.
107 C2 **Clovis** U.S.A.
97 D2 **Cluff Lake Mine** Can.
78 B1 **Cluj-Napoca** Romania
19 C2 **Cluny** Austr.
73 D2 **Cluses** France
22 A3 **Clutha** *r.* N.Z.
64 B3 **Clyde** *r.* U.K.
64 B3 **Clyde, Firth of** *est.* U.K.
64 B3 **Clydebank** U.K.
95 G2 **Clyde River** Can.
112 B3 **Coalcomán** Mex.
103 C3 **Coaldale** U.S.A.
96 B2 **Coal River** Can.
118 B2 **Coari** Brazil
118 B2 **Coari** *r.* Brazil
110 B2 **Coastal Plain** U.S.A.
96 B2 **Coast Mountains** Can.
103 B2 **Coast Ranges** *mts* U.S.A.
64 B3 **Coatbridge** U.K.
95 F2 **Coats Island** Can.
23 R2 **Coats Land** *reg.* Antarctica
113 C3 **Coatzacoalcos** Mex.
114 A3 **Cobán** Guat.
21 C2 **Cobar** Austr.
65 B3 **Cóbh** Rep. of Ireland
120 B2 **Cobija** Bol.
109 D2 **Cobourg** Can.
18 C1 **Cobourg Peninsula** Austr.
21 C3 **Cobram** Austr.
69 E2 **Coburg** Ger.
120 B2 **Cochabamba** Bol.
68 C2 **Cochem** Ger.
41 B4 **Cochin** India
96 C2 **Cochrane** *Alta* Can.
98 B2 **Cochrane** *Ont.* Can.
121 A5 **Cochrane** Chile
20 B2 **Cockburn** Austr.
Cockburn Town Turks and Caicos Is *see* **Grand Turk**
66 B1 **Cockermouth** U.K.
90 B3 **Cockscomb** *mt.* S. Africa
114 B3 **Coco** *r.* Hond./Nic.
16 A4 **Cocos Islands** *terr.* Indian Ocean
112 B2 **Cocula** Mex.
118 A1 **Cocuy, Sierra Nevada del** *mt.* Col.
109 E2 **Cod, Cape** U.S.A.
76 B2 **Codigoro** Italy
99 D1 **Cod Island** Can.
119 D2 **Codó** Brazil
104 B2 **Cody** U.S.A.
19 D1 **Coen** Austr.
68 C2 **Coesfeld** Ger.
102 C1 **Coeur d'Alene** U.S.A.
91 C3 **Coffee Bay** S. Africa
105 D3 **Coffeyville** U.S.A.
21 D2 **Coffs Harbour** Austr.
72 B2 **Cognac** France
86 A2 **Cogo** Equat. Guinea
20 B3 **Cohuna** Austr.
114 B4 **Coiba, Isla** *i.* Panama
121 A5 **Coihaique** Chile
41 B3 **Coimbatore** India
74 B1 **Coimbra** Port.
20 B3 **Colac** Austr.
123 D1 **Colatina** Brazil
104 C3 **Colby** U.S.A.
67 D3 **Colchester** U.K.
97 C2 **Cold Lake** *l.* Can.
64 C3 **Coldstream** U.K.
107 D2 **Coleman** U.S.A.
20 B3 **Coleraine** Austr.
65 C1 **Coleraine** U.K.
91 C3 **Colesberg** S. Africa
112 B3 **Colima** Mex.
112 B3 **Colima, Nevado de** *vol.* Mex.
64 A2 **Coll** *i.* U.K.
21 C1 **Collarenebri** Austr.
18 B1 **Collier Bay** Austr.
22 B2 **Collingwood** N.Z.
65 B1 **Collooney** Rep. of Ireland
73 D2 **Colmar** France
68 C2 **Cologne** Ger.
122 C2 **Colômbia** Brazil
118 A1 **Colombia** *country* S. America
41 B4 **Colombo** Sri Lanka
72 C3 **Colomiers** France
120 C4 **Colón** Arg.
114 C4 **Colón** Panama
77 C3 **Colonna, Capo** *c.* Italy
64 A2 **Colonsay** *i.* U.K.
121 B4 **Colorado** *r.* Arg.
106 A2 **Colorado** *r.* Mex./U.S.A.
107 D3 **Colorado** *r.* U.S.A.
104 B3 **Colorado** *state* U.S.A.
103 E3 **Colorado Plateau** U.S.A.
104 C3 **Colorado Springs** U.S.A.
112 B2 **Colotlán** Mex.
104 B1 **Colstrip** U.S.A.
105 E3 **Columbia** *MO* U.S.A.
111 D2 **Columbia** *SC* U.S.A.
110 C1 **Columbia** *TN* U.S.A.
102 B1 **Columbia** *r.* U.S.A.
96 C2 **Columbia, Mount** Can.
102 D1 **Columbia Falls** U.S.A.
96 B2 **Columbia Mountains** Can.
102 C1 **Columbia Plateau** U.S.A.
111 D2 **Columbus** *GA* U.S.A.
108 B3 **Columbus** *IN* U.S.A.
110 C2 **Columbus** *MS* U.S.A.
105 D2 **Columbus** *NE* U.S.A.
106 B2 **Columbus** *NM* U.S.A.
108 C3 **Columbus** *OH* U.S.A.
102 C1 **Colville** U.S.A.
94 A2 **Colville** *r.* U.S.A.
94 C2 **Colville Lake** Can.
66 B2 **Colwyn Bay** U.K.
76 B2 **Comacchio** Italy
113 C3 **Comalcalco** Mex.
78 C1 **Comănești** Romania
98 C1 **Comencho, Lac** *l.* Can.
65 C2 **Comeragh Mountains** Rep. of Ireland
43 D2 **Comilla** Bangl.
76 A2 **Comino, Capo** *c. Sardinia* Italy
113 C3 **Comitán de Domínguez** Mex.
72 C2 **Commentry** France
107 D2 **Commerce** U.S.A.
95 F2 **Committee Bay** Can.
76 A1 **Como** Italy
76 A1 **Como, Lago di** *l.* Italy
121 B5 **Comodoro Rivadavia** Arg.
89 D2 **Comoros** *country* Africa
72 C2 **Compiègne** France
112 B2 **Compostela** Mex.

58 B2 **Comrat** Moldova
82 A4 **Conakry** Guinea
123 E1 **Conceição da Barra** Brazil
119 D2 **Conceição do Araguaia** Brazil
123 D1 **Conceição do Mato Dentro** Brazil
120 B3 **Concepción** Arg.
121 A4 **Concepción** Chile
112 B2 **Concepción** Mex.
114 B4 **Concepción** Panama
103 B4 **Conception, Point** U.S.A.
122 C2 **Conchas** Brazil
106 C1 **Conchas Lake** U.S.A.
112 B2 **Conchos** *r. Chihuahua* Mex.
113 C2 **Conchos** *r. Nuevo León/ Tamaulipas* Mex.
103 B3 **Concord** *CA* U.S.A.
109 E2 **Concord** *NH* U.S.A.
120 C4 **Concordia** Arg.
90 A2 **Concordia** S. Africa
105 D3 **Concordia** U.S.A.
21 C2 **Condobolin** Austr.
72 C3 **Condom** France
102 B1 **Condon** U.S.A.
76 B1 **Conegliano** Italy
72 C2 **Confolens** France
86 B3 **Congo** *country* Africa
86 B3 **Congo** *r.* Congo/ Dem. Rep. Congo
86 C3 **Congo, Democratic Republic of** *country* Africa
97 C2 **Conklin** Can.
65 B1 **Conn, Lough** *l.* Rep. of Ireland
65 B2 **Connaught** *reg.* Rep. of Ireland
109 E2 **Connecticut** *r.* U.S.A.
109 E2 **Connecticut** *state* U.S.A.
65 B2 **Connemara** *reg.* Rep. of Ireland
102 D1 **Conrad** U.S.A.
107 D2 **Conroe** U.S.A.
123 D2 **Conselheiro Lafaiete** Brazil
123 D1 **Conselheiro Pena** Brazil
66 C1 **Consett** U.K.
31 B3 **Côn Sơn** *i.* Vietnam
78 C2 **Constanța** Romania
74 B2 **Constantina** Spain
83 C1 **Constantine** Alg.
102 D2 **Contact** U.S.A.
118 A2 **Contamana** Peru
121 A6 **Contreras, Isla** *i.* Chile
94 D2 **Contwoyto Lake** Can.
110 B1 **Conway** *AR* U.S.A.
109 E2 **Conway** *NH* U.S.A.
19 C2 **Coober Pedy** Austr.
22 B2 **Cook, Mount** *mt.* N.Z.
111 C1 **Cookeville** U.S.A.
17 G4 **Cook Islands** S. Pacific Ocean
99 E1 **Cook's Harbour** Can.
65 C1 **Cookstown** U.K.
22 B2 **Cook Strait** N.Z.
19 D1 **Cooktown** Austr.
21 C2 **Coolabah** Austr.
21 C2 **Coolamon** Austr.
21 D1 **Coolangatta** Austr.
18 B3 **Coolgardie** Austr.
21 C3 **Cooma** Austr.
20 B2 **Coombah** Austr.
21 C2 **Coonabarabran** Austr.
20 A3 **Coonalpyn** Austr.
21 C2 **Coonamble** Austr.
20 A1 **Cooper Creek** *watercourse* Austr.
102 B2 **Coos Bay** U.S.A.
21 C2 **Cootamundra** Austr.
113 C3 **Copainalá** Mex.
113 C3 **Copala** Mex.
61 F4 **Copenhagen** Denmark
77 C2 **Copertino** Italy
120 A3 **Copiapó** Chile
108 B1 **Copper Harbor** U.S.A.
Coppermine Can. *see* **Kugluktuk**
94 D2 **Coppermine** *r.* Can.
90 B2 **Copperton** S. Africa
120 A3 **Coquimbo** Chile
78 B2 **Corabia** Romania
123 D1 **Coração de Jesus** Brazil
118 A3 **Coracora** Peru
21 D1 **Coraki** Austr.
18 A2 **Coral Bay** Austr.
95 F2 **Coral Harbour** Can.
124 D7 **Coral Sea** S. Pacific Ocean
20 B3 **Corangamite, Lake** Austr.
67 C2 **Corby** U.K.
121 A5 **Corcovado, Golfo de** *sea chan.* Chile
111 D2 **Cordele** U.S.A.
32 B1 **Cordilleras Range** *mts* Phil.
120 B4 **Córdoba** Arg.
113 C3 **Córdoba** Mex.
74 C2 **Córdoba** Spain
121 B4 **Córdoba, Sierras de** *mts* Arg.
79 A3 **Corfu** *i.* Greece
74 B2 **Coria** Spain
110 C2 **Corinth** U.S.A.
123 D1 **Corinto** Brazil
65 B3 **Cork** Rep. of Ireland
79 C2 **Çorlu** Turkey
122 B2 **Cornélio Procópio** Brazil
99 E2 **Corner Brook** Can.
21 C3 **Corner Inlet** *b.* Austr.
103 B3 **Corning** *CA* U.S.A.
109 D2 **Corning** *NY* U.S.A.
Corn Islands *is* Nic. *see* **Maíz, Islas del**
76 B2 **Corno, Monte** *mt.* Italy
98 C2 **Cornwall** Can.
94 E1 **Cornwallis Island** Can.
118 B1 **Coro** Venez.
122 C1 **Coromandel** Brazil
22 C1 **Coromandel Peninsula** N.Z.
97 C2 **Coronation** Can.
94 D2 **Coronation Gulf** Can.
120 C3 **Coronel Oviedo** Para.
121 B4 **Coronel Suárez** Arg.
77 D2 **Çorovodë** Albania
107 D3 **Corpus Christi** U.S.A.
120 B2 **Corque** Bol.
119 D3 **Corrente** Brazil
122 B1 **Correntes** Brazil
119 D3 **Correntina** Brazil
65 B2 **Corrib, Lough** *l.* Rep. of Ireland
120 C3 **Corrientes** Arg.
112 B2 **Corrientes, Cabo** *c.* Mex.
21 C3 **Corryong** Austr.
Corse *i.* France *see* **Corsica**
73 D3 **Corse, Cap** *c. Corsica* France
73 D3 **Corsica** *i.* France
107 D2 **Corsicana** U.S.A.
73 D3 **Corte** *Corsica* France
74 B2 **Cortegana** Spain
104 B3 **Cortez** U.S.A.
76 B1 **Cortina d'Ampezzo** Italy
109 D2 **Cortland** U.S.A.
76 B2 **Cortona** Italy
74 B2 **Coruche** Port.
Çoruh Turkey *see* **Artvin**
48 B1 **Çorum** Turkey
120 C2 **Corumbá** Brazil
122 C1 **Corumbá** *r.* Brazil
102 B2 **Corvallis** U.S.A.
66 B2 **Corwen** U.K.
112 B2 **Cosalá** Mex.
113 C3 **Cosamaloapan** Mex.
77 C3 **Cosenza** Italy
73 C2 **Cosne-Cours-sur-Loire** France
75 C2 **Costa Blanca** *coastal area* Spain
75 D1 **Costa Brava** *coastal area* Spain
74 B2 **Costa del Sol** *coastal area* Spain
114 B3 **Costa de Mosquitos** *coastal area* Nic.
122 B1 **Costa Rica** Brazil
114 B3 **Costa Rica** *country* Central America
112 B2 **Costa Rica** Mex.
78 B2 **Costești** Romania
32 B2 **Cotabato** Phil.
82 B4 **Côte d'Ivoire** *country* Africa
118 A2 **Cotopaxi, Volcán** *vol.* Ecuador
67 B3 **Cotswold Hills** U.K.
102 B2 **Cottage Grove** U.S.A.
70 C1 **Cottbus** Ger.
20 A3 **Couëdic, Cape de** Austr.
105 D2 **Council Bluffs** U.S.A.
56 B2 **Courland Lagoon** *b.* Lith./Rus. Fed.
96 B3 **Courtenay** Can.
72 B2 **Coutances** France
72 B2 **Coutras** France
68 B2 **Couvin** Belgium
67 C2 **Coventry** U.K.
108 C3 **Covington** U.S.A.
18 B3 **Cowan, Lake** *salt flat* Austr.
64 C2 **Cowdenbeath** U.K.
21 C3 **Cowes** Austr.
102 B1 **Cowlitz** *r.* U.S.A.
21 C2 **Cowra** Austr.
122 B1 **Coxim** Brazil
43 D2 **Cox's Bazar** Bangl.
113 B3 **Coyuca de Benitez** Mex.
113 D2 **Cozumel** Mex.
113 D2 **Cozumel, Isla de** *i.* Mex.
91 C3 **Cradock** S. Africa
104 B2 **Craig** U.S.A.
70 C2 **Crailsheim** Ger.
78 B2 **Craiova** Romania
97 D2 **Cranberry Portage** Can.
21 C3 **Cranbourne** Austr.
96 C3 **Cranbrook** Can.
119 D2 **Crateús** Brazil
119 E2 **Crato** Brazil
104 C2 **Crawford** U.S.A.
108 B2 **Crawfordsville** U.S.A.
67 C3 **Crawley** U.K.
102 D1 **Crazy Mountains** U.S.A.
97 D2 **Cree** *r.* Can.
112 B2 **Creel** Mex.
97 D2 **Cree Lake** Can.
68 B1 **Creil** Neth.
73 D2 **Crema** Italy
76 B1 **Cremona** Italy
76 B2 **Cres** *i.* Croatia
102 B2 **Crescent City** U.S.A.
96 C3 **Creston** Can.
105 E2 **Creston** U.S.A.
110 C2 **Crestview** U.S.A.
79 B3 **Crete** *i.* Greece
75 D1 **Creus, Cap de** *c.* Spain
66 B2 **Crewe** U.K.
64 B2 **Crianlarich** U.K.
120 D3 **Criciúma** Brazil
64 C2 **Crieff** U.K.
76 B1 **Crikvenica** Croatia
59 C2 **Crimea** *pen.* Ukr.
69 F2 **Crimmitschau** Ger.
122 C1 **Cristalina** Brazil
69 E1 **Crivitz** Ger.
Crna Gora *aut. rep.* Yugo. *see* **Montenegro**

77 D2 **Crni Vrh** *mt.* Yugo.
77 C1 **Črnomelj** Slovenia
65 B2 **Croagh Patrick** *hill* Rep. of Ireland
77 C1 **Croatia** *country* Europe
29 C1 **Crocker, Banjaran** *mts* Malaysia
107 D2 **Crockett** U.S.A.
27 C3 **Croker Island** Austr.
67 D2 **Cromer** U.K.
22 A3 **Cromwell** N.Z.
105 D1 **Crookston** U.S.A.
21 C2 **Crookwell** Austr.
104 C1 **Crosby** U.S.A.
111 D3 **Cross City** U.S.A.
110 B2 **Crossett** U.S.A.
66 B1 **Cross Fell** *hill* U.K.
97 E2 **Cross Lake** *l.* Can.
111 C1 **Crossville** U.S.A.
77 C3 **Crotone** Italy
67 D3 **Crowborough** U.K.
110 B2 **Crowley** U.S.A.
96 C3 **Crowsnest Pass** Can.
114 C3 **Cruz, Cabo** *c.* Cuba
120 C3 **Cruz Alta** Brazil
120 B4 **Cruz del Eje** Arg.
123 D2 **Cruzeiro** Brazil
118 A2 **Cruzeiro do Sul** Brazil
20 A2 **Crystal Brook** Austr.
107 D3 **Crystal City** U.S.A.
108 B1 **Crystal Falls** U.S.A.
71 E2 **Csongrád** Hungary
88 B2 **Cuando** *r.* Angola/Zambia
88 A2 **Cuangar** Angola
86 B3 **Cuango** *r.* Angola/ Dem. Rep. Congo
88 A1 **Cuanza** *r.* Angola
107 C3 **Cuatro Ciénegas** Mex.
112 B2 **Cuauhtémoc** Mex.
113 C3 **Cuautla** Mex.
114 B2 **Cuba** *country* West Indies
88 A2 **Cubal** Angola
88 B2 **Cubango** *r.* Angola/Namibia
118 A1 **Cúcuta** Col.
41 B3 **Cuddalore** India
41 B3 **Cuddapah** India
74 C1 **Cuéllar** Spain
88 A2 **Cuemba** Angola
118 A2 **Cuenca** Ecuador
75 C1 **Cuenca** Spain
75 C1 **Cuenca, Serranía de** *mts* Spain
113 C3 **Cuernavaca** Mex.
107 D3 **Cuero** U.S.A.
119 C3 **Cuiabá** Brazil
119 C3 **Cuiabá** *r.* Brazil
64 A2 **Cuillin Sound** *sea chan.* U.K.
88 A1 **Cuilo** Angola
88 B2 **Cuito** *r.* Angola
88 A2 **Cuito Cuanavale** Angola
28 B1 **Cukai** Malaysia
21 C3 **Culcairn** Austr.
21 C1 **Culgoa** Austr.
112 B2 **Culiacán** Mex.
32 A1 **Culion** *i.* Phil.
75 C2 **Cullera** Spain
110 C2 **Cullman** U.S.A.
65 C1 **Cullybackey** U.K.
119 C3 **Culuene** *r.* Brazil
118 B1 **Cumaná** Venez.
109 D3 **Cumberland** U.S.A.
108 B3 **Cumberland** *r.* U.S.A.
97 D2 **Cumberland Lake** Can.
95 G2 **Cumberland Peninsula** Can.
110 C1 **Cumberland Plateau** U.S.A.
95 G2 **Cumberland Sound** *sea chan.* Can.
64 C3 **Cumbernauld** U.K.
103 B3 **Cummings** U.S.A.
64 B3 **Cumnock** U.K.
112 B1 **Cumpas** Mex.
113 C3 **Cunduacán** Mex.
88 A2 **Cunene** *r.* Angola
76 A2 **Cuneo** Italy
21 C1 **Cunnamulla** Austr.
76 A1 **Cuorgnè** Italy
64 C2 **Cupar** U.K.
115 D3 **Curaçao** *i.* Neth. Antilles
118 A2 **Curaray** *r.* Ecuador
121 A4 **Curicó** Chile
122 C3 **Curitiba** Brazil
20 A2 **Curnamona** Austr.
19 D3 **Currie** Austr.
19 E2 **Curtis Island** Austr.
119 C2 **Curuá** *r.* Brazil
28 B2 **Curup** Indon.
119 D2 **Cururupu** Brazil
123 D1 **Curvelo** Brazil
118 A3 **Cusco** Peru
107 D1 **Cushing** U.S.A.
102 D1 **Cut Bank** U.S.A.
43 C2 **Cuttack** India
69 D1 **Cuxhaven** Ger.
32 B1 **Cuyo Islands** Phil.
Cuzco Peru *see* **Cusco**
87 C3 **Cyangugu** Rwanda
79 B3 **Cyclades** *is* Greece
97 C3 **Cypress Hills** Can.
48 B2 **Cyprus** *country* Asia
70 C2 **Czech Republic** *country* Europe
71 D1 **Czersk** Pol.
71 D1 **Częstochowa** Pol.

D

30 B1 **Đa, Sông** *r.* Vietnam
37 D2 **Daban** China
82 A3 **Dabola** Guinea
Dacca Bangl. *see* **Dhaka**
70 C2 **Dachau** Ger.
38 A2 **Dachuan** China
42 A2 **Dadu** Pak.
32 B1 **Daet** Phil.
82 A3 **Dagana** Senegal
32 B1 **Dagupan** Phil.
42 B3 **Dahanu** India
37 D2 **Da Hinggan Ling** *mts* China
84 C3 **Dahlak Archipelago** *is* Eritrea
68 C2 **Dahlem** Ger.
28 B2 **Daik** Indon.
74 C2 **Daimiel** Spain
19 C2 **Dajarra** Austr.
82 A3 **Dakar** Senegal
85 C4 **Daketa Shet'** *watercourse* Eth.
84 A2 **Dakhla Oasis** Egypt
31 A3 **Dakoank** India
56 C3 **Dakol'ka** *r.* Belarus
77 D2 **Đakovica** Yugo.
77 C1 **Đakovo** Croatia
88 B2 **Dala** Angola
36 C2 **Dalain Hob** China
61 G3 **Dalälven** *r.* Sweden
79 C3 **Dalaman** Turkey
79 C3 **Dalaman** *r.* Turkey
37 C2 **Dalandzadgad** Mongolia
31 B2 **Đa Lat** Vietnam
42 A2 **Dalbandin** Pak.
64 C3 **Dalbeattie** U.K.
19 E2 **Dalby** Austr.
111 C1 **Dale Hollow Lake** U.S.A.
21 C3 **Dalgety** Austr.
107 C1 **Dalhart** U.S.A.
99 D2 **Dalhousie** Can.
30 B1 **Dali** China
38 C2 **Dalian** China
64 C3 **Dalkeith** U.K.
107 D2 **Dallas** U.S.A.
96 A2 **Dall Island** U.S.A.
77 C2 **Dalmacija** *reg.* Croatia
34 C2 **Dal'negorsk** Rus. Fed.
34 B1 **Dal'nerechensk** Rus. Fed.
82 B4 **Daloa** Côte d'Ivoire
19 D2 **Dalrymple, Mount** Austr.
60 □A3 **Dalsmynni** Iceland
43 C2 **Daltenganj** India
111 D2 **Dalton** U.S.A.
28 B1 **Daludalu** Indon.
60 □B2 **Dalvík** Iceland
18 C1 **Daly** *r.* Austr.
19 C1 **Daly Waters** Austr.
42 B2 **Damān** India
48 B2 **Damanhûr** Egypt
27 C3 **Damar** *i.* Indon.
48 B2 **Damascus** Syria
83 D3 **Damaturu** Nigeria
44 B3 **Damāvand, Qolleh-ye** *mt.* Iran
49 D2 **Damghan** Iran
69 D1 **Damme** Ger.
43 B2 **Damoh** India
82 B4 **Damongo** Ghana
27 C3 **Dampir, Selat** *sea chan.* Indon.
43 D1 **Damxung** China
82 B4 **Danané** Côte d'Ivoire
31 B2 **Đa Năng** Vietnam
109 E2 **Danbury** U.S.A.
33 A1 **Dandong** China
114 B3 **Dangriga** Belize
38 B2 **Dangshan** China
57 F2 **Danilov** Rus. Fed.
57 E2 **Danilovskaya Vozvyshennost'** *hills* Rus. Fed.
38 B2 **Danjiangkou** China
57 E3 **Dankov** Rus. Fed.
114 B3 **Danlí** Hond.
69 E1 **Dannenberg (Elbe)** Ger.
22 C2 **Dannevirke** N.Z.
30 B2 **Dan Sai** Thai.
Dantu China *see* **Zhenjiang**
78 A1 **Danube** *r.* Europe
108 B2 **Danville** *IL* U.S.A.
108 C3 **Danville** *KY* U.S.A.
109 D3 **Danville** *VA* U.S.A.
39 A4 **Danzhou** China
39 B3 **Daoxian** China
82 C3 **Dapaong** Togo
36 C2 **Da Qaidam Zhen** China
37 E1 **Daqing** China
48 B2 **Dar'ā** Syria
47 C2 **Dārāb** Iran
83 D1 **Daraj** Libya
49 D2 **Dārān** Iran
31 B2 **Đa Răng, Sông** *r.* Vietnam
43 C2 **Darbhanga** India
87 D3 **Dar es Salaam** Tanz.
42 B1 **Dargai** Pak.
22 B1 **Dargaville** N.Z.
21 C3 **Dargo** Austr.
37 D1 **Darhan** Mongolia
118 A1 **Darién, Golfo del** *g.* Col.
43 C2 **Dārjiling** India
36 C2 **Darlag** China
20 B2 **Darling** *r.* Austr.
21 C1 **Darling Downs** *hills* Austr.
18 A3 **Darling Range** *hills* Austr.
66 C1 **Darlington** U.K.
21 C2 **Darlington Point** Austr.
71 D1 **Darłowo** Pol.
69 D3 **Darmstadt** Ger.
83 E1 **Darnah** Libya
20 B2 **Darnick** Austr.
75 C1 **Daroca** Spain

67 D3 **Dartford** U.K.
67 A3 **Dartmoor** *hills* U.K.
99 D2 **Dartmouth** Can.
67 B3 **Dartmouth** U.K.
65 B1 **Darty Mountains** Rep. of Ireland
27 D3 **Daru** P.N.G.
18 C1 **Darwin** Austr.
44 B2 **Dashkhovuz** Turkm.
42 A2 **Dasht** *r.* Pak.
29 C1 **Datadian** Indon.
79 C3 **Datça** Turkey
38 B1 **Datong** China
32 B2 **Datu Piang** Phil.
42 B1 **Daud Khel** Pak.
56 B2 **Daugava** *r.* Latvia
56 C2 **Daugavpils** Latvia
68 C2 **Daun** Ger.
97 D2 **Dauphin** Can.
97 E2 **Dauphin Lake** Can.
41 B3 **Davangere** India
32 B2 **Davao** Phil.
32 B2 **Davao Gulf** Phil.
105 E2 **Davenport** U.S.A.
67 C2 **Daventry** U.K.
91 C2 **Daveyton** S. Africa
114 B4 **David** Panama
97 D2 **Davidson** Can.
94 E3 **Davidson Lake** Can.
103 B3 **Davis** U.S.A.
99 D1 **Davis Inlet** Can.
127 F3 **Davis Sea** Antarctica
128 P3 **Davis Strait** Can./Greenland
73 D2 **Davos** Switz.
47 C3 **Dawqah** Oman
94 B2 **Dawson** Can.
111 D2 **Dawson** U.S.A.
96 B2 **Dawson Creek** Can.
96 B2 **Dawsons Landing** Can.
36 C2 **Dawu** China
Dawukou China *see* Shizuishan
72 B3 **Dax** France
36 C2 **Daxue Shan** *mts* China
39 B3 **Dayong** China
48 C2 **Dayr az Zawr** Syria
108 C3 **Dayton** U.S.A.
111 D3 **Daytona Beach** U.S.A.
90 B3 **De Aar** S. Africa
48 B2 **Dead Sea** *salt l.* Asia
39 B3 **De'an** China
120 B4 **Deán Funes** Arg.
96 B2 **Dease Lake** Can.
94 D2 **Dease Strait** Can.
103 C3 **Death Valley** *depr.* U.S.A.
72 C2 **Deauville** France
29 C1 **Debak** *Sarawak* Malaysia
77 D2 **Debar** Macedonia
71 E2 **Debrecen** Hungary
85 B3 **Debre Markos** Eth.
85 B3 **Debre Tabor** Eth.
85 B4 **Debre Zeyit** Eth.
110 C2 **Decatur** *AL* U.S.A.
108 B3 **Decatur** *IL* U.S.A.
41 B3 **Deccan** *plat.* India
70 C1 **Děčín** Czech Rep.
105 E2 **Decorah** U.S.A.
56 C2 **Dedovichi** Rus. Fed.
89 C2 **Dedza** Malawi
66 B2 **Dee** *r. England/Wales* U.K.
64 C2 **Dee** *r. Scotland* U.K.
21 D1 **Deepwater** Austr.
99 E2 **Deer Lake** Can.
102 D1 **Deer Lodge** U.S.A.
108 C2 **Defiance** U.S.A.
36 C2 **Dêgê** China
85 C4 **Degeh Bur** Eth.
70 C2 **Deggendorf** Ger.
59 E2 **Degtevo** Rus. Fed.
43 B1 **Dehra Dun** India
43 C2 **Dehri** India
37 E2 **Dehui** China
68 A2 **Deinze** Belgium
78 B1 **Dej** Romania
108 B2 **De Kalb** U.S.A.
46 A3 **Dekemhare** Eritrea
86 C3 **Dekese** Dem. Rep. Congo
103 C3 **Delano** U.S.A.
103 D3 **Delano Peak** U.S.A.
42 A1 **Delārām** Afgh.
91 C2 **Delareyville** S. Africa
97 D2 **Delaronde Lake** Can.
108 C2 **Delaware** U.S.A.
109 D3 **Delaware** *r.* U.S.A.
109 D3 **Delaware** *state* U.S.A.
109 D3 **Delaware Bay** U.S.A.
21 C3 **Delegate** Austr.
73 D2 **Delémont** Switz.
68 B1 **Delft** Neth.
68 C1 **Delfzijl** Neth.
89 D2 **Delgado, Cabo** *c.* Moz.
43 B2 **Delhi** India
28 B2 **Deli** *i.* Indon.
94 C2 **Déline** Can.
69 F2 **Delitzsch** Ger.
75 D2 **Dellys** Alg.
69 D1 **Delmenhorst** Ger.
77 B1 **Delnice** Croatia
51 L1 **De-Longa, Ostrova** *is* Rus. Fed.
97 D3 **Deloraine** Can.
79 B3 **Delphi** Greece
107 C3 **Del Rio** U.S.A.
104 B3 **Delta** *CO* U.S.A.
103 D3 **Delta** *UT* U.S.A.
94 B2 **Delta Junction** U.S.A.
77 D3 **Delvinë** Albania
86 C3 **Demba** Dem. Rep. Congo
85 B4 **Dembī Dolo** Eth.
57 D2 **Demidov** Rus. Fed.
106 B2 **Deming** U.S.A.
79 C3 **Demirci** Turkey
78 C2 **Demirköy** Turkey
70 C1 **Demmin** Ger.
110 C2 **Demopolis** U.S.A.
28 B2 **Dempo, Gunung** *vol.* Indon.
57 D2 **Demyansk** Rus. Fed.
90 B3 **De Naawte** S. Africa
85 C3 **Denakil** *reg.* Eritrea/Eth.
68 B1 **Den Burg** Neth.
68 B2 **Dendermonde** Belgium
38 A1 **Dengkou** China
38 B2 **Dengzhou** China
Den Haag Neth. *see* The Hague
18 A2 **Denham** Austr.
68 B1 **Den Helder** Neth.
20 B3 **Deniliquin** Austr.
102 C2 **Denio** U.S.A.
105 D2 **Denison** *IA* U.S.A.
107 D2 **Denison** *TX* U.S.A.
79 C3 **Denizli** Turkey
21 D2 **Denman** Austr.
18 A3 **Denmark** Austr.
61 F4 **Denmark** *country* Europe
128 Q3 **Denmark Strait** Greenland/Iceland
29 C2 **Denpasar** Indon.
107 D2 **Denton** U.S.A.
18 A3 **D'Entrecasteaux, Point** Austr.
104 B3 **Denver** U.S.A.
43 C2 **Deogarh** *Orissa* India
42 B2 **Deogarh** *Rajasthan* India
43 C2 **Deoghar** India
51 K2 **Deputatskiy** Rus. Fed.
36 C3 **Dêqên** China
110 B2 **De Queen** U.S.A.
42 A2 **Dera Bugti** Pak.
42 B1 **Dera Ghazi Khan** Pak.
42 B1 **Dera Ismail Khan** Pak.
55 D4 **Derbent** Rus. Fed.
18 B1 **Derby** Austr.
67 C2 **Derby** U.K.
65 B2 **Derg, Lough** *l.* Rep. of Ireland
59 D1 **Derhachi** Ukr.
110 B2 **De Ridder** U.S.A.
59 D2 **Derkul** *r.* Rus. Fed./Ukr.
84 B3 **Derudeb** Sudan
90 B3 **De Rust** S. Africa
77 C2 **Derventa** Bos.-Herz.
66 C2 **Derwent** *r.* U.K.
66 B1 **Derwent Water** *l.* U.K.
45 C1 **Derzhavinsk** Kazakh.
120 B2 **Desaguadero** *r.* Bol.
97 D2 **Deschambault Lake** Can.
102 B1 **Deschutes** *r.* U.S.A.
85 B3 **Desē** Eth.
121 B5 **Deseado** Arg.
121 B5 **Deseado** *r.* Arg.
105 E2 **Des Moines** U.S.A.
105 E2 **Des Moines** *r.* U.S.A.
59 C1 **Desna** *r.* Rus. Fed./Ukr.
57 D3 **Desnogorsk** Rus. Fed.
69 F2 **Dessau** Ger.
94 B2 **Destruction Bay** Can.
96 C1 **Detah** Can.
69 D2 **Detmold** Ger.
108 C2 **Detroit** U.S.A.
105 D1 **Detroit Lakes** U.S.A.
68 B2 **Deurne** Neth.
78 B1 **Deva** Romania
68 C1 **Deventer** Neth.
64 C2 **Deveron** *r.* U.K.
71 D2 **Devét Skal** *hill* Czech Rep.
105 D1 **Devil's Lake** U.S.A.
96 A2 **Devil's Paw** *mt.* U.S.A.
67 C3 **Devizes** U.K.
42 B2 **Devli** India
78 C2 **Devnya** Bulg.
96 C2 **Devon** Can.
94 E1 **Devon Island** Can.
19 D4 **Devonport** Austr.
42 B2 **Dewas** India
105 F3 **Dexter** U.S.A.
38 A2 **Deyang** China
27 D3 **Deyong, Tanjung** *pt* Indon.
49 C2 **Dezfūl** Iran
38 B2 **Dezhou** China
47 C2 **Dhahran** Saudi Arabia
43 D2 **Dhaka** Bangl.
46 B3 **Dhamār** Yemen
43 C2 **Dhamtari** India
43 C2 **Dhanbad** India
43 C2 **Dhankuta** Nepal
30 A1 **Dharmanagar** India
43 C2 **Dharmjaygarh** India
41 B3 **Dhārwād** India
42 B2 **Dhasa** India
46 B3 **Dhubāb** Yemen
42 B2 **Dhule** India
112 A1 **Diablo, Picacho del** *mt.* Mex.
19 C2 **Diamantina** *watercourse* Austr.
123 D1 **Diamantina** Brazil
119 D3 **Diamantina, Chapada** *plat.* Brazil
119 C3 **Diamantino** Brazil
39 B3 **Dianbai** China
119 D3 **Dianópolis** Brazil
82 B4 **Dianra** Côte d'Ivoire
82 C3 **Diapaga** Burkina
47 C2 **Dibā al Ḥiṣn** U.A.E.
86 C3 **Dibaya** Dem. Rep. Congo
30 A1 **Dibrugarh** India
104 C1 **Dickinson** U.S.A.
110 C1 **Dickson** U.S.A.
Dicle *r.* Turkey *see* **Tigris**

73 D3 **Die** France
97 D2 **Diefenbaker, Lake** Can.
82 B3 **Diéma** Mali
30 B2 **Diên Châu** Vietnam
69 D1 **Diepholz** Ger.
72 C2 **Dieppe** France
83 D3 **Diffa** Niger
99 D2 **Digby** Can.
73 D3 **Digne-les-Bains** France
73 C2 **Digoin** France
32 B2 **Digos** Phil.
27 D3 **Digul** *r.* Indon.
Dihang *r.* China/India *see* Brahmaputra
73 D2 **Dijon** France
85 C3 **Dikhil** Djibouti
79 C3 **Dikili** Turkey
68 A2 **Diksmuide** Belgium
83 D3 **Dikwa** Nigeria
85 B4 **Dīla** Eth.
27 C3 **Dili** East Timor
69 D2 **Dillenburg** Ger.
102 D1 **Dillon** U.S.A.
86 C4 **Dilolo** Dem. Rep. Congo
30 A1 **Dimapur** India
Dimashq Syria *see* Damascus
20 B3 **Dimboola** Austr.
78 C2 **Dimitrovgrad** Bulg.
55 D3 **Dimitrovgrad** Rus. Fed.
32 B1 **Dinagat** *i.* Phil.
72 B2 **Dinan** France
68 B2 **Dinant** Belgium
79 D3 **Dinar** Turkey
49 D2 **Dīnār, Kūh-e** *mt.* Iran
41 B3 **Dindigul** India
91 D1 **Dindiza** Moz.
69 E2 **Dingelstädt** Ger.
65 A2 **Dingle** Rep. of Ireland
65 A2 **Dingle Bay** Rep. of Ireland
64 B2 **Dingwall** U.K.
38 A2 **Dingxi** China
43 C2 **Dinngyê** China
122 B3 **Dionísio Cerqueira** Brazil
82 A3 **Diourbel** Senegal
32 B2 **Dipolog** Phil.
42 B1 **Dir** Pak.
19 D1 **Direction, Cape** Austr.
85 C4 **Dirē Dawa** Eth.
88 B2 **Dirico** Angola
18 A2 **Dirk Hartog Island** Austr.
21 C1 **Dirranbandi** Austr.
46 B3 **Ḑirs** Saudi Arabia
18 B2 **Disappointment, Lake** *salt flat* Austr.
20 B3 **Discovery Bay** *b.* Austr.
111 E1 **Dismal Swamp** U.S.A.
67 D2 **Diss** U.K.
76 B3 **Dittaino** *r. Sicily* Italy
42 B2 **Diu** India
123 D2 **Divinópolis** Brazil
55 D4 **Divnoye** Rus. Fed.
82 B4 **Divo** Côte d'Ivoire
48 B2 **Divriği** Turkey
108 B2 **Dixon** U.S.A.
96 A2 **Dixon Entrance** *sea chan.* Can./U.S.A.
48 C2 **Diyarbakır** Turkey
42 A2 **Diz** Pak.
83 D2 **Djado** Niger
83 D2 **Djado, Plateau du** Niger
86 B3 **Djambala** Congo
83 C2 **Djanet** Alg.
83 C1 **Djelfa** Alg.
87 C2 **Djéma** C.A.R.
82 B3 **Djenné** Mali
82 B3 **Djibo** Burkina
85 C3 **Djibouti** *country* Africa
85 C3 **Djibouti** Djibouti
82 C4 **Djougou** Benin
60 □C3 **Djúpivogur** Iceland
59 E1 **Dmitriyevka** Rus. Fed.
57 E3 **Dmitriyev-L'govskiy** Rus. Fed.
57 E2 **Dmitrov** Rus. Fed.
Dnepr *r.* Rus. Fed. *see* Dnieper
57 D3 **Dnieper** *r.* Europe
Dnipro *r.* Ukr. *see* **Dnieper**
59 C2 **Dniprodzerzhyns'k** Ukr.
59 D2 **Dnipropetrovs'k** Ukr.
59 C2 **Dniprorudne** Ukr.
58 B2 **Dnister** *r.* Ukr.
56 C2 **Dno** Rus. Fed.
83 D4 **Doba** Chad
56 B2 **Dobele** Latvia
69 F2 **Döbeln** Ger.
27 C3 **Doberai, Jazirah** *pen.* Indon.
27 C3 **Dobo** Indon.
77 C2 **Doboj** Bos.-Herz.
78 C2 **Dobrich** Bulg.
57 F3 **Dobrinka** Rus. Fed.
57 E3 **Dobroye** Rus. Fed.
57 D3 **Dobrush** Belarus
123 E1 **Doce** *r.* Brazil
113 B2 **Doctor Arroyo** Mex.
112 B2 **Doctor Belisario Domínguez** Mex.
79 C3 **Dodecanese** *is* Greece
Dodekanisos *is* Greece *see* Dodecanese
104 C3 **Dodge City** U.S.A.
87 D3 **Dodoma** Tanz.
68 C2 **Doetinchem** Neth.
27 C3 **Dofa** Indon.
43 C1 **Dogai Coring** *salt l.* China
96 B2 **Dog Creek** Can.
35 B3 **Dōgo** *i.* Japan
83 C3 **Dogondoutchi** Niger
49 C2 **Doğubeyazıt** Turkey
47 C2 **Doha** Qatar
30 A2 **Doi Saket** Thai.
68 B1 **Dokkum** Neth.
56 C3 **Dokshytsy** Belarus
59 D2 **Dokuchayevs'k** Ukr.
27 D3 **Dolak, Pulau** *i.* Indon.
99 C2 **Dolbeau** Can.
72 B2 **Dol-de-Bretagne** France
73 D2 **Dole** France
67 B2 **Dolgellau** U.K.
57 E3 **Dolgorukovo** Rus. Fed.
57 E3 **Dolgoye** Rus. Fed.
76 B1 **Dolomiti** *mts* Italy
85 C4 **Dolo Odo** Eth.
94 D2 **Dolphin and Union Strait** Can.
58 A2 **Dolyna** Ukr.
70 C2 **Domažlice** Czech Rep.
61 E3 **Dombås** Norway
71 D2 **Dombóvár** Hungary
96 B2 **Dome Creek** Can.
115 D3 **Dominica** *country* West Indies
115 C3 **Dominican Republic** *country* West Indies
57 E2 **Domodedovo** Rus. Fed.
79 B3 **Domokos** Greece
29 C2 **Dompu** Indon.
57 E3 **Don** *r.* Rus. Fed.
64 C2 **Don** *r.* U.K.
65 D1 **Donaghadee** U.K.
20 B3 **Donald** Austr.
Donau *r.* Austria/Ger. *see* Danube
70 C2 **Donauwörth** Ger.
74 B2 **Don Benito** Spain
66 C2 **Doncaster** U.K.
88 A1 **Dondo** Angola
89 C2 **Dondo** Moz.
41 C4 **Dondra Head** *hd* Sri Lanka
65 B1 **Donegal** Rep. of Ireland
65 B1 **Donegal Bay** Rep. of Ireland
59 D2 **Donets'k** Ukr.
59 D2 **Donets'kyy Kryazh** *hills* Rus. Fed./Ukr.
18 A2 **Dongara** Austr.
30 B1 **Dongchuan** China
39 A4 **Dongfang** China
34 B1 **Dongfanghong** China
26 B3 **Donggala** Indon.
33 A2 **Donggang** China
30 B2 **Đông Ha** Vietnam
Dong Hai *sea* N. Pacific Ocean *see* **East China Sea**
30 B2 **Đông Hôi** Vietnam
86 B2 **Dongou** Congo
31 B2 **Dong Phraya Fai** *mts* Thai.
39 B3 **Dongshan** China
38 B2 **Dongsheng** China
38 C2 **Dongtai** China
39 B3 **Dongting Hu** *l.* China
Dong Ujimqin Qi China *see* Uliastai
38 B2 **Dongying** China
22 B1 **Donnellys Crossing** N.Z.
75 C1 **Donostia-San Sebastián** Spain
85 C4 **Dooxo Nugaaleed** *val.* Somalia
67 B3 **Dorchester** U.K.
90 A1 **Dordabis** Namibia
72 B2 **Dordogne** *r.* France
68 B2 **Dordrecht** Neth.
91 C3 **Dordrecht** S. Africa
97 D2 **Doré Lake** *l.* Can.
69 D1 **Dorfmark** Ger.
82 B3 **Dori** Burkina
90 A3 **Doring** *r.* S. Africa
64 B2 **Dornoch** U.K.
64 B2 **Dornoch Firth** *est.* U.K.
57 D3 **Dorogobuzh** Rus. Fed.
58 B2 **Dorohoi** Romania
36 C1 **Döröö Nuur** *salt l.* Mongolia
60 G3 **Dorotea** Sweden
18 A2 **Dorre Island** Austr.
21 D2 **Dorrigo** Austr.
86 B2 **Dorsale Camerounaise** *slope* Cameroon/Nigeria
68 C2 **Dortmund** Ger.
68 C2 **Dortmund-Ems-Kanal** *canal* Ger.
121 B5 **Dos Bahías, Cabo** *c.* Arg.
69 F1 **Dosse** *r.* Ger.
82 C3 **Dosso** Niger
111 C2 **Dothan** U.S.A.
69 D1 **Dötlingen** Ger.
73 C1 **Douai** France
86 A2 **Douala** Cameroon
72 B2 **Douarnenez** France
82 B3 **Douentza** Mali
66 A1 **Douglas** Isle of Man
90 B2 **Douglas** S. Africa
96 A2 **Douglas** *AK* U.S.A.
106 B2 **Douglas** *AZ* U.S.A.
111 D2 **Douglas** *GA* U.S.A.
104 B2 **Douglas** *WY* U.S.A.
72 C1 **Doullens** France
122 B1 **Dourada, Serra** *hills* Brazil
122 B2 **Dourados** Brazil
122 B2 **Dourados, Serra dos** *hills* Brazil
74 B1 **Douro** *r.* Port.
67 D3 **Dover** U.K.
109 D3 **Dover** U.S.A.
63 D3 **Dover, Strait of** France/U.K.
109 F1 **Dover-Foxcroft** U.S.A.
47 C2 **Dowlatābād** *Būshehr* Iran
47 C2 **Dowlatābād** *Kermān* Iran

65 D1 **Downpatrick** U.K.
45 C3 **Dowshī** Afgh.
35 B3 **Dōzen** *is* Japan
98 C2 **Dozois, Réservoir** *resr* Can.
122 B2 **Dracena** Brazil
68 C1 **Drachten** Neth.
78 B2 **Drăgănești-Olt** Romania
78 B2 **Drăgășani** Romania
56 C3 **Drahichyn** Belarus
91 C2 **Drakensberg** *mts* Lesotho/S. Africa
91 C2 **Drakensberg** *mts* S. Africa
126 B8 **Drake Passage** S. Atlantic Ocean
79 B2 **Drama** Greece
61 F4 **Drammen** Norway
77 C1 **Drava** *r.* Europe
96 C2 **Drayton Valley** Can.
69 D2 **Dreieich** Ger.
79 B3 **Drepano, Akra** *pt* Greece
70 C1 **Dresden** Ger.
72 C2 **Dreux** France
68 B1 **Driemond** Neth.
77 C2 **Drina** *r.* Bos.-Herz./Yugo.
77 C2 **Drniš** Croatia
78 B2 **Drobeta-Turnu Severin** Romania
69 D1 **Drochtersen** Ger.
65 C2 **Drogheda** Rep. of Ireland
58 A2 **Drohobych** Ukr.
65 C1 **Dromore** U.K.
42 B1 **Drosh** Pak.
21 C3 **Drouin** Austr.
96 C2 **Drumheller** Can.
108 C1 **Drummond Island** U.S.A.
99 C2 **Drummondville** Can.
56 B3 **Druskininkai** Lith.
59 D2 **Druzhkivka** Ukr.
56 D2 **Druzhnaya Gorka** Rus. Fed.
98 A2 **Dryden** Can.
18 B1 **Drysdale** *r.* Austr.
46 A2 **Ḍubā** Saudi Arabia
47 C2 **Dubai** U.A.E.
97 D1 **Dubawnt Lake** Can.
Dubayy U.A.E. *see* **Dubai**
46 A2 **Dubbagh, Jabal ad** *mt.* Saudi Arabia
21 C2 **Dubbo** Austr.
65 C2 **Dublin** Rep. of Ireland
111 D2 **Dublin** U.S.A.
58 B1 **Dubno** Ukr.
109 D2 **Du Bois** U.S.A.
82 A4 **Dubréka** Guinea
77 C2 **Dubrovnik** Croatia
58 B1 **Dubrovytsya** Ukr.
57 D3 **Dubrowna** Belarus
105 E2 **Dubuque** U.S.A.
97 D2 **Duck Bay** Can.
31 B2 **Đưc Trong** Vietnam
69 E2 **Duderstadt** Ger.
50 G2 **Dudinka** Rus. Fed.
67 B2 **Dudley** U.K.
74 B1 **Duero** *r.* Spain
64 C2 **Dufftown** U.K.
99 C1 **Dufreboy, Lac** *l.* Can.
77 C2 **Dugi Rat** Croatia
68 C2 **Duisburg** Ger.
91 D1 **Duiwelskloof** S. Africa
91 C3 **Dukathole** S. Africa
47 C2 **Dukhān** Qatar
57 D2 **Dukhovshchina** Rus. Fed.
Dukou China *see* Panzhihua
56 C2 **Dūkštas** Lith.
36 C2 **Dulan** China
120 B4 **Dulce** *r.* Arg.
68 C2 **Dülmen** Ger.
78 C2 **Dulovo** Bulg.
105 E1 **Duluth** U.S.A.
32 B2 **Dumaguete** Phil.
28 B1 **Dumai** Indon.
32 B1 **Dumaran** *i.* Phil.
110 B2 **Dumas** *AR* U.S.A.
107 C1 **Dumas** *TX* U.S.A.
64 B3 **Dumbarton** U.K.
71 D2 **Ďumbier** *mt.* Slovakia
64 C3 **Dumfries** U.K.
23 J3 **Dumont d'Urville Sea** Antarctica
84 B1 **Dumyâṭ** Egypt
Duna *r.* Hungary *see* **Danube**
Dunaj *r.* Slovakia *see* **Danube**
71 D2 **Dunakeszi** Hungary
Dunărea *r.* Romania *see* Danube
78 C1 **Dunării, Delta** Romania
71 D2 **Dunaújváros** Hungary
Dunav *r.* Yugo. *see* **Danube**
58 B2 **Dunayivtsi** Ukr.
64 C2 **Dunbar** U.K.
64 C1 **Dunbeath** U.K.
96 B3 **Duncan** Can.
107 D2 **Duncan** U.S.A.
64 C1 **Duncansby Head** *hd* U.K.
65 C1 **Dundalk** Rep. of Ireland
65 C2 **Dundalk Bay** Rep. of Ireland
Dundas Greenland *see* Uummannaq
91 D2 **Dundee** S. Africa
64 C2 **Dundee** U.K.
65 D1 **Dundrum Bay** U.K.
22 B3 **Dunedin** N.Z.
64 C2 **Dunfermline** U.K.
65 C1 **Dungannon** U.K.
42 B2 **Dungarpur** India
65 C2 **Dungarvan** Rep. of Ireland
67 D3 **Dungeness** *hd* U.K.
65 C1 **Dungiven** U.K.
21 D2 **Dungog** Austr.
87 C2 **Dungu** Dem. Rep. Congo
28 B1 **Dungun** Malaysia
84 B2 **Dungunab** Sudan
37 E2 **Dunhua** China
36 C2 **Dunhuang** China
Dunkerque France *see* Dunkirk
72 C1 **Dunkirk** France
109 D2 **Dunkirk** U.S.A.
65 C2 **Dún Laoghaire** Rep. of Ireland
65 D1 **Dunmurry** U.K.
64 C1 **Dunnet Head** *hd* U.K.
64 C3 **Duns** U.K.
67 C3 **Dunstable** U.K.
78 B2 **Dupnitsa** Bulg.
104 C1 **Dupree** U.S.A.
18 B1 **Durack** *r.* Austr.
112 B2 **Durango** Mex.
74 C1 **Durango** Spain
104 B3 **Durango** U.S.A.
107 D2 **Durant** U.S.A.
121 C4 **Durazno** Uru.
91 D2 **Durban** S. Africa
72 C3 **Durban-Corbières** France
90 A3 **Durbanville** S. Africa
68 B2 **Durbuy** Belgium
68 C2 **Düren** Ger.
66 C1 **Durham** U.K.
111 E1 **Durham** U.S.A.
28 B1 **Duri** Indon.
77 C2 **Durmitor** *mt.* Yugo.
64 B1 **Durness** U.K.
77 C2 **Durrës** Albania
65 A3 **Dursey Island** Rep. of Ireland
79 C3 **Dursunbey** Turkey
27 D3 **D'Urville, Tanjung** *pt* Indon.
22 B2 **D'Urville Island** N.Z.
39 A3 **Dushan** China
45 C3 **Dushanbe** Tajik.
68 C2 **Düsseldorf** Ger.
39 A3 **Duyun** China
59 D2 **Dvorichna** Ukr.
42 A2 **Dwarka** India
91 C1 **Dwarsberg** S. Africa
102 C1 **Dworshak Reservoir** U.S.A.
57 D3 **Dyat'kovo** Rus. Fed.
64 C2 **Dyce** U.K.
95 G2 **Dyer, Cape** Can.
110 C1 **Dyersburg** U.S.A.
71 D2 **Dyje** *r.* Austria/Czech Rep.
71 D1 **Dylewska Góra** *hill* Pol.
59 D2 **Dymytrov** Ukr.
91 C3 **Dyoki** S. Africa
55 E3 **Dyurtyuli** Rus. Fed.
37 D2 **Dzamïn Üüd** Mongolia
Dzhambul Kazakh. *see* **Taraz**
49 D1 **Dzhanga** Turkm.
44 B2 **Dzhangala** Kazakh.
59 C2 **Dzhankoy** Ukr.
55 D4 **Dzhanybek** Kazakh.
45 C2 **Dzhizak** Uzbek.
59 D3 **Dzhubga** Rus. Fed.
51 K3 **Dzhugdzhur, Khrebet** *mts* Rus. Fed.
45 D2 **Dzhungarskiy Alatau, Khrebet** *mts* China/Kazakh.
44 C2 **Dzhusaly** Kazakh.
71 E1 **Działdowo** Pol.
37 D1 **Dzuunmod** Mongolia
56 C3 **Dzyarzhynsk** Belarus
56 C3 **Dzyatlavichy** Belarus

E

99 E1 **Eagle** *r.* Can.
102 C1 **Eagle Cap** *mt.* U.S.A.
98 A2 **Eagle Lake** *l.* Can.
102 B2 **Eagle Lake** *l.* U.S.A.
107 C3 **Eagle Pass** U.S.A.
94 B2 **Eagle Plain** Can.
98 A1 **Ear Falls** Can.
67 D3 **Eastbourne** U.K.
124 C3 **East China Sea** N. Pacific Ocean
22 B1 **East Coast Bays** N.Z.
67 D2 **East Dereham** U.K.
97 D3 **Eastend** Can.
116 B4 **Easter Island** *i.* S. Pacific Ocean
91 C3 **Eastern Cape** *prov.* S. Africa
84 B2 **Eastern Desert** Egypt
41 B3 **Eastern Ghats** *mts* India
97 E2 **Easterville** Can.
121 C6 **East Falkland** *i.* Falkland Is
68 C1 **East Frisian Islands** *is* Ger.
64 B3 **East Kilbride** U.K.
67 C3 **Eastleigh** U.K.
108 C2 **East Liverpool** U.S.A.
91 C3 **East London** S. Africa
98 C1 **Eastmain** Can.
98 C1 **Eastmain** *r.* Can.
111 D2 **Eastman** U.S.A.
108 A3 **East St Louis** U.S.A.
East Siberian Sea Rus. Fed. *see* **Vostochno-Sibirskoye More**
27 C3 **East Timor** *terr.* Asia
108 A2 **Eau Claire** U.S.A.
27 D2 **Eauripik** *atoll* Micronesia
113 C2 **Ebano** Mex.
67 B3 **Ebbw Vale** U.K.
70 C1 **Eberswalde-Finow** Ger.
77 C2 **Eboli** Italy
86 B2 **Ebolowa** Cameroon
75 D1 **Ebro** *r.* Spain
75 C1 **Echarri-Aranaz** Spain
82 C1 **Ech Chélif** Alg.

112 A2 **Echeverria, Pico** *mt.* Mex.
97 E2 **Echoing** *r.* Can.
68 C3 **Echternach** Lux.
20 B3 **Echuca** Austr.
74 B2 **Écija** Spain
70 B1 **Eckernförde** Ger.
95 F2 **Eclipse Sound** *sea chan.* Can.
118 A2 **Ecuador** *country* S. America
84 C3 **Ed** Eritrea
85 A3 **Ed Da'ein** Sudan
85 B3 **Ed Damazin** Sudan
84 B3 **Ed Damer** Sudan
84 B3 **Ed Debba** Sudan
84 B3 **Ed Dueim** Sudan
19 D4 **Eddystone Point** Austr.
68 B1 **Ede** Neth.
86 B2 **Edéa** Cameroon
122 C1 **Edéia** Brazil
21 C3 **Eden** Austr.
66 B1 **Eden** *r.* U.K.
91 C2 **Edenburg** S. Africa
65 C2 **Edenderry** Rep. of Ireland
20 B3 **Edenhope** Austr.
79 B2 **Edessa** Greece
50 C1 **Edgeøya** *i.* Svalbard
107 D3 **Edinburg** U.S.A.
64 C3 **Edinburgh** U.K.
78 C2 **Edirne** Turkey
96 C2 **Edmonton** Can.
99 D2 **Edmundston** Can.
79 C3 **Edremit** Turkey
79 C3 **Edremit Körfezi** *b.* Turkey
96 C2 **Edson** Can.
87 C3 **Edward, Lake** Dem. Rep. Congo/Uganda
107 C2 **Edwards Plateau** *TX* U.S.A.
68 C1 **Eenrum** Neth.
108 B3 **Effingham** U.S.A.
103 C3 **Egan Range** *mts* U.S.A.
71 E2 **Eger** Hungary
61 E4 **Egersund** Norway
60 □C2 **Egilsstaðir** Iceland
48 B2 **Eğirdir** Turkey
48 B2 **Eğirdir Gölü** *l.* Turkey
72 C2 **Égletons** France
22 B1 **Egmont, Mt** *vol.* N.Z.
51 M2 **Egvekinot** Rus. Fed.
84 B2 **Egypt** *country* Africa
36 C2 **Ehen Hudag** China
68 C1 **Eibergen** Neth.
68 C2 **Eifel** *hills* Ger.
64 A2 **Eigg** *i.* U.K.
41 B4 **Eight Degree Channel** India/Maldives
18 B1 **Eighty Mile Beach** Austr.
69 F2 **Eilenburg** Ger.
69 D2 **Einbeck** Ger.
68 B2 **Eindhoven** Neth.
118 B2 **Eirunepé** Brazil
88 B2 **Eiseb** *watercourse* Namibia
69 E2 **Eisenach** Ger.
70 C1 **Eisenhüttenstadt** Ger.
71 D2 **Eisenstadt** Austria
69 E2 **Eisleben Lutherstadt** Ger.
75 D2 **Eivissa** Spain
Eivissa *i.* Spain *see* **Ibiza**
75 C1 **Ejea de los Caballeros** Spain
89 □D3 **Ejeda** Madag.
Ejin Qi China *see* **Dalain Hob**
61 H4 **Ekenäs** Fin.
45 D1 **Ekibastuz** Kazakh.
60 J2 **Ekostrovskaya Imandra, Ozero** *l.* Rus. Fed.
61 F4 **Eksjö** Sweden
90 A2 **Eksteenfontein** S. Africa
98 B1 **Ekwan** *r.* Can.
30 A2 **Ela** Myanmar
48 A2 **El 'Amirîya** Egypt
91 C2 **Elandsdoorn** S. Africa
48 B2 **El 'Arîsh** Egypt
79 B3 **Elassona** Greece
48 B3 **Elat** Israel
48 B2 **Elazığ** Turkey
76 B2 **Elba, Isola d'** *i.* Italy
118 A1 **El Banco** Col.
77 D2 **Elbasan** Albania
118 B1 **El Baúl** Venez.
82 C1 **El Bayadh** Alg.
69 D1 **Elbe** *r.* Ger.
104 B3 **Elbert, Mount** U.S.A.
111 D2 **Elberton** U.S.A.
72 C2 **Elbeuf** France
48 B2 **Elbistan** Turkey
71 D1 **Elbląg** Pol.
55 D4 **Elbrus** *mt.* Rus. Fed.
Elburz Mountains Iran *see* Alborz, Reshteh-ye
118 B1 **El Callao** Venez.
107 D3 **El Campo** U.S.A.
103 C4 **El Centro** U.S.A.
120 B2 **El Cerro** Bol.
75 C2 **Elche** Spain
75 C2 **Elda** Spain
105 E3 **Eldon** U.S.A.
112 B2 **El Dorado** Mex.
110 B2 **El Dorado** *AR* U.S.A.
105 D3 **El Dorado** *KS* U.S.A.
82 B2 **El Eglab** *plat.* Alg.
74 C2 **El Ejido** Spain
57 E2 **Elektrostal'** Rus. Fed.
118 A2 **El Encanto** Col.
114 C2 **Eleuthera** *i.* Bahamas
85 A3 **El Fasher** Sudan
112 B2 **El Fuerte** Mex.
85 A3 **El Geneina** Sudan
84 B3 **El Geteina** Sudan
64 C2 **Elgin** U.K.
108 B2 **Elgin** U.S.A.
84 B2 **El Gîza** Egypt
83 C1 **El Goléa** Alg.
112 A1 **El Golfo de Santa Clara** Mex.
87 D2 **Elgon, Mount** Uganda
48 A2 **El Hammâm** Egypt
82 A2 **El Ḥammâmi** *reg.* Maur.
82 A2 **El Hierro** *i.* Canary Is
113 C2 **El Higo** Mex.
82 C2 **El Homr** Alg.
El Iskandarîya Egypt *see* Alexandria
55 D4 **Elista** Rus. Fed.
109 E2 **Elizabeth** U.S.A.
111 E1 **Elizabeth City** U.S.A.
108 B3 **Elizabethtown** U.S.A.
82 B1 **El Jadida** Morocco
71 E1 **Ełk** Pol.
107 D1 **Elk City** U.S.A.
96 C2 **Elkford** Can.
84 B2 **El Khârga** Egypt
108 B2 **Elkhart** U.S.A.
El Khartum Sudan *see* Khartoum
78 C2 **Elkhovo** Bulg.
108 D3 **Elkins** U.S.A.
96 C3 **Elko** Can.
102 C2 **Elko** U.S.A.
97 C2 **Elk Point** Can.
94 E1 **Ellef Ringnes Island** Can.
105 D1 **Ellendale** U.S.A.
102 B1 **Ellensburg** U.S.A.
22 B2 **Ellesmere, Lake** N.Z.
95 F1 **Ellesmere Island** Can.
66 B2 **Ellesmere Port** U.K.
23 O2 **Ellesworth Mountains** Antarctica
94 E2 **Ellice** *r.* Can.
Ellice Islands *country* S. Pacific Ocean *see* **Tuvalu**
91 C3 **Elliotdale** S. Africa
91 C1 **Ellisras** S. Africa
64 C2 **Ellon** U.K.
109 F2 **Ellsworth** U.S.A.
79 C3 **Elmalı** Turkey
48 B2 **El Mansûra** Egypt
83 C1 **El Meghaïer** Alg.
84 B2 **El Minya** Egypt
109 D2 **Elmira** U.S.A.
75 C2 **El Moral** Spain
69 D1 **Elmshorn** Ger.
85 A3 **El Muglad** Sudan
32 A1 **El Nido** Phil.
85 B3 **El Obeid** Sudan
112 B2 **El Oro** Mex.
83 C1 **El Oued** Alg.
106 A2 **Eloy** U.S.A.
106 B2 **El Paso** U.S.A.
112 B1 **El Porvenir** Mex.
75 D1 **El Prat de Llobregat** Spain
El Qâhira Egypt *see* **Cairo**
107 D1 **El Reno** U.S.A.
113 B2 **El Salado** Mex.
112 B2 **El Salto** Mex.
114 B3 **El Salvador** *country* Central America
113 B2 **El Salvador** Mex.
106 B3 **El Sauz** Mex.
112 A1 **El Socorro** Mex.
El Suweis Egypt *see* **Suez**
113 C2 **El Temascal** Mex.
118 B1 **El Tigre** Venez.
115 D4 **El Tocuyo** Venez.
84 B2 **El Tûr** Egypt
El Uqṣur Egypt *see* **Luxor**
56 C2 **Elva** Estonia
74 B2 **Elvas** Port.
61 F3 **Elverum** Norway
87 E2 **El Wak** Kenya
67 D2 **Ely** U.K.
105 E1 **Ely** *MN* U.S.A.
103 D3 **Ely** *NV* U.S.A.
49 D2 **Emāmrūd** Iran
61 G4 **Emån** *r.* Sweden
44 B2 **Emba** Kazakh.
44 B2 **Emba** *r.* Kazakh.
91 C2 **Embalenhle** S. Africa
122 C1 **Emborcação, Represa de** *resr* Brazil
87 D3 **Embu** Kenya
68 C1 **Emden** Ger.
19 D2 **Emerald** Austr.
97 E3 **Emerson** Can.
79 C3 **Emet** Turkey
91 D2 **eMgwenya** S. Africa
91 D2 **eMijindini** S. Africa
83 D3 **Emi Koussi** *mt.* Chad
78 C2 **Emine, Nos** *pt* Bulg.
48 B2 **Emirdağ** Turkey
56 B2 **Emmaste** Estonia
68 B1 **Emmeloord** Neth.
68 C2 **Emmelshausen** Ger.
68 C1 **Emmen** Neth.
107 C3 **Emory Peak** U.S.A.
112 A2 **Empalme** Mex.
91 D2 **Empangeni** S. Africa
76 B2 **Empoli** Italy
105 D3 **Emporia** *KS* U.S.A.
109 D3 **Emporia** *VA* U.S.A.
68 C1 **Ems** *r.* Ger.
68 C1 **Emsdetten** Ger.
91 C2 **Emzinoni** S. Africa
27 D3 **Enarotali** Indon.
112 B2 **Encarnación** Mex.
120 C3 **Encarnación** Para.
123 D1 **Encruzilhada** Brazil
28 B1 **Endau** Malaysia
26 C3 **Endeh** Indon.
94 A2 **Endicott Mountains** U.S.A.
59 C2 **Enerhodar** Ukr.

55 D3 **Engel's** Rus. Fed.
28 B2 **Enggano** *i.* Indon.
66 C2 **England** *admin. div.* U.K.
98 A1 **English** *r.* Can.
63 C4 **English Channel** France/U.K.
107 D1 **Enid** U.S.A.
68 B1 **Enkhuizen** Neth.
61 G4 **Enköping** Sweden
76 B3 **Enna** *Sicily* Italy
97 D1 **Ennadai Lake** Can.
85 A3 **En Nahud** Sudan
83 E3 **Ennedi, Massif** *mts* Chad
21 C1 **Enngonia** Austr.
65 B2 **Ennis** Rep. of Ireland
107 D2 **Ennis** U.S.A.
65 C2 **Enniscorthy** Rep. of Ireland
65 C1 **Enniskillen** U.K.
65 B2 **Ennistymon** Rep. of Ireland
70 C2 **Enns** *r.* Austria
60 H2 **Enontekiö** Fin.
21 C3 **Ensay** Austr.
68 C1 **Enschede** Neth.
112 A1 **Ensenada** Mex.
38 A2 **Enshi** China
96 C1 **Enterprise** Can.
110 C2 **Enterprise** *AL* U.S.A.
102 C1 **Enterprise** *OR* U.S.A.
120 B3 **Entre Ríos** Bol.
74 B2 **Entroncamento** Port.
83 C4 **Enugu** Nigeria
118 A2 **Envira** Brazil
103 D3 **Ephraim** U.S.A.
102 C1 **Ephrata** U.S.A.
73 D2 **Épinal** France
67 C3 **Epsom** U.K.
86 A2 **Equatorial Guinea** *country* Africa
69 F3 **Erbendorf** Ger.
68 C3 **Erbeskopf** *hill* Ger.
49 C2 **Erciş** Turkey
33 B1 **Erdao Jiang** *r.* China
79 C2 **Erdek** Turkey
48 B2 **Erdemli** Turkey
120 C3 **Erechim** Brazil
37 D1 **Ereentsav** Mongolia
48 B2 **Ereğli** *Konya* Turkey
48 B1 **Ereğli** *Zonguldak* Turkey
37 D2 **Erenhot** China
Erevan Armenia *see* **Yerevan**
69 E2 **Erfurt** Ger.
48 B2 **Ergani** Turkey
82 B2 **'Erg Chech** *des.* Alg./Mali
83 D3 **Erg du Ténéré** *des.* Niger
79 C2 **Ergene** *r.* Turkey
108 C2 **Erie** U.S.A.
108 C2 **Erie, Lake** Can./U.S.A.
34 D2 **Erimo-misaki** *c.* Japan
84 B3 **Eritrea** *country* Africa
69 E3 **Erlangen** Ger.
18 C2 **Erldunda** Austr.
91 C2 **Ermelo** S. Africa
48 B2 **Ermenek** Turkey
79 B3 **Ermoupoli** Greece
41 B4 **Ernakulam** India
41 B3 **Erode** India
68 B2 **Erp** Neth.
82 B1 **Er Rachidia** Morocco
85 B3 **Er Rahad** Sudan
65 B1 **Errigal** *hill* Rep. of Ireland
65 A1 **Erris Head** *hd* Rep. of Ireland
77 D2 **Ersekë** Albania
59 E1 **Ertil'** Rus. Fed.
69 D2 **Erwitte** Ger.
69 F2 **Erzgebirge** *mts* Czech Rep./Ger.
48 B2 **Erzincan** Turkey
49 C2 **Erzurum** Turkey
61 E4 **Esbjerg** Denmark
103 D3 **Escalante** U.S.A.
112 B2 **Escalón** Mex.
108 B1 **Escanaba** U.S.A.
113 C3 **Escárcega** Mex.
75 C1 **Escatrón** Spain
68 A2 **Escaut** *r.* Belgium
69 E1 **Eschede** Ger.
68 B3 **Esch-sur-Alzette** Lux.
69 E2 **Eschwege** Ger.
68 C2 **Eschweiler** Ger.
103 C4 **Escondido** U.S.A.
112 B2 **Escuinapa** Mex.
79 C3 **Eşen** Turkey
49 D2 **Eşfahān** Iran
91 D2 **Esikhawini** S. Africa
66 B1 **Esk** *r.* U.K.
99 D1 **Esker** Can.
60 □C2 **Eskifjörður** Iceland
61 G4 **Eskilstuna** Sweden
Eskimo Point Can. *see* Arviat
48 B2 **Eskişehir** Turkey
49 C2 **Eslāmābād-e Gharb** Iran
79 C3 **Esler Dağı** *mt.* Turkey
79 C3 **Eşme** Turkey
118 A1 **Esmeraldas** Ecuador
47 D2 **Espakeh** Iran
72 C3 **Espalion** France
98 B2 **Espanola** Can.
106 B1 **Espanola** U.S.A.
18 B3 **Esperance** Austr.
112 B2 **Esperanza** Mex.
74 B2 **Espichel, Cabo** *c.* Port.
123 D1 **Espinhaço, Serra do** *mts* Brazil
119 D3 **Espinosa** Brazil
112 A2 **Espíritu Santo, Isla** *i.* Mex.
61 H3 **Espoo** Fin.
121 A5 **Esquel** Arg.
82 B1 **Essaouira** Morocco
82 A2 **Es Semara** Western Sahara
68 C2 **Essen** Ger.
118 C1 **Essequibo** *r.* Guyana
51 L3 **Esso** Rus. Fed.
47 C2 **Eşţahbān** Iran
119 E3 **Estância** Brazil
91 C2 **Estcourt** S. Africa
75 C1 **Estella** Spain
74 B2 **Estepona** Spain
74 C1 **Esteras de Medinaceli** Spain
97 D2 **Esterhazy** Can.
120 B3 **Esteros** Para.
104 B2 **Estes Park** U.S.A.
97 D3 **Estevan** Can.
105 E2 **Estherville** U.S.A.
97 D2 **Eston** Can.
56 C2 **Estonia** *country* Europe
74 B1 **Estrela, Serra da** *mts* Port.
74 B2 **Estremoz** Port.
20 A1 **Etadunna** Austr.
72 C2 **Étampes** France
72 C1 **Étaples** France
43 B2 **Etawah** India
91 D2 **eThandakukhanya** S. Africa
85 B4 **Ethiopia** *country* Africa
77 C3 **Etna, Monte** *vol. Sicily* Italy
96 A2 **Etolin Island** *i.* U.S.A.
88 A2 **Etosha Pan** *salt pan* Namibia
68 C3 **Ettelbruck** Lux.
68 B2 **Etten-Leur** Neth.
67 D3 **Eu** France
21 C2 **Euabalong** Austr.
Euboea *i.* Greece *see* **Evvoia**
18 B3 **Eucla** Austr.
111 C2 **Eufaula** U.S.A.
107 D1 **Eufaula Lake** *resr* U.S.A.
102 B2 **Eugene** U.S.A.
112 A2 **Eugenia, Punta** *pt* Mex.
21 C1 **Eulo** Austr.
21 C2 **Eumungerie** Austr.
107 C2 **Eunice** U.S.A.
48 C2 **Euphrates** *r.* Asia
102 B2 **Eureka** *CA* U.S.A.
102 C1 **Eureka** *MT* U.S.A.
103 C3 **Eureka** *NV* U.S.A.
20 B2 **Euriowie** Austr.
21 C3 **Euroa** Austr.
74 B2 **Europa Point** Gibraltar
96 B2 **Eutsuk Lake** Can.
98 C1 **Evans, Lac** *l.* Can.
21 D1 **Evans Head** Austr.
95 F2 **Evans Strait** Can.
104 A2 **Evanston** U.S.A.
108 B3 **Evansville** U.S.A.
91 C2 **Evaton** S. Africa
47 C2 **Evaz** Iran
51 L2 **Evensk** Rus. Fed.
18 C2 **Everard Range** *hills* Austr.
43 C2 **Everest, Mount** China/Nepal
102 B1 **Everett** U.S.A.
68 A2 **Evergem** Belgium
111 D3 **Everglades** *swamp* U.S.A.
86 B2 **Evinayong** Equat. Guinea
61 E4 **Evje** Norway
74 B2 **Évora** Port.
72 C2 **Évreux** France
79 C2 **Evros** *r.* Greece/Turkey
79 B3 **Evrotas** *r.* Greece
48 B2 **Evrychou** Cyprus
79 B3 **Evvoia** *i.* Greece
87 E2 **Ewaso Ngiro** *r.* Kenya
120 B2 **Exaltación** Bol.
67 B3 **Exe** *r.* U.K.
67 B3 **Exeter** U.K.
67 B3 **Exmoor** *hills* U.K.
67 B3 **Exmouth** U.K.
18 A2 **Exmouth Gulf** Austr.
114 C2 **Exuma Cays** *is* Bahamas
59 D2 **Eya** *r.* Rus. Fed.
85 C4 **Eyl** Somalia
20 A1 **Eyre, Lake** *salt flat* Austr.
19 C3 **Eyre Peninsula** Austr.
62 B1 **Eysturoy** *i.* Faroe Is
91 D2 **Ezakheni** S. Africa
91 C2 **Ezenzeleni** S. Africa
38 B2 **Ezhou** China
54 E2 **Ezhva** Rus. Fed.
79 C3 **Ezine** Turkey

F

106 B2 **Fabens** U.S.A.
76 B2 **Fabriano** Italy
83 D3 **Fachi** Niger
82 C3 **Fada-Ngourma** Burkina
76 B2 **Faenza** Italy
27 C3 **Fafanlap** Indon.
78 B1 **Făgăraş** Romania
61 E3 **Fagernes** Norway
61 G4 **Fagersta** Sweden
121 B6 **Fagnano, Lago** *l.* Arg./Chile
82 B3 **Faguibine, Lac** *l.* Mali
60 □B3 **Fagurhólsmýri** Iceland
94 B2 **Fairbanks** U.S.A.
105 D2 **Fairbury** U.S.A.
103 B3 **Fairfield** U.S.A.
64 □ **Fair Isle** *i.* U.K.
105 E2 **Fairmont** *MN* U.S.A.
108 C3 **Fairmont** *WV* U.S.A.
96 C2 **Fairview** Can.
27 D2 **Fais** *i.* Micronesia
42 B1 **Faisalabad** Pak.
104 C1 **Faith** U.S.A.
43 C2 **Faizabad** India
27 C3 **Fakfak** Indon.
33 A1 **Faku** China
82 A4 **Falaba** Sierra Leone

107 D3 **Falcon Lake** *l.* Mex./U.S.A.
107 D3 **Falfurrias** U.S.A.
96 C2 **Falher** Can.
69 F2 **Falkenberg** Ger.
61 F4 **Falkenberg** Sweden
69 F1 **Falkensee** Ger.
64 C3 **Falkirk** U.K.
121 C6 **Falkland Islands** *terr.* S. Atlantic Ocean
61 F4 **Falköping** Sweden
103 C3 **Fallon** U.S.A.
109 E2 **Fall River** U.S.A.
105 D2 **Falls City** U.S.A.
67 A3 **Falmouth** U.K.
90 A3 **False Bay** S. Africa
61 F4 **Falster** *i.* Denmark
78 C1 **Fălticeni** Romania
61 G3 **Falun** Sweden
39 A3 **Fangchenggang** China
39 C3 **Fangshan** Taiwan
37 E1 **Fangzheng** China
76 B2 **Fano** Italy
30 B1 **Fan Si Pan** *mt.* Vietnam
87 C2 **Faradje** Dem. Rep. Congo
89 □D3 **Farafangana** Madag.
84 A2 **Farafra Oasis** Egypt
44 C3 **Farāh** Afgh.
82 A3 **Faranah** Guinea
46 B3 **Farasān, Jazā'ir** *is* Saudi Arabia
22 B2 **Farewell, Cape** N.Z.
105 D1 **Fargo** U.S.A.
105 E2 **Faribault** U.S.A.
109 E2 **Farmington** *ME* U.S.A.
106 B1 **Farmington** *NM* U.S.A.
109 D3 **Farmville** U.S.A.
67 C3 **Farnborough** U.K.
96 C2 **Farnham, Mount** Can.
96 A1 **Faro** Can.
74 B2 **Faro** Port.
56 A2 **Fårö** *i.* Sweden
62 B1 **Faroe Islands** *terr.* N. Atlantic Ocean
47 C2 **Farrāshband** Iran
Farvel, Kap *c.* Greenland *see* **Nunap Isua**
107 C2 **Farwell** U.S.A.
47 C2 **Fāryāb** Iran
47 C2 **Fasā** Iran
77 C2 **Fasano** Italy
58 B1 **Fastiv** Ukr.
43 B2 **Fatehgarh** India
43 C2 **Fatehpur** India
60 G2 **Fauske** Norway
60 □A3 **Faxaflói** *b.* Iceland
60 G3 **Faxälven** *r.* Sweden
83 D3 **Faya** Chad
110 B1 **Fayetteville** *AR* U.S.A.
111 E1 **Fayetteville** *NC* U.S.A.
110 C1 **Fayetteville** *TN* U.S.A.
42 B1 **Fazilka** India
82 A2 **Fdérik** Maur.
111 E2 **Fear, Cape** U.S.A.
22 C2 **Featherston** N.Z.
72 C2 **Fécamp** France
16 D2 **Federated States of Micronesia** *country* N. Pacific Ocean
70 C1 **Fehmarn** *i.* Ger.
69 F1 **Fehrbellin** Ger.
123 D2 **Feia, Lagoa** *lag.* Brazil
118 A2 **Feijó** Brazil
22 C2 **Feilding** N.Z.
119 E3 **Feira de Santana** Brazil
75 D2 **Felanitx** Spain
113 D3 **Felipe C. Puerto** Mex.
123 D1 **Felixlândia** Brazil
67 D3 **Felixstowe** U.K.
69 D2 **Felsberg** Ger.
76 B1 **Feltre** Italy
61 F3 **Femunden** *l.* Norway
39 B3 **Fengcheng** *Jiangxi* China
33 A1 **Fengcheng** *Liaoning* China
30 A1 **Fengqing** China
38 B2 **Fengxian** China
Fengxiang China *see* Lincang
39 C3 **Fengyüan** Taiwan
38 B1 **Fengzhen** China
73 D3 **Feno, Capo di** *c. Corsica* France
89 □D2 **Fenoarivo Atsinanana** Madag.
59 D2 **Feodosiya** Ukr.
76 A3 **Fer, Cap de** *c.* Alg.
45 D2 **Fergana** Uzbek.
105 D1 **Fergus Falls** U.S.A.
82 B4 **Ferkessédougou** Côte d'Ivoire
76 B2 **Fermo** Italy
99 D1 **Fermont** Can.
74 B1 **Fermoselle** Spain
65 B2 **Fermoy** Rep. of Ireland
111 D2 **Fernandina Beach** U.S.A.
122 B2 **Fernandópolis** Brazil
96 C3 **Fernie** Can.
76 B2 **Ferrara** Italy
122 B2 **Ferreiros** Brazil
76 A2 **Ferro, Capo** *c. Sardinia* Italy
74 B1 **Ferrol** Spain
68 B1 **Ferwerd** Neth.
82 B1 **Fès** Morocco
86 B3 **Feshi** Dem. Rep. Congo
105 E3 **Festus** U.S.A.
79 C3 **Fethiye** Turkey
64 □ **Fetlar** *i.* U.K.
98 C1 **Feuilles, Rivière aux** *r.* Can.
45 D3 **Feyzābād** Afgh.
Fez Morocco *see* **Fès**
89 □D3 **Fianarantsoa** Madag.
85 B4 **Fichē** Eth.
77 C2 **Fier** Albania
64 C2 **Fife Ness** *pt* U.K.
72 C3 **Figeac** France
74 B1 **Figueira da Foz** Port.
75 D1 **Figueres** Spain
82 B1 **Figuig** Morocco
17 F4 **Fiji** *country* S. Pacific Ocean
120 B3 **Filadélfia** Para.
23 Q2 **Filchner Ice Shelf** Antarctica
66 C1 **Filey** U.K.
79 B3 **Filippiada** Greece
61 F4 **Filipstad** Sweden
64 C2 **Findhorn** *r.* U.K.
108 C2 **Findlay** U.S.A.
19 D4 **Fingal** Austr.
109 D2 **Finger Lakes** U.S.A.
60 I3 **Finland** *country* Europe
61 H4 **Finland, Gulf of** Europe
96 B2 **Finlay** *r.* Can.
21 C3 **Finley** Austr.
69 E2 **Finne** *ridge* Ger.
60 H2 **Finnmarksvidda** *reg.* Norway
60 G2 **Finnsnes** Norway
61 G4 **Finspång** Sweden
64 A2 **Fionnphort** U.K.
Fırat *r.* Turkey *see* **Euphrates**
Firenze Italy *see* **Florence**
73 C2 **Firminy** France
43 B2 **Firozabad** India
42 B1 **Firozpur** India
47 C2 **Fīrūzābād** Iran
90 A2 **Fish** *watercourse* Namibia
90 B3 **Fish** *r.* S. Africa
67 A3 **Fishguard** U.K.
73 C2 **Fismes** France
74 B1 **Fisterra, Cabo** *c.* Spain
109 E2 **Fitchburg** U.S.A.
97 C2 **Fitzgerald** Can.
18 B1 **Fitzroy Crossing** Austr.
76 B2 **Fivizzano** Italy
87 C3 **Fizi** Dem. Rep. Congo
60 G3 **Fjällsjöälven** *r.* Sweden
91 C3 **Flagstaff** S. Africa
106 A1 **Flagstaff** U.S.A.
98 C1 **Flaherty Island** Can.
66 C1 **Flamborough Head** *hd* U.K.
69 F1 **Fläming** *hills* Ger.
104 B2 **Flaming Gorge Reservoir** U.S.A.
102 D1 **Flathead** *r.* U.S.A.
102 D1 **Flathead Lake** U.S.A.
19 D1 **Flattery, Cape** Austr.
102 B1 **Flattery, Cape** U.S.A.
66 B2 **Fleetwood** U.K.
61 E4 **Flekkefjord** Norway
70 B1 **Flensburg** Ger.
72 B2 **Flers** France
19 D1 **Flinders** *r.* Austr.
18 A3 **Flinders Bay** Austr.
19 D3 **Flinders Island** Austr.
20 A2 **Flinders Ranges** *mts* Austr.
97 D2 **Flin Flon** Can.
108 C2 **Flint** U.S.A.
75 D1 **Florac** France
76 B2 **Florence** Italy
110 C2 **Florence** *AL* U.S.A.
106 A2 **Florence** *AZ* U.S.A.
102 B2 **Florence** *OR* U.S.A.
111 E2 **Florence** *SC* U.S.A.
118 A1 **Florencia** Col.
114 B3 **Flores** Guat.
26 C3 **Flores** *i.* Indon.
26 B3 **Flores Sea** Indon.
119 E2 **Floresta** Brazil
107 D3 **Floresville** U.S.A.
119 D2 **Floriano** Brazil
120 D3 **Florianópolis** Brazil
121 C4 **Florida** Uru.
111 D2 **Florida** *state* U.S.A.
111 D4 **Florida, Straits of** Bahamas/U.S.A.
111 D4 **Florida Keys** *is* U.S.A.
79 B2 **Florina** Greece
61 E3 **Florø** Norway
97 D2 **Foam Lake** Can.
77 C2 **Foča** Bos.-Herz.
78 C1 **Focşani** Romania
39 B3 **Fogang** China
77 C2 **Foggia** Italy
99 E2 **Fogo Island** Can.
72 C3 **Foix** France
57 D3 **Fokino** Rus. Fed.
98 B2 **Foleyet** Can.
76 B2 **Foligno** Italy
67 D3 **Folkestone** U.K.
111 D2 **Folkston** U.S.A.
76 B2 **Follonica** Italy
97 D2 **Fond-du-Lac** Can.
97 D2 **Fond du Lac** *r.* Can.
108 B2 **Fond du Lac** U.S.A.
74 B1 **Fondevila** Spain
76 B2 **Fondi** Italy
114 B3 **Fonseca, Golfo de** *b.* Central America
118 B2 **Fonte Boa** Brazil
72 B2 **Fontenay-le-Comte** France
60 □C2 **Fontur** *pt* Iceland
21 C2 **Forbes** Austr.
69 E3 **Forchheim** Ger.
61 E3 **Førde** Norway
21 C1 **Fords Bridge** Austr.
110 B2 **Fordyce** U.S.A.
110 C2 **Forest** U.S.A.
21 C3 **Forest Hill** Austr.
99 D2 **Forestville** Can.
64 C2 **Forfar** U.K.
102 B1 **Forks** U.S.A.
76 B2 **Forlì** Italy
75 D2 **Formentera** *i.* Spain

75 D2 **Formentor, Cap de** *c.* Spain
123 C2 **Formiga** Brazil
120 C3 **Formosa** Arg.
Formosa *country* Asia *see* Taiwan
122 C1 **Formosa** Brazil
64 C2 **Forres** U.K.
18 B3 **Forrest** Austr.
110 B1 **Forrest City** U.S.A.
19 D1 **Forsayth** Austr.
61 H3 **Forssa** Fin.
21 D2 **Forster** Austr.
104 B1 **Forsyth** U.S.A.
42 B2 **Fort Abbas** Pak.
98 B1 **Fort Albany** Can.
119 E2 **Fortaleza** Brazil
96 C2 **Fort Assiniboine** Can.
64 B2 **Fort Augustus** U.K.
102 D1 **Fort Benton** U.S.A.
103 B3 **Fort Bragg** U.S.A.
Fort Chimo Can. *see* Kuujjuaq
97 C2 **Fort Chipewyan** Can.
104 B2 **Fort Collins** U.S.A.
115 D3 **Fort-de-France** Martinique
105 E2 **Fort Dodge** U.S.A.
98 A2 **Fort Frances** Can.
Fort Franklin Can. *see* **Déline**
98 C1 **Fort George** Can.
94 C2 **Fort Good Hope** Can.
64 C2 **Forth** *r.* U.K.
64 C2 **Forth, Firth of** *est.* U.K.
120 C3 **Fortín Madrejón** Para.
111 D3 **Fort Lauderdale** U.S.A.
96 B1 **Fort Liard** Can.
97 C2 **Fort Mackay** Can.
97 C2 **Fort McMurray** Can.
94 C2 **Fort McPherson** Can.
104 C2 **Fort Morgan** U.S.A.
111 D3 **Fort Myers** U.S.A.
96 B2 **Fort Nelson** Can.
96 B2 **Fort Nelson** *r.* Can.
110 C2 **Fort Payne** U.S.A.
104 B1 **Fort Peck** U.S.A.
104 B1 **Fort Peck Reservoir** U.S.A.
111 D3 **Fort Pierce** U.S.A.
96 C1 **Fort Providence** Can.
97 D2 **Fort Qu'Appelle** Can.
96 C1 **Fort Resolution** Can.
98 C1 **Fort Rupert** Can.
96 B2 **Fort St James** Can.
96 B2 **Fort St John** Can.
96 C2 **Fort Saskatchewan** Can.
105 E3 **Fort Scott** U.S.A.
98 B1 **Fort Severn** Can.
44 B2 **Fort-Shevchenko** Kazakh.
96 B1 **Fort Simpson** Can.
97 C1 **Fort Smith** Can.
110 B1 **Fort Smith** U.S.A.
107 C2 **Fort Stockton** U.S.A.
106 C2 **Fort Sumner** U.S.A.
96 C2 **Fort Vermilion** Can.
110 C2 **Fort Walton Beach** U.S.A.
108 B2 **Fort Wayne** U.S.A.
64 B2 **Fort William** U.K.
107 D2 **Fort Worth** U.S.A.
94 B2 **Fort Yukon** U.S.A.
39 B3 **Foshan** China
60 F3 **Fosna** *pen.* Norway
61 E3 **Fosnavåg** Norway
60 □B3 **Foss** Iceland
60 □A2 **Fossá** Iceland
76 A2 **Fossano** Italy
21 C3 **Foster** Austr.
72 B2 **Fougères** France
64 □ **Foula** *i.* U.K.
79 C3 **Fournoi** *i.* Greece
82 A3 **Fouta Djallon** *reg.* Guinea
22 A3 **Foveaux Strait** N.Z.
104 C3 **Fowler** U.S.A.
18 C3 **Fowlers Bay** Austr.
96 C2 **Fox Creek** Can.
95 F2 **Foxe Basin** *g.* Can.
95 F2 **Foxe Channel** Can.
95 F2 **Foxe Peninsula** Can.
22 B2 **Fox Glacier** N.Z.
96 C2 **Fox Lake** Can.
96 A1 **Fox Mountain** Can.
22 C2 **Foxton** N.Z.
97 D2 **Fox Valley** Can.
65 C1 **Foyle, Lough** *b.* Rep. of Ireland/U.K.
122 B3 **Foz de Areia, Represa de** *resr* Brazil
88 A2 **Foz do Cunene** Angola
122 B3 **Foz do Iguaçu** Brazil
75 D1 **Fraga** Spain
122 C2 **Franca** Brazil
77 C2 **Francavilla Fontana** Italy
72 C2 **France** *country* Europe
86 B3 **Franceville** Gabon
105 D2 **Francis Case, Lake** U.S.A.
123 D1 **Francisco Sá** Brazil
88 B3 **Francistown** Botswana
96 B2 **François Lake** Can.
69 D2 **Frankenberg (Eder)** Ger.
69 D3 **Frankenthal (Pfalz)** Ger.
69 E2 **Frankenwald** *mts* Ger.
108 C3 **Frankfort** U.S.A.
69 D2 **Frankfurt am Main** Ger.
70 C1 **Frankfurt an der Oder** Ger.
69 E3 **Fränkische Schweiz** *reg.* Ger.
109 D2 **Franklin** U.S.A.
94 C2 **Franklin Bay** Can.
102 C1 **Franklin D. Roosevelt Lake** U.S.A.
96 B1 **Franklin Mountains** Can.
94 E2 **Franklin Strait** Can.
22 B2 **Franz Josef Glacier** N.Z.
Franz Josef Land *is* Rus. Fed. *see* **Zemlya Frantsa-Iosifa**
76 A3 **Frasca, Capo della** *c.* *Sardinia* Italy
96 B3 **Fraser** *r. B.C.* Can.
99 D1 **Fraser** *r. Nfld.* Can.
90 B3 **Fraserburg** S. Africa
64 C2 **Fraserburgh** U.K.
98 B2 **Fraserdale** Can.
19 E2 **Fraser Island** Austr.
96 B2 **Fraser Lake** Can.
121 C4 **Fray Bentos** Uru.
61 E4 **Fredericia** Denmark
107 D2 **Frederick** U.S.A.
107 D2 **Fredericksburg** *TX* U.S.A.
109 D3 **Fredericksburg** *VA* U.S.A.
96 A2 **Frederick Sound** *sea chan.* U.S.A.
99 D2 **Fredericton** Can.
61 F4 **Frederikshavn** Denmark
Fredrikshamn Fin. *see* Hamina
61 F4 **Fredrikstad** Norway
108 B2 **Freeport** *IL* U.S.A.
107 D3 **Freeport** *TX* U.S.A.
114 C2 **Freeport City** Bahamas
107 D3 **Freer** U.S.A.
91 C2 **Free State** *prov.* S. Africa
82 A4 **Freetown** Sierra Leone
74 B2 **Fregenal de la Sierra** Spain
72 B2 **Fréhel, Cap** *c.* France
70 B2 **Freiburg im Breisgau** Ger.
70 C2 **Freising** Ger.
70 C2 **Freistadt** Austria
73 D3 **Fréjus** France
18 A3 **Fremantle** Austr.
105 D2 **Fremont** *NE* U.S.A.
108 C2 **Fremont** *OH* U.S.A.
119 C1 **French Guiana** *terr.* S. America
102 E1 **Frenchman** *r.* U.S.A.
17 H5 **French Polynesia** *terr.* S. Pacific Ocean
112 B2 **Fresnillo** Mex.
103 C3 **Fresno** U.S.A.
75 D2 **Freu, Cap des** *c.* Spain
73 D2 **Freyming-Merlebach** France
82 A3 **Fria** Guinea
120 B3 **Frias** Arg.
70 B2 **Friedrichshafen** Ger.
69 F1 **Friesack** Ger.
68 C1 **Friesoythe** Ger.
Frobisher Bay Can. *see* Iqaluit
95 G2 **Frobisher Bay** *b.* Can.
69 F2 **Frohburg** Ger.
55 D4 **Frolovo** Rus. Fed.
20 A2 **Frome, Lake** *salt flat* Austr.
20 A2 **Frome Downs** Austr.
68 C2 **Fröndenberg** Ger.
113 C3 **Frontera** Mex.
112 B1 **Fronteras** Mex.
76 B2 **Frosinone** Italy
60 E3 **Frøya** *i.* Norway
Frunze Kyrg. *see* **Bishkek**
73 D2 **Frutigen** Switz.
71 D2 **Frýdek-Místek** Czech Rep.
39 B3 **Fu'an** China
74 C1 **Fuenlabrada** Spain
120 C3 **Fuerte Olimpo** Para.
82 A2 **Fuerteventura** *i.* Canary Is
32 B1 **Fuga** *i.* Phil.
47 C2 **Fujairah** U.A.E.
35 C3 **Fuji** Japan
39 B3 **Fujian** *prov.* China
35 C3 **Fujinomiya** Japan
35 C3 **Fuji-san** *vol.* Japan
35 C3 **Fukui** Japan
35 B4 **Fukuoka** Japan
35 D3 **Fukushima** Japan
69 D2 **Fulda** Ger.
69 D2 **Fulda** *r.* Ger.
38 A3 **Fuling** China
105 E3 **Fulton** U.S.A.
73 C2 **Fumay** France
82 A1 **Funchal** Madeira
74 B1 **Fundão** Port.
99 D2 **Fundy, Bay of** *g.* Can.
38 B2 **Funing** *Jiangsu* China
39 A3 **Funing** *Yunnan* China
83 C3 **Funtua** Nigeria
47 C2 **Fūrgun, Kūh-e** *mt.* Iran
57 F2 **Furmanov** Rus. Fed.
123 C2 **Furnas, Represa** *resr* Brazil
19 D4 **Furneaux Group** *is* Austr.
68 C1 **Fürstenau** Ger.
69 E3 **Fürth** Ger.
34 D3 **Furukawa** Japan
95 F2 **Fury and Hecla Strait** Can.
33 A1 **Fushun** China
33 B1 **Fusong** China
47 C2 **Fuwayriṭ** Qatar
38 B2 **Fuyang** China
37 E1 **Fuyu** China
36 B1 **Fuyun** China
39 B3 **Fuzhou** China
Fuzhou China *see* **Linchuan**
61 F4 **Fyn** *i.* Denmark
F.Y.R.O.M. *country* Europe *see* **Macedonia**

G

85 C4 **Gaalkacyo** Somalia
88 A2 **Gabela** Angola
83 D1 **Gabès** Tunisia
83 D1 **Gabès, Golfe de** *g.* Tunisia

86 B3 **Gabon** *country* Africa
91 C1 **Gaborone** Botswana
78 C2 **Gabrovo** Bulg.
82 A3 **Gabú** Guinea-Bissau
41 B3 **Gadag** India
69 E1 **Gadebusch** Ger.
110 C2 **Gadsden** U.S.A.
78 C2 **Găești** Romania
76 B2 **Gaeta** Italy
111 D1 **Gaffney** U.S.A.
83 C1 **Gafsa** Tunisia
57 E2 **Gagarin** Rus. Fed.
82 B4 **Gagnoa** Côte d'Ivoire
99 D1 **Gagnon** Can.
49 C1 **Gagra** Georgia
90 A2 **Gaiab** *watercourse* Namibia
72 C3 **Gaillac** France
111 D3 **Gainesville** *FL* U.S.A.
111 D2 **Gainesville** *GA* U.S.A.
107 D2 **Gainesville** *TX* U.S.A.
66 C2 **Gainsborough** U.K.
20 A2 **Gairdner, Lake** *salt flat* Austr.
64 B2 **Gairloch** U.K.
87 E3 **Galana** *r.* Kenya
71 D2 **Galanta** Slovakia
116 C2 **Galapagos Islands** *is* Pacific Ocean
64 C3 **Galashiels** U.K.
78 C1 **Galați** Romania
61 E3 **Galdhøpiggen** *mt.* Norway
113 B2 **Galeana** Mex.
96 C2 **Galena Bay** Can.
108 A2 **Galesburg** U.S.A.
90 B2 **Galeshewe** S. Africa
54 D3 **Galich** Rus. Fed.
84 B1 **Galilee, Sea of** Israel
110 C1 **Gallatin** U.S.A.
41 C4 **Galle** Sri Lanka
118 A1 **Gallinas, Punta** *pt* Col.
77 C2 **Gallipoli** Italy
Gallipoli Turkey *see* **Gelibolu**
60 H2 **Gällivare** Sweden
106 B1 **Gallup** U.S.A.
82 A2 **Galtat Zemmour** Western Sahara
65 B2 **Galtymore** *hill* Rep. of Ireland
107 E3 **Galveston** U.S.A.
107 E3 **Galveston Bay** U.S.A.
65 B2 **Galway** Rep. of Ireland
65 B2 **Galway Bay** Rep. of Ireland
122 C1 **Gamá** Brazil
91 D3 **Gamalakhe** S. Africa
Gambia *country* Africa *see* **The Gambia**
20 A3 **Gambier Islands** Austr.
99 E2 **Gambo** Can.
86 B3 **Gamboma** Congo
106 B1 **Ganado** U.S.A.
49 D3 **Ganāveh** Iran
49 C1 **Gäncä** Azer.
Gand Belgium *see* **Gent**
29 C2 **Gandadiwata, Bukit** *mt.* Indon.
86 C3 **Gandajika** Dem. Rep. Congo
99 E2 **Gander** Can.
99 E2 **Gander** *r.* Can.
69 D1 **Ganderkesee** Ger.
75 D1 **Gandesa** Spain
42 B2 **Gāndhīdhām** India
42 B2 **Gandhinagar** India
42 B2 **Gāndhi Sāgar** *resr* India
75 C2 **Gandía** Spain
Ganga *r.* Bangl./India *see* Ganges
121 B5 **Gangán** Arg.
42 B2 **Ganganagar** India
36 C2 **Gangca** China
43 C1 **Gangdisê Shan** *mts* China
43 D2 **Ganges** *r.* Bangl./India
73 C3 **Ganges** France
43 C2 **Ganges, Mouths of the** Bangl./India
43 C2 **Gangtok** India
73 C2 **Gannat** France
104 B2 **Gannett Peak** U.S.A.
90 A3 **Gansbaai** S. Africa
38 A1 **Gansu** *prov.* China
39 B3 **Ganzhou** China
82 B3 **Gao** Mali
82 A3 **Gaoual** Guinea
38 B2 **Gaoyou** China
38 B2 **Gaoyou Hu** *l.* China
73 D3 **Gap** France
75 C2 **Gap Carbon** *hd* Alg.
43 C1 **Gar** China
21 C1 **Garah** Austr.
119 E2 **Garanhuns** Brazil
103 B2 **Garberville** U.S.A.
69 D1 **Garbsen** Ger.
122 C2 **Garça** Brazil
76 B1 **Garda, Lago di** *l.* Italy
69 E1 **Gardelegen** Ger.
104 C3 **Garden City** U.S.A.
97 E2 **Garden Hill** Can.
45 C3 **Gardez** Afgh.
56 B2 **Gargždai** Lith.
43 C2 **Garhchiroli** India
91 C3 **Gariep Dam** *resr* S. Africa
90 A3 **Garies** S. Africa
87 D3 **Garissa** Kenya
70 C2 **Garmisch-Partenkirchen** Ger.
45 D3 **Garmo, Qullai** *mt.* Tajik.
20 B2 **Garnpung Lake** *imp. l.* Austr.
72 B3 **Garonne** *r.* France
85 C4 **Garoowe** Somalia
42 B2 **Garoth** India
86 B2 **Garoua** Cameroon
64 B2 **Garry** *r.* U.K.
94 E2 **Garry Lake** Can.
87 E3 **Garsen** Kenya
29 B2 **Garut** Indon.
108 B2 **Gary** U.S.A.
113 B2 **Garza García** Mex.
36 C2 **Garzê** China
72 B3 **Gascony, Gulf of** *g.* France/Spain
18 A2 **Gascoyne** *r.* Austr.
83 D3 **Gashua** Nigeria
99 D2 **Gaspé** Can.
99 D2 **Gaspé, Péninsule de** *pen.* Can.
111 E1 **Gaston, Lake** U.S.A.
111 D1 **Gastonia** U.S.A.
75 C2 **Gata, Cabo de** *c.* Spain
56 D2 **Gatchina** Rus. Fed.
66 C1 **Gateshead** U.K.
107 D2 **Gatesville** U.S.A.
98 C2 **Gatineau** *r.* Can.
21 D1 **Gatton** Austr.
97 E2 **Gauer Lake** Can.
61 E4 **Gausta** *mt.* Norway
91 C2 **Gauteng** *prov.* S. Africa
47 C2 **Gāvbandī** Iran
79 B3 **Gavdos** *i.* Greece
61 G3 **Gävle** Sweden
57 F2 **Gavrilov Posad** Rus. Fed.
57 E2 **Gavrilov-Yam** Rus. Fed.
30 A1 **Gawai** Myanmar
20 A2 **Gawler** Austr.
20 A2 **Gawler Ranges** *hills* Austr.
43 C2 **Gaya** India
82 C3 **Gaya** Niger
108 C1 **Gaylord** U.S.A.
54 E2 **Gayny** Rus. Fed.
84 B1 **Gaza** *terr.* Asia
48 B2 **Gaza** Gaza
44 C2 **Gaz-Achak** Turkm.
44 B3 **Gazandzhyk** Turkm.
48 B2 **Gaziantep** Turkey
82 B4 **Gbarnga** Liberia
71 D1 **Gdańsk** Pol.
56 A3 **Gdańsk, Gulf of** Pol./Rus. Fed.
56 C2 **Gdov** Rus. Fed.
71 D1 **Gdynia** Pol.
84 B3 **Gedaref** Sudan
69 D2 **Gedern** Ger.
79 C3 **Gediz** *r.* Turkey
70 C1 **Gedser** Denmark
68 B2 **Geel** Belgium
20 B3 **Geelong** Austr.
69 E1 **Geesthacht** Ger.
43 C1 **Gê'gyai** China
97 D2 **Geikie** *r.* Can.
61 E3 **Geilo** Norway
30 B1 **Gejiu** China
76 B3 **Gela** *Sicily* Italy
59 D3 **Gelendzhik** Rus. Fed.
79 C2 **Gelibolu** Turkey
68 C2 **Gelsenkirchen** Ger.
86 B2 **Gemena** Dem. Rep. Congo
79 C2 **Gemlik** Turkey
85 C4 **Genalē Wenz** *r.* Eth.
121 B4 **General Acha** Arg.
121 B4 **General Alvear** Arg.
121 C4 **General Belgrano** Arg.
112 B2 **General Cepeda** Mex.
121 B4 **General Pico** Arg.
121 B4 **General Roca** Arg.
32 B2 **General Santos** Phil.
109 D2 **Genesee** *r.* U.S.A.
109 D2 **Geneseo** U.S.A.
Geneva Switz. *see* **Genève**
109 D2 **Geneva** U.S.A.
Geneva, Lake France/Switz. *see* **Léman, Lac**
73 D2 **Genève** Switz.
74 B2 **Genil** *r.* Spain
68 B2 **Genk** Belgium
76 A2 **Genoa** Italy
Genova Italy *see* **Genoa**
68 A2 **Gent** Belgium
69 F1 **Genthin** Ger.
18 A3 **Geographe Bay** Austr.
99 D1 **George** *r.* Can.
90 B3 **George** S. Africa
111 D3 **George, Lake** *FL* U.S.A.
109 E2 **George, Lake** *NY* U.S.A.
82 A3 **Georgetown** Gambia
119 C1 **Georgetown** Guyana
28 B1 **George Town** Malaysia
111 E2 **Georgetown** *SC* U.S.A.
107 D2 **Georgetown** *TX* U.S.A.
23 I2 **George V Land** *reg.* Antarctica
49 C1 **Georgia** *country* Asia
111 D2 **Georgia** *state* U.S.A.
98 B2 **Georgian Bay** Can.
19 C2 **Georgina** *watercourse* Austr.
45 E2 **Georgiyevka** Kazakh.
55 D4 **Georgiyevsk** Rus. Fed.
69 F2 **Gera** Ger.
119 D3 **Geral de Goiás, Serra** *hills* Brazil
22 B2 **Geraldine** N.Z.
18 A2 **Geraldton** Austr.
48 B1 **Gerede** Turkey
42 A1 **Gereshk** Afgh.
103 C2 **Gerlach** U.S.A.
71 E2 **Gerlachovský štít** *mt.* Slovakia
70 C1 **Germany** *country* Europe
68 C2 **Gerolstein** Ger.
69 E3 **Gerolzhofen** Ger.
69 D2 **Gersfeld (Rhön)** Ger.
43 C1 **Gêrzê** China
109 D3 **Gettysburg** *PA* U.S.A.
104 D1 **Gettysburg** *SD* U.S.A.

77 D2 **Gevgelija** Macedonia
74 C1 **Gexto** Spain
79 C3 **Geyikli** Turkey
90 B2 **Ghaap Plateau** S. Africa
83 C1 **Ghadāmis** Libya
43 C2 **Ghaghara** *r.* India
82 B4 **Ghana** *country* Africa
88 B3 **Ghanzi** Botswana
83 C1 **Ghardaïa** Alg.
83 D1 **Gharyān** Libya
43 C2 **Ghatal** India
43 C2 **Ghazipur** India
45 C3 **Ghaznī** Afgh.
Ghent Belgium *see* **Gent**
78 B1 **Gherla** Romania
73 D3 **Ghisonaccia** *Corsica* France
42 B2 **Ghotāru** India
42 A2 **Ghotki** Pak.
59 E3 **Giaginskaya** Rus. Fed.
65 C1 **Giant's Causeway** *lava field* U.K.
29 C2 **Gianyar** Indon.
77 C3 **Giarre** *Sicily* Italy
76 A1 **Giaveno** Italy
90 A2 **Gibeon** Namibia
74 B2 **Gibraltar** *terr.* Europe
74 B2 **Gibraltar, Strait of** Morocco/Spain
18 B2 **Gibson Desert** Austr.
36 C1 **Gichgeniyn Nuruu** *mts* Mongolia
85 B4 **Gīdolē** Eth.
72 C2 **Gien** France
69 D2 **Gießen** Ger.
69 E1 **Gifhorn** Ger.
96 C2 **Gift Lake** Can.
35 C3 **Gifu** Japan
64 B3 **Gigha** *i.* U.K.
74 B1 **Gijón** Spain
106 A2 **Gila** *r.* U.S.A.
106 A2 **Gila Bend** U.S.A.
19 D1 **Gilbert** *r.* Austr.
Gilbert Islands *country* Pacific Ocean *see* **Kiribati**
102 D1 **Gildford** U.S.A.
84 A2 **Gilf Kebir Plateau** Egypt
21 C2 **Gilgandra** Austr.
42 B1 **Gilgit** Jammu and Kashmir
42 B1 **Gilgit** *r.* Jammu and Kashmir
21 C2 **Gilgunnia** Austr.
97 E2 **Gillam** Can.
104 B2 **Gillette** U.S.A.
67 D3 **Gillingham** U.K.
97 E2 **Gimli** Can.
85 C4 **Gīnīr** Eth.
77 C2 **Ginosa** Italy
21 C3 **Gippsland** *reg.* Austr.
42 A2 **Girdar Dhor** *r.* Pak.
48 B1 **Giresun** Turkey
75 D1 **Girona** Spain
64 B3 **Girvan** U.K.
22 C1 **Gisborne** N.Z.
61 F4 **Gislaved** Sweden
87 C3 **Gitarama** Rwanda
76 B2 **Giulianova** Italy
78 C2 **Giurgiu** Romania
78 C1 **Giuvala, Pasul** *pass* Romania
73 C2 **Givors** France
91 D1 **Giyani** S. Africa
48 B3 **Giza Pyramids** Egypt
77 D2 **Gjirokastër** Albania
94 E2 **Gjoa Haven** Can.
61 F3 **Gjøvik** Norway
99 E2 **Glace Bay** Can.
102 B1 **Glacier Peak** *vol.* U.S.A.
19 E2 **Gladstone** Austr.
60 □A2 **Gláma** *mts* Iceland
77 C2 **Glamoč** Bos.-Herz.
68 C3 **Glan** *r.* Ger.
65 B2 **Glanaruddery Mountains** Rep. of Ireland
64 B3 **Glasgow** U.K.
108 B3 **Glasgow** *KY* U.S.A.
104 B1 **Glasgow** *MT* U.S.A.
67 B3 **Glastonbury** U.K.
69 F2 **Glauchau** Ger.
54 E3 **Glazov** Rus. Fed.
57 E3 **Glazunovka** Rus. Fed.
64 B2 **Glen Coe** *val.* U.K.
106 A2 **Glendale** U.S.A.
21 D2 **Glen Davis** Austr.
104 C1 **Glendive** U.S.A.
20 B3 **Glenelg** *r.* Austr.
21 D1 **Glen Innes** Austr.
94 B2 **Glennallen** U.S.A.
64 C2 **Glenrothes** U.K.
109 E2 **Glens Falls** U.S.A.
64 C2 **Glen Shee** *val.* U.K.
65 B1 **Glenties** Rep. of Ireland
106 B2 **Glenwood** U.S.A.
104 B3 **Glenwood Springs** U.S.A.
71 D1 **Gliwice** Pol.
106 A2 **Globe** U.S.A.
71 D1 **Głogów** Pol.
60 F2 **Glomfjord** Norway
61 F4 **Glomma** *r.* Norway
21 D2 **Gloucester** Austr.
67 B3 **Gloucester** U.K.
69 F1 **Glöwen** Ger.
45 E1 **Glubokoye** Kazakh.
69 D1 **Glückstadt** Ger.
71 C2 **Gmünd** Austria
70 C2 **Gmunden** Austria
69 D1 **Gnarrenburg** Ger.
71 D1 **Gniezno** Pol.
77 D2 **Gnjilane** Yugo.
43 D2 **Goalpara** India
64 B3 **Goat Fell** *hill* U.K.
85 C4 **Goba** Eth.
88 A3 **Gobabis** Namibia
121 A5 **Gobernador Gregores** Arg.
37 D2 **Gobi** *des.* China/Mongolia
68 C2 **Goch** Ger.
90 A1 **Gochas** Namibia
42 B3 **Godavari** *r.* India
41 C3 **Godavari, Mouths of the** India
98 B2 **Goderich** Can.
42 B2 **Godhra** India
97 E2 **Gods** *r.* Can.
97 E2 **Gods Lake** *l.* Can.
Godwin-Austen, Mount China/Jammu and Kashmir *see* **K2**
98 C2 **Goéland, Lac au** *l.* Can.
99 D1 **Goélands, Lac aux** *l.* Can.
68 A2 **Goes** Neth.
122 C1 **Goiandira** Brazil
122 C1 **Goiânia** Brazil
122 B1 **Goiás** Brazil
122 B2 **Goio-Erê** Brazil
79 C2 **Gökçeada** *i.* Turkey
79 C3 **Gökçedağ** Turkey
89 B2 **Gokwe** Zimbabwe
61 E3 **Gol** Norway
30 A1 **Golaghat** India
79 C2 **Gölcük** Turkey
71 E1 **Gołdap** Pol.
69 F1 **Goldberg** Ger.
21 D1 **Gold Coast** Austr.
82 B4 **Gold Coast** *coastal area* Ghana
96 C2 **Golden** Can.
22 B2 **Golden Bay** N.Z.
96 B3 **Golden Hinde** *mt.* Can.
65 B2 **Golden Vale** *lowland* Rep. of Ireland
103 C3 **Goldfield** U.S.A.
96 B3 **Gold River** Can.
111 E1 **Goldsboro** U.S.A.
103 C4 **Goleta** U.S.A.
36 C2 **Golmud** China
49 D2 **Golpāyegān** Iran
64 C2 **Golspie** U.K.
87 C3 **Goma** Dem. Rep. Congo
43 C2 **Gomati** *r.* India
83 D3 **Gombe** Nigeria
83 D3 **Gombi** Nigeria
Gomel' Belarus *see* **Homyel'**
112 B2 **Gómez Palacio** Mex.
115 C3 **Gonaïves** Haiti
49 D2 **Gonbad-e Kavus** Iran
85 B3 **Gonder** Eth.
43 C2 **Gondia** India
79 C2 **Gönen** Turkey
83 D4 **Gongola** *r.* Nigeria
21 C2 **Gongolgon** Austr.
113 C2 **Gonzáles** Mex.
107 D3 **Gonzales** U.S.A.
90 A3 **Good Hope, Cape of** S. Africa
102 D2 **Gooding** U.S.A.
104 C3 **Goodland** U.S.A.
21 C1 **Goodooga** Austr.
66 C2 **Goole** U.K.
21 C2 **Goolgowi** Austr.
20 A3 **Goolwa** Austr.
21 D1 **Goondiwindi** Austr.
102 B2 **Goose Lake** U.S.A.
70 B2 **Göppingen** Ger.
43 C2 **Gorakhpur** India
77 C2 **Goražde** Bos.-Herz.
79 C3 **Gördes** Turkey
57 D3 **Gordeyevka** Rus. Fed.
19 D4 **Gordon, Lake** Austr.
83 D4 **Goré** Chad
85 B4 **Gorē** Eth.
22 A3 **Gore** N.Z.
65 C2 **Gorey** Rep. of Ireland
49 D2 **Gorgān** Iran
49 C1 **Gori** Georgia
76 B1 **Gorizia** Italy
Gor'kiy Rus. Fed. *see* **Nizhniy Novgorod**
71 E2 **Gorlice** Pol.
71 C1 **Görlitz** Ger.
77 D2 **Gornji Milanovac** Yugo.
77 C2 **Gornji Vakuf** Bos.-Herz.
45 E1 **Gorno-Altaysk** Rus. Fed.
78 C2 **Gornotrakiyska Nizina** *lowland* Bulg.
45 E1 **Gornyak** Rus. Fed.
27 D3 **Goroka** P.N.G.
82 B3 **Gorom Gorom** Burkina
27 C2 **Gorontalo** Indon.
57 E3 **Gorshechnoye** Rus. Fed.
65 B2 **Gorumna Island** Rep. of Ireland
59 D3 **Goryachiy Klyuch** Rus. Fed.
71 D1 **Gorzów Wielkopolski** Pol.
21 D2 **Gosford** Austr.
34 D2 **Goshogawara** Japan
69 E2 **Goslar** Ger.
77 C2 **Gospić** Croatia
67 C3 **Gosport** U.K.
77 D2 **Gostivar** Macedonia
Göteborg Sweden *see* Gothenburg
69 E2 **Gotha** Ger.
61 F4 **Gothenburg** Sweden
104 C2 **Gothenburg** U.S.A.
61 G4 **Gotland** *i.* Sweden
79 B2 **Gotse Delchev** Bulg.
61 G4 **Gotska Sandön** *i.* Sweden
35 B4 **Gōtsu** Japan
69 D2 **Göttingen** Ger.
96 B2 **Gott Peak** Can.
68 B1 **Gouda** Neth.
82 A3 **Goudiri** Senegal

83 D3 **Goudoumaria** Niger
98 C2 **Gouin, Réservoir** *resr* Can.
21 C2 **Goulburn** Austr.
21 B3 **Goulburn** *r. N.S.W.* Austr.
21 B3 **Goulburn** *r. Vic.* Austr.
82 B3 **Goundam** Mali
75 D2 **Gouraya** Alg.
72 C3 **Gourdon** France
83 D3 **Gouré** Niger
90 B3 **Gourits** *r.* S. Africa
82 B3 **Gourma-Rharous** Mali
21 C3 **Gourock Range** *mts* Austr.
123 D1 **Governador Valadares** Brazil
111 E3 **Governor's Harbour** Bahamas
36 C2 **Govĭ Altayn Nuruu** *mts* Mongolia
43 C2 **Govind Ballash Pant Sāgar** *resr* India
67 A3 **Gower** *pen.* U.K.
120 C3 **Goya** Arg.
49 C1 **Göyçay** Azer.
43 C1 **Gozha Co** *salt l.* China
90 B3 **Graaf-Reinet** S. Africa
69 E1 **Grabow** Ger.
77 C2 **Gračac** Croatia
55 E3 **Grachevka** Rus. Fed.
69 F2 **Gräfenhainichen** Ger.
21 D1 **Grafton** Austr.
105 D1 **Grafton** U.S.A.
107 D2 **Graham** U.S.A.
96 A2 **Graham Island** Can.
23 P3 **Graham Land** *reg.* Antarctica
91 C3 **Grahamstown** S. Africa
119 D2 **Grajaú** Brazil
79 B2 **Grámmos** *mt.* Greece
64 B2 **Grampian Mountains** U.K.
114 B3 **Granada** Nic.
74 C2 **Granada** Spain
109 E1 **Granby** Can.
82 A2 **Gran Canaria** *i.* Canary Is
120 B3 **Gran Chaco** *reg.* Arg./Para.
104 C1 **Grand** *r.* U.S.A.
114 C2 **Grand Bahama** *i.* Bahamas
99 E2 **Grand Bank** Can.
126 C2 **Grand Banks of Newfoundland** *sea feature* N. Atlantic Ocean
106 A1 **Grand Canyon** U.S.A.
106 A1 **Grand Canyon** *gorge* U.S.A.
114 B3 **Grand Cayman** *i.* Cayman Is
97 C2 **Grand Centre** Can.
102 C1 **Grand Coulee** U.S.A.
120 B2 **Grande** *r.* Bol.
122 B2 **Grande** *r.* Brazil
121 B6 **Grande, Bahía** *b.* Arg.
123 D2 **Grande, Ilha** *i.* Brazil
96 C2 **Grande Cache** Can.
89 D2 **Grande Comore** *i.* Comoros
96 C2 **Grande Prairie** Can.
83 D3 **Grand Erg de Bilma** *des.* Niger
82 B1 **Grand Erg Occidental** *des.* Alg.
83 C2 **Grand Erg Oriental** *des.* Alg.
99 D2 **Grande-Rivière** Can.
98 C1 **Grande Rivière de la Baleine** *r.* Can.
120 B4 **Grandes, Salinas** *salt marsh* Arg.
99 D2 **Grand Falls** *N.B.* Can.
99 E2 **Grand Falls** *Nfld.* Can.
96 C3 **Grand Forks** Can.
105 D1 **Grand Forks** U.S.A.
96 C1 **Grandin, Lac** *l.* Can.
105 D2 **Grand Island** U.S.A.
110 B3 **Grand Isle** U.S.A.
104 B3 **Grand Junction** U.S.A.
82 B4 **Grand-Lahou** Côte d'Ivoire
99 D2 **Grand Lake** *l. N.B.* Can.
99 E2 **Grand Lake** *l. Nfld.* Can.
105 E1 **Grand Marais** U.S.A.
74 B2 **Grândola** Port.
97 E2 **Grand Rapids** Can.
108 B2 **Grand Rapids** *MI* U.S.A.
105 E1 **Grand Rapids** *MN* U.S.A.
104 A2 **Grand Teton** *mt.* U.S.A.
115 C2 **Grand Turk** Turks and Caicos Is
102 C1 **Grangeville** U.S.A.
96 B2 **Granisle** Can.
102 E1 **Granite Peak** U.S.A.
76 B3 **Granitola, Capo** *c. Sicily* Italy
61 F4 **Gränna** Sweden
69 F1 **Gransee** Ger.
67 C2 **Grantham** U.K.
64 C2 **Grantown-on-Spey** U.K.
106 B1 **Grants** U.S.A.
102 B2 **Grants Pass** U.S.A.
72 B2 **Granville** France
97 D2 **Granville Lake** Can.
123 D1 **Grão Mogol** Brazil
91 D1 **Graskop** S. Africa
73 D3 **Grasse** France
75 D1 **Graus** Spain
60 F2 **Gravdal** Norway
72 B2 **Grave, Pointe de** *pt* France
97 D3 **Gravelbourg** Can.
98 C2 **Gravenhurst** Can.
21 D1 **Gravesend** Austr.
67 D3 **Gravesend** U.K.
73 D2 **Gray** France
71 D2 **Graz** Austria
114 C2 **Great Abaco** *i.* Bahamas
18 B3 **Great Australian Bight** *g.* Austr.
22 C1 **Great Barrier Island** N.Z.
19 D1 **Great Barrier Reef** Austr.
103 C3 **Great Basin** U.S.A.
94 D2 **Great Bear Lake** Can.
105 D3 **Great Bend** U.S.A.
31 A2 **Great Coco Island** Cocos Is
21 B3 **Great Dividing Range** *mts* Austr.
114 B2 **Greater Antilles** *is* Caribbean Sea
102 D1 **Great Falls** U.S.A.
91 C3 **Great Fish** *r.* S. Africa
91 C3 **Great Fish Point** S. Africa
115 C2 **Great Inagua** *i.* Bahamas
90 B3 **Great Karoo** *plat.* S. Africa
91 C3 **Great Kei** *r.* S. Africa
67 B2 **Great Malvern** U.K.
31 A3 **Great Nicobar** *i.* India
67 D2 **Great Ouse** *r.* U.K.
87 D3 **Great Rift Valley** Africa
87 D3 **Great Ruaha** *r.* Tanz.
102 D2 **Great Salt Lake** U.S.A.
103 D2 **Great Salt Lake Desert** U.S.A.
84 A2 **Great Sand Sea** *des.* Egypt/Libya
18 B2 **Great Sandy Desert** Austr.
96 C1 **Great Slave Lake** Can.
111 D1 **Great Smoky Mountains** U.S.A.
18 B2 **Great Victoria Desert** Austr.
38 B1 **Great Wall** China
67 D2 **Great Yarmouth** U.K.
74 B1 **Gredos, Sierra de** *mts* Spain
79 B3 **Greece** *country* Europe
104 C2 **Greeley** U.S.A.
50 F1 **Greem-Bell, Ostrov** *i.* Rus. Fed.
108 B3 **Green** *r. KY* U.S.A.
104 B3 **Green** *r. WY* U.S.A.
108 B2 **Green Bay** U.S.A.
108 B1 **Green Bay** *b.* U.S.A.
108 C3 **Greenbrier** *r.* U.S.A.
108 B3 **Greencastle** U.S.A.
111 D1 **Greeneville** U.S.A.
109 E2 **Greenfield** U.S.A.
97 D2 **Green Lake** Can.
95 I2 **Greenland** *terr.* N. America
128 R2 **Greenland Sea** Greenland/Svalbard
64 B3 **Greenock** U.K.
103 D3 **Green River** *UT* U.S.A.
104 B2 **Green River** *WY* U.S.A.
108 B3 **Greensburg** *IN* U.S.A.
109 D2 **Greensburg** *PA* U.S.A.
111 E2 **Green Swamp** U.S.A.
106 A2 **Green Valley** U.S.A.
82 B4 **Greenville** Liberia
110 C2 **Greenville** *AL* U.S.A.
110 B2 **Greenville** *MS* U.S.A.
111 E1 **Greenville** *NC* U.S.A.
111 D2 **Greenville** *SC* U.S.A.
107 D2 **Greenville** *TX* U.S.A.
21 D2 **Greenwell Point** Austr.
111 D2 **Greenwood** U.S.A.
18 B2 **Gregory, Lake** *salt flat* Austr.
19 D1 **Gregory Range** *hills* Austr.
70 C1 **Greifswald** Ger.
69 F2 **Greiz** Ger.
61 F4 **Grenå** Denmark
110 C2 **Grenada** U.S.A.
115 D3 **Grenada** *country* West Indies
72 C3 **Grenade** France
21 C2 **Grenfell** Austr.
97 D2 **Grenfell** Can.
73 D2 **Grenoble** France
19 D1 **Grenville, Cape** Austr.
102 B1 **Gresham** U.S.A.
68 C1 **Greven** Ger.
79 B2 **Grevena** Greece
68 C2 **Grevenbroich** Ger.
69 E1 **Grevesmühlen** Ger.
104 B2 **Greybull** U.S.A.
96 A1 **Grey Hunter Peak** Can.
99 E1 **Grey Islands** Can.
22 B2 **Greymouth** N.Z.
20 B1 **Grey Range** *hills* Austr.
65 C2 **Greystones** Rep. of Ireland
111 D2 **Griffin** U.S.A.
21 C2 **Griffith** Austr.
69 F2 **Grimma** Ger.
70 C1 **Grimmen** Ger.
66 C2 **Grimsby** U.K.
96 C2 **Grimshaw** Can.
60 □B2 **Grímsstaðir** Iceland
61 E4 **Grimstad** Norway
105 E2 **Grinnell** U.S.A.
91 C3 **Griqualand East** *reg.* S. Africa
90 B2 **Griqualand West** *reg.* S. Africa
95 F1 **Grise Fiord** Can.
64 C1 **Gritley** U.K.
91 C2 **Groblersdal** S. Africa
90 B2 **Groblershoop** S. Africa
Grodno Belarus *see* **Hrodna**
72 B2 **Groix, Île de** *i.* France
68 C1 **Gronau (Westfalen)** Ger.
60 F3 **Grong** Norway
68 C1 **Groningen** Neth.
90 B2 **Grootdrink** S. Africa
19 C1 **Groote Eylandt** *i.* Austr.
88 A2 **Grootfontein** Namibia
90 A2 **Groot Karas Berg** *plat.* Namibia
90 B3 **Groot Swartberge** *mts* S. Africa
91 C3 **Groot Winterberg** *mt.* S. Africa
69 D2 **Großenlüder** Ger.
70 C2 **Großer Rachel** *mt.* Ger.
71 C2 **Grosser Speikkogel** *mt.* Austria

76 B2 **Grosseto** Italy
69 D3 **Groß-Gerau** Ger.
70 C2 **Großglockner** *mt.* Austria
68 C1 **Groß-Hesepe** Ger.
69 E2 **Großlohra** Ger.
90 A1 **Gross Ums** Namibia
99 E1 **Groswater Bay** Can.
98 B2 **Groundhog** *r.* Can.
103 B3 **Grover Beach** U.S.A.
109 E2 **Groveton** U.S.A.
55 D4 **Groznyy** Rus. Fed.
77 C1 **Grubišno Polje** Croatia
71 D1 **Grudziądz** Pol.
90 A2 **Grünau** Namibia
60 □A3 **Grundarfjörður** Iceland
57 E3 **Gryazi** Rus. Fed.
57 F2 **Gryazovets** Rus. Fed.
71 D1 **Gryfice** Pol.
70 C1 **Gryfino** Pol.
114 C2 **Guacanayabo, Golfo de** *b.* Cuba
112 B2 **Guadalajara** Mex.
75 C1 **Guadalope** *r.* Spain
74 B2 **Guadalquivir** *r.* Spain
100 B4 **Guadalupe** *i.* Mex.
74 B2 **Guadalupe, Sierra de** *mts* Spain
106 C2 **Guadalupe Peak** U.S.A.
112 B2 **Guadalupe Victoria** Mex.
112 B2 **Guadalupe y Calvo** Mex.
74 C1 **Guadarrama, Sierra de** *mts* Spain
115 D3 **Guadeloupe** *terr.* West Indies
74 B2 **Guadiana** *r.* Port./Spain
74 C2 **Guadix** Spain
122 B2 **Guaíra** Brazil
115 C3 **Guajira, Península de la** *pen.* Col.
118 A2 **Gualaceo** Ecuador
27 D2 **Guam** *terr.* N. Pacific Ocean
112 B2 **Guamúchil** Mex.
112 B2 **Guanacevi** Mex.
119 D3 **Guanambi** Brazil
118 B1 **Guanare** Venez.
114 B2 **Guane** Cuba
38 A2 **Guang'an** China
39 B3 **Guangchang** China
39 B3 **Guangdong** *prov.* China
39 A3 **Guangxi** *aut. reg.* China
38 A2 **Guangyuan** China
39 B3 **Guangzhou** China
123 D1 **Guanhães** Brazil
115 D4 **Guanipa** *r.* Venez.
39 A3 **Guanling** China
33 A1 **Guanshui** China
Guansuo China *see* **Guanling**
115 C2 **Guantánamo** Cuba
118 B3 **Guaporé** *r.* Bol./Brazil
122 B3 **Guarapuava** Brazil
122 C3 **Guaraqueçaba** Brazil
123 C2 **Guaratinguetá** Brazil
74 B1 **Guarda** Port.
122 C1 **Guarda Mor** Brazil
74 C1 **Guardo** Spain
123 C2 **Guarujá** Brazil
112 B2 **Guasave** Mex.
114 A3 **Guatemala** *country* Central America
114 A3 **Guatemala** Guat.
118 B1 **Guaviare** *r.* Col.
123 C2 **Guaxupé** Brazil
118 A2 **Guayaquil** Ecuador
118 B3 **Guayaramerín** Bol.
112 A2 **Guaymas** Mex.
85 B3 **Guba** Eth.
54 E1 **Guba Dolgaya** Rus. Fed.
57 E3 **Gubkin** Rus. Fed.
83 C1 **Guelma** Alg.
82 A2 **Guelmine** Morocco
98 B2 **Guelph** Can.
113 C2 **Guémez** Mex.
72 C2 **Guéret** France
63 C4 **Guernsey** *terr.* Channel Is
112 A2 **Guerrero Negro** Mex.
99 D1 **Guers, Lac** *l.* Can.
38 B2 **Guichi** China
86 B2 **Guider** Cameroon
76 B2 **Guidonia-Montecelio** Italy
39 A3 **Guigang** China
68 A3 **Guignicourt** France
91 D1 **Guija** Moz.
67 C3 **Guildford** U.K.
39 B3 **Guilin** China
98 C1 **Guillaume-Delisle, Lac** *l.* Can.
74 B1 **Guimarães** Port.
82 A3 **Guinea** *country* Africa
80 D3 **Guinea, Gulf of** Africa
82 A3 **Guinea-Bissau** *country* Africa
72 B2 **Guingamp** France
72 B2 **Guipavas** France
122 B1 **Guiratinga** Brazil
118 B1 **Güiria** Venez.
68 A3 **Guise** France
32 B1 **Guiuan** Phil.
39 A3 **Guiyang** China
39 A3 **Guizhou** *prov.* China
42 B1 **Gujranwala** Pak.
42 B1 **Gujrat** Pak.
59 D2 **Gukovo** Rus. Fed.
44 B2 **Gulabie** Uzbek.
21 C2 **Gulargambone** Austr.
41 B3 **Gulbarga** India
56 C2 **Gulbene** Latvia
110 C2 **Gulfport** U.S.A.
37 E1 **Gulian** China
45 C2 **Gulistan** Uzbek.
Gulja China *see* **Yining**
97 D2 **Gull Lake** Can.
79 C3 **Güllük** Turkey
87 D2 **Gulu** Uganda
88 B2 **Gumare** Botswana
44 B3 **Gumdag** Turkm.
43 C2 **Gumla** India
68 C2 **Gummersbach** Ger.
42 B2 **Guna** India
21 C3 **Gundagai** Austr.
79 C3 **Güney** Turkey
86 B3 **Gungu** Dem. Rep. Congo
97 E2 **Gunisao** *r.* Can.
21 D2 **Gunnedah** Austr.
104 B3 **Gunnison** *CO* U.S.A.
103 D3 **Gunnison** *UT* U.S.A.
104 B3 **Gunnison** *r.* U.S.A.
41 B3 **Guntakal** India
28 A1 **Gunungsitoli** Indon.
28 A1 **Gunungtua** Indon.
70 C2 **Günzburg** Ger.
70 C2 **Gunzenhausen** Ger.
42 B2 **Gurgaon** India
119 D2 **Gurgueia** *r.* Brazil
118 B1 **Guri, Embalse de** *resr* Venez.
122 C1 **Gurinhatã** Brazil
119 D2 **Gurupi** *r.* Brazil
42 B2 **Guru Sikhar** *mt.* India
Gur'yev Kazakh. *see* **Atyrau**
83 C3 **Gusau** Nigeria
33 A2 **Gushan** China
42 A1 **Gushgy** Turkm.
38 B2 **Gushi** China
51 I3 **Gusinoozersk** Rus. Fed.
57 F2 **Gus'-Khrustal'nyy** Rus. Fed.
76 A3 **Guspini** *Sardinia* Italy
96 A2 **Gustavus** U.S.A.
69 F1 **Güstrow** Ger.
69 D2 **Gütersloh** Ger.
89 C2 **Gutu** Zimbabwe
43 D2 **Guwahati** India
118 C1 **Guyana** *country* S. America
Guyi China *see* **Sanjiang**
107 C1 **Guymon** U.S.A.
21 D2 **Guyra** Austr.
38 A2 **Guyuan** China
112 B1 **Guzmán** Mex.
112 B1 **Guzmán, Lago de** *l.* Mex.
42 A2 **Gwadar** Pak.
43 B2 **Gwalior** India
89 B3 **Gwanda** Zimbabwe
65 B1 **Gweebarra Bay** Rep. of Ireland
65 B1 **Gweedore** Rep. of Ireland
89 B2 **Gweru** Zimbabwe
83 D3 **Gwoza** Nigeria
21 D2 **Gwydir** *r.* Austr.
43 C2 **Gyangzê** China
43 C1 **Gyaring Co** *l.* China
36 C2 **Gyaring Hu** *l.* China
54 G1 **Gydanskiy Poluostrov** *pen.* Rus. Fed.
Gyêgu China *see* **Yushu**
19 E2 **Gympie** Austr.
71 D2 **Gyöngyös** Hungary
71 D2 **Győr** Hungary
97 E2 **Gypsumville** Can.
71 E2 **Gyula** Hungary
49 C1 **Gyumri** Armenia
44 B3 **Gyzylarbat** Turkm.

H

56 B2 **Haapsalu** Estonia
68 B1 **Haarlem** Neth.
69 C2 **Haarstrang** *ridge* Ger.
22 A2 **Haast** N.Z.
46 B3 **Habbān** Yemen
49 C2 **Ḩabbānīyah, Hawr al** *l.* Iraq
35 C4 **Hachijō-jima** *i.* Japan
34 D2 **Hachinohe** Japan
89 C3 **Hacufera** Moz.
47 C2 **Ḩadd, Ra's al** *pt* Oman
64 C3 **Haddington** U.K.
83 D3 **Hadejia** Nigeria
61 E4 **Haderslev** Denmark
59 C1 **Hadyach** Ukr.
33 B2 **Haeju** N. Korea
33 B2 **Haeju-man** *b.* N. Korea
33 B3 **Haenam** S. Korea
46 B2 **Ḩafar al Bāţin** Saudi Arabia
30 A1 **Hāflong** India
60 □A3 **Hafnarfjörður** Iceland
46 A3 **Hagar Nish Plateau** Eritrea
68 C2 **Hagen** Ger.
69 E1 **Hagenow** Ger.
96 B2 **Hagensborg** Can.
109 D3 **Hagerstown** U.S.A.
61 F3 **Hagfors** Sweden
35 B4 **Hagi** Japan
30 B1 **Ha Giang** Vietnam
65 B2 **Hag's Head** *hd* Rep. of Ireland
87 D3 **Hai** Tanz.
30 B1 **Hai Dương** Vietnam
39 B3 **Haifeng** China
39 B3 **Haikou** China
46 B2 **Ḩā'il** Saudi Arabia
37 D1 **Hailar** China
60 H2 **Hailuoto** *i.* Fin.
37 D3 **Hainan** *i.* China
39 A4 **Hainan** *prov.* China
96 A2 **Haines** U.S.A.
96 A1 **Haines Junction** Can.
69 E2 **Hainich** *ridge* Ger.
69 E2 **Hainleite** *ridge* Ger.
30 B1 **Hai Phong** Vietnam
115 C3 **Haiti** *country* West Indies

84 B3 **Haiya** Sudan
71 E2 **Hajdúböszörmény** Hungary
46 B3 **Ḥajjah** Yemen
47 C2 **Ḥājjīābād** Iran
30 A1 **Haka** Myanmar
49 C2 **Hakkârı** Turkey
34 D2 **Hakodate** Japan
Ḥalab Syria *see* **Aleppo**
46 B2 **Ḥalabān** Saudi Arabia
49 C2 **Halabja** Iraq
84 B2 **Halaib** Sudan
46 A2 **Halaib Triangle** *terr.* Egypt/Sudan
47 C3 **Ḥalāniyāt, Juzur al** *is* Oman
46 A2 **Ḥālat 'Ammār** Saudi Arabia
36 C1 **Halban** Mongolia
69 E2 **Halberstadt** Ger.
32 B1 **Halcon, Mount** Phil.
61 F4 **Halden** Norway
69 E1 **Haldensleben** Ger.
43 B2 **Haldwani** India
47 C2 **Hāleh** Iran
22 A3 **Halfmoon Bay** N.Z.
99 D2 **Halifax** Can.
66 C2 **Halifax** U.K.
109 D3 **Halifax** U.S.A.
33 B3 **Halla-san** *mt.* S. Korea
95 F2 **Hall Beach** Can.
68 B2 **Halle** Belgium
70 C2 **Hallein** Austria
69 E2 **Halle (Saale)** Ger.
105 D1 **Hallock** U.S.A.
95 G2 **Hall Peninsula** Can.
18 B1 **Halls Creek** Austr.
27 C2 **Halmahera** *i.* Indon.
61 F4 **Halmstad** Sweden
35 B4 **Hamada** Japan
49 C2 **Hamadān** Iran
48 B2 **Ḥamāh** Syria
35 C4 **Hamamatsu** Japan
61 F3 **Hamar** Norway
84 B2 **Hamâta, Gebel** *mt.* Egypt
41 C4 **Hambantota** Sri Lanka
69 D1 **Hamburg** Ger.
46 A2 **Ḥamḍ, Wādī al** *watercourse* Saudi Arabia
46 B3 **Ḥamḍah** Saudi Arabia
61 H3 **Hämeenlinna** Fin.
69 D1 **Hameln** Ger.
18 A2 **Hamersley Range** *mts* Austr.
33 B2 **Hamhŭng** N. Korea
36 C2 **Hami** China
84 B2 **Hamid** Sudan
20 B3 **Hamilton** Austr.
98 C2 **Hamilton** Can.
22 C1 **Hamilton** N.Z.
64 B3 **Hamilton** U.K.
110 C2 **Hamilton** *AL* U.S.A.
102 D1 **Hamilton** *MT* U.S.A.
108 C3 **Hamilton** *OH* U.S.A.
61 I3 **Hamina** Fin.
68 C2 **Hamm** Ger.
82 B2 **Hammada du Drâa** *plat.* Alg.
49 C2 **Ḥammār, Hawr al** *imp. l.* Iraq
69 D2 **Hammelburg** Ger.
60 G3 **Hammerdal** Sweden
60 H1 **Hammerfest** Norway
110 B2 **Hammond** U.S.A.
109 E3 **Hammonton** U.S.A.
47 C2 **Hāmūn-e Jaz Mūrīān** *salt marsh* Iran
42 A2 **Hamun-i-Lora** *dry lake* Pak.
42 A2 **Hamun-i-Mashkel** *salt flat* Pak.
46 A2 **Ḥanalc** Saudi Arabia
34 D3 **Hanamaki** Japan
69 D2 **Hanau** Ger.
38 B2 **Hancheng** China
108 B1 **Hancock** U.S.A.
38 B2 **Handan** China
103 C3 **Hanford** U.S.A.
36 C1 **Hangayn Nuruu** *mts* Mongolia
Hanggin Houqi China *see* **Xamba**
Hangö Fin. *see* **Hanko**
38 C2 **Hangzhou** China
38 C2 **Hangzhou Wan** *b.* China
Hanjia China *see* **Pengshui**
Hanjiang China *see* **Yangzhou**
61 H4 **Hanko** Fin.
103 D3 **Hanksville** U.S.A.
22 B2 **Hanmer Springs** N.Z.
97 C2 **Hanna** Can.
105 E3 **Hannibal** U.S.A.
69 D1 **Hannover** Ger.
69 D2 **Hannoversch Münden** Ger.
61 F4 **Hanöbukten** *b.* Sweden
30 B1 **Ha Nôi** Vietnam
Hanoi Vietnam *see* **Ha Nôi**
98 B2 **Hanover** Can.
90 B3 **Hanover** S. Africa
61 E4 **Hanstholm** Denmark
56 C3 **Hantsavichy** Belarus
43 C2 **Hanumana** India
42 B2 **Hanumangarh** India
38 A2 **Hanzhong** China
60 H2 **Haparanda** Sweden
68 B2 **Hapert** Neth.
99 D1 **Happy Valley-Goose Bay** Can.
46 A2 **Ḥaql** Saudi Arabia
47 B2 **Ḥaraḍh** Saudi Arabia
56 C2 **Haradok** Belarus
46 B3 **Ḥarajā** Saudi Arabia
89 C2 **Harare** Zimbabwe
47 C3 **Ḥarāsīs, Jiddat al** *des.* Oman
37 D1 **Har-Ayrag** Mongolia
37 E1 **Harbin** China
99 E2 **Harbour Breton** Can.
42 B2 **Harda Khās** India
68 C1 **Hardenberg** Neth.
68 B1 **Harderwijk** Neth.
90 A3 **Hardeveld** *mts* S. Africa
102 E1 **Hardin** U.S.A.
96 C1 **Hardisty Lake** Can.
68 C1 **Haren (Ems)** Ger.
85 C4 **Hārer** Eth.
85 C4 **Hargeysa** Somalia
78 C1 **Harghita-Mădăraș, Vârful** *mt.* Romania
36 C2 **Har Hu** *l.* China
42 B1 **Haripur** Pak.
42 A1 **Hari Rūd** *r.* Afgh./Iran
68 B1 **Harlingen** Neth.
107 D3 **Harlingen** U.S.A.
67 D3 **Harlow** U.K.
102 E1 **Harlowton** U.S.A.
102 C2 **Harney Basin** U.S.A.
102 C2 **Harney Lake** U.S.A.
61 G3 **Härnösand** Sweden
37 E1 **Har Nur** China
36 C1 **Har Nuur** *l.* Mongolia
82 B4 **Harper** Liberia
69 D1 **Harpstedt** Ger.
98 C1 **Harricanaw** *r.* Can.
21 D2 **Harrington** Austr.
99 E1 **Harrington Harbour** Can.
64 A2 **Harris, Sound of** *sea chan.* U.K.
108 B3 **Harrisburg** *IL* U.S.A.
109 D2 **Harrisburg** *PA* U.S.A.
91 C2 **Harrismith** S. Africa
110 B1 **Harrison** U.S.A.
99 E1 **Harrison, Cape** Can.
94 A2 **Harrison Bay** U.S.A.
109 D3 **Harrisonburg** U.S.A.
105 E3 **Harrisonville** U.S.A.
66 C2 **Harrogate** U.K.
78 C2 **Hârșova** Romania
60 G2 **Harstad** Norway
90 B2 **Hartbees** *watercourse* S. Africa
71 D2 **Hartberg** Austria
109 E2 **Hartford** U.S.A.
67 A3 **Hartland Point** U.K.
66 C1 **Hartlepool** U.K.
96 B2 **Hartley Bay** Can.
91 B2 **Harts** *r.* S. Africa
111 D2 **Hartwell Reservoir** U.S.A.
36 C1 **Har Us Nuur** *l.* Mongolia
104 C1 **Harvey** U.S.A.
67 D3 **Harwich** U.K.
69 E2 **Harz** *hills* Ger.
41 B3 **Hassan** India
68 B2 **Hasselt** Belgium
83 C1 **Hassi Messaoud** Alg.
61 F4 **Hässleholm** Sweden
68 B2 **Hastière-Lavaux** Belgium
21 C3 **Hastings** Austr.
22 C1 **Hastings** N.Z.
67 D3 **Hastings** U.K.
105 E2 **Hastings** *MN* U.S.A.
105 D2 **Hastings** *NE* U.S.A.
Hatay Turkey *see* **Antakya**
97 D2 **Hatchet Lake** Can.
20 B2 **Hatfield** Austr.
36 C1 **Hatgal** Mongolia
30 B2 **Ha Tĩnh** Vietnam
111 E1 **Hatteras, Cape** U.S.A.
110 C2 **Hattiesburg** U.S.A.
68 C2 **Hattingen** Ger.
31 B3 **Hat Yai** Thai.
85 C4 **Haud** *reg.* Eth.
61 E4 **Haugesund** Norway
61 E4 **Haukeligrend** Norway
60 I2 **Haukipudas** Fin.
22 C1 **Hauraki Gulf** N.Z.
82 B1 **Haut Atlas** *mts* Morocco
99 D2 **Hauterive** Can.
82 B1 **Hauts Plateaux** *plat.* Alg.
114 B2 **Havana** Cuba
67 C3 **Havant** U.K.
69 E1 **Havel** *r.* Ger.
69 F1 **Havelberg** Ger.
22 B2 **Havelock** N.Z.
22 C1 **Havelock North** N.Z.
67 A3 **Haverfordwest** U.K.
71 D2 **Havlíčkův Brod** Czech Rep.
60 H1 **Havøysund** Norway
79 C3 **Havran** Turkey
102 E1 **Havre** U.S.A.
99 D2 **Havre Aubert** Can.
99 D1 **Havre-St-Pierre** Can.
92 A5 **Hawaiian Islands** N. Pacific Ocean
66 B2 **Hawarden** U.K.
22 A2 **Hawea, Lake** N.Z.
22 B1 **Hawera** N.Z.
66 B1 **Hawes** U.K.
64 C3 **Hawick** U.K.
22 C1 **Hawke Bay** N.Z.
20 A2 **Hawker** Austr.
20 B1 **Hawkers Gate** Austr.
103 C3 **Hawthorne** U.S.A.
20 B2 **Hay** Austr.
96 C1 **Hay** *r.* Can.
102 C1 **Hayden** U.S.A.
97 E2 **Hayes** *r. Man.* Can.
94 E2 **Hayes** *r. Nunavut* Can.
47 C3 **Haymā'** Oman
79 C2 **Hayrabolu** Turkey
96 C1 **Hay River** Can.
105 D3 **Hays** U.S.A.
46 B3 **Ḥays** Yemen
58 B2 **Haysyn** Ukr.
67 C3 **Haywards Heath** U.K.
42 A1 **Hazarajat** *reg.* Afgh.

108 C3 **Hazard** U.S.A.
43 C2 **Hazāribāg** India
43 C2 **Hazaribagh Range** *mts* India
96 B2 **Hazelton** Can.
109 D2 **Hazleton** U.S.A.
21 C3 **Healesville** Austr.
98 B2 **Hearst** Can.
38 B2 **Hebei** *prov.* China
21 C1 **Hebel** Austr.
110 B1 **Heber Springs** U.S.A.
38 B2 **Hebi** China
99 D1 **Hebron** Can.
96 A2 **Hecate Strait** Can.
39 A3 **Hechi** China
61 F3 **Hede** Sweden
68 B1 **Heerenveen** Neth.
68 B1 **Heerhugowaard** Neth.
68 B2 **Heerlen** Neth.
48 B2 **Ḥefa** Israel
38 B2 **Hefei** China
38 B3 **Hefeng** China
37 E1 **Hegang** China
70 B1 **Heide** Ger.
90 A1 **Heide** Namibia
70 B2 **Heidelberg** Ger.
90 B3 **Heidelberg** S. Africa
70 B2 **Heilbronn** Ger.
37 E1 **Heilong Jiang** *r.* China/Rus. Fed.
61 I3 **Heinola** Fin.
60 F3 **Helagsfjället** *mt.* Sweden
110 B2 **Helena** *AR* U.S.A.
102 D1 **Helena** *MT* U.S.A.
64 B2 **Helensburgh** U.K.
70 B1 **Helgoland** *i.* Ger.
70 B1 **Helgoländer Bucht** *b.* Ger.
Helixi China *see* **Ningguo**
60 □A3 **Hella** Iceland
68 B2 **Hellevoetsluis** Neth.
75 C2 **Hellín** Spain
44 C3 **Helmand** *r.* Afgh.
69 E2 **Helmbrechts** Ger.
90 A2 **Helmeringhausen** Namibia
68 B2 **Helmond** Neth.
64 C1 **Helmsdale** U.K.
64 C1 **Helmsdale** *r.* U.K.
69 E1 **Helmstedt** Ger.
33 B1 **Helong** China
61 F4 **Helsingborg** Sweden
Helsingfors Fin. *see* **Helsinki**
61 F4 **Helsingør** Denmark
61 H3 **Helsinki** Fin.
65 C2 **Helvick Head** *hd* Rep. of Ireland
67 C3 **Hemel Hempstead** U.K.
69 D1 **Hemmoor** Ger.
60 F2 **Hemnesberget** Norway
38 B2 **Henan** *prov.* China
108 B3 **Henderson** *KY* U.S.A.
111 E1 **Henderson** *NC* U.S.A.
103 D3 **Henderson** *NV* U.S.A.
107 E2 **Henderson** *TX* U.S.A.
111 D1 **Hendersonville** U.S.A.
67 C3 **Hendon** U.K.
30 A1 **Hengduan Shan** *mts* China
68 C1 **Hengelo** Neth.
Hengnan China *see* Hengyang
38 B2 **Hengshui** China
39 A3 **Hengxian** China
39 B3 **Hengyang** China
Hengzhou China *see* Hengxian
59 C2 **Heniches'k** Ukr.
68 C2 **Hennef (Sieg)** Ger.
98 B1 **Henrietta Maria, Cape** Can.
107 D1 **Henryetta** U.S.A.
95 G2 **Henry Kater, Cape** Can.
69 D1 **Henstedt-Ulzburg** Ger.
88 A3 **Hentiesbaai** Namibia
30 A2 **Henzada** Myanmar
39 A3 **Hepu** China
44 C3 **Herāt** Afgh.
97 D2 **Herbert** Can.
69 D2 **Herbstein** Ger.
67 B2 **Hereford** U.K.
107 C2 **Hereford** U.S.A.
69 D1 **Herford** Ger.
68 C2 **Herkenbosch** Neth.
64 □ **Herma Ness** *hd* U.K.
90 A3 **Hermanus** S. Africa
21 C2 **Hermidale** Austr.
102 C1 **Hermiston** U.S.A.
27 D3 **Hermit Islands** P.N.G.
112 A2 **Hermosillo** Mex.
122 B3 **Hernandarias** Para.
68 C2 **Herne** Ger.
61 E4 **Herning** Denmark
67 C3 **Hertford** U.K.
91 C2 **Hertzogville** S. Africa
19 E2 **Hervey Bay** *b.* Austr.
69 F2 **Herzberg** Ger.
69 E3 **Herzogenaurach** Ger.
39 A3 **Heshan** China
96 A1 **Hess** *r.* Can.
69 D2 **Hessisch Lichtenau** Ger.
104 C1 **Hettinger** U.S.A.
69 E2 **Hettstedt** Ger.
66 B1 **Hexham** U.K.
39 B3 **Hexian** China
39 B3 **Heyuan** China
20 B3 **Heywood** Austr.
38 B2 **Heze** China
105 D3 **Hiawatha** U.S.A.
105 E1 **Hibbing** U.S.A.
22 C1 **Hicks Bay** N.Z.
34 D2 **Hidaka-sanmyaku** *mts* Japan
113 C2 **Hidalgo** Mex.
112 B2 **Hidalgo del Parral** Mex.
122 C1 **Hidrolândia** Brazil
High Atlas *mts* Morocco *see* Haut Atlas
102 B2 **High Desert** U.S.A.
96 C2 **High Level** Can.
111 E1 **High Point** U.S.A.
96 C2 **High Prairie** Can.
96 C2 **High River** Can.
97 D2 **Highrock Lake** Can.
67 C3 **High Wycombe** U.K.
56 B2 **Hiiumaa** *i.* Estonia
46 A2 **Hijaz** *reg.* Saudi Arabia
22 C1 **Hikurangi** *mt.* N.Z.
69 E2 **Hildburghausen** Ger.
69 E2 **Hilders** Ger.
69 D1 **Hildesheim** Ger.
68 C2 **Hillesheim** Ger.
108 C3 **Hillsboro** *OH* U.S.A.
107 D2 **Hillsboro** *TX* U.S.A.
21 C2 **Hillston** Austr.
111 D2 **Hilton Head Island** U.S.A.
68 B1 **Hilversum** Neth.
36 B2 **Himalaya** *mts* Asia
35 B4 **Himeji** Japan
91 C2 **Himeville** S. Africa
48 B2 **Ḥimş** Syria
19 D1 **Hinchinbrook Island** Austr.
42 A1 **Hindu Kush** *mts* Afgh./Pak.
111 D2 **Hinesville** U.S.A.
43 B2 **Hinganghat** India
49 C2 **Hınıs** Turkey
60 G2 **Hinnøya** *i.* Norway
74 B2 **Hinojosa del Duque** Spain
96 C2 **Hinton** Can.
43 C2 **Hirakud Reservoir** India
34 D2 **Hirosaki** Japan
35 B4 **Hiroshima** Japan
69 E3 **Hirschaid** Ger.
69 E2 **Hirschberg** Ger.
73 C2 **Hirson** France
61 E4 **Hirtshals** Denmark
42 B2 **Hisar** India
115 C2 **Hispaniola** *i.* Caribbean Sea
49 C2 **Hīt** Iraq
35 D3 **Hitachi** Japan
35 D3 **Hitachinaka** Japan
60 E3 **Hitra** *i.* Norway
61 G4 **Hjälmaren** *l.* Sweden
97 D1 **Hjalmar Lake** Can.
61 F4 **Hjørring** Denmark
91 D2 **Hlabisa** S. Africa
60 □B2 **Hlíð** Iceland
59 C2 **Hlobyne** Ukr.
91 C2 **Hlohlowane** S. Africa
91 C2 **Hlotse** Lesotho
59 C1 **Hlukhiv** Ukr.
56 C2 **Hlybokaye** Belarus
82 C4 **Ho** Ghana
90 A1 **Hoachanas** Namibia
19 D4 **Hobart** Austr.
107 D1 **Hobart** U.S.A.
107 C2 **Hobbs** U.S.A.
61 E4 **Hobro** Denmark
85 C4 **Hobyo** Somalia
31 B2 **Hô Chi Minh** Vietnam
82 B3 **Ḥôḍ** *reg.* Maur.
71 E2 **Hódmezővásárhely** Hungary
68 B2 **Hoek van Holland** Neth.
33 B2 **Hoeyang** N. Korea
69 E2 **Hof** Ger.
69 E2 **Hofheim in Unterfranken** Ger.
60 □B3 **Höfn** *Austurland* Iceland
60 □A2 **Höfn** *Vestfirðir* Iceland
60 □B3 **Hofsjökull** *ice cap* Iceland
35 B4 **Hōfu** Japan
83 C2 **Hoggar** *plat.* Alg.
61 G4 **Högsby** Sweden
69 D2 **Hohe Rhön** *mts* Ger.
68 C2 **Hohe Venn** *moorland* Belgium
43 C1 **Hoh Xil Shan** *mts* China
31 B2 **Hôi An** Vietnam
30 A1 **Hojai** India
22 B2 **Hokitika** N.Z.
34 D2 **Hokkaidō** *i.* Japan
96 B2 **Holberg** Can.
106 A2 **Holbrook** U.S.A.
105 D2 **Holdrege** U.S.A.
114 C2 **Holguín** Cuba
60 □B2 **Hóll** Iceland
Holland *country* Europe *see* Netherlands
108 B2 **Holland** U.S.A.
68 B1 **Hollum** Neth.
110 C2 **Holly Springs** U.S.A.
111 D3 **Hollywood** U.S.A.
60 F2 **Holm** Norway
94 D2 **Holman** Can.
60 H3 **Holmsund** Sweden
90 A2 **Holoog** Namibia
61 E4 **Holstebro** Denmark
111 D1 **Holston** *r.* U.S.A.
66 A2 **Holyhead** U.K.
66 C1 **Holy Island** *England* U.K.
66 A2 **Holy Island** *Wales* U.K.
104 C2 **Holyoke** U.S.A.
69 D2 **Holzminden** Ger.
30 A1 **Homalin** Myanmar
69 D2 **Homberg (Efze)** Ger.
82 B3 **Hombori** Mali
95 G2 **Home Bay** Can.
111 D3 **Homestead** U.S.A.
60 F3 **Hommelvik** Norway
57 D3 **Homyel'** Belarus
90 A3 **Hondeklipbaai** S. Africa
113 D3 **Hondo** *r.* Belize/Mex.
107 D3 **Hondo** U.S.A.
114 B3 **Honduras** *country* Central America

61 F3 **Hønefoss** Norway
103 B2 **Honey Lake** U.S.A.
72 C2 **Honfleur** France
30 B1 **Hông Gai** Vietnam
38 B3 **Honghu** China
39 A3 **Hongjiang** China
37 D3 **Hong Kong** China
39 B3 **Hong Kong** *special admin. reg.* China
33 B1 **Hongwŏn** N. Korea
38 B2 **Hongze Hu** *l.* China
34 D3 **Honjō** Japan
60 I1 **Honningsvåg** Norway
35 B3 **Honshū** *i.* Japan
102 B1 **Hood, Mount** *vol.* U.S.A.
18 A3 **Hood Point** Austr.
102 B1 **Hood River** U.S.A.
68 C1 **Hoogeveen** Neth.
68 C1 **Hoogezand-Sappemeer** Neth.
68 C2 **Hoog-Keppel** Neth.
Hook of Holland Neth. *see* **Hoek van Holland**
96 A2 **Hoonah** U.S.A.
68 B1 **Hoorn** Neth.
96 B3 **Hope** Can.
110 B2 **Hope** U.S.A.
51 N2 **Hope, Point** U.S.A.
99 D1 **Hopedale** Can.
99 D1 **Hope Mountains** Can.
20 B3 **Hopetoun** Austr.
90 B2 **Hopetown** S. Africa
109 D3 **Hopewell** U.S.A.
98 C1 **Hopewell Islands** Can.
18 B2 **Hopkins, Lake** *salt flat* Austr.
108 B3 **Hopkinsville** U.S.A.
102 B1 **Hoquiam** U.S.A.
49 C1 **Horasan** Turkey
61 F4 **Hörby** Sweden
57 D3 **Horki** Belarus
59 D2 **Horlivka** Ukr.
47 D2 **Hormak** Iran
47 C2 **Hormuz, Strait of** Iran/Oman
71 D2 **Horn** Austria
60 □A2 **Horn** *c.* Iceland
121 B6 **Horn, Cape** *c.* Chile
109 D2 **Hornell** U.S.A.
98 B2 **Hornepayne** Can.
66 C2 **Hornsea** U.K.
58 B2 **Horodenka** Ukr.
59 C1 **Horodnya** Ukr.
58 B2 **Horodok** *Khmel'nyts'ka Oblast'* Ukr.
58 A2 **Horodok** *L'vivs'ka Oblast'* Ukr.
58 A1 **Horokhiv** Ukr.
Horqin Youyi Qianqi China *see* **Ulanhot**
99 E1 **Horse Islands** Can.
20 B3 **Horsham** Austr.
94 C2 **Horton** *r.* Can.
85 B4 **Hosa'ina** Eth.
42 A2 **Hoshab** Pak.
42 B1 **Hoshiarpur** India
45 E3 **Hotan** China
90 B2 **Hotazel** S. Africa
110 B2 **Hot Springs** *AR* U.S.A.
104 C2 **Hot Springs** *SD* U.S.A.
96 C1 **Hottah Lake** Can.
68 B2 **Houffalize** Belgium
38 B2 **Houma** China
110 B3 **Houma** U.S.A.
96 B2 **Houston** Can.
107 D3 **Houston** U.S.A.
90 B3 **Houwater** S. Africa
36 C1 **Hovd** Mongolia
67 C3 **Hove** U.K.
36 C1 **Hövsgöl Nuur** *l.* Mongolia
36 C2 **Hövüün** Mongolia
84 A3 **Howar, Wadi** *watercourse* Sudan
21 C3 **Howe, Cape** Austr.
21 C3 **Howlong** Austr.
69 D2 **Höxter** Ger.
64 C1 **Hoy** *i.* U.K.
61 E3 **Høyanger** Norway
70 C1 **Hoyerswerda** Ger.
71 D1 **Hradec Králové** Czech Rep.
77 C2 **Hrasnica** Bos.-Herz.
59 C1 **Hrebinka** Ukr.
56 B3 **Hrodna** Belarus
30 A1 **Hsi-hseng** Myanmar
39 C3 **Hsinchu** Taiwan
39 C3 **Hsinying** Taiwan
30 A1 **Hsipaw** Myanmar
38 A2 **Huachi** China
118 A3 **Huacho** Peru
37 D2 **Huade** China
33 B1 **Huadian** China
38 B2 **Huaibei** China
39 A3 **Huaihua** China
38 B2 **Huainan** China
38 B2 **Huaiyang** China
38 B2 **Huaiyin** China
113 C3 **Huajuápan de León** Mex.
27 C3 **Huaki** Indon.
39 C3 **Hualien** Taiwan
118 A2 **Huallaga** *r.* Peru
88 A2 **Huambo** Angola
118 A3 **Huancayo** Peru
Huangcaoba China *see* **Xingyi**
38 B2 **Huangchuan** China
Huang Hai *sea* N. Pacific Ocean *see* **Yellow Sea**
38 B2 **Huang He** *r.* China
39 A4 **Huangliu** China
38 B3 **Huangshan** China
38 B2 **Huangshi** China
38 A2 **Huangtu Gaoyuan** *plat.* China
39 C3 **Huangyan** China
33 B1 **Huanren** China
118 A2 **Huánuco** Peru
120 B2 **Huanuni** Bol.
118 A2 **Huaráz** Peru
118 A3 **Huarmey** Peru
120 A3 **Huasco** Chile
120 A3 **Huasco** *r.* Chile
112 B2 **Huatabampo** Mex.
113 C3 **Huatusco** Mex.
39 A3 **Huayuan** China
38 B2 **Hubei** *prov.* China
41 B3 **Hubli** India
68 C2 **Hückelhoven** Ger.
66 C2 **Hucknall** U.K.
66 C2 **Huddersfield** U.K.
61 G3 **Hudiksvall** Sweden
109 E2 **Hudson** *r.* U.S.A.
97 D2 **Hudson Bay** Can.
95 F3 **Hudson Bay** *sea* Can.
96 B2 **Hudson's Hope** Can.
95 G2 **Hudson Strait** Can.
31 B2 **Huê** Vietnam
114 A3 **Huehuetenango** Guat.
112 B2 **Huehueto, Cerro** *mt.* Mex.
113 C2 **Huejutla** Mex.
74 B2 **Huelva** Spain
75 C2 **Huércal-Overa** Spain
75 C1 **Huesca** Spain
74 C2 **Huéscar** Spain
18 B3 **Hughes** Austr.
107 D2 **Hugo** U.S.A.
38 B1 **Huhhot** China
90 B2 **Huhudi** S. Africa
90 A2 **Huib-Hoch Plateau** Namibia
39 B3 **Huichang** China
33 B1 **Huich'ŏn** N. Korea
39 B3 **Huilai** China
88 A2 **Huíla Plateau** Angola
30 B1 **Huili** China
33 B1 **Huinan** China
61 H3 **Huittinen** Fin.
113 C3 **Huixtla** Mex.
Huiyang China *see* **Huizhou**
30 B1 **Huize** China
39 B3 **Huizhou** China
37 C1 **Hujirt** Mongolia
46 B2 **Hujr** Saudi Arabia
90 B1 **Hukuntsi** Botswana
46 B2 **Ḥulayfah** Saudi Arabia
34 B1 **Hulin** China
98 C2 **Hull** Can.
Hulun China *see* **Hailar**
37 D1 **Hulun Nur** *l.* China
59 D2 **Hulyaypole** Ukr.
37 E1 **Huma** China
118 B2 **Humaitá** Brazil
90 B3 **Humansdorp** S. Africa
66 C2 **Humber** *est.* U.K.
94 D3 **Humboldt** Can.
110 C1 **Humboldt** U.S.A.
103 C2 **Humboldt** *r.* U.S.A.
71 E2 **Humenné** Slovakia
21 C3 **Hume Reservoir** Austr.
106 A1 **Humphreys Peak** U.S.A.
33 A1 **Hun** *r.* China
60 □A2 **Húnaflói** *b.* Iceland
39 B3 **Hunan** *prov.* China
33 C1 **Hunchun** China
78 B1 **Hunedoara** Romania
69 D2 **Hünfeld** Ger.
71 D2 **Hungary** *country* Europe
20 B1 **Hungerford** Austr.
33 B2 **Hŭngnam** N. Korea
67 D2 **Hunstanton** U.K.
69 D1 **Hunte** *r.* Ger.
19 D4 **Hunter Islands** Austr.
67 C2 **Huntingdon** U.K.
108 B2 **Huntington** *IN* U.S.A.
108 C3 **Huntington** *WV* U.S.A.
22 C1 **Huntly** N.Z.
64 C2 **Huntly** U.K.
98 C2 **Huntsville** Can.
110 C2 **Huntsville** *AL* U.S.A.
107 D2 **Huntsville** *TX* U.S.A.
27 D3 **Huon Peninsula** P.N.G.
38 B2 **Huozhou** China
84 B2 **Hurghada** Egypt
105 D2 **Huron** U.S.A.
108 C2 **Huron, Lake** Can./U.S.A.
103 D3 **Hurricane** U.S.A.
60 □B2 **Húsavík** Iceland
78 C1 **Huşi** Romania
94 A2 **Huslia** U.S.A.
46 B3 **Ḥusn Āl 'Abr** Yemen
70 B1 **Husum** Ger.
37 C1 **Hutag** Mongolia
28 A1 **Hutanopan** Indon.
105 D3 **Hutchinson** U.S.A.
38 C2 **Huzhou** China
60 □C3 **Hvalnes** Iceland
60 □B3 **Hvannadalshnúkur** *vol.* Iceland
77 C2 **Hvar** *i.* Croatia
88 B2 **Hwange** Zimbabwe
104 C2 **Hyannis** U.S.A.
36 C1 **Hyargas Nuur** *salt l.* Mongolia
18 A3 **Hyden** Austr.
41 B3 **Hyderabad** India
42 A2 **Hyderabad** Pak.
73 D3 **Hyères** France
73 D3 **Hyères, Îles d'** *is* France
33 B1 **Hyesan** N. Korea
96 B2 **Hyland Post** Can.
35 B3 **Hyōno-sen** *mt.* Japan
67 D3 **Hythe** U.K.
61 H3 **Hyvinkää** Fin.

I

118 B2 **Iaco** *r.* Brazil
78 C2 **Ialomiţa** *r.* Romania
78 C1 **Ianca** Romania
78 C1 **Iaşi** Romania
32 A1 **Iba** Phil.
83 C4 **Ibadan** Nigeria
118 A1 **Ibagué** Col.
118 A1 **Ibarra** Ecuador
46 B3 **Ibb** Yemen
68 C1 **Ibbenbüren** Ger.
83 C4 **Ibi** Nigeria
123 C1 **Ibiá** Brazil
123 D1 **Ibiraçu** Brazil
75 D2 **Ibiza** *i.* Spain
119 D3 **Ibotirama** Brazil
47 C2 **Ibrā'** Oman
47 C2 **Ibrī** Oman
118 A3 **Ica** Peru
48 B2 **İçel** Turkey
60 □A3 **Iceland** *country* Europe
34 D3 **Ichinoseki** Japan
59 C1 **Ichnya** Ukr.
33 B2 **Ich'ŏn** N. Korea
107 E2 **Idabel** U.S.A.
102 D2 **Idaho** *state* U.S.A.
102 D2 **Idaho Falls** U.S.A.
68 C3 **Idar-Oberstein** Ger.
84 B2 **Idfu** Egypt
83 D2 **Idhān Awbārī** *des.* Libya
83 D2 **Idhān Murzūq** *des.* Libya
86 B3 **Idiofa** Dem. Rep. Congo
48 B2 **Idlib** Syria
122 B2 **Iepê** Brazil
68 A2 **Ieper** Belgium
87 D3 **Ifakara** Tanz.
89 □D3 **Ifanadiana** Madag.
83 C4 **Ife** Nigeria
82 C3 **Ifôghas, Adrar des** *hills* Mali
29 C1 **Igan** *Sarawak* Malaysia
122 C2 **Igarapava** Brazil
50 G2 **Igarka** Rus. Fed.
42 B3 **Igatpuri** India
49 C2 **Iğdır** Turkey
76 A3 **Iglesias** *Sardinia* Italy
95 F2 **Igloolik** Can.
Igluligaarjuk Can. *see* **Chesterfield Inlet**
98 A2 **Ignace** Can.
56 C2 **Ignalina** Lith.
78 C2 **İğneada** Turkey
79 B3 **Igoumenitsa** Greece
54 E3 **Igra** Rus. Fed.
54 F2 **Igrim** Rus. Fed.
122 B3 **Iguaçu** *r.* Brazil
122 B3 **Iguaçu Falls** Arg./Brazil
113 C3 **Iguala** Mex.
75 D1 **Igualada** Spain
122 C2 **Iguape** Brazil
122 B2 **Iguatemi** Brazil
122 B2 **Iguatemi** *r.* Brazil
119 E2 **Iguatu** Brazil
86 A3 **Iguéla** Gabon
87 D3 **Igunga** Tanz.
37 D2 **Ihbulag** Mongolia
89 □D3 **Ihosy** Madag.
60 I3 **Iisalmi** Fin.
83 C4 **Ijebu-Ode** Nigeria
68 B1 **IJmuiden** Neth.
68 B1 **IJssel** *r.* Neth.
68 B1 **IJsselmeer** *l.* Neth.
91 C2 **Ikageng** S. Africa
79 C3 **Ikaria** *i.* Greece
86 C3 **Ikela** Dem. Rep. Congo
78 B2 **Ikhtiman** Bulg.
35 A4 **Iki** *i.* Japan
89 □D3 **Ikongo** Madag.
33 B2 **Iksan** S. Korea
32 B1 **Ilagan** Phil.
49 C2 **Īlām** Iran
43 C2 **Ilam** Nepal
71 D1 **Iława** Pol.
47 C2 **Ilazārān, Kūh-e** *mt.* Iran
97 D2 **Île-à-la-Crosse** Can.
97 D2 **Île-à-la-Crosse, Lac** *l.* Can.
86 C3 **Ilebo** Dem. Rep. Congo
87 D2 **Ileret** Kenya
67 D3 **Ilford** U.K.
67 A3 **Ilfracombe** U.K.
123 D2 **Ilha Grande, Baía da** *b.* Brazil
122 B2 **Ilha Grande, Represa** *resr* Brazil
122 B2 **Ilha Solteíra, Represa** *resr* Brazil
74 B1 **Ílhavo** Port.
119 E3 **Ilhéus** Brazil
32 B2 **Iligan** Phil.
120 A4 **Illapel** Chile
58 C2 **Illichivs'k** Ukr.
108 A3 **Illinois** *r.* U.S.A.
108 B3 **Illinois** *state* U.S.A.
58 B2 **Illintsi** Ukr.
83 C2 **Illizi** Alg.
57 D2 **Il'men', Ozero** *l.* Rus. Fed.
69 E2 **Ilmenau** Ger.
118 A3 **Ilo** Peru
32 B1 **Iloilo** Phil.
60 J3 **Ilomantsi** Fin.
83 C4 **Ilorin** Nigeria
21 D1 **Iluka** Austr.
95 H2 **Ilulissat** Greenland
35 A4 **Imari** Japan
85 C4 **Īmī** Eth.
76 B2 **Imola** Italy
119 D2 **Imperatriz** Brazil
104 C2 **Imperial** U.S.A.
86 B2 **Impfondo** Congo
30 A1 **Imphal** India
79 C2 **İmroz** Turkey
118 B3 **Inambari** *r.* Peru
83 C2 **In Aménas** Alg.
27 C3 **Inanwatan** Indon.
60 I2 **Inari** Fin.
60 I2 **Inarijärvi** *l.* Fin.
35 D3 **Inawashiro-ko** *l.* Japan
48 B1 **İnce Burun** *pt* Turkey
33 B2 **Inch'ŏn** S. Korea
91 D2 **Incomati** *r.* Moz.
46 A3 **Inda Silasē** Eth.
112 B2 **Indé** Mex.
103 C3 **Independence** *CA* U.S.A.
105 E2 **Independence** *IA* U.S.A.
105 D3 **Independence** *KS* U.S.A.
105 E3 **Independence** *MO* U.S.A.
102 C2 **Independence Mountains** U.S.A.
44 B2 **Inderborskiy** Kazakh.
40 D2 **India** *country* Asia
109 D2 **Indiana** U.S.A.
108 B2 **Indiana** *state* U.S.A.
108 B3 **Indianapolis** U.S.A.
97 D2 **Indian Head** Can.
127 **Indian Ocean**
105 E2 **Indianola** *IA* U.S.A.
110 B2 **Indianola** *MS* U.S.A.
103 C3 **Indian Springs** U.S.A.
54 D2 **Indiga** Rus. Fed.
51 K2 **Indigirka** *r.* Rus. Fed.
77 D1 **Inđija** Yugo.
103 C4 **Indio** U.S.A.
26 B3 **Indonesia** *country* Asia
42 B2 **Indore** India
29 B2 **Indramayu, Tanjung** *pt* Indon.
72 C2 **Indre** *r.* France
42 A2 **Indus** *r.* China/Pak.
42 A2 **Indus, Mouths of the** Pak.
48 B1 **İnebolu** Turkey
79 C2 **İnegöl** Turkey
112 B3 **Infiernillo, Presa** *resr* Mex.
21 D1 **Inglewood** Austr.
70 C2 **Ingolstadt** Ger.
43 C2 **Ingrāj Bāzār** India
91 D2 **Inhaca** Moz.
89 C3 **Inhambane** Moz.
65 A2 **Inishbofin** *i.* Rep. of Ireland
65 B2 **Inishmore** *i.* Rep. of Ireland
65 C1 **Inishowen** *pen.* Rep. of Ireland
22 B2 **Inland Kaikoura Range** *mts* N.Z.
70 C2 **Inn** *r.* Europe
95 G1 **Innaanganeq** *c.* Greenland
Inner Mongolia *aut. reg.* China *see* **Nei Mongol Zizhiqu**
64 B2 **Inner Sound** *sea chan.* U.K.
19 D1 **Innisfail** Austr.
70 C2 **Innsbruck** Austria
122 B1 **Inocência** Brazil
86 B3 **Inongo** Dem. Rep. Congo
71 D1 **Inowrocław** Pol.
82 C2 **In Salah** Alg.
30 A2 **Insein** Myanmar
54 F2 **Inta** Rus. Fed.
105 E1 **International Falls** U.S.A.
98 C1 **Inukjuak** Can.
94 C2 **Inuvik** Can.
64 B2 **Inveraray** U.K.
22 A3 **Invercargill** N.Z.
21 D1 **Inverell** Austr.
64 B2 **Invergordon** U.K.
96 C2 **Invermere** Can.
99 D2 **Inverness** Can.
64 B2 **Inverness** U.K.
64 C2 **Inverurie** U.K.
20 A3 **Investigator Strait** Austr.
45 E1 **Inya** Rus. Fed.
87 D3 **Inyonga** Tanz.
55 D3 **Inza** Rus. Fed.
79 B3 **Ioannina** Greece
105 D3 **Iola** U.S.A.
64 A2 **Iona** *i.* U.K.
79 B3 **Ionian Islands** Greece
77 C3 **Ionian Sea** Greece/Italy
Ionoi Nisoi *is* Greece *see* **Ionian Islands**
79 C3 **Ios** *i.* Greece
105 E2 **Iowa** *state* U.S.A.
105 E2 **Iowa City** U.S.A.
122 C1 **Ipameri** Brazil
123 D1 **Ipatinga** Brazil
49 C1 **Ipatovo** Rus. Fed.
91 C2 **Ipelegeng** S. Africa
118 A1 **Ipiales** Col.
122 B3 **Ipiranga** Brazil
28 B1 **Ipoh** Malaysia
122 B1 **Iporá** Brazil
86 C2 **Ippy** C.A.R.
79 C2 **İpsala** Turkey
21 D1 **Ipswich** Austr.
67 D2 **Ipswich** U.K.
95 G2 **Iqaluit** Can.
120 A3 **Iquique** Chile
118 A2 **Iquitos** Peru
79 C3 **Irakleio** Greece
Iraklion Greece *see* **Irakleio**
44 B3 **Iran** *country* Asia
29 C1 **Iran, Pegunungan** *mts* Indon.
47 D2 **Īrānshahr** Iran
112 B2 **Irapuato** Mex.
49 C2 **Iraq** *country* Asia
122 B3 **Irati** Brazil
48 B2 **Irbid** Jordan

54 F3 **Irbit** Rus. Fed.
119 D3 **Irecê** Brazil
65 C2 **Ireland, Republic of** *country* Europe
86 C3 **Irema** Dem. Rep. Congo
44 C2 **Irgiz** Kazakh.
82 B3 **Irîgui** *reg.* Mali/Maur.
87 D3 **Iringa** Tanz.
119 C2 **Iriri** *r.* Brazil
63 B3 **Irish Sea** Rep. of Ireland/U.K.
37 C1 **Irkutsk** Rus. Fed.
20 A2 **Iron Knob** Austr.
108 B1 **Iron Mountain** U.S.A.
108 A1 **Ironwood** U.S.A.
32 B1 **Irosin** Phil.
35 C4 **Irō-zaki** *pt* Japan
58 C1 **Irpin'** Ukr.
30 A2 **Irrawaddy** *r.* Myanmar
31 A2 **Irrawaddy, Mouths of the** Myanmar
54 F2 **Irtysh** *r.* Kazakh./Rus. Fed.
75 C1 **Irún** Spain
64 B3 **Irvine** U.K.
32 B2 **Isabela** Phil.
114 B3 **Isabelia, Cordillera** *mts* Nic.
60 □A2 **Ísafjörður** Iceland
35 B4 **Isahaya** Japan
73 E2 **Isar** *r.* Ger.
64 □ **Isbister** U.K.
76 B2 **Ischia, Isola d'** *i.* Italy
35 C4 **Ise** Japan
86 C2 **Isengi** Dem. Rep. Congo
68 C2 **Iserlohn** Ger.
69 D1 **Isernhagen** Ger.
35 C4 **Ise-wan** *b.* Japan
82 C4 **Iseyin** Nigeria
Isfahan Iran *see* **Eşfahān**
34 D2 **Ishikari-wan** *b.* Japan
45 D1 **Ishim** *r.* Kazakh./Rus. Fed.
34 D3 **Ishinomaki** Japan
35 D3 **Ishioka** Japan
108 B1 **Ishpeming** U.S.A.
79 C2 **Işıklar Dağı** *mts* Turkey
79 C3 **Işıklı** Turkey
91 D2 **Isipingo** S. Africa
87 C2 **Isiro** Dem. Rep. Congo
48 B2 **İskenderun** Turkey
50 G3 **Iskitim** Rus. Fed.
78 B2 **Iskŭr** *r.* Bulg.
96 A2 **Iskut** *r.* Can.
42 B1 **Islamabad** Pak.
20 A2 **Island Lagoon** *salt flat* Austr.
97 E2 **Island Lake** *l.* Can.
22 B1 **Islands, Bay of** N.Z.
64 A3 **Islay** *i.* U.K.
66 A1 **Isle of Man** *i.* Irish Sea
84 B1 **Ismâ'ilîya** Egypt
77 C3 **Isola di Capo Rizzuto** Italy
78 C2 **Isperikh** Bulg.
48 B2 **Israel** *country* Asia
73 C2 **Issoire** France
79 C2 **İstanbul** Turkey
İstanbul Boğazı *str.* Turkey *see* **Bosporus**
79 B3 **Istiaia** Greece
76 B1 **Istra** *pen.* Croatia
73 C3 **Istres** France
119 D3 **Itaberaba** Brazil
123 D1 **Itabira** Brazil
123 D2 **Itabirito** Brazil
119 E3 **Itabuna** Brazil
118 C2 **Itacoatiara** Brazil
122 B2 **Itaguajé** Brazil
122 C2 **Itaí** Brazil
122 B3 **Itaipu, Represa de** *resr* Brazil
119 C2 **Itaituba** Brazil
120 D3 **Itajaí** Brazil
123 C2 **Itajubá** Brazil
76 B2 **Italy** *country* Europe
123 D1 **Itamarandiba** Brazil
123 D1 **Itambacuri** Brazil
123 D1 **Itambé, Pico de** *mt.* Brazil
30 A1 **Itanagar** India
122 C2 **Itanhaém** Brazil
123 D1 **Itanhém** Brazil
123 D1 **Itaobím** Brazil
122 C1 **Itapajipe** Brazil
123 E1 **Itapebi** Brazil
123 D2 **Itapemirim** Brazil
123 D2 **Itaperuna** Brazil
123 D1 **Itapetinga** Brazil
122 C2 **Itapetininga** Brazil
122 C2 **Itapeva** Brazil
119 E3 **Itapicuru** *r.* Brazil
119 D2 **Itapicuru Mirim** Brazil
122 C1 **Itapuranga** Brazil
122 C2 **Itararé** Brazil
43 B2 **Itarsi** India
122 B1 **Itarumã** Brazil
123 D2 **Itaúna** Brazil
39 C3 **Itbayat** *i.* Phil.
109 D2 **Ithaca** U.S.A.
69 D1 **Ith Hils** *ridge* Ger.
86 C2 **Itimbiri** *r.* Dem. Rep. Congo
123 D1 **Itinga** Brazil
122 B1 **Itiquira** Brazil
122 A1 **Itiquira** *r.* Brazil
35 C4 **Itō** Japan
122 C2 **Itu** Brazil
118 A2 **Ituí** *r.* Brazil
122 C1 **Ituiutaba** Brazil
87 C3 **Itula** Dem. Rep. Congo
122 C1 **Itumbiara** Brazil
122 B1 **Iturama** Brazil
69 D1 **Itzehoe** Ger.
51 N2 **Iul'tin** Rus. Fed.
122 B2 **Ivaí** *r.* Brazil
60 I2 **Ivalo** Fin.
56 C3 **Ivanava** Belarus
20 B2 **Ivanhoe** Austr.
58 B1 **Ivankiv** Ukr.
58 A2 **Ivano-Frankivs'k** Ukr.
57 F2 **Ivanovo** Rus. Fed.
56 C3 **Ivatsevichy** Belarus
79 C2 **Ivaylovgrad** Bulg.
54 F2 **Ivdel'** Rus. Fed.
122 B2 **Ivinheima** Brazil
122 B2 **Ivinheima** *r.* Brazil
95 H2 **Ivittuut** Greenland
Ivory Coast *country* Africa *see* **Côte d'Ivoire**
76 A1 **Ivrea** Italy
79 C3 **İvrindi** Turkey
95 F2 **Ivujivik** Can.
35 D3 **Iwaki** Japan
35 B4 **Iwakuni** Japan
34 D2 **Iwamizawa** Japan
34 D2 **Iwanai** Japan
56 C3 **Iwye** Belarus
91 D3 **Ixopo** S. Africa
112 B2 **Ixtlán** Mex.
113 D2 **Izamal** Mex.
49 C1 **Izberbash** Rus. Fed.
54 E3 **Izhevsk** Rus. Fed.
54 E2 **Izhma** Rus. Fed.
57 E3 **Izmalkovo** Rus. Fed.
58 B2 **Izmayil** Ukr.
79 C3 **İzmir** Turkey
79 C2 **İznik Gölü** *l.* Turkey
113 C3 **Izúcar de Matamoros** Mex.
35 B3 **Izumo** Japan
58 B1 **Izyaslav** Ukr.
59 D2 **Izyum** Ukr.

J

Jabal, Bahr el *r.* Sudan/Uganda *see* **White Nile**
74 C2 **Jabalón** *r.* Spain
43 B2 **Jabalpur** India
18 C1 **Jabiru** Austr.
77 C2 **Jablanica** Bos.-Herz.
119 E2 **Jaboatão** Brazil
122 C2 **Jaboticabal** Brazil
75 C1 **Jaca** Spain
113 C2 **Jacala** Mex.
119 C2 **Jacareacanga** Brazil
123 C2 **Jacareí** Brazil
123 D1 **Jacinto** Brazil
109 E1 **Jackman** U.S.A.
110 C2 **Jackson** *AL* U.S.A.
108 C2 **Jackson** *MI* U.S.A.
110 B2 **Jackson** *MS* U.S.A.
110 C1 **Jackson** *TN* U.S.A.
104 A2 **Jackson** *WY* U.S.A.
22 A2 **Jackson Head** *hd* N.Z.
110 B2 **Jacksonville** *AR* U.S.A.
111 D2 **Jacksonville** *FL* U.S.A.
108 A3 **Jacksonville** *IL* U.S.A.
111 E2 **Jacksonville** *NC* U.S.A.
107 D2 **Jacksonville** *TX* U.S.A.
115 C3 **Jacmel** Haiti
42 A2 **Jacobabad** Pak.
119 D3 **Jacobina** Brazil
99 D2 **Jacques Cartier, Mont** *mt.* Can.
119 D2 **Jacunda** Brazil
122 C2 **Jacupiranga** Brazil
77 C2 **Jadovnik** *mt.* Bos.-Herz.
118 A2 **Jaén** Peru
74 C2 **Jaén** Spain
20 A3 **Jaffa, Cape** Austr.
41 B4 **Jaffna** Sri Lanka
41 C3 **Jagdalpur** India
91 C2 **Jagersfontein** S. Africa
47 C2 **Jaghīn** Iran
122 C2 **Jaguariaíva** Brazil
47 C2 **Jahrom** Iran
42 B2 **Jaipur** India
42 B2 **Jaisalmer** India
43 C2 **Jajarkot** Nepal
77 C2 **Jajce** Bos.-Herz.
28 B2 **Jakarta** Indon.
96 A1 **Jakes Corner** Can.
60 G2 **Jäkkvik** Sweden
60 H3 **Jakobstad** Fin.
45 D3 **Jalālābād** Afgh.
45 D2 **Jalal-Abad** Kyrg.
42 B1 **Jalandhar** India
113 C3 **Jalapa Enríquez** Mex.
122 B2 **Jales** Brazil
42 B2 **Jalgaon** India
83 D4 **Jalingo** Nigeria
42 B3 **Jalna** India
112 B2 **Jalpa** Mex.
43 C2 **Jalpaiguri** India
113 C2 **Jalpan** Mex.
83 E2 **Jālū** Libya
114 C3 **Jamaica** *country* West Indies
114 C3 **Jamaica Channel** Haiti/Jamaica
43 C2 **Jamalpur** Bangl.
28 B2 **Jambi** Indon.
105 D2 **James** *r. ND/SD* U.S.A.
109 D3 **James** *r. VA* U.S.A.
98 B1 **James Bay** Can.
20 A2 **Jamestown** Austr.
105 D1 **Jamestown** *ND* U.S.A.
109 D2 **Jamestown** *NY* U.S.A.
42 B1 **Jammu** Jammu and Kashmir
42 B1 **Jammu and Kashmir** *terr.* Asia

42 B2 **Jamnagar** India
61 I3 **Jämsä** Fin.
43 C2 **Jamshedpur** India
43 C2 **Jamuna** *r.* Bangl.
43 C2 **Janakpur** Nepal
123 D1 **Janaúba** Brazil
49 D2 **Jandaq** Iran
108 B2 **Janesville** U.S.A.
52 C2 **Jan Mayen** *i.* Arctic Ocean
90 B3 **Jansenville** S. Africa
123 D1 **Januária** Brazil
42 B2 **Jaora** India
35 B4 **Japan** *country* Asia
124 C3 **Japan, Sea of** N. Pacific Ocean
118 B2 **Japurá** *r.* Brazil
122 C1 **Jaraguá** Brazil
122 B2 **Jaraguari** Brazil
120 C3 **Jardim** Brazil
37 D1 **Jargalant** Mongolia
71 D1 **Jarocin** Pol.
71 E1 **Jarosław** Pol.
60 F3 **Järpen** Sweden
38 A2 **Jartai** China
118 B3 **Jarú** Brazil
Jarud China *see* **Lubei**
47 C2 **Jäsk** Iran
71 E2 **Jasło** Pol.
96 C2 **Jasper** Can.
108 B3 **Jasper** *IN* U.S.A.
107 E2 **Jasper** *TX* U.S.A.
71 D2 **Jastrzębie-Zdrój** Pol.
71 D2 **Jászberény** Hungary
122 B1 **Jataí** Brazil
42 A2 **Jati** Pak.
122 C2 **Jaú** Brazil
118 B2 **Jaú** *r.* Brazil
113 C2 **Jaumave** Mex.
43 C2 **Jaunpur** India
122 B1 **Jauru** Brazil
29 B2 **Java** *i.* Indon.
37 D1 **Javarthushuu** Mongolia
127 D4 **Java Sea** Indon.
Jawa *i.* Indon. *see* **Java**
85 C4 **Jawhar** Somalia
71 D1 **Jawor** Pol.
71 D1 **Jaworzno** Pol.
27 D3 **Jaya, Puncak** *mt.* Indon.
27 D3 **Jayapura** Indon.
96 B1 **Jean Marie River** Can.
99 D1 **Jeannin, Lac** *l.* Can.
84 A3 **Jebel Abyad Plateau** Sudan
64 C3 **Jedburgh** U.K.
46 A2 **Jeddah** Saudi Arabia
69 E1 **Jeetze** *r.* Ger.
103 C3 **Jefferson, Mount** U.S.A.
105 E3 **Jefferson City** U.S.A.
56 C2 **Jēkabpils** Latvia
71 D1 **Jelenia Góra** Pol.
56 B2 **Jelgava** Latvia
29 C2 **Jember** Indon.
69 E2 **Jena** Ger.
48 B2 **Jenīn** West Bank
110 B2 **Jennings** U.S.A.
119 D3 **Jequié** Brazil
123 D1 **Jequitaí** Brazil
123 D1 **Jequitinhonha** Brazil
123 E1 **Jequitinhonha** *r.* Brazil
115 C3 **Jérémie** Haiti
112 B2 **Jerez** Mex.
74 B2 **Jerez de la Frontera** Spain
77 D3 **Jergucat** Albania
83 C1 **Jerid, Chott el** *salt l.* Tunisia
102 D2 **Jerome** U.S.A.
63 C4 **Jersey** *terr.* Channel Is
119 D2 **Jerumenha** Brazil
48 B2 **Jerusalem** Israel/West Bank
21 D3 **Jervis Bay Territory** *admin. div.* Austr.
76 B1 **Jesenice** Slovenia
76 B2 **Jesi** Italy
69 F2 **Jessen** Ger.
43 C2 **Jessore** Bangl.
111 D2 **Jesup** U.S.A.
113 C3 **Jesús Carranza** Mex.
42 B2 **Jhalawar** India
42 B1 **Jhang** Pak.
43 B2 **Jhansi** India
43 C2 **Jharsuguda** India
42 B1 **Jhelum** Pak.
38 C2 **Jiading** China
37 E1 **Jiamusi** China
39 B3 **Ji'an** *Jiangxi* China
33 B1 **Ji'an** *Jilin* China
30 A1 **Jianchuan** China
Jiandaoyu China *see* **Zigui**
38 B2 **Jiangsu** *prov.* China
39 B3 **Jiangxi** *prov.* China
38 A2 **Jiangyou** China
38 B3 **Jianli** China
39 B3 **Jianyang** *Fujian* China
38 A2 **Jianyang** *Sichuan* China
38 B2 **Jiaohe** China
38 C2 **Jiaozhou** China
38 B2 **Jiaozuo** China
38 C2 **Jiaxing** China
36 C2 **Jiayuguan** China
Jiddah Saudi Arabia *see* **Jeddah**
60 G2 **Jiehkkevarri** *mt.* Norway
38 B2 **Jiexiu** China
36 C2 **Jigzhi** China
71 D2 **Jihlava** Czech Rep.
85 C4 **Jijiga** Eth.
85 C4 **Jilib** Somalia
37 E2 **Jilin** China
33 B1 **Jilin** *prov.* China
33 A1 **Jilin Handa Ling** *mts* China
85 B4 **Jīma** Eth.
112 B2 **Jiménez** *Chihuahua* Mex.
113 C2 **Jiménez** *Tamaulipas* Mex.
38 B2 **Jinan** China
38 B2 **Jincheng** China
21 C3 **Jindabyne** Austr.
71 D2 **Jindřichův Hradec** Czech Rep.
39 B3 **Jingdezhen** China
39 B3 **Jinggangshan** China
30 B1 **Jinghong** China
38 B2 **Jingmen** China
38 A2 **Jingning** China
38 B2 **Jingsha** China
38 A2 **Jingtai** China
39 A3 **Jingxi** China
33 B1 **Jingyu** China
38 A2 **Jingyuan** China
38 B2 **Jingzhou** China
39 B3 **Jinhua** China
38 B1 **Jining** *Nei Mongol* China
38 B2 **Jining** *Shandong* China
87 D2 **Jinja** Uganda
85 B4 **Jinka** Eth.
114 B3 **Jinotepe** Nic.
39 A3 **Jinping** China
Jinsha Jiang *r.* China *see* Yangtze
38 B3 **Jinshi** China
38 C1 **Jinzhou** China
118 B2 **Jiparaná** *r.* Brazil
47 C2 **Jīroft** Iran
39 A3 **Jishou** China
78 B2 **Jiu** *r.* Romania
38 A2 **Jiuding Shan** *mt.* China
38 B3 **Jiujiang** China
34 B1 **Jixi** China
46 B3 **Jīzān** Saudi Arabia
119 E2 **João Pessoa** Brazil
123 C1 **João Pinheiro** Brazil
42 B2 **Jodhpur** India
60 I3 **Joensuu** Fin.
35 C3 **Jōetsu** Japan
89 C3 **Jofane** Moz.
56 C2 **Jõgeva** Estonia
91 C2 **Johannesburg** S. Africa
102 C2 **John Day** U.S.A.
102 B1 **John Day** *r.* U.S.A.
96 C2 **John d'Or Prairie** Can.
111 E1 **John H. Kerr Reservoir** U.S.A.
64 C1 **John o'Groats** U.K.
111 D1 **Johnson City** U.S.A.
96 A1 **Johnson's Crossing** Can.
17 G2 **Johnston Atoll** N. Pacific Ocean
64 B3 **Johnstone** U.K.
109 D2 **Johnstown** U.S.A.
28 B1 **Johor Bahru** Malaysia
56 C2 **Jõhvi** Estonia
122 C3 **Joinville** Brazil
73 D2 **Joinville** France
60 G2 **Jokkmokk** Sweden
60 □B2 **Jökulsá á Fjöllum** *r.* Iceland
108 B2 **Joliet** U.S.A.
98 C2 **Joliette** Can.
32 B2 **Jolo** Phil.
32 B2 **Jolo** *i.* Phil.
29 C2 **Jombang** Indon.
43 C2 **Jomsom** Nepal
56 B2 **Jonava** Lith.
110 B1 **Jonesboro** *AR* U.S.A.
110 B2 **Jonesboro** *LA* U.S.A.
95 F1 **Jones Sound** *sea chan.* Can.
61 F4 **Jönköping** Sweden
99 C2 **Jonquière** Can.
113 C3 **Jonuta** Mex.
105 E3 **Joplin** U.S.A.
48 B2 **Jordan** *country* Asia
48 B2 **Jordan** *r.* Asia
104 B1 **Jordan** U.S.A.
102 C2 **Jordan Valley** U.S.A.
30 A1 **Jorhat** India
61 E4 **Jørpeland** Norway
83 C4 **Jos** Nigeria
113 C3 **José Cardel** Mex.
99 D1 **Joseph, Lac** *l.* Can.
18 B1 **Joseph Bonaparte Gulf** Austr.
83 C4 **Jos Plateau** Nigeria
61 E3 **Jotunheimen** *mts* Norway
90 B3 **Joubertina** S. Africa
91 C2 **Jouberton** S. Africa
61 I3 **Joutseno** Fin.
102 B1 **Juan de Fuca Strait** Can./U.S.A.
Juanshui China *see* Tongcheng
113 B2 **Juaréz** Mex.
119 D2 **Juàzeiro** Brazil
119 E2 **Juàzeiro do Norte** Brazil
85 B4 **Juba** Sudan
85 C5 **Jubba** *r.* Somalia
46 B2 **Jubbah** Saudi Arabia
75 C2 **Júcar** *r.* Spain
113 C3 **Juchitán** Mex.
70 C2 **Judenburg** Austria
69 D2 **Jühnde** Ger.
114 B3 **Juigalpa** Nic.
68 C1 **Juist** *i.* Ger.
123 D2 **Juiz de Fora** Brazil
118 A3 **Juliaca** Peru
43 C2 **Jumla** Nepal
42 B2 **Junagadh** India
107 D2 **Junction** U.S.A.
105 D3 **Junction City** U.S.A.
122 C2 **Jundiaí** Brazil
96 A2 **Juneau** U.S.A.
21 C2 **Junee** Austr.
73 D2 **Jungfrau** *mt.* Switz.
109 D2 **Juniata** *r.* U.S.A.

60 G3 **Junsele** Sweden
102 C2 **Juntura** U.S.A.
122 C2 **Juquiá** Brazil
85 A4 **Jur** *r.* Sudan
73 D2 **Jura** *mts* France/Switz.
64 B2 **Jura** *i.* U.K.
64 B3 **Jura, Sound of** *sea chan.* U.K.
56 B2 **Jurbarkas** Lith.
56 B2 **Jūrmala** Latvia
118 B2 **Juruá** *r.* Brazil
118 C2 **Juruena** *r.* Brazil
118 B2 **Jutaí** *r.* Brazil
69 F2 **Jüterbog** Ger.
38 B2 **Juxian** China
47 C2 **Jūyom** Iran
90 B1 **Jwaneng** Botswana
61 I3 **Jyväskylä** Fin.

K

42 B1 **K2** *mt.* China/ Jammu and Kashmir
87 C3 **Kabalo** Dem. Rep. Congo
87 C3 **Kabare** Dem. Rep. Congo
98 B2 **Kabinakagami Lake** Can.
86 C3 **Kabinda** Dem. Rep. Congo
86 B2 **Kabo** C.A.R.
88 B2 **Kabompo** Zambia
87 C3 **Kabongo** Dem. Rep. Congo
45 C3 **Kābul** Afgh.
32 B2 **Kaburuang** *i.* Indon.
89 B2 **Kabwe** Zambia
42 A2 **Kachchh, Gulf of** India
51 I3 **Kachug** Rus. Fed.
49 C1 **Kaçkar Dağı** *mt.* Turkey
79 C2 **Kadıköy** Turkey
82 B3 **Kadiolo** Mali
41 B3 **Kadmat** *i.* India
89 B2 **Kadoma** Zimbabwe
31 A2 **Kadonkani** Myanmar
85 A3 **Kadugli** Sudan
83 C3 **Kaduna** Nigeria
57 E2 **Kaduy** Rus. Fed.
54 E2 **Kadzherom** Rus. Fed.
82 A3 **Kaédi** Maur.
86 B1 **Kaélé** Cameroon
33 B2 **Kaesŏng** N. Korea
82 A3 **Kaffrine** Senegal
79 B3 **Kafireas, Akra** *pt* Greece
89 B2 **Kafue** Zambia
88 B2 **Kafue** *r.* Zambia
86 B2 **Kaga Bandoro** C.A.R.
59 E2 **Kagal'nitskaya** Rus. Fed.
35 B4 **Kagoshima** Japan
58 C2 **Kaharlyk** Ukr.
29 C2 **Kahayan** *r.* Indon.
86 B3 **Kahemba** Dem. Rep. Congo
69 E2 **Kahla** Ger.
47 C2 **Kahnūj** Iran
60 H2 **Kahperusvaarat** *mts* Fin.
48 B2 **Kahramanmaraş** Turkey
47 C2 **Kahūrak** Iran
27 C3 **Kai, Kepulauan** *is* Indon.
83 C4 **Kaiama** Nigeria
22 B2 **Kaiapoi** N.Z.
27 C3 **Kai Besar** *i.* Indon.
38 B2 **Kaifeng** China
Kaihua China *see* **Wenshan**
90 B2 **Kaiingveld** *reg.* S. Africa
27 C3 **Kai Kecil** *i.* Indon.
22 B2 **Kaikoura** N.Z.
82 A4 **Kailahun** Sierra Leone
39 A3 **Kaili** China
27 C3 **Kaimana** Indon.
22 C1 **Kaimanawa Mountains** N.Z.
35 C4 **Kainan** Japan
83 C3 **Kainji Reservoir** Nigeria
22 B1 **Kaipara Harbour** N.Z.
42 B2 **Kairana** India
83 D1 **Kairouan** Tunisia
22 B1 **Kaitaia** N.Z.
22 C1 **Kaitawa** N.Z.
Kaitong China *see* **Tongyu**
27 C3 **Kaiwatu** Indon.
33 A1 **Kaiyuan** *Liaoning* China
30 B1 **Kaiyuan** *Yunnan* China
60 I3 **Kajaani** Fin.
19 D2 **Kajabbi** Austr.
90 B2 **Kakamas** S. Africa
87 D2 **Kakamega** Kenya
59 C2 **Kakhovka** Ukr.
59 C2 **Kakhovs'ke Vodoskhovyshche** *resr* Ukr.
41 C3 **Kākināda** India
96 C1 **Kakisa** Can.
87 C3 **Kakoswa** Dem. Rep. Congo
94 B2 **Kaktovik** U.S.A.
Kalaallit Nunaat *terr.* N. America *see* **Greenland**
27 C3 **Kalabahi** Indon.
88 B2 **Kalabo** Zambia
59 E1 **Kalach** Rus. Fed.
87 D2 **Kalacha Dida** Kenya
30 A1 **Kaladan** *r.* India/Myanmar
88 B3 **Kalahari Desert** Africa
60 H3 **Kalajoki** Fin.
79 B2 **Kalamaria** Greece
79 B3 **Kalamata** Greece
108 B2 **Kalamazoo** U.S.A.
79 B3 **Kalampaka** Greece
59 C2 **Kalanchak** Ukr.
31 B2 **Kalasin** Thai.
47 C2 **Kalāt** Iran
42 A2 **Kalat** Pak.
18 A2 **Kalbarri** Austr.
79 C3 **Kale** Turkey
48 B1 **Kalecik** Turkey
86 C3 **Kalema** Dem. Rep. Congo
87 C3 **Kalémié** Dem. Rep. Congo
30 A1 **Kalemyo** Myanmar
Kalgan China *see* **Zhangjiakou**
18 B3 **Kalgoorlie** Austr.
77 C2 **Kali** Croatia
78 C2 **Kaliakra, Nos** *pt* Bulg.
87 C3 **Kalima** Dem. Rep. Congo
29 C2 **Kalimantan** *reg.* Indon.
56 B3 **Kaliningrad** Rus. Fed.
59 D2 **Kalininskaya** Rus. Fed.
56 C3 **Kalinkavichy** Belarus
102 D1 **Kalispell** U.S.A.
71 D1 **Kalisz** Pol.
59 E2 **Kalitva** *r.* Rus. Fed.
60 H2 **Kalix** Sweden
60 H2 **Kalixälven** *r.* Sweden
79 C3 **Kalkan** Turkey
88 A3 **Kalkfeld** Namibia
60 I3 **Kallavesi** *l.* Fin.
60 F3 **Kallsjön** *l.* Sweden
61 G4 **Kalmar** Sweden
61 G4 **Kalmarsund** *sea chan.* Sweden
88 B2 **Kalomo** Zambia
96 B2 **Kalone Peak** Can.
43 B1 **Kalpa** India
41 B3 **Kalpeni** *i.* India
43 B2 **Kalpi** India
69 D1 **Kaltenkirchen** Ger.
57 E3 **Kaluga** Rus. Fed.
61 F4 **Kalundborg** Denmark
58 A2 **Kalush** Ukr.
57 E2 **Kalyazin** Rus. Fed.
79 C3 **Kalymnos** *i.* Greece
87 C3 **Kama** Dem. Rep. Congo
30 A2 **Kama** Myanmar
54 E3 **Kama** *r.* Rus. Fed.
34 D3 **Kamaishi** Japan
48 B2 **Kaman** Turkey
88 A2 **Kamanjab** Namibia
46 B3 **Kamarān** *i.* Yemen
42 A2 **Kamarod** Pak.
18 B3 **Kambalda** Austr.
87 C4 **Kambove** Dem. Rep. Congo
78 C2 **Kamchiya** *r.* Bulg.
78 B2 **Kamenitsa** *mt.* Bulg.
76 B2 **Kamenjak, Rt** *pt* Croatia
44 B1 **Kamenka** Kazakh.
59 D1 **Kamenka** Rus. Fed.
59 E2 **Kamenolomni** Rus. Fed.
51 M2 **Kamenskoye** Rus. Fed.
59 E2 **Kamensk-Shakhtinskiy** Rus. Fed.
54 F3 **Kamensk-Ural'skiy** Rus. Fed.
90 A3 **Kamiesberge** *mts* S. Africa
90 A3 **Kamieskroon** S. Africa
97 D1 **Kamilukuak Lake** *l.* Can.
87 C3 **Kamina** Dem. Rep. Congo
97 E1 **Kaminak Lake** Can.
58 A1 **Kamin'-Kashyrs'kyy** Ukr.
96 B2 **Kamloops** Can.
86 C3 **Kamonia** Dem. Rep. Congo
87 D2 **Kampala** Uganda
28 B1 **Kampar** *r.* Indon.
28 B1 **Kampar** Malaysia
68 B1 **Kampen** Neth.
87 C3 **Kampene** Dem. Rep. Congo
31 A2 **Kamphaeng Phet** Thai.
31 B2 **Kâmpóng Cham** Cambodia
31 B2 **Kâmpóng Chhnăng** Cambodia
Kâmpóng Saôm Cambodia *see* **Sihanoukville**
31 B2 **Kâmpóng Spoe** Cambodia
31 B2 **Kâmpôt** Cambodia
97 D2 **Kamsack** Can.
54 E3 **Kamskoye Vodokhranilishche** *resr* Rus. Fed.
85 C4 **Kamsuuma** Somalia
58 B2 **Kam"yanets'-Podil's'kyy** Ukr.
58 A1 **Kam"yanka-Buz'ka** Ukr.
56 B3 **Kamyanyets** Belarus
59 D2 **Kamyshevatskaya** Rus. Fed.
55 D3 **Kamyshin** Rus. Fed.
103 D3 **Kanab** U.S.A.
86 C3 **Kananga** Dem. Rep. Congo
55 D3 **Kanash** Rus. Fed.
108 C3 **Kanawha** *r.* U.S.A.
35 C3 **Kanazawa** Japan
30 A1 **Kanbalu** Myanmar
31 A2 **Kanchanaburi** Thai.
41 B3 **Kanchipuram** India
45 C3 **Kandahār** Afgh.
54 C2 **Kandalaksha** Rus. Fed.
29 C2 **Kandangan** Indon.
42 A2 **Kandhkot** Pak.
82 C3 **Kandi** Benin
42 B2 **Kandla** India
21 C2 **Kandos** Austr.
89 □D2 **Kandreho** Madag.
41 C4 **Kandy** Sri Lanka
44 B2 **Kandyagash** Kazakh.
95 G1 **Kane Bassin** *b.* Greenland
59 D2 **Kanevskaya** Rus. Fed.
90 B1 **Kang** Botswana
95 H2 **Kangaatsiaq** Greenland
82 B3 **Kangaba** Mali
48 B2 **Kangal** Turkey
47 C2 **Kangān** Iran
28 B1 **Kangar** Malaysia
20 A3 **Kangaroo Island** Austr.
43 C2 **Kangchenjunga** *mt.* India/Nepal
33 B2 **Kangdong** N. Korea

29 C2 **Kangean, Kepulauan** *is* Indon.
95 I2 **Kangeq** *c.* Greenland
95 I2 **Kangerlussuaq** *inlet* Greenland
95 H2 **Kangerlussuaq** *inlet* Greenland
95 H2 **Kangersuatsiaq** Greenland
33 B1 **Kanggye** N. Korea
99 D1 **Kangiqsualujjuaq** Can.
95 G2 **Kangiqsujuaq** Can.
95 G2 **Kangirsuk** Can.
43 C2 **Kangmar** China
33 B2 **Kangnŭng** S. Korea
33 A1 **Kangping** China
40 D2 **Kangto** *mt.* China/India
30 A1 **Kani** Myanmar
54 D2 **Kanin, Poluostrov** *pen.* Rus. Fed.
54 D2 **Kanin Nos** Rus. Fed.
59 C2 **Kaniv** Ukr.
61 H3 **Kankaanpää** Fin.
108 B2 **Kankakee** U.S.A.
82 B3 **Kankan** Guinea
43 C2 **Kanker** India
Kannur India *see* **Cannanore**
83 C3 **Kano** Nigeria
90 B3 **Kanonpunt** *pt* S. Africa
35 B4 **Kanoya** Japan
43 C2 **Kanpur** India
104 C3 **Kansas** *r.* U.S.A.
105 D3 **Kansas** *state* U.S.A.
105 E3 **Kansas City** U.S.A.
51 H3 **Kansk** Rus. Fed.
82 C3 **Kantchari** Burkina
59 D2 **Kantemirovka** Rus. Fed.
65 B2 **Kanturk** Rep. of Ireland
91 D2 **KaNyamazane** S. Africa
91 C1 **Kanye** Botswana
39 C3 **Kaohsiung** Taiwan
88 A2 **Kaokoveld** *plat.* Namibia
82 A3 **Kaolack** Senegal
88 B2 **Kaoma** Zambia
86 C3 **Kapanga** Dem. Rep. Congo
45 D2 **Kapchagay** Kazakh.
45 D2 **Kapchagayskoye Vodokhranilishche** *resr* Kazakh.
68 B2 **Kapellen** Belgium
89 B2 **Kapiri Mposhi** Zambia
95 H2 **Kapisillit** Greenland
98 B1 **Kapiskau** *r.* Can.
29 C1 **Kapit** *Sarawak* Malaysia
31 A3 **Kapoe** Thai.
85 B4 **Kapoeta** Sudan
71 D2 **Kaposvár** Hungary
29 B2 **Kapuas** *r.* Indon.
20 A2 **Kapunda** Austr.
98 B2 **Kapuskasing** Can.
71 D2 **Kapuvár** Hungary
56 C3 **Kapyl'** Belarus
82 C4 **Kara** Togo
79 C3 **Kara Ada** *i.* Turkey
44 B2 **Kara-Bogaz Gol, Zaliv** *b.* Turkm.
44 B2 **Karabogazkel'** Turkm.
48 B1 **Karabük** Turkey
44 C2 **Karabutak** Kazakh.
78 C2 **Karacaköy** Turkey
57 D3 **Karachev** Rus. Fed.
42 A2 **Karachi** Pak.
Kara Deniz *sea* Asia/Europe *see* **Black Sea**
45 D2 **Karaganda** Kazakh.
45 D2 **Karagayly** Kazakh.
51 L3 **Karaginskiy Zaliv** *b.* Rus. Fed.
49 D2 **Karaj** Iran
44 B2 **Karakalpakiya** Kazakh.
32 B2 **Karakelong** *i.* Indon.
45 D2 **Kara-Köl** Kyrg.
45 D2 **Karakol** Kyrg.
42 B1 **Karakoram** *mts* Asia
44 B2 **Karakum, Peski** *des.* Kazakh.
Karakum Desert Turkm. *see* **Karakumy, Peski**
44 C3 **Karakumy, Peski** *des.* Turkm.
48 B2 **Karaman** Turkey
45 E2 **Karamay** China
22 B2 **Karamea** N.Z.
22 B2 **Karamea Bight** *b.* N.Z.
48 B2 **Karapınar** Turkey
90 A2 **Karasburg** Namibia
54 F1 **Kara Sea** *sea* Rus. Fed.
60 I2 **Karasjok** Norway
50 G3 **Karasuk** Rus. Fed.
45 D2 **Karatau** Kazakh.
45 C2 **Karatau, Khrebet** *mts* Kazakh.
54 F2 **Karatayka** Rus. Fed.
35 A4 **Karatsu** Japan
28 B2 **Karawang** Indon.
49 C2 **Karbalā'** Iraq
71 E2 **Karcag** Hungary
79 B3 **Karditsa** Greece
56 B2 **Kärdla** Estonia
90 B3 **Kareeberge** *mts* S. Africa
43 B2 **Kareli** India
42 B1 **Kargil** Jammu and Kashmir
89 B2 **Kariba** Zimbabwe
89 B2 **Kariba, Lake** *resr* Zambia/Zimbabwe
29 B2 **Karimata, Pulau-pulau** *is* Indon.
29 B2 **Karimata, Selat** *str.* Indon.
41 B3 **Karimnagar** India
29 C2 **Karimunjawa, Pulau-pulau** *is* Indon.
59 C2 **Karkinits'ka Zatoka** *g.* Ukr.
59 D2 **Karlivka** Ukr.
77 C1 **Karlovac** Croatia
70 C1 **Karlovy Vary** Czech Rep.
61 F4 **Karlshamn** Sweden
61 G4 **Karlskrona** Sweden
70 B2 **Karlsruhe** Ger.
61 F4 **Karlstad** Sweden
69 D3 **Karlstadt** Ger.
57 D3 **Karma** Belarus
61 E4 **Karmøy** *i.* Norway
43 D2 **Karnafuli Reservoir** Bangl.
42 B2 **Karnal** India
78 C2 **Karnobat** Bulg.
42 A2 **Karodi** Pak.
89 B2 **Karoi** Zimbabwe
89 C1 **Karonga** Malawi
84 B3 **Karora** Eritrea
79 C3 **Karpathos** *i.* Greece
79 B3 **Karpenisi** Greece
54 D2 **Karpogory** Rus. Fed.
18 A2 **Karratha** Austr.
49 C1 **Kars** Turkey
56 C2 **Kārsava** Latvia
45 C3 **Karshi** Uzbek.
54 E2 **Karskiye Vorota, Proliv** *str.* Rus. Fed.
Karskoye More *sea* Rus. Fed. *see* **Kara Sea**
69 E1 **Karstädt** Ger.
79 C2 **Kartal** Turkey
55 F3 **Kartaly** Rus. Fed.
41 B3 **Karwar** India
51 I3 **Karymskoye** Rus. Fed.
79 B3 **Karystos** Greece
48 A2 **Kaş** Turkey
98 B1 **Kasabonika Lake** Can.
86 C4 **Kasaji** Dem. Rep. Congo
89 C2 **Kasama** Zambia
88 B2 **Kasane** Botswana
86 B3 **Kasangulu** Dem. Rep. Congo
41 B3 **Kasaragod** India
97 D1 **Kasba Lake** Can.
88 B2 **Kasempa** Zambia
87 C4 **Kasenga** Dem. Rep. Congo
87 C3 **Kasese** Dem. Rep. Congo
87 D2 **Kasese** Uganda
49 D2 **Kāshān** Iran
Kashgar China *see* **Kashi**
45 D3 **Kashi** China
35 D3 **Kashima-nada** *b.* Japan
57 E2 **Kashin** Rus. Fed.
57 E3 **Kashira** Rus. Fed.
57 E3 **Kashirskoye** Rus. Fed.
35 C3 **Kashiwazaki** Japan
44 B3 **Kāshmar** Iran
42 A2 **Kashmor** Pak.
87 C3 **Kashyukulu** Dem. Rep. Congo
55 D3 **Kasimov** Rus. Fed.
61 H3 **Kaskinen** Fin.
87 C3 **Kasongo** Dem. Rep. Congo
86 B3 **Kasongo-Lunda** Dem. Rep. Congo
79 C3 **Kasos** *i.* Greece
Kaspiyskoye More *sea* Asia/Europe *see* **Caspian Sea**
84 B3 **Kassala** Sudan
69 D2 **Kassel** Ger.
48 B1 **Kastamonu** Turkey
79 B3 **Kastelli** Greece
79 B2 **Kastoria** Greece
57 D3 **Kastsyukovichy** Belarus
87 D3 **Kasulu** Tanz.
89 C2 **Kasungu** Malawi
109 F1 **Katahdin, Mount** U.S.A.
86 C3 **Katako-Kombe** Dem. Rep. Congo
18 A3 **Katanning** Austr.
31 A3 **Katchall** *i.* India
79 B2 **Katerini** Greece
96 A2 **Kate's Needle** *mt.* Can./U.S.A.
89 C2 **Katete** Zambia
30 A1 **Katha** Myanmar
46 A2 **Katherîna, Gebel** *mt.* Egypt
18 C1 **Katherine** *r.* Austr.
42 B2 **Kathiawar** *pen.* India
91 C2 **Kathlehong** S. Africa
43 C2 **Kathmandu** Nepal
90 B2 **Kathu** S. Africa
42 B1 **Kathua** Jammu and Kashmir
82 B3 **Kati** Mali
43 C2 **Katihar** India
22 C1 **Katikati** N.Z.
91 C3 **Kati-Kati** S. Africa
88 B2 **Katima Mulilo** Namibia
79 B3 **Kato Achaïa** Greece
21 D2 **Katoomba** Austr.
71 D1 **Katowice** Pol.
61 G4 **Katrineholm** Sweden
83 C3 **Katsina** Nigeria
83 C4 **Katsina-Ala** Nigeria
45 C3 **Kattakurgan** Uzbek.
61 F4 **Kattegat** *str.* Denmark/Sweden
68 B1 **Katwijk aan Zee** Neth.
61 H3 **Kauhajoki** Fin.
56 B3 **Kaunas** Lith.
83 C3 **Kaura-Namoda** Nigeria
60 H2 **Kautokeino** Norway
79 B2 **Kavala** Greece
34 C2 **Kavalerovo** Rus. Fed.
41 C3 **Kavali** India
41 B3 **Kavaratti** *i.* India
78 C2 **Kavarna** Bulg.
49 D2 **Kavīr, Dasht-e** *des.* Iran
35 C3 **Kawagoe** Japan
22 B1 **Kawakawa** N.Z.
35 C3 **Kawanishi** Japan

98 C2 Kawartha Lakes Can.
35 C3 Kawasaki Japan
22 C1 Kawerau N.Z.
31 A2 Kawkareik Myanmar
30 A1 Kawlin Myanmar
31 A2 Kawmapyin Myanmar
31 A2 Kawthaung Myanmar
45 D3 Kaxgar He *r.* China
82 B3 Kaya Burkina
29 C1 Kayan *r.* Indon.
104 B2 Kaycee U.S.A.
106 A1 Kayenta U.S.A.
82 A3 Kayes Mali
45 D2 Kaynar Kazakh.
48 B2 Kayseri Turkey
45 D1 Kazakhskiy Melkosopochnik *plain* Kazakh.
49 D1 Kazakhskiy Zaliv *b.* Kazakh.
44 E2 Kazakhstan *country* Asia
55 D3 Kazan' Rus. Fed.
48 A1 Kazanlŭk Bulg.
44 A2 Kazbek *mt.* Georgia/Rus. Fed.
49 D3 Kāzerūn Iran
71 E2 Kazincbarcika Hungary
54 F2 Kazymskiy Mys Rus. Fed.
79 B3 Kea *i.* Greece
65 C1 Keady U.K.
105 D2 Kearney U.S.A.
106 A2 Kearny U.S.A.
84 A3 Kebkabiya Sudan
60 G2 Kebnekaise *mt.* Sweden
85 C4 K'ebrī Dehar Eth.
29 B2 Kebumen Indon.
96 B2 Kechika *r.* Can.
71 D2 Kecskemét Hungary
56 B2 Kėdainiai Lith.
82 A3 Kédougou Senegal
71 D1 Kędzierzyn-Koźle Pol.
96 B1 Keele *r.* Can.
96 A1 Keele Peak Can.
109 E2 Keene U.S.A.
90 A2 Keetmanshoop Namibia
97 E3 Keewatin Can.
79 B3 Kefallonia *i.* Greece
27 C3 Kefamenanu Indon.
60 □A3 Keflavík Iceland
45 D2 Kegen Kazakh.
96 C2 Keg River Can.
30 A1 Kehsi Mansam Myanmar
66 C2 Keighley U.K.
56 B2 Keila Estonia
90 B2 Keimoes S. Africa
60 I3 Keitele *l.* Fin.
20 B3 Keith Austr.
71 E2 Kékes *mt.* Hungary
28 B1 Kelang Malaysia
60 J2 Keles-Uayv, Gora *hill* Rus. Fed.
70 C2 Kelheim Ger.
48 B1 Kelkit *r.* Turkey
96 B1 Keller Lake *l.* Can.
102 C1 Kellogg U.S.A.
60 I2 Kelloselkä Fin.
65 C2 Kells Rep. of Ireland
56 B2 Kelmė Lith.
83 D4 Kelo Chad
96 C3 Kelowna Can.
64 C3 Kelso U.K.
102 B1 Kelso U.S.A.
28 B1 Keluang Malaysia
97 D2 Kelvington Can.
54 C2 Kem' Rus. Fed.
96 B2 Kemano Can.
79 C3 Kemer Turkey
50 G3 Kemerovo Rus. Fed.
60 H2 Kemi Fin.
60 I2 Kemijärvi Fin.
60 I2 Kemijoki *r.* Fin.
104 A2 Kemmerer U.S.A.
60 I3 Kempele Fin.
21 D2 Kempsey Austr.
98 C2 Kempt, Lac *l.* Can.
70 C2 Kempten (Allgäu) Ger.
91 C2 Kempton Park S. Africa
29 C2 Kemujan *i.* Indon.
97 D2 Kenaston Can.
66 B1 Kendal U.K.
27 C3 Kendari Indon.
29 C2 Kendawangan Indon.
83 D3 Kendégué Chad
43 C2 Kendujhargarh India
82 A4 Kenema Sierra Leone
86 B3 Kenge Dem. Rep. Congo
30 A1 Kengtung Myanmar
90 B2 Kenhardt S. Africa
82 B1 Kénitra Morocco
65 B3 Kenmare Rep. of Ireland
104 C1 Kenmare U.S.A.
65 A3 Kenmare River *inlet* Rep. of Ireland
68 C3 Kenn Ger.
105 E3 Kennett U.S.A.
102 C1 Kennewick U.S.A.
97 E3 Kenora Can.
108 B2 Kenosha U.S.A.
106 C2 Kent U.S.A.
45 C2 Kentau Kazakh.
108 B3 Kentucky *r.* U.S.A.
108 C3 Kentucky *state* U.S.A.
108 B3 Kentucky Lake U.S.A.
87 D2 Kenya *country* Africa
Kenya, Mount Kenya *see* Kirinyaga
105 E2 Keokuk U.S.A.
79 C3 Kepsut Turkey
20 B3 Kerang Austr.
59 D2 Kerch Ukr.
27 D3 Kerema P.N.G.
96 C3 Keremeos Can.
84 B3 Keren Eritrea
49 C2 Kerend Iran
127 C7 Kerguelen Plateau *sea feature* Indian Ocean
87 D3 Kericho Kenya
22 B1 Kerikeri N.Z.
28 B2 Kerinci, Gunung *vol.* Indon.
45 C3 Kerki Turkm.
79 A3 Kerkyra Greece
Kerkyra *i.* Greece *see* Corfu
84 B3 Kerma Sudan
47 C1 Kermān Iran
49 C2 Kermānshāh Iran
107 C2 Kermit U.S.A.
103 C3 Kern *r.* U.S.A.
68 C2 Kerpen Ger.
97 D2 Kerrobert Can.
107 D2 Kerrville U.S.A.
48 B2 Keryneia Cyprus
98 B1 Kesagami Lake Can.
79 C2 Keşan Turkey
34 D3 Kesennuma Japan
42 B2 Keshod India
68 C2 Kessel Neth.
66 B1 Keswick U.K.
71 D2 Keszthely Hungary
50 G3 Ket' *r.* Rus. Fed.
29 C2 Ketapang Indon.
96 A2 Ketchikan U.S.A.
102 D2 Ketchum U.S.A.
67 C2 Kettering U.K.
61 H3 Keuruu Fin.
68 C2 Kevelaer Ger.
108 B2 Kewanee U.S.A.
108 B1 Keweenaw Bay *b.* U.S.A.
108 B1 Keweenaw Peninsula U.S.A.
111 D3 Key Largo U.S.A.
109 D3 Keyser U.S.A.
111 D4 Key West U.S.A.
91 C2 Kgotsong S. Africa
37 E1 Khabarovsk Rus. Fed.
48 B1 Khadyzhensk Rus. Fed.
30 A1 Khagrachari Bangl.
42 A2 Khairpur Pak.
90 B1 Khakhea Botswana
36 C1 Khamar-Daban, Khrebet *mts* Rus. Fed.
42 B2 Khambhat India
41 B3 Khambhat, Gulf of India
42 B2 Khamgaon India
46 B3 Khamir Yemen
46 B3 Khamis Mushayţ Saudi Arabia
45 C3 Khānābād Afgh.
42 B2 Khandwa India
51 K2 Khandyga Rus. Fed.
42 B1 Khanewal Pak.
34 B2 Khanka, Lake China/Rus. Fed.
42 B2 Khanpur Pak.
45 D2 Khantau Kazakh.
51 H2 Khantayskoye, Ozero *l.* Rus. Fed.
54 F2 Khanty-Mansiysk Rus. Fed.
31 A3 Khao Chum Thong Thai.
31 A2 Khao Laem Reservoir Thai.
55 D4 Kharabali Rus. Fed.
43 C2 Kharagpur India
47 C2 Khārān *r.* Iran
59 D2 Kharkiv Ukr.
Khar'kov Ukr. *see* Kharkiv
78 C2 Kharmanli Bulg.
54 D3 Kharovsk Rus. Fed.
84 B3 Khartoum Sudan
55 D4 Khasav'yurt Rus. Fed.
47 D2 Khāsh Iran
46 A3 Khashm el Girba Sudan
46 A3 Khashm el Girba Dam Sudan
42 D2 Khāsi Hills India
78 C2 Khaskovo Bulg.
91 C3 Khayamnandi S. Africa
45 C2 Khayatbashi, Gora *mt.* Uzbek.
46 A2 Khaybar Saudi Arabia
90 A3 Khayelitsha S. Africa
30 B2 Khê Bo Vietnam
75 D2 Khemis Miliana Alg.
31 B2 Khemmarat Thai.
83 C1 Khenchela Alg.
49 D3 Kherāmeh Iran
59 C2 Kherson Ukr.
51 H2 Kheta *r.* Rus. Fed.
37 D1 Khilok Rus. Fed.
57 E2 Khimki Rus. Fed.
57 E3 Khlevnoye Rus. Fed.
58 B2 Khmel'nyts'kyy Ukr.
44 B1 Khobda Kazakh.
44 B2 Khodzheyli Uzbek.
57 E3 Khokhol'skiy Rus. Fed.
42 B2 Khokhropar Pak.
57 D2 Kholm Rus. Fed.
57 D2 Kholm-Zhirkovskiy Rus. Fed.
47 C2 Khonj Iran
31 B2 Khon Kaen Thai.
30 A1 Khonsa India
51 K2 Khonuu Rus. Fed.
54 E2 Khoreyver Rus. Fed.
37 D1 Khorinsk Rus. Fed.
59 C2 Khorol Ukr.
49 C2 Khorramābād Iran
49 C2 Khorramshahr Iran
45 D3 Khorugh Tajik.
42 A1 Khowst Afgh.
30 A1 Khreum Myanmar
44 B1 Khromtau Kazakh.
58 B2 Khrystynivka Ukr.
45 C2 Khŭjand Tajik.

31 B2 **Khu Khan** Thai.
46 A2 **Khulays** Saudi Arabia
43 C2 **Khulna** Bangl.
49 D2 **Khunsar** Iran
47 B2 **Khurayş** Saudi Arabia
42 B1 **Khushab** Pak.
58 A2 **Khust** Ukr.
91 C2 **Khutsong** S. Africa
42 A2 **Khuzdar** Pak.
47 C2 **Khvormūj** Iran
49 C2 **Khvoy** Iran
57 D2 **Khvoynaya** Rus. Fed.
45 D3 **Khyber Pass** Afgh./Pak.
21 D2 **Kiama** Austr.
87 C3 **Kiambi** Dem. Rep. Congo
87 D3 **Kibiti** Tanz.
77 D2 **Kičevo** Macedonia
82 C3 **Kidal** Mali
67 B2 **Kidderminster** U.K.
82 A3 **Kidira** Senegal
22 C1 **Kidnappers, Cape** N.Z.
70 C1 **Kiel** Ger.
71 E1 **Kielce** Pol.
66 B1 **Kielder Water** *resr* U.K.
58 C1 **Kiev** Ukr.
82 A3 **Kiffa** Maur.
87 D3 **Kigali** Rwanda
87 C3 **Kigoma** Tanz.
56 B2 **Kihnu** *i.* Estonia
60 I2 **Kiiminki** Fin.
35 B4 **Kii-suidō** *sea chan.* Japan
77 D1 **Kikinda** Yugo.
87 C3 **Kikondja** Dem. Rep. Congo
27 D3 **Kikori** P.N.G.
27 D3 **Kikori** *r.* P.N.G.
86 B3 **Kikwit** Dem. Rep. Congo
33 B1 **Kilchu** N. Korea
86 B3 **Kilembe** Dem. Rep. Congo
107 E2 **Kilgore** U.S.A.
87 D3 **Kilimanjaro** *vol.* Tanz.
48 B2 **Kilis** Turkey
58 B2 **Kiliya** Ukr.
65 B2 **Kilkee** Rep. of Ireland
65 D1 **Kilkeel** U.K.
65 C2 **Kilkenny** Rep. of Ireland
79 B2 **Kilkis** Greece
65 B1 **Killala Bay** Rep. of Ireland
65 B2 **Killaloe** Rep. of Ireland
97 C2 **Killam** Can.
65 B2 **Killarney** Rep. of Ireland
107 D2 **Killeen** U.S.A.
64 B2 **Killin** U.K.
65 B1 **Killybegs** Rep. of Ireland
64 B3 **Kilmarnock** U.K.
21 B3 **Kilmore** Austr.
87 D3 **Kilosa** Tanz.
65 B2 **Kilrush** Rep. of Ireland
87 C3 **Kilwa** Dem. Rep. Congo
87 D3 **Kilwa Masoko** Tanz.
87 D3 **Kimambi** Tanz.
20 A2 **Kimba** Austr.
104 C2 **Kimball** U.S.A.
96 C3 **Kimberley** Can.
90 B2 **Kimberley** S. Africa
18 B1 **Kimberley Plateau** Austr.
33 B1 **Kimch'aek** N. Korea
33 B2 **Kimch'ŏn** S. Korea
95 G2 **Kimmirut** Can.
57 E3 **Kimovsk** Rus. Fed.
86 B3 **Kimpese** Dem. Rep. Congo
57 E2 **Kimry** Rus. Fed.
29 C1 **Kinabalu, Gunung** *mt. Sabah* Malaysia
96 C2 **Kinbasket Lake** Can.
98 B2 **Kincardine** Can.
30 A1 **Kinchang** Myanmar
87 C3 **Kinda** Dem. Rep. Congo
66 C2 **Kinder Scout** *hill* U.K.
97 D2 **Kindersley** Can.
82 A3 **Kindia** Guinea
87 C3 **Kindu** Dem. Rep. Congo
54 D3 **Kineshma** Rus. Fed.
19 E2 **Kingaroy** Austr.
103 B3 **King City** U.S.A.
98 C1 **King George Islands** Can.
56 C2 **Kingisepp** Rus. Fed.
19 D3 **King Island** Austr.
18 B1 **King Leopold Ranges** *hills* Austr.
106 A1 **Kingman** U.S.A.
103 B3 **Kings** *r.* U.S.A.
20 A3 **Kingscote** Austr.
67 D2 **King's Lynn** U.K.
18 B1 **King Sound** *b.* Austr.
102 D2 **Kings Peak** U.S.A.
111 D1 **Kingsport** U.S.A.
98 C2 **Kingston** Can.
114 C3 **Kingston** Jamaica
109 E2 **Kingston** U.S.A.
20 A3 **Kingston South East** Austr.
66 C2 **Kingston upon Hull** U.K.
115 D3 **Kingstown** St Vincent
107 D3 **Kingsville** U.S.A.
67 B3 **Kingswood** U.K.
64 B2 **Kingussie** U.K.
94 E2 **King William Island** Can.
91 C3 **King William's Town** S. Africa
35 D3 **Kinka-san** *i.* Japan
61 F4 **Kinna** Sweden
65 B3 **Kinsale** Rep. of Ireland
86 B3 **Kinshasa** Dem. Rep. Congo
111 E1 **Kinston** U.S.A.
56 B2 **Kintai** Lith.
82 B4 **Kintampo** Ghana
64 B3 **Kintyre** *pen.* U.K.
98 C2 **Kipawa, Lac** *l.* Can.
51 I3 **Kirensk** Rus. Fed.
57 E3 **Kireyevsk** Rus. Fed.
Kirghizia *country* Asia *see* Kyrgyzstan
17 G3 **Kiribati** *country* Pacific Ocean
48 B2 **Kırıkkale** Turkey
57 E2 **Kirillov** Rus. Fed.
Kirin China *see* **Jilin**
87 D3 **Kirinyaga** *mt.* Kenya
57 D2 **Kirishi** Rus. Fed.
79 C3 **Kırkağaç** Turkey
64 C2 **Kirkcaldy** U.K.
64 B3 **Kirkcudbright** U.K.
60 J2 **Kirkenes** Norway
98 B2 **Kirkland Lake** Can.
105 E2 **Kirksville** U.S.A.
49 C2 **Kirkūk** Iraq
64 C1 **Kirkwall** U.K.
57 D3 **Kirov** *Kaluzhskaya Oblast'* Rus. Fed.
54 D3 **Kirov** Rus. Fed.
Kirovabad Azer. *see* **Gäncä**
54 E3 **Kirovo-Chepetsk** Rus. Fed.
59 C2 **Kirovohrad** Ukr.
54 C2 **Kirovsk** Rus. Fed.
59 D2 **Kirovs'ke** Ukr.
64 C2 **Kirriemuir** U.K.
54 E3 **Kirs** Rus. Fed.
42 A2 **Kirthar Range** *mts* Pak.
60 H2 **Kiruna** Sweden
35 C3 **Kiryū** Japan
57 E2 **Kirzhach** Rus. Fed.
87 C2 **Kisangani** Dem. Rep. Congo
86 B3 **Kisantu** Dem. Rep. Congo
28 A1 **Kisaran** Indon.
50 G3 **Kiselevsk** Rus. Fed.
43 C2 **Kishanganj** India
Kishinev Moldova *see* **Chişinău**
45 D1 **Kishkenekol'** Kazakh.
42 B1 **Kishtwar** Jammu and Kashmir
83 C4 **Kisi** Nigeria
87 D3 **Kisii** Kenya
71 D2 **Kiskunfélegyháza** Hungary
71 D2 **Kiskunhalas** Hungary
55 D4 **Kislovodsk** Rus. Fed.
85 C5 **Kismaayo** Somalia
87 C3 **Kisoro** Uganda
82 A4 **Kissidougou** Guinea
111 D3 **Kissimmee** U.S.A.
111 D3 **Kissimmee, Lake** U.S.A.
97 D2 **Kississing Lake** Can.
87 D3 **Kisumu** Kenya
82 B3 **Kita** Mali
34 D3 **Kitakami** Japan
34 D3 **Kitakami-gawa** *r.* Japan
35 B4 **Kita-Kyūshū** Japan
87 D2 **Kitale** Kenya
34 D2 **Kitami** Japan
98 B2 **Kitchener** Can.
61 J3 **Kitee** Fin.
87 D2 **Kitgum** Uganda
96 B2 **Kitimat** Can.
86 B3 **Kitona** Dem. Rep. Congo
87 D3 **Kitunda** Tanz.
96 B2 **Kitwanga** Can.
89 B2 **Kitwe** Zambia
69 E3 **Kitzingen** Ger.
60 I3 **Kiuruvesi** Fin.
87 C3 **Kivu, Lake** Dem. Rep. Congo/Rwanda
78 C2 **Kıyıköy** Turkey
54 E3 **Kizel** Rus. Fed.
79 C3 **Kızılca Dağ** *mt.* Turkey
55 D4 **Kizlyar** Rus. Fed.
60 I1 **Kjøllefjord** Norway
60 G2 **Kjøpsvik** Norway
70 C1 **Kladno** Czech Rep.
70 C2 **Klagenfurt** Austria
56 B2 **Klaipėda** Lith.
62 B1 **Klaksvík** Faroe Is
102 B2 **Klamath** *r.* U.S.A.
102 B2 **Klamath Falls** U.S.A.
102 B2 **Klamath Mountains** U.S.A.
70 C2 **Klatovy** Czech Rep.
90 A3 **Klawer** S. Africa
96 A2 **Klawock** U.S.A.
96 B2 **Kleena Kleene** Can.
90 B2 **Kleinbegin** S. Africa
90 A2 **Kleinsee** S. Africa
91 C2 **Klerksdorp** S. Africa
57 D3 **Kletnya** Rus. Fed.
68 C2 **Kleve** Ger.
57 D3 **Klimavichy** Belarus
57 D3 **Klimovo** Rus. Fed.
57 E2 **Klimovsk** Rus. Fed.
57 E2 **Klin** Rus. Fed.
69 F2 **Klínovec** *mt.* Czech Rep.
61 G4 **Klintehamn** Sweden
57 D3 **Klintsy** Rus. Fed.
77 C2 **Ključ** Bos.-Herz.
71 D1 **Kłodzko** Pol.
68 C1 **Kloosterhaar** Neth.
71 D2 **Klosterneuburg** Austria
69 E1 **Klötze (Altmark)** Ger.
71 D1 **Kluczbork** Pol.
96 A2 **Klukwan** U.S.A.
42 A2 **Klupro** Pak.
56 C3 **Klyetsk** Belarus
66 C1 **Knaresborough** U.K.
61 F3 **Knästen** *hill* Sweden
97 E2 **Knee Lake** Can.
69 E1 **Knesebeck** Ger.
69 E3 **Knetzgau** Ger.
77 C2 **Knin** Croatia
71 C2 **Knittelfeld** Austria
77 D2 **Knjaževac** Yugo.
68 A2 **Knokke-Heist** Belgium
111 D1 **Knoxville** U.S.A.
90 B3 **Knysna** S. Africa
35 C4 **Kōbe** Japan

København Denmark *see* **Copenhagen**
68 C2 **Koblenz** Ger.
27 C3 **Kobroör** *i.* Indon.
56 B3 **Kobryn** Belarus
77 D2 **Kočani** Macedonia
79 C2 **Kocasu** *r.* Turkey
43 C2 **Koch Bihār** India
Kochi India *see* **Cochin**
35 B4 **Kōchi** Japan
55 D4 **Kochubey** Rus. Fed.
43 C2 **Kodarma** India
85 B4 **Kodok** Sudan
58 B2 **Kodyma** Ukr.
79 C2 **Kodzhaele** *mt.* Bulg./Greece
90 A2 **Koës** Namibia
90 C2 **Koffiefontein** S. Africa
82 B4 **Koforidua** Ghana
35 C3 **Kōfu** Japan
99 D1 **Kogaluk** *r.* Can.
42 B1 **Kohat** Pak.
30 A1 **Kohima** India
56 C2 **Kohtla-Järve** Estonia
45 D2 **Kokand** Uzbek.
60 H3 **Kokkola** Fin.
108 B2 **Kokomo** U.S.A.
45 E2 **Kokpekti** Kazakh.
45 C1 **Kokshetau** Kazakh.
99 D1 **Koksoak** *r.* Can.
91 C3 **Kokstad** S. Africa
Koktokay China *see* **Fuyun**
26 C3 **Kolaka** Indon.
54 C2 **Kola Peninsula** *pen.* Rus. Fed.
60 H2 **Kolari** Fin.
82 A3 **Kolda** Senegal
61 E4 **Kolding** Denmark
75 D2 **Koléa** Alg.
54 D2 **Kolguyev, Ostrov** *i.* Rus. Fed.
41 B3 **Kolhapur** India
56 B2 **Kolkasrags** *pt* Latvia
Kollam India *see* **Quilon**
Köln Ger. *see* **Cologne**
71 D1 **Kołobrzeg** Pol.
82 B3 **Kolokani** Mali
57 E2 **Kolomna** Rus. Fed.
58 B2 **Kolomyya** Ukr.
90 B2 **Kolonkwane** Botswana
50 G3 **Kolpashevo** Rus. Fed.
57 E3 **Kolpny** Rus. Fed.
Kol'skiy Poluostrov *pen.* Rus. Fed. *see* **Kola Peninsula**
46 B3 **Koluli** Eritrea
87 C4 **Kolwezi** Dem. Rep. Congo
51 L2 **Kolyma** *r.* Rus. Fed.
51 L2 **Kolymskaya Nizmennost'** *lowland* Rus. Fed.
51 M2 **Kolymskiy, Khrebet** *mts* Rus. Fed.
35 C3 **Komaki** Japan
51 M3 **Komandorskiye Ostrova** *is* Rus. Fed.
71 D2 **Komárno** Slovakia
91 D2 **Komati** *r.* Swaziland
35 C3 **Komatsu** Japan
87 C3 **Kombe** Dem. Rep. Congo
58 C2 **Kominternivs'ke** Ukr.
77 C2 **Komiža** Croatia
71 D2 **Komló** Hungary
86 B3 **Komono** Congo
79 C2 **Komotini** Greece
90 B3 **Komsberg** *mts* S. Africa
44 C1 **Komsomolets** Kazakh.
51 H1 **Komsomolets, Ostrov** *i.* Rus. Fed.
57 F2 **Komsomol'sk** Rus. Fed.
59 C2 **Komsomol's'k** Ukr.
55 D4 **Komsomol'skiy** Rus. Fed.
37 E1 **Komsomol'sk-na-Amure** Rus. Fed.
57 E2 **Konakovo** Rus. Fed.
43 C3 **Kondagaon** India
54 F2 **Kondinskoye** Rus. Fed.
87 D3 **Kondoa** Tanz.
54 C2 **Kondopoga** Rus. Fed.
57 E3 **Kondrovo** Rus. Fed.
95 I2 **Kong Christian IX Land** *reg.* Greenland
95 I2 **Kong Frederik VI Kyst** *coastal area* Greenland
33 B2 **Kongju** S. Korea
87 C3 **Kongolo** Dem. Rep. Congo
61 E4 **Kongsberg** Norway
61 F3 **Kongsvinger** Norway
45 D3 **Kongur Shan** *mt.* China
68 C2 **Königswinter** Ger.
71 D1 **Konin** Pol.
77 C2 **Konjic** Bos.-Herz.
90 A2 **Konkiep** *watercourse* Namibia
59 C1 **Konotop** Ukr.
70 B2 **Konstanz** Ger.
83 C3 **Kontagora** Nigeria
31 B2 **Kon Tum** Vietnam
31 B2 **Kontum, Plateau du** Vietnam
48 B2 **Konya** Turkey
68 C3 **Konz** Ger.
54 E3 **Konzhakovskiy Kamen', Gora** *mt.* Rus. Fed.
102 C1 **Kooskia** U.S.A.
96 C3 **Kootenay Lake** Can.
90 B3 **Kootjieskolk** S. Africa
60 □B2 **Kópasker** Iceland
76 B1 **Koper** Slovenia
61 G4 **Kopparberg** Sweden
77 C1 **Koprivnica** Croatia
57 F3 **Korablino** Rus. Fed.
41 C3 **Koraput** India
69 D2 **Korbach** Ger.
77 D2 **Korçë** Albania
77 C2 **Korčula** Croatia
77 C2 **Korčula** *i.* Croatia
33 B1 **Korea, North** *country* Asia
33 B2 **Korea, South** *country* Asia
38 C2 **Korea Bay** *g.* China/N. Korea
33 B3 **Korea Strait** Japan/S. Korea
59 D2 **Korenovsk** Rus. Fed.
58 B1 **Korets'** Ukr.
79 C2 **Körfez** Turkey
82 B4 **Korhogo** Côte d'Ivoire
79 B3 **Korinthiakos Kolpos** *sea chan.* Greece
79 B3 **Korinthos** Greece
71 D2 **Kőris-hegy** *hill* Hungary
77 D2 **Koritnik** *mt.* Albania
35 D3 **Kōriyama** Japan
48 B2 **Korkuteli** Turkey
36 B2 **Korla** China
71 D2 **Körmend** Hungary
82 B3 **Koro** Mali
99 D1 **Koroc** *r.* Can.
59 D1 **Korocha** Rus. Fed.
87 D3 **Korogwe** Tanz.
27 C2 **Koror** Palau
58 B1 **Korosten'** Ukr.
58 B1 **Korostyshiv** Ukr.
83 D3 **Koro Toro** Chad
59 C2 **Korsun'-Shevchenkivs'kyy** Ukr.
71 E1 **Korsze** Pol.
68 A2 **Kortrijk** Belgium
51 L3 **Koryakskaya, Sopka** *vol.* Rus. Fed.
51 M2 **Koryakskiy Khrebet** *mts* Rus. Fed.
54 D2 **Koryazhma** Rus. Fed.
33 B2 **Koryŏng** S. Korea
59 C1 **Koryukivka** Ukr.
79 C3 **Kos** Greece
79 C3 **Kos** *i.* Greece
33 B2 **Kosan** N. Korea
71 D1 **Kościan** Pol.
21 C3 **Kosciuszko, Mount** Austr.
36 B1 **Kosh-Agach** Rus. Fed.
71 E2 **Košice** Slovakia
60 H2 **Koskullskulle** Sweden
33 B2 **Kosŏng** N. Korea
77 D2 **Kosovo** *prov.* Yugo.
77 D2 **Kosovska Mitrovica** Yugo.
82 B4 **Kossou, Lac de** *l.* Côte d'Ivoire
44 C1 **Kostanay** Kazakh.
78 B2 **Kostenets** Bulg.
91 C2 **Koster** S. Africa
85 B3 **Kosti** Sudan
54 C2 **Kostomuksha** Rus. Fed.
58 B1 **Kostopil'** Ukr.
57 F2 **Kostroma** Rus. Fed.
57 F2 **Kostroma** *r.* Rus. Fed.
70 C1 **Kostrzyn** Pol.
71 D1 **Koszalin** Pol.
71 D2 **Kőszeg** Hungary
42 B2 **Kota** India
28 B2 **Kotaagung** Indon.
29 C2 **Kotabaru** Indon.
29 C1 **Kota Belud** *Sabah* Malaysia
28 B1 **Kota Bharu** Malaysia
28 B2 **Kotabumi** Indon.
29 C1 **Kota Kinabalu** *Sabah* Malaysia
29 C1 **Kota Samarahan** *Sarawak* Malaysia
54 D3 **Kotel'nich** Rus. Fed.
55 D4 **Kotel'nikovo** Rus. Fed.
51 K1 **Kotel'nyy, Ostrov** *i.* Rus. Fed.
69 E2 **Köthen (Anhalt)** Ger.
61 I3 **Kotka** Fin.
54 D2 **Kotlas** Rus. Fed.
77 C2 **Kotor Varoš** Bos.-Herz.
55 D3 **Kotovo** Rus. Fed.
59 E1 **Kotovsk** Rus. Fed.
58 B2 **Kotovs'k** Ukr.
51 H2 **Kotuy** *r.* Rus. Fed.
68 A2 **Koudekerke** Neth.
82 B3 **Koudougou** Burkina
90 B3 **Kougaberge** *mts* S. Africa
86 B3 **Koulamoutou** Gabon
82 B3 **Koulikoro** Mali
82 A3 **Koundâra** Guinea
119 C1 **Kourou** Fr. Guiana
82 B3 **Kouroussa** Guinea
83 D3 **Kousséri** Cameroon
82 B3 **Koutiala** Mali
61 I3 **Kouvola** Fin.
60 J2 **Kovdor** Rus. Fed.
58 A1 **Kovel'** Ukr.
57 F2 **Kovrov** Rus. Fed.
22 B2 **Kowhitirangi** N.Z.
39 B3 **Kowloon** China
79 C3 **Köyceğiz** Turkey
54 D2 **Koyda** Rus. Fed.
94 A2 **Koyukuk** *r.* U.S.A.
79 B2 **Kozani** Greece
58 C1 **Kozelets'** Ukr.
57 E3 **Kozel'sk** Rus. Fed.
Kozhikode India *see* **Calicut**
58 B2 **Kozyatyn** Ukr.
31 A3 **Krabi** Thai.
31 A2 **Kra Buri** Thai.
31 B2 **Krâchéh** Cambodia
61 E4 **Kragerø** Norway
68 B1 **Kraggenburg** Neth.
77 D2 **Kragujevac** Yugo.
28 B2 **Krakatau** *i.* Indon.
71 D1 **Kraków** Pol.
59 D2 **Kramators'k** Ukr.

61 G3 **Kramfors** Sweden
79 B3 **Kranidi** Greece
54 E1 **Krasino** Rus. Fed.
56 C2 **Krāslava** Latvia
69 F2 **Kraslice** Czech Rep.
57 D3 **Krasnapollye** Belarus
57 D3 **Krasnaya Gora** Rus. Fed.
55 D3 **Krasnoarmeysk** Rus. Fed.
59 D2 **Krasnoarmiys'k** Ukr.
54 D2 **Krasnoborsk** Rus. Fed.
59 D2 **Krasnodar** Rus. Fed.
59 D2 **Krasnodon** Ukr.
56 C2 **Krasnogorodskoye** Rus. Fed.
59 D1 **Krasnogvardeyskoye** Rus. Fed.
59 D2 **Krasnohrad** Ukr.
59 C2 **Krasnohvardiys'ke** Ukr.
54 E3 **Krasnokamsk** Rus. Fed.
57 D2 **Krasnomayskiy** Rus. Fed.
59 C2 **Krasnoperekops'k** Ukr.
55 D3 **Krasnoslobodsk** Rus. Fed.
54 E3 **Krasnoufimsk** Rus. Fed.
51 H3 **Krasnoyarsk** Rus. Fed.
57 F2 **Krasnoye-na-Volge** Rus. Fed.
57 D3 **Krasnyy** Rus. Fed.
57 E2 **Krasnyy Kholm** Rus. Fed.
55 C4 **Krasnyy Luch** Ukr.
59 E2 **Krasnyy Sulin** Rus. Fed.
58 B2 **Krasyliv** Ukr.
68 C2 **Krefeld** Ger.
59 C2 **Kremenchuk** Ukr.
59 C2 **Kremenchuts'ka Vodoskhovyshche** *resr* Ukr.
71 D2 **Křemešník** *hill* Czech Rep.
59 D2 **Kreminna** Ukr.
71 D2 **Krems an der Donau** Austria
57 D2 **Kresttsy** Rus. Fed.
56 B2 **Kretinga** Lith.
68 C2 **Kreuzau** Ger.
68 C2 **Kreuztal** Ger.
86 A2 **Kribi** Cameroon
79 B3 **Krikellos** Greece
34 D1 **Kril'on, Mys** *c.* Rus. Fed.
41 C3 **Krishna** *r.* India
41 C3 **Krishna, Mouths of the** India
43 C2 **Krishnanagar** India
61 E4 **Kristiansand** Norway
61 F4 **Kristianstad** Sweden
60 E3 **Kristiansund** Norway
61 F4 **Kristinehamn** Sweden
Kriti *i.* Greece *see* **Crete**
Krivoy Rog Ukr. *see* **Kryvyy Rih**
77 C1 **Križevci** Croatia
76 B1 **Krk** *i.* Croatia
60 F3 **Krokom** Sweden
59 C1 **Krolevets'** Ukr.
69 E2 **Kronach** Ger.
31 B2 **Krŏng Kaôh Kŏng** Cambodia
95 I2 **Kronprins Frederik Bjerge** *nunataks* Greenland
91 C2 **Kroonstad** S. Africa
55 D4 **Kropotkin** Rus. Fed.
71 E2 **Krosno** Pol.
71 D1 **Krotoszyn** Pol.
28 B2 **Krui** Indon.
77 C2 **Krujë** Albania
79 C2 **Krumovgrad** Bulg.
Krung Thep Thai. *see* **Bangkok**
56 C3 **Krupki** Belarus
77 D2 **Kruševac** Yugo.
69 E2 **Krušné Hory** *mts* Czech Rep.
96 A2 **Kruzof Island** U.S.A.
57 D3 **Krychaw** Belarus
59 D3 **Krymsk** Rus. Fed.
79 C3 **Krytiko Pelagos** *sea* Greece
59 C2 **Kryvyy Rih** Ukr.
82 B2 **Ksabi** Alg.
75 D2 **Ksar el Boukhari** Alg.
82 B1 **Ksar el Kebir** Morocco
57 E3 **Kshenskiy** Rus. Fed.
46 B2 **Kū', Jabal al** *hill* Saudi Arabia
29 C1 **Kuala Belait** Brunei
28 B1 **Kuala Kerai** Malaysia
28 B1 **Kuala Lipis** Malaysia
28 B1 **Kuala Lumpur** Malaysia
29 C2 **Kualapembuang** Indon.
28 B1 **Kuala Terengganu** Malaysia
28 B2 **Kualatungal** Indon.
29 C1 **Kuamut** *Sabah* Malaysia
33 A1 **Kuandian** China
28 B1 **Kuantan** Malaysia
59 D2 **Kuban'** *r.* Rus. Fed.
57 E2 **Kubenskoye, Ozero** *l.* Rus. Fed.
78 C2 **Kubrat** Bulg.
29 C1 **Kubuang** Indon.
29 C1 **Kuching** *Sarawak* Malaysia
77 C2 **Kuçovë** Albania
29 C2 **Kudus** Indon.
70 C2 **Kufstein** Austria
94 D2 **Kugluktuk** Can.
60 I3 **Kuhmo** Fin.
47 C2 **Kūhrān, Kūh-e** *mt.* Iran
Kuitin China *see* **Kuytun**
88 A2 **Kuito** Angola
96 A2 **Kuiu Island** U.S.A.
60 I2 **Kuivaniemi** Fin.
33 B2 **Kujang** N. Korea
77 D2 **Kukës** Albania
79 C3 **Kula** Turkey
43 D2 **Kula Kangri** *mt.* Bhutan
44 B2 **Kulandy** Kazakh.
56 B2 **Kuldīga** Latvia
90 B1 **Kule** Botswana
69 E2 **Kulmbach** Ger.
45 C3 **Kŭlob** Tajik.
44 B2 **Kul'sary** Kazakh.
45 D1 **Kulunda** Rus. Fed.
95 I2 **Kulusuk** Greenland
35 C3 **Kumagaya** Japan
35 B4 **Kumamoto** Japan
77 D2 **Kumanovo** Macedonia
82 B4 **Kumasi** Ghana
86 A2 **Kumba** Cameroon
46 B2 **Kumdah** Saudi Arabia
55 E3 **Kumertau** Rus. Fed.
33 B2 **Kumi** S. Korea
61 G4 **Kumla** Sweden
83 D3 **Kumo** Nigeria
30 A1 **Kumon Range** *mts* Myanmar
30 B2 **Kumphawapi** Thai.
Kumul China *see* **Hami**
34 D2 **Kunashir, Ostrov** *i.* Rus. Fed.
29 C1 **Kundat** *Sabah* Malaysia
45 D2 **Kungei Alatau** *mts* Kazakh./Kyrg.
44 B2 **Kungrad** Uzbek.
61 F4 **Kungsbacka** Sweden
86 B2 **Kungu** Dem. Rep. Congo
54 E3 **Kungur** Rus. Fed.
30 A1 **Kunhing** Myanmar
45 D3 **Kunlun Shan** *mts* China
30 B1 **Kunming** China
33 B2 **Kunsan** S. Korea
18 B1 **Kununurra** Austr.
60 I3 **Kuopio** Fin.
77 C1 **Kupa** *r.* Croatia/Slovenia
27 C3 **Kupang** Indon.
56 B2 **Kupiškis** Lith.
79 C2 **Küplü** Turkey
96 A2 **Kupreanof Island** U.S.A.
59 D2 **Kup"yans'k** Ukr.
45 E2 **Kuqa** China
35 B4 **Kurashiki** Japan
35 B3 **Kurayoshi** Japan
57 E3 **Kurchatov** Rus. Fed.
78 C2 **Kŭrdzhali** Bulg.
35 B4 **Kure** Japan
56 B2 **Kuressaare** Estonia
54 F3 **Kurgan** Rus. Fed.
61 H3 **Kurikka** Fin.
124 C3 **Kuril Trench** *sea feature* N. Pacific Ocean
57 E3 **Kurkino** Rus. Fed.
85 B3 **Kurmuk** Sudan
41 B3 **Kurnool** India
35 D3 **Kuroiso** Japan
21 D2 **Kurri Kurri** Austr.
57 E3 **Kursk** Rus. Fed.
90 B2 **Kuruman** S. Africa
90 B2 **Kuruman** *watercourse* S. Africa
35 B4 **Kurume** Japan
51 I3 **Kurumkan** Rus. Fed.
41 C4 **Kurunegala** Sri Lanka
79 C3 **Kuşadası** Turkey
79 C3 **Kuşadası Körfezi** *b.* Turkey
79 C2 **Kuş Gölü** *l.* Turkey
59 D2 **Kushchevskaya** Rus. Fed.
34 D2 **Kushiro** Japan
44 C1 **Kushmurun** Kazakh.
43 C2 **Kushtia** Bangl.
34 D2 **Kussharo-ko** *l.* Japan
79 C3 **Kütahya** Turkey
49 C1 **K'ut'aisi** Georgia
77 C1 **Kutjevo** Croatia
71 D1 **Kutno** Pol.
86 B3 **Kutu** Dem. Rep. Congo
94 D2 **Kuujjua** *r.* Can.
99 D1 **Kuujjuaq** Can.
Kuujjuarapik Can. *see* **Poste-de-la-Baleine**
60 I2 **Kuusamo** Fin.
88 A2 **Kuvango** Angola
57 D2 **Kuvshinovo** Rus. Fed.
46 B2 **Kuwait** *country* Asia
47 B2 **Kuwait** Kuwait
Kuybyshev Rus. Fed. *see* **Samara**
59 D2 **Kuybysheve** Ukr.
55 D3 **Kuybyshevskoye Vodokhranilishche** *resr* Rus. Fed.
45 E2 **Kuytun** China
79 C3 **Kuyucak** Turkey
55 D3 **Kuznetsk** Rus. Fed.
58 B1 **Kuznetsovs'k** Ukr.
60 H1 **Kvalsund** Norway
91 D2 **KwaMashu** S. Africa
33 B2 **Kwangju** S. Korea
33 B1 **Kwanmo-bong** *mt.* N. Korea
91 C3 **Kwanobuhle** S. Africa
90 B3 **Kwanonzame** S. Africa
91 C3 **Kwatinidubu** S. Africa
91 C2 **KwaZamokhule** S. Africa
91 D2 **Kwazulu-Natal** *prov.* S. Africa
89 B2 **Kwekwe** Zimbabwe
86 B3 **Kwenge** *r.* Dem. Rep. Congo
71 D1 **Kwidzyn** Pol.
86 B3 **Kwilu** *r.* Angola/Dem. Rep. Congo
27 C3 **Kwoka** *mt.* Indon.
21 C3 **Kyabram** Austr.
30 A2 **Kyaikto** Myanmar
37 D1 **Kyakhta** Rus. Fed.
20 A2 **Kyancutta** Austr.
30 A1 **Kyaukpadaung** Myanmar
30 A2 **Kyaukpyu** Myanmar
56 B3 **Kybartai** Lith.
30 A2 **Kyebogyi** Myanmar
30 A2 **Kyeintali** Myanmar
42 B1 **Kyelang** India
Kyiv Ukr. *see* **Kiev**

Kyklades *is* Greece *see* **Cyclades**
97 D2 **Kyle** Can.
64 B2 **Kyle of Lochalsh** U.K.
68 C3 **Kyll** *r.* Ger.
79 B3 **Kyllini** *mt.* Greece
79 B3 **Kymi** Greece
20 B3 **Kyneton** Austr.
87 D2 **Kyoga, Lake** Uganda
21 D1 **Kyogle** Austr.
33 B2 **Kyŏngju** S. Korea
35 C4 **Kyōto** Japan
79 B3 **Kyparissia** Greece
45 C1 **Kypshak, Ozero** *salt l.* Kazakh.
79 B3 **Kyra Panagia** *i.* Greece
45 D2 **Kyrgyzstan** *country* Asia
69 F1 **Kyritz** Ger.
60 H3 **Kyrönjoki** *r.* Fin.
54 E2 **Kyrta** Rus. Fed.
54 D2 **Kyssa** Rus. Fed.
79 B3 **Kythira** *i.* Greece
79 B3 **Kythnos** *i.* Greece
96 B2 **Kyuquot** Can.
35 B4 **Kyūshū** *i.* Japan
78 B2 **Kyustendil** Bulg.
58 C1 **Kyyivs'ke Vodoskhovyshche** *resr* Ukr.
60 H3 **Kyyjärvi** Fin.
36 C1 **Kyzyl** Rus. Fed.
44 C2 **Kyzylkum Desert** Kazakh./Uzbek.
45 C2 **Kyzylorda** Kazakh.

L

113 C3 **La Angostura, Presa de** *resr* Mex.
85 C4 **Laascaanood** Somalia
113 D3 **La Ascensión, Bahía de** *b.* Mex.
118 B1 **La Asunción** Venez.
82 A2 **Laâyoune** Western Sahara
55 D4 **Laba** *r.* Rus. Fed.
112 B2 **La Babia** Mex.
114 B3 **La Bahía, Islas de** *is* Hond.
99 D1 **La Baleine, Rivière à** *r.* Can.
120 B3 **La Banda** Arg.
72 B2 **La Baule-Escoublac** France
82 A3 **Labé** Guinea
96 A1 **Laberge, Lake** Can.
97 C2 **La Biche, Lac** *l.* Can.
55 D4 **Labinsk** Rus. Fed.
32 B1 **Labo** Phil.
112 B2 **La Boquilla, Presa de** *resr* Mex.
72 B3 **Labouheyre** France
99 D1 **Labrador** *reg.* Can.
99 D1 **Labrador City** Can.
126 C1 **Labrador Sea** Can./Greenland
118 B2 **Lábrea** Brazil
29 C1 **Labuan** Malaysia
28 B1 **Labuhanbilik** Indon.
27 C3 **Labuna** Indon.
31 A2 **Labutta** Myanmar
54 F2 **Labytnangi** Rus. Fed.
77 C2 **Laç** Albania
74 B1 **La Cabrera, Sierra de** *mts* Spain
73 C2 **La Capelle** France
99 D1 **Lac Bermen** *l.* Can.
41 B3 **Laccadive Islands** India
97 E2 **Lac du Bonnet** Can.
114 B3 **La Ceiba** Hond.
21 B2 **Lachlan** *r.* Austr.
109 E1 **Lachute** Can.
73 D3 **La Ciotat** France
97 C2 **Lac La Biche** Can.
Lac la Martre Can. *see* **Wha Ti**
96 C2 **Lacombe** Can.
113 C3 **La Concordia** Mex.
76 A3 **Laconi** *Sardinia* Italy
109 E2 **Laconia** U.S.A.
72 B2 **La Coubre, Pointe de** *pt* France
96 C2 **La Crete** Can.
108 A2 **La Crosse** U.S.A.
112 B2 **La Cruz** Mex.
112 B2 **La Cuesta** Mex.
43 B1 **Ladakh Range** *mts* India
74 C1 **La Demanda, Sierra de** *mts* Spain
90 B3 **Ladismith** S. Africa
47 D2 **Lādīz** Iran
54 C2 **Ladoga, Lake** Rus. Fed.
Ladozhskoye Ozero *l.* Rus. Fed. *see* **Ladoga, Lake**
96 B3 **Ladysmith** Can.
91 C2 **Ladysmith** S. Africa
27 D3 **Lae** P.N.G.
30 B2 **Lae** Thai.
120 B3 **La Esmeralda** Bol.
61 F4 **Læsø** *i.* Denmark
111 C2 **La Fayette** U.S.A.
108 B2 **Lafayette** *IN* U.S.A.
110 B2 **Lafayette** *LA* U.S.A.
83 C4 **Lafia** Nigeria
72 B2 **La Flèche** France
111 D1 **La Follette** U.S.A.
99 C1 **Laforge** Can.
47 C2 **Lāft** Iran
76 A3 **La Galite** *i.* Tunisia
49 C1 **Lagan'** Rus. Fed.
86 B2 **Lagdo, Lac de** *l.* Cameroon
83 C1 **Laghouat** Alg.
123 D1 **Lagoa Santa** Brazil
82 A2 **La Gomera** *i.* Canary Is
115 C3 **La Gonâve, Île de** *i.* Haiti
82 C4 **Lagos** Nigeria
74 B2 **Lagos** Port.
102 C1 **La Grande** U.S.A.
98 C1 **La Grande 2, Réservoir** *resr* Can.
98 C1 **La Grande 3, Réservoir** *resr* Can.
99 C1 **La Grande 4, Réservoir** *resr* Can.
18 B1 **Lagrange** Austr.
111 C2 **La Grange** U.S.A.
118 B1 **La Gran Sabana** *plat.* Venez.
120 D3 **Laguna** Brazil
118 A2 **Lagunas** Peru
La Habana Cuba *see* **Havana**
29 C1 **Lahad Datu** *Sabah* Malaysia
72 B2 **La Hague, Cap de** *c.* France
28 B2 **Lahat** Indon.
46 B3 **Laḥij** Yemen
49 C2 **Lāhījān** Iran
68 C2 **Lahnstein** Ger.
42 B1 **Lahore** Pak.
42 A2 **Lahri** Pak.
61 I3 **Lahti** Fin.
83 D4 **Laï** Chad
21 D1 **Laidley** Austr.
72 C2 **L'Aigle** France
90 B3 **Laingsburg** S. Africa
60 H2 **Lainioälven** *r.* Sweden
83 C3 **L'Aïr, Massif de** *mts* Niger
64 B1 **Lairg** U.K.
76 B1 **Laives** Italy
38 B2 **Laiwu** China
38 C2 **Laiyang** China
38 B2 **Laizhou** China
38 B2 **Laizhou Wan** *b.* China
18 C1 **Lajamanu** Austr.
120 C3 **Lajes** Brazil
112 B2 **La Junta** Mex.
104 C3 **La Junta** U.S.A.
114 B2 **La Juventud, Isla de** *i.* Cuba
21 C2 **Lake Cargelligo** Austr.
21 D2 **Lake Cathie** Austr.
110 B2 **Lake Charles** U.S.A.
111 D2 **Lake City** *FL* U.S.A.
111 E2 **Lake City** *SC* U.S.A.
96 B3 **Lake Cowichan** Can.
106 A2 **Lake Havasu City** U.S.A.
111 D3 **Lakeland** U.S.A.
96 C2 **Lake Louise** Can.
102 B1 **Lake Oswego** U.S.A.
22 A2 **Lake Paringa** N.Z.
110 B2 **Lake Providence** U.S.A.
22 B2 **Lake Pukaki** N.Z.
21 C3 **Lakes Entrance** Austr.
102 B2 **Lakeview** U.S.A.
104 B3 **Lakewood** *CO* U.S.A.
109 E2 **Lakewood** *NJ* U.S.A.
111 D3 **Lake Worth** U.S.A.
42 A2 **Lakhpat** India
42 B1 **Lakki** Pak.
79 B3 **Lakonikos Kolpos** *b.* Greece
60 H1 **Lakselv** Norway
112 A2 **La Laguna, Picacho de** *mt.* Mex.
113 C3 **La Libertad** Guat.
74 B1 **Lalín** Spain
43 B2 **Lalitpur** India
97 D2 **La Loche** Can.
68 B2 **La Louvière** Belgium
76 A2 **La Maddalena** *Sardinia* Italy
99 D2 **La Madeleine, Îles de** *is* Can.
29 C1 **Lamag** *Sabah* Malaysia
La Manche *str.* France/U.K. *see* **English Channel**
104 C3 **Lamar** U.S.A.
47 C2 **Lamard** Iran
76 A3 **La Marmora, Punta** *mt.* *Sardinia* Italy
96 C1 **La Martre, Lac** *l.* Can.
86 B3 **Lambaréné** Gabon
90 A3 **Lambert's Bay** S. Africa
60 □A2 **Lambeyri** Iceland
74 B1 **Lamego** Port.
118 A3 **La Merced** Peru
20 B3 **Lameroo** Austr.
107 C2 **Lamesa** U.S.A.
79 B3 **Lamia** Greece
105 E2 **Lamoni** U.S.A.
30 A2 **Lampang** Thai.
107 D2 **Lampasas** U.S.A.
113 B2 **Lampazos** Mex.
67 A2 **Lampeter** U.K.
30 A2 **Lamphun** Thai.
87 E3 **Lamu** Kenya
75 D2 **La Nao, Cabo de** *c.* Spain
64 C3 **Lanark** U.K.
31 A2 **Lanbi Kyun** *i.* Myanmar
Lancang Jiang *r.* China *see* **Mekong**
66 B1 **Lancaster** U.K.
103 C4 **Lancaster** *CA* U.S.A.
108 C3 **Lancaster** *OH* U.S.A.
109 D2 **Lancaster** *PA* U.S.A.
111 D2 **Lancaster** *SC* U.S.A.
95 F2 **Lancaster Sound** *str.* Can.
Lanchow China *see* **Lanzhou**
70 C2 **Landeck** Austria
104 B2 **Lander** U.S.A.
67 A3 **Land's End** *pt* U.K.
70 C2 **Landshut** Ger.
61 F4 **Landskrona** Sweden
111 C2 **Lanett** U.S.A.
90 B2 **Langberg** *mts* S. Africa
105 D1 **Langdon** U.S.A.
61 F4 **Langeland** *i.* Denmark

69 D1 **Langen** Ger.
68 C1 **Langeoog** Ger.
68 C1 **Langeoog** *i.* Ger.
69 D2 **Langgöns** Ger.
60 □A3 **Langjökull** *ice cap* Iceland
73 C3 **Langogne** France
72 B3 **Langon** France
73 D2 **Langres** France
28 A1 **Langsa** Indon.
30 B1 **Lang Sơn** Vietnam
69 D1 **Langwedel** Ger.
97 D2 **Lanigan** Can.
121 A4 **Lanín, Volcán** *vol.* Arg./Chile
49 C2 **Länkäran** Azer.
72 B2 **Lannion** France
108 C2 **Lansing** U.S.A.
39 B3 **Lanxi** China
85 B4 **Lanya** Sudan
82 A2 **Lanzarote** *i.* Canary Is
38 A2 **Lanzhou** China
32 B1 **Laoag** Phil.
30 B1 **Lao Cai** Vietnam
38 B2 **Laohekou** China
Laojunmiao China *see* **Yumen**
33 B1 **Laoling** China
33 B1 **Lao Ling** *mts* China
73 C2 **Laon** France
30 B2 **Laos** *country* Asia
33 B1 **Laotougou** China
122 C3 **Lapa** Brazil
82 A2 **La Palma** *i.* Canary Is
114 C4 **La Palma** Panama
118 B1 **La Paragua** Venez.
120 B2 **La Paz** Bol.
112 A2 **La Paz** Mex.
118 B2 **La Pedrera** Col.
34 D1 **La Pérouse Strait** Japan/Rus. Fed.
113 C2 **La Pesca** Mex.
112 B2 **La Piedad** Mex.
121 C4 **La Plata** Arg.
121 C4 **La Plata, Río de** *sea chan.* Arg./Uru.
61 I3 **Lappeenranta** Fin.
60 G2 **Lappland** *reg.* Europe
79 C2 **Lâpseki** Turkey
51 J1 **Laptev Sea** Rus. Fed.
Laptevykh, More *sea* Rus. Fed. *see* **Laptev Sea**
60 H3 **Lapua** Fin.
120 B3 **La Quiaca** Arg.
47 C2 **Lār** Iran
82 B1 **Larache** Morocco
104 B2 **Laramie** U.S.A.
104 B2 **Laramie Mountains** U.S.A.
122 B3 **Laranjeiras do Sul** Brazil
27 C3 **Larantuka** Indon.
27 C3 **Larat** *i.* Indon.
75 D2 **Larba** Alg.
68 B3 **L'Ardenne, Plateau de** Belgium
74 C1 **Laredo** Spain
107 D3 **Laredo** U.S.A.
64 B3 **Largs** U.K.
120 B3 **La Rioja** Arg.
79 B3 **Larisa** Greece
42 A2 **Larkana** Pak.
48 B2 **Larnaka** Cyprus
65 D1 **Larne** U.K.
68 B2 **La Roche-en-Ardenne** Belgium
72 B2 **La Rochelle** France
72 B2 **La Roche-sur-Yon** France
75 C2 **La Roda** Spain
115 D3 **La Romana** Dom. Rep.
97 D2 **La Ronge** Can.
97 D2 **La Ronge, Lac** *l.* Can.
18 C1 **Larrimah** Austr.
23 P3 **Larsen Ice Shelf** Antarctica
61 F4 **Larvik** Norway
98 C2 **La Sarre** Can.
106 B2 **Las Cruces** U.S.A.
121 C4 **Las Flores** Arg.
121 B4 **Las Heras** Arg.
30 A1 **Lashio** Myanmar
77 C3 **La Sila** *reg.* Italy
120 B3 **Las Lomitas** Arg.
112 B2 **Las Nieves** Mex.
82 A2 **Las Palmas de Gran Canaria** Canary Is
76 A2 **La Spezia** Italy
118 B3 **Las Piedras, Río de** *r.* Peru
121 B5 **Las Plumas** Arg.
97 D2 **Last Mountain Lake** Can.
120 B3 **Las Tórtolas, Cerro** *mt.* Chile
86 B3 **Lastoursville** Gabon
77 C2 **Lastovo** *i.* Croatia
112 A2 **Las Tres Vírgenes, Volcán** *vol.* Mex.
114 C2 **Las Tunas** Cuba
112 B2 **Las Varas** *Chihuahua* Mex.
112 B2 **Las Varas** *Nayarit* Mex.
106 B1 **Las Vegas** *NM* U.S.A.
103 C3 **Las Vegas** *NV* U.S.A.
99 E1 **La Tabatière** Can.
72 B3 **La Teste** France
76 B2 **Latina** Italy
115 D3 **La Tortuga, Isla** *i.* Venez.
32 B1 **La Trinidad** Phil.
57 E2 **Latskoye** Rus. Fed.
99 C2 **La Tuque** Can.
56 B2 **Latvia** *country* Europe
70 C1 **Lauchhammer** Ger.
69 E3 **Lauf an der Pegnitz** Ger.
73 D2 **Laufen** Switz.
19 D4 **Launceston** Austr.
67 A3 **Launceston** U.K.
30 A1 **Launggyaung** Myanmar
19 D1 **Laura** Austr.
110 C2 **Laurel** *MS* U.S.A.
102 E1 **Laurel** *MT* U.S.A.
77 C2 **Lauria** Italy
111 E2 **Laurinburg** U.S.A.
73 D2 **Lausanne** Switz.
29 C2 **Laut** *i.* Indon.
69 D2 **Lautersbach (Hessen)** Ger.
29 C2 **Laut Kecil, Kepulauan** *is* Indon.
68 C1 **Lauwersmeer** *l.* Neth.
72 B2 **Laval** France
18 B2 **Laverton** Austr.
123 D2 **Lavras** Brazil
91 D2 **Lavumisa** Swaziland
29 C1 **Lawas** *Sarawak* Malaysia
30 A1 **Lawksawk** Myanmar
82 B3 **Lawra** Ghana
105 D3 **Lawrence** U.S.A.
110 C1 **Lawrenceburg** U.S.A.
107 D2 **Lawton** U.S.A.
46 A2 **Lawz, Jabal al** *mt.* Saudi Arabia
61 F4 **Laxå** Sweden
77 D2 **Lazarevac** Yugo.
112 A1 **Lázaro Cárdenas** *Baja California Norte* Mex.
112 B3 **Lázaro Cárdenas** *Michoacán* Mex.
56 B3 **Lazdijai** Lith.
104 C2 **Lead** U.S.A.
97 D2 **Leader** Can.
104 B3 **Leadville** U.S.A.
Leaf Bay Can. *see* **Tasiujaq**
97 D2 **Leaf Rapids** Can.
65 B2 **Leane, Lough** *l.* Rep. of Ireland
98 C1 **L'Eau Claire, Lac à** *l.* Can.
105 E3 **Leavenworth** U.S.A.
48 B2 **Lebanon** *country* Asia
105 E3 **Lebanon** *MO* U.S.A.
109 E2 **Lebanon** *NH* U.S.A.
102 B2 **Lebanon** *OR* U.S.A.
109 D2 **Lebanon** *PA* U.S.A.
57 E3 **Lebedyan'** Rus. Fed.
59 C1 **Lebedyn** Ukr.
72 C2 **Le Blanc** France
71 D1 **Lębork** Pol.
91 C1 **Lebowakgomo** S. Africa
74 B2 **Lebrija** Spain
121 A4 **Lebu** Chile
72 C3 **Le Bugue** France
77 C2 **Lecce** Italy
76 A1 **Lecco** Italy
70 C2 **Lech** *r.* Austria/Ger.
79 B3 **Lechaina** Greece
39 B3 **Lechang** China
70 B1 **Leck** Ger.
72 C3 **Lectoure** France
74 B1 **Ledesma** Spain
72 C2 **Le Dorat** France
96 C2 **Leduc** Can.
105 E1 **Leech Lake** U.S.A.
66 C2 **Leeds** U.K.
66 B2 **Leek** U.K.
68 C1 **Leer (Ostfriesland)** Ger.
111 D3 **Leesburg** U.S.A.
110 B2 **Leesville** U.S.A.
21 C2 **Leeton** Austr.
90 B3 **Leeu-Gamka** S. Africa
68 B1 **Leeuwarden** Neth.
18 A3 **Leeuwin, Cape** Austr.
115 D3 **Leeward Islands** Caribbean Sea
79 B3 **Lefkada** Greece
79 B3 **Lefkada** *i.* Greece
Lefkosia Cyprus *see* **Nicosia**
32 B1 **Legaspi** Phil.
76 B1 **Legnago** Italy
71 D1 **Legnica** Pol.
43 B1 **Leh** Jammu and Kashmir
72 C2 **Le Havre** France
90 B1 **Lehututu** Botswana
42 B1 **Leiah** Pak.
71 D2 **Leibnitz** Austria
67 C2 **Leicester** U.K.
19 C1 **Leichhardt** *r.* Austr.
68 B1 **Leiden** Neth.
68 A2 **Leie** *r.* Belgium
20 A2 **Leigh Creek** Austr.
61 E3 **Leikanger** Norway
69 D1 **Leine** *r.* Ger.
65 C2 **Leinster** *reg.* Rep. of Ireland
69 F2 **Leipzig** Ger.
60 F2 **Leiranger** Norway
74 B2 **Leiria** Port.
61 E4 **Leirvik** Norway
65 C2 **Leixlip** Rep. of Ireland
39 B3 **Leiyang** China
39 B3 **Leizhou** China
39 A3 **Leizhou Bandao** *pen.* China
86 B3 **Lékana** Congo
68 B1 **Lelystad** Neth.
121 B6 **Le Maire, Estrecho de** *sea chan.* Arg.
73 D2 **Léman, Lac** *l.* France/Switz.
72 C2 **Le Mans** France
105 D2 **Le Mars** U.S.A.
122 C2 **Leme** Brazil
48 B2 **Lemesos** Cyprus
69 D1 **Lemförde** Ger.
95 G2 **Lemieux Islands** Can.
104 C1 **Lemmon** U.S.A.
103 C3 **Lemoore** U.S.A.
99 D1 **Le Moyne, Lac** *l.* Can.
30 A1 **Lemro** *r.* Myanmar
77 C2 **Le Murge** *hills* Italy
51 J2 **Lena** *r.* Rus. Fed.
68 C1 **Lengerich** Ger.
39 B3 **Lengshuijiang** China

39 B3 **Lengshuitan** China
59 D2 **Lenine** Ukr.
Leningrad Rus. Fed. *see* **St Petersburg**
59 D2 **Leningradskaya** Rus. Fed.
45 D3 **Lenin Peak** Kyrg./Tajik.
Leninsk Kazakh. *see* **Baykonur**
57 E3 **Leninskiy** Rus. Fed.
68 A2 **Lens** Belgium
72 C1 **Lens** France
51 I2 **Lensk** Rus. Fed.
71 D2 **Lenti** Hungary
77 C3 **Lentini** *Sicily* Italy
82 B3 **Léo** Burkina
71 D2 **Leoben** Austria
67 B2 **Leominster** U.K.
112 B2 **León** Mex.
114 B3 **León** Nic.
74 B1 **León** Spain
90 A1 **Leonardville** Namibia
18 B2 **Leonora** Austr.
123 D2 **Leopoldina** Brazil
91 C1 **Lephepe** Botswana
91 C3 **Lephoi** S. Africa
39 B3 **Leping** China
73 C2 **Le-Puy-en-Velay** France
118 A2 **Lérida** Col.
74 C1 **Lerma** Spain
79 C3 **Leros** *i.* Greece
98 C1 **Le Roy, Lac** *l.* Can.
61 F4 **Lerum** Sweden
64 □ **Lerwick** U.K.
79 C3 **Lesbos** *i.* Greece
115 C3 **Les Cayes** Haiti
37 C3 **Leshan** China
54 D2 **Leshukonskoye** Rus. Fed.
77 D2 **Leskovac** Yugo.
72 B2 **Lesneven** France
57 E2 **Lesnoye** Rus. Fed.
51 H3 **Lesosibirsk** Rus. Fed.
91 C2 **Lesotho** *country* Africa
34 B1 **Lesozavodsk** Rus. Fed.
72 B2 **Les Sables-d'Olonne** France
115 D3 **Lesser Antilles** *is* Caribbean Sea
96 C2 **Lesser Slave Lake** Can.
73 C3 **Les Vans** France
Lesvos *i.* Greece *see* **Lesbos**
71 D1 **Leszno** Pol.
67 C3 **Letchworth** U.K.
96 C3 **Lethbridge** Can.
118 C1 **Lethem** Guyana
27 C3 **Leti, Kepulauan** *is* Indon.
118 B2 **Leticia** Col.
91 C1 **Letlhakeng** Botswana
72 C1 **Le Touquet-Paris-Plage** France
67 D3 **Le Tréport** France
31 A2 **Letsok-aw Kyun** *i.* Myanmar
91 C2 **Letsopa** S. Africa
65 C1 **Letterkenny** Rep. of Ireland
73 C3 **Leucate, Étang de** *l.* France
28 A1 **Leuser, Gunung** *mt.* Indon.
68 B2 **Leuven** Belgium
79 B3 **Levadeia** Greece
60 F3 **Levanger** Norway
107 C2 **Levelland** U.S.A.
18 B1 **Lévêque, Cape** Austr.
68 C2 **Leverkusen** Ger.
71 D2 **Levice** Slovakia
22 C2 **Levin** N.Z.
99 C2 **Lévis** Can.
57 E3 **Lev Tolstoy** Rus. Fed.
67 D3 **Lewes** U.K.
64 A1 **Lewis, Isle of** *i.* U.K.
108 C3 **Lewisburg** U.S.A.
102 D1 **Lewis Range** *mts* U.S.A.
102 C1 **Lewiston** *ID* U.S.A.
109 E2 **Lewiston** *ME* U.S.A.
102 E1 **Lewistown** U.S.A.
108 C3 **Lexington** *KY* U.S.A.
104 D2 **Lexington** *NE* U.S.A.
109 D3 **Lexington** *VA* U.S.A.
32 B1 **Leyte** *i.* Phil.
77 C2 **Lezhë** Albania
57 F2 **Lezhnevo** Rus. Fed.
57 E3 **L'gov** Rus. Fed.
43 C2 **Lhagoi Kangri** *mt.* China
43 D2 **Lhasa** China
43 C2 **Lhazê** China
28 A1 **Lhokseumawe** Indon.
35 B3 **Liancourt Rocks** *i.* N. Pacific Ocean
39 B3 **Lianhua** China
39 B3 **Lianjiang** China
38 C1 **Lianshan** China
38 B2 **Lianyungang** China
39 B3 **Lianzhou** China
33 A1 **Liao** *r.* China
38 B2 **Liaocheng** China
38 C1 **Liaodong Bandao** *pen.* China
38 C1 **Liaodong Wan** *b.* China
38 C1 **Liaoning** *prov.* China
33 A1 **Liaoyang** China
33 B1 **Liaoyuan** China
96 B1 **Liard** *r.* Can.
102 C1 **Libby** U.S.A.
86 B2 **Libenge** Dem. Rep. Congo
104 C3 **Liberal** U.S.A.
71 D1 **Liberec** Czech Rep.
82 B4 **Liberia** *country* Africa
114 B3 **Liberia** Costa Rica
118 B1 **Libertad** Venez.
68 B3 **Libin** Belgium
72 B3 **Libourne** France
86 A2 **Libreville** Gabon
83 D2 **Libya** *country* Africa
83 E2 **Libyan Desert** Egypt/Libya
84 A1 **Libyan Plateau** Egypt
76 B3 **Licata** *Sicily* Italy
Licheng China *see* **Lipu**
89 C2 **Lichinga** Moz.
69 E2 **Lichte** Ger.
91 C2 **Lichtenburg** S. Africa
69 E2 **Lichtenfels** Ger.
56 C3 **Lida** Belarus
61 F4 **Lidköping** Sweden
18 C2 **Liebig, Mount** Austr.
73 D2 **Liechtenstein** *country* Europe
68 B2 **Liège** Belgium
60 J3 **Lieksa** Fin.
87 C2 **Lienart** Dem. Rep. Congo
70 C2 **Lienz** Austria
56 B2 **Liepāja** Latvia
70 C2 **Liezen** Austria
65 C2 **Liffey** *r.* Rep. of Ireland
65 C1 **Lifford** Rep. of Ireland
21 C1 **Lightning Ridge** Austr.
89 C2 **Ligonha** *r.* Moz.
73 D3 **Ligurian Sea** France/Italy
87 C4 **Likasi** Dem. Rep. Congo
96 B2 **Likely** Can.
57 E2 **Likhoslavl'** Rus. Fed.
29 B1 **Liku** Indon.
73 D3 **L'Île-Rousse** *Corsica* France
39 B3 **Liling** China
61 F4 **Lilla Edet** Sweden
68 B2 **Lille** Belgium
73 C1 **Lille** France
61 E4 **Lille Bælt** *sea chan.* Denmark
61 F3 **Lillehammer** Norway
61 F4 **Lillestrøm** Norway
96 B2 **Lillooet** Can.
89 C2 **Lilongwe** Malawi
32 B2 **Liloy** Phil.
118 A3 **Lima** Peru
108 C2 **Lima** U.S.A.
65 C1 **Limavady** U.K.
121 B4 **Limay** *r.* Arg.
69 F2 **Limbach-Oberfrohna** Ger.
56 B2 **Limbaži** Latvia
86 A2 **Limbe** Cameroon
68 D2 **Limburg an der Lahn** Ger.
90 B2 **Lime Acres** S. Africa
122 C2 **Limeira** Brazil
65 B2 **Limerick** Rep. of Ireland
61 E4 **Limfjorden** *sea chan.* Denmark
79 C3 **Limnos** *i.* Greece
72 C2 **Limoges** France
114 B3 **Limón** Costa Rica
104 C3 **Limon** U.S.A.
72 C3 **Limoux** France
89 C3 **Limpopo** *r.* S. Africa/Zimbabwe
32 A1 **Linapacan** *i.* Phil.
121 A4 **Linares** Chile
113 C2 **Linares** Mex.
74 C2 **Linares** Spain
30 B1 **Lincang** China
39 B3 **Linchuan** China
66 C2 **Lincoln** U.K.
108 B2 **Lincoln** *IL* U.S.A.
109 F1 **Lincoln** *ME* U.S.A.
105 D2 **Lincoln** *NE* U.S.A.
119 C1 **Linden** Guyana
110 C1 **Linden** U.S.A.
87 C2 **Lindi** *r.* Dem. Rep. Congo
87 D3 **Lindi** Tanz.
Lindisfarne *i.* U.K. *see* **Holy Island**
48 A2 **Lindos** Greece
109 D2 **Lindsay** Can.
38 B2 **Linfen** China
32 B1 **Lingayen** Phil.
38 B2 **Lingbao** China
91 C3 **Lingelethu** S. Africa
91 C3 **Lingelihle** S. Africa
68 C1 **Lingen (Ems)** Ger.
28 B2 **Lingga, Kepulauan** *is* Indon.
39 A3 **Lingshan** China
39 A4 **Lingshui** China
82 A3 **Linguère** Senegal
123 D1 **Linhares** Brazil
38 A1 **Linhe** China
61 G4 **Linköping** Sweden
34 B1 **Linkou** China
38 B2 **Linqing** China
122 C2 **Lins** Brazil
104 C1 **Linton** U.S.A.
37 D2 **Linxi** China
37 C2 **Linxia** China
38 B2 **Linyi** *Shandong* China
38 B2 **Linyi** *Shandong* China
38 B2 **Linying** China
70 C2 **Linz** Austria
73 C3 **Lion, Golfe du** *g.* France
77 B3 **Lipari** Italy
76 B3 **Lipari, Isole** *is* Italy
57 E3 **Lipetsk** Rus. Fed.
78 B1 **Lipova** Romania
69 D2 **Lippstadt** Ger.
39 B3 **Lipu** China
44 C1 **Lisakovsk** Kazakh.
86 C2 **Lisala** Dem. Rep. Congo
Lisboa Port. *see* **Lisbon**
74 B2 **Lisbon** Port.
65 C1 **Lisburn** U.K.
65 B2 **Liscannor Bay** Rep. of Ireland
72 C2 **Lisieux** France
67 A3 **Liskeard** U.K.
57 E3 **Liski** Rus. Fed.
21 D1 **Lismore** Austr.
65 C1 **Lisnaskea** U.K.
65 B2 **Listowel** Rep. of Ireland

39 A3 **Litang** *Guangxi* China
36 C2 **Litang** *Sichuan* China
108 B3 **Litchfield** *IL* U.S.A.
105 E1 **Litchfield** *MN* U.S.A.
21 D2 **Lithgow** Austr.
79 B3 **Lithino, Akra** *pt* Greece
56 B2 **Lithuania** *country* Europe
70 C1 **Litoměřice** Czech Rep.
114 C2 **Little Abaco** *i.* Bahamas
41 D3 **Little Andaman** *i.* India
105 E1 **Little Falls** U.S.A.
107 C2 **Littlefield** U.S.A.
67 C3 **Littlehampton** U.K.
90 A2 **Little Karas Berg** *plat.* Namibia
90 B3 **Little Karoo** *plat.* S. Africa
64 A2 **Little Minch** *sea chan.* U.K.
104 C1 **Little Missouri** *r.* U.S.A.
31 A3 **Little Nicobar** *i.* India
110 B2 **Little Rock** U.S.A.
109 E2 **Littleton** U.S.A.
38 B2 **Liujiachang** China
39 A3 **Liuzhou** China
56 C2 **Līvāni** Latvia
18 B1 **Liveringa** Austr.
106 C2 **Livermore, Mount** U.S.A.
99 D2 **Liverpool** Can.
66 B2 **Liverpool** U.K.
95 F2 **Liverpool, Cape** Can.
21 C2 **Liverpool Range** *mts* Austr.
102 D1 **Livingston** *MT* U.S.A.
107 E2 **Livingston** *TX* U.S.A.
107 D2 **Livingston, Lake** U.S.A.
88 B2 **Livingstone** Zambia
77 C2 **Livno** Bos.-Herz.
57 E3 **Livny** Rus. Fed.
108 C2 **Livonia** U.S.A.
76 B2 **Livorno** Italy
87 D3 **Liwale** Tanz.
38 B2 **Liyang** China
67 A4 **Lizard Point** U.K.
76 B1 **Ljubljana** Slovenia
61 G3 **Ljungan** *r.* Sweden
61 F4 **Ljungby** Sweden
61 G3 **Ljusdal** Sweden
61 G3 **Ljusnan** *r.* Sweden
67 B3 **Llandeilo** U.K.
67 B3 **Llandovery** U.K.
67 B2 **Llandrindod Wells** U.K.
66 B2 **Llandudno** U.K.
67 A3 **Llanelli** U.K.
66 A2 **Llangefni** U.K.
107 C2 **Llano Estacado** *plain* U.S.A.
118 B1 **Llanos** *reg.* Col./Venez.
75 D1 **Lleida** Spain
67 A2 **Lleyn Peninsula** U.K.
75 C2 **Lliria** Spain
74 C1 **Llodio** Spain
96 B2 **Lloyd George, Mount** Can.
97 D2 **Lloyd Lake** Can.
97 C2 **Lloydminster** Can.
120 B3 **Llullaillaco, Volcán** *vol.* Chile
91 C2 **Lobatse** Botswana
88 A2 **Lobito** Angola
69 F1 **Loburg** Ger.
64 B2 **Lochaber** *mts* U.K.
72 C2 **Loches** France
64 B2 **Lochgilphead** U.K.
64 B1 **Lochinver** U.K.
64 A2 **Lochmaddy** U.K.
64 C2 **Lochnagar** *mt.* U.K.
20 A2 **Lock** Austr.
64 C3 **Lockerbie** U.K.
107 D3 **Lockhart** U.S.A.
109 D2 **Lock Haven** U.S.A.
109 D2 **Lockport** U.S.A.
31 B2 **Lôc Ninh** Vietnam
73 C3 **Lodève** France
54 C2 **Lodeynoye Pole** Rus. Fed.
60 F2 **Løding** Norway
87 D2 **Lodwar** Kenya
71 D1 **Łódź** Pol.
30 B2 **Loei** Thai.
90 A3 **Loeriesfontein** S. Africa
60 F2 **Lofoten** *is* Norway
102 D2 **Logan** U.S.A.
94 B2 **Logan, Mount** Can.
108 B2 **Logansport** U.S.A.
76 B1 **Logatec** Slovenia
74 C1 **Logroño** Spain
69 D1 **Löhne** Ger.
69 D1 **Lohne (Oldenburg)** Ger.
30 A2 **Loikaw** Myanmar
30 A2 **Loi Lan** *mt.* Myanmar/Thai.
72 B2 **Loire** *r.* France
118 A2 **Loja** Ecuador
74 C2 **Loja** Spain
60 I2 **Lokan tekojärvi** *l.* Fin.
68 B2 **Lokeren** Belgium
90 B1 **Lokgwabe** Botswana
87 D2 **Lokichar** Kenya
87 D2 **Lokichokio** Kenya
57 D2 **Loknya** Rus. Fed.
83 C4 **Lokoja** Nigeria
57 D3 **Lokot'** Rus. Fed.
56 C2 **Loksa** Estonia
95 G2 **Loks Land** *i.* Can.
61 F4 **Lolland** *i.* Denmark
87 D3 **Lollondo** Tanz.
90 B2 **Lolwane** S. Africa
78 B2 **Lom** Bulg.
61 E3 **Lom** Norway
87 C2 **Lomami** *r.* Dem. Rep. Congo
121 C4 **Lomas de Zamora** Arg.
29 C2 **Lombok** *i.* Indon.
29 C2 **Lombok, Selat** *sea chan.* Indon.
82 C4 **Lomé** Togo
86 C3 **Lomela** *r.* Dem. Rep. Congo
68 B2 **Lommel** Belgium
64 B2 **Lomond, Loch** *l.* U.K.
56 C2 **Lomoriosov** Rus. Fed.
103 B4 **Lompoc** U.S.A.
31 B2 **Lom Sak** Thai.
71 E1 **Łomża** Pol.
98 B2 **London** Can.
67 C3 **London** U.K.
108 C3 **London** U.S.A.
65 C1 **Londonderry** U.K.
18 B1 **Londonderry, Cape** Austr.
122 B2 **Londrina** Brazil
51 M2 **Longa, Proliv** *sea chan.* Rus. Fed.
29 C1 **Long Akah** *Sarawak* Malaysia
111 E2 **Long Bay** *b.* U.S.A.
103 C4 **Long Beach** U.S.A.
65 C2 **Longford** Rep. of Ireland
29 C2 **Longiram** Indon.
115 C2 **Long Island** Bahamas
98 C1 **Long Island** Can.
27 D3 **Long Island** P.N.G.
109 E2 **Long Island** U.S.A.
98 B2 **Longlac** Can.
98 B2 **Long Lake** *l.* Can.
104 B2 **Longmont** U.S.A.
108 C2 **Long Point** *pt* Can.
39 B3 **Longquan** China
99 E2 **Long Range Mountains** Can.
19 D2 **Longreach** Austr.
67 D2 **Long Stratton** U.K.
66 B1 **Longtown** U.K.
73 D2 **Longuyon** France
107 E2 **Longview** *TX* U.S.A.
102 B1 **Longview** *WA* U.S.A.
38 A2 **Longxi** China
39 B3 **Longxi Shan** *mt.* China
31 B2 **Long Xuyên** Vietnam
39 B3 **Longyan** China
50 C1 **Longyearbyen** Svalbard
70 B2 **Lons-le-Saunier** France
111 E2 **Lookout, Cape** U.S.A.
65 B2 **Loop Head** *hd* Rep. of Ireland
31 B2 **Lop Buri** Thai.
32 B1 **Lopez** Phil.
36 C2 **Lop Nur** *salt l.* China
86 B2 **Lopori** *r.* Dem. Rep. Congo
74 B2 **Lora del Río** Spain
108 C2 **Lorain** U.S.A.
42 A1 **Loralai** Pak.
75 C2 **Lorca** Spain
106 B2 **Lordsburg** U.S.A.
123 C2 **Lorena** Brazil
120 B2 **Loreto** Bol.
112 A2 **Loreto** Mex.
72 B2 **Lorient** France
64 B2 **Lorn, Firth of** *est.* U.K.
20 B3 **Lorne** Austr.
106 B1 **Los Alamos** U.S.A.
107 D3 **Los Aldamas** Mex.
121 A4 **Los Angeles** Chile
103 C4 **Los Angeles** U.S.A.
103 B3 **Los Banos** U.S.A.
120 B3 **Los Blancos** Arg.
121 A5 **Los Chonos, Archipiélago de** *is* Chile
59 E1 **Losevo** Rus. Fed.
76 B2 **Lošinj** *i.* Croatia
112 B2 **Los Mochis** Mex.
114 B4 **Los Mosquitos, Golfo de** *b.* Panama
86 B2 **Losombo** Dem. Rep. Congo
115 D3 **Los Roques, Islas** *is* Venez.
64 C2 **Lossiemouth** U.K.
118 B1 **Los Teques** Venez.
120 A4 **Los Vilos** Chile
72 C3 **Lot** *r.* France
64 C1 **Loth** U.K.
87 D2 **Lotikipi Plain** Kenya
86 C3 **Loto** Dem. Rep. Congo
57 E2 **Lotoshino** Rus. Fed.
30 B1 **Louang Namtha** Laos
30 B2 **Louangphrabang** Laos
86 B3 **Loubomo** Congo
72 B2 **Loudéac** France
39 B3 **Loudi** China
86 B3 **Loudima** Congo
82 A3 **Louga** Senegal
67 C2 **Loughborough** U.K.
65 B2 **Loughrea** Rep. of Ireland
73 D2 **Louhans** France
65 B2 **Louisburgh** Rep. of Ireland
110 B2 **Louisiana** *state* U.S.A.
91 C1 **Louis Trichardt** S. Africa
108 B3 **Louisville** *KY* U.S.A.
110 C2 **Louisville** *MS* U.S.A.
54 C2 **Loukhi** Rus. Fed.
98 C1 **Loups-Marins, Lacs des** *lakes* Can.
72 B3 **Lourdes** France
119 C1 **Lourenço** Brazil
74 B1 **Lousã** Port.
21 C2 **Louth** Austr.
66 C2 **Louth** U.K.
Louvain Belgium *see* **Leuven**
57 D2 **Lovat'** *r.* Rus. Fed.
78 B2 **Lovech** Bulg.
104 B2 **Loveland** U.S.A.
103 C2 **Lovelock** U.S.A.
107 C2 **Lovington** U.S.A.
87 C3 **Lowa** Dem. Rep. Congo
109 E2 **Lowell** U.S.A.
96 C3 **Lower Arrow Lake** Can.
22 B2 **Lower Hutt** N.Z.
65 C1 **Lower Lough Erne** *l.* U.K.
96 B2 **Lower Post** Can.
67 D2 **Lowestoft** U.K.
71 D1 **Łowicz** Pol.
109 D2 **Lowville** U.S.A.

20 B2 **Loxton** Austr.
Loyang China *see* **Luoyang**
57 D3 **Loyew** Belarus
60 F2 **Løypskardtinden** *mt.* Norway
77 C2 **Loznica** Yugo.
59 D2 **Lozova** Ukr.
88 B2 **Luacano** Angola
38 B2 **Lu'an** China
88 A1 **Luanda** Angola
89 C2 **Luangwa** *r.* Zambia
89 B2 **Luanshya** Zambia
74 B1 **Luarca** Spain
88 B2 **Luau** Angola
71 E1 **Lubaczów** Pol.
88 A2 **Lubango** Angola
87 C3 **Lubao** Dem. Rep. Congo
71 E1 **Lubartów** Pol.
69 D1 **Lübbecke** Ger.
107 C2 **Lubbock** U.S.A.
69 E1 **Lübeck** Ger.
37 E2 **Lubei** China
55 E3 **Lubenka** Kazakh.
71 D1 **Lubin** Pol.
71 E1 **Lublin** Pol.
59 C1 **Lubny** Ukr.
29 C1 **Lubok Antu** *Sarawak* Malaysia
69 E1 **Lübow** Ger.
69 E1 **Lübtheen** Ger.
87 C3 **Lubudi** Dem. Rep. Congo
28 B2 **Lubuklinggau** Indon.
87 C4 **Lubumbashi** Dem. Rep. Congo
88 B2 **Lubungu** Zambia
87 C3 **Lubutu** Dem. Rep. Congo
88 A1 **Lucala** Angola
65 C2 **Lucan** Rep. of Ireland
88 B1 **Lucapa** Angola
64 B3 **Luce Bay** U.K.
122 B2 **Lucélia** Brazil
32 B1 **Lucena** Phil.
74 C2 **Lucena** Spain
71 D2 **Lučenec** Slovakia
77 C2 **Lucera** Italy
Lucerne Switz. *see* **Luzern**
69 E1 **Lüchow** Ger.
88 A2 **Lucira** Angola
69 F1 **Luckenwalde** Ger.
90 B2 **Luckhoff** S. Africa
43 C2 **Lucknow** India
88 B2 **Lucusse** Angola
Lüda China *see* **Dalian**
68 C2 **Lüdenscheid** Ger.
69 E1 **Lüder** Ger.
88 A3 **Lüderitz** Namibia
42 B1 **Ludhiana** India
108 B2 **Ludington** U.S.A.
67 B2 **Ludlow** U.K.
103 C4 **Ludlow** U.S.A.
78 C2 **Ludogorie** *reg.* Bulg.
61 G3 **Ludvika** Sweden
70 B2 **Ludwigsburg** Ger.
69 F1 **Ludwigsfelde** Ger.
69 D3 **Ludwigshafen am Rhein** Ger.
69 E1 **Ludwigslust** Ger.
56 C2 **Ludza** Latvia
86 C3 **Luebo** Dem. Rep. Congo
88 A2 **Luena** Angola
38 A2 **Lüeyang** China
39 B3 **Lufeng** China
107 E2 **Lufkin** U.S.A.
56 C2 **Luga** Rus. Fed.
56 C2 **Luga** *r.* Rus. Fed.
73 D2 **Lugano** Switz.
89 C2 **Lugenda** *r.* Moz.
74 B1 **Lugo** Spain
78 B1 **Lugoj** Romania
59 D2 **Luhans'k** Ukr.
87 D3 **Luhombero** Tanz.
58 B1 **Luhyny** Ukr.
88 B2 **Luiana** Angola
86 C3 **Luilaka** *r.* Dem. Rep. Congo
73 D2 **Luino** Italy
60 I2 **Luiro** *r.* Fin.
86 C3 **Luiza** Dem. Rep. Congo
38 B2 **Lujiang** China
77 C2 **Lukavac** Bos.-Herz.
86 B3 **Lukenie** *r.* Dem. Rep. Congo
106 A2 **Lukeville** U.S.A.
57 E3 **Lukhovitsy** Rus. Fed.
71 E1 **Łuków** Pol.
88 B2 **Lukulu** Zambia
60 H2 **Luleå** Sweden
60 H2 **Luleälven** *r.* Sweden
78 C2 **Lüleburgaz** Turkey
38 B2 **Lüliang Shan** *mts* China
29 C2 **Lumajang** Indon.
43 C1 **Lumajangdong Co** *salt l.* China
88 B2 **Lumbala Kaquengue** Angola
88 B2 **Lumbala N'guimbo** Angola
111 E2 **Lumberton** U.S.A.
29 C1 **Lumbis** Indon.
74 B1 **Lumbrales** Spain
31 B2 **Lumphăt** Cambodia
97 D2 **Lumsden** Can.
22 A3 **Lumsden** N.Z.
61 F4 **Lund** Sweden
89 C2 **Lundazi** Zambia
67 A3 **Lundy Island** U.K.
69 E1 **Lüneburg** Ger.
69 E1 **Lüneburger Heide** *reg.* Ger.
68 C2 **Lünen** Ger.
73 D2 **Lunéville** France
88 B2 **Lunga** *r.* Zambia
82 A4 **Lungi** Sierra Leone
30 A1 **Lunglei** India
65 C2 **Lungnaquilla Mountain** Rep. of Ireland
88 B2 **Lungwebungu** *r.* Zambia
42 B2 **Luni** *r.* India
56 C3 **Luninyets** Belarus
82 A4 **Lunsar** Sierra Leone
45 E2 **Luntai** China
39 A3 **Luodian** China
39 B3 **Luoding** China
38 B2 **Luohe** China
38 B2 **Luoyang** China
89 B2 **Lupane** Zimbabwe
39 A3 **Lupanshui** China
78 B1 **Lupeni** Romania
89 C2 **Lupilichi** Moz.
69 F2 **Luppa** Ger.
89 D2 **Lúrio** Moz.
89 D2 **Lurio** *r.* Moz.
89 B2 **Lusaka** Zambia
86 C3 **Lusambo** Dem. Rep. Congo
77 C2 **Lushnjë** Albania
104 C2 **Lusk** U.S.A.
44 B3 **Lut, Dasht-e** *des.* Iran
69 F2 **Lutherstadt Wittenberg** Ger.
67 C3 **Luton** U.K.
29 C1 **Lutong** *Sarawak* Malaysia
97 C1 **Łutselk'e** Can.
58 B1 **Luts'k** Ukr.
90 B2 **Lutzputs** S. Africa
90 A3 **Lutzville** S. Africa
85 C4 **Luuq** Somalia
105 D2 **Luverne** U.S.A.
87 C3 **Luvua** *r.* Dem. Rep. Congo
91 D1 **Luvuvhu** *r.* S. Africa
87 D2 **Luwero** Uganda
27 C3 **Luwuk** Indon.
68 C3 **Luxembourg** *country* Europe
68 C3 **Luxembourg** Lux.
73 D2 **Luxeuil-les-Bains** France
91 C3 **Luxolweni** S. Africa
84 B2 **Luxor** Egypt
68 B2 **Luyksgestel** Neth.
54 D2 **Luza** Rus. Fed.
73 D2 **Luzern** Switz.
39 A3 **Luzhou** China
122 C1 **Luziânia** Brazil
119 D2 **Luzilândia** Brazil
32 B1 **Luzon** *i.* Phil.
39 C3 **Luzon Strait** Phil.
58 A2 **L'viv** Ukr.
L'vov Ukr. *see* **L'viv**
56 C3 **Lyakhavichy** Belarus
60 G3 **Lycksele** Sweden
91 D2 **Lydenburg** S. Africa
56 C3 **Lyel'chytsy** Belarus
56 C3 **Lyepyel'** Belarus
67 B3 **Lyme Bay** U.K.
109 D3 **Lynchburg** U.S.A.
97 D2 **Lynn Lake** Can.
97 D1 **Lynx Lake** Can.
73 C2 **Lyon** France
57 D2 **Lyozna** Belarus
54 E3 **Lys'va** Rus. Fed.
59 D2 **Lysychans'k** Ukr.
66 B2 **Lytham St Anne's** U.K.
56 C3 **Lyuban'** Belarus
57 E2 **Lyubertsy** Rus. Fed.
58 B1 **Lyubeshiv** Ukr.
57 F2 **Lyubim** Rus. Fed.
59 D2 **Lyubotyn** Ukr.
57 D2 **Lyubytino** Rus. Fed.
57 D3 **Lyudinovo** Rus. Fed.

M

48 B2 **Ma'an** Jordan
56 C2 **Maardu** Estonia
68 B2 **Maas** *r.* Neth.
68 B2 **Maaseik** Belgium
32 B1 **Maasin** Phil.
68 B2 **Maastricht** Neth.
89 C3 **Mabalane** Moz.
118 C1 **Mabaruma** Guyana
89 C3 **Mabote** Moz.
90 B2 **Mabule** Botswana
90 B1 **Mabutsane** Botswana
123 D2 **Macaé** Brazil
89 C2 **Macaloge** Moz.
94 E2 **MacAlpine Lake** Can.
119 C1 **Macapá** Brazil
118 A2 **Macará** Ecuador
123 D1 **Macarani** Brazil
29 C2 **Macassar Strait** Indon.
119 E2 **Macau** Brazil
39 B3 **Macau** *special admin. reg.* China
91 D1 **Maccaretane** Moz.
66 B2 **Macclesfield** U.K.
18 B2 **Macdonald, Lake** *salt flat* Austr.
18 C2 **Macdonnell Ranges** *mts* Austr.
98 A1 **MacDowell Lake** Can.
74 B1 **Macedo de Cavaleiros** Port.
20 B3 **Macedon** *mt.* Austr.
77 D2 **Macedonia** *country* Europe
119 E2 **Maceió** Brazil
76 B2 **Macerata** Italy
20 A2 **Macfarlane, Lake** *salt flat* Austr.
65 B3 **Macgillycuddy's Reeks** *mts* Rep. of Ireland
42 A2 **Mach** Pak.
123 C2 **Machado** Brazil
89 C3 **Machaila** Moz.
87 D3 **Machakos** Kenya
118 A2 **Machala** Ecuador
89 C3 **Machanga** Moz.

38 B2 **Macheng** China
109 F2 **Machias** U.S.A.
118 A1 **Machiques** Venez.
118 A3 **Machupicchu** Peru
91 D2 **Macia** Moz.
78 C1 **Măcin** Romania
21 D1 **Macintyre** *r.* Austr.
19 D2 **Mackay** Austr.
18 B2 **Mackay, Lake** *salt flat* Austr.
97 C1 **MacKay Lake** Can.
96 B2 **Mackenzie** Can.
96 A1 **Mackenzie** *r.* Can.
94 B2 **Mackenzie Bay** Can.
94 D1 **Mackenzie King Island** Can.
96 A1 **Mackenzie Mountains** Can.
97 D2 **Macklin** Can.
21 D2 **Macksville** Austr.
21 D1 **Maclean** Austr.
18 A2 **MacLeod, Lake** *imp. l.* Austr.
108 A2 **Macomb** U.S.A.
76 A2 **Macomer** *Sardinia* Italy
73 C2 **Mâcon** France
111 D2 **Macon** *GA* U.S.A.
105 E3 **Macon** *MO* U.S.A.
21 C2 **Macquarie** *r.* Austr.
21 C2 **Macquarie Marshes** Austr.
65 B3 **Macroom** Rep. of Ireland
20 A1 **Macumba** *watercourse* Austr.
113 C3 **Macuspana** Mex.
112 B2 **Macuzari, Presa** *resr* Mex.
91 D2 **Madadeni** S. Africa
89 □D3 **Madagascar** *country* Africa
83 D2 **Madama** Niger
79 B2 **Madan** Bulg.
27 D3 **Madang** P.N.G.
118 C2 **Madeira** *r.* Brazil
82 A1 **Madeira** *terr.* N. Atlantic Ocean
112 B2 **Madera** Mex.
103 B3 **Madera** U.S.A.
41 B3 **Madgaon** India
86 B3 **Madingou** Congo
108 B3 **Madison** *IN* U.S.A.
105 D2 **Madison** *SD* U.S.A.
108 B2 **Madison** *WI* U.S.A.
108 C3 **Madison** *WV* U.S.A.
102 D1 **Madison** *r.* U.S.A.
108 B3 **Madisonville** U.S.A.
29 C2 **Madiun** Indon.
87 D2 **Mado Gashi** Kenya
36 C2 **Madoi** China
56 C2 **Madona** Latvia
46 A2 **Madrakah** Saudi Arabia
41 C3 **Madras** India
102 B2 **Madras** U.S.A.
113 C2 **Madre, Laguna** *lag.* Mex.
113 B3 **Madre del Sur, Sierra** *mts* Mex.
112 B2 **Madre Occidental, Sierra** *mts* Mex.
113 B2 **Madre Oriental, Sierra** *mts* Mex.
74 C1 **Madrid** Spain
74 C2 **Madridejos** Spain
29 C2 **Madura** *i.* Indon.
29 C2 **Madura, Selat** *sea chan.* Indon.
41 B4 **Madurai** India
35 C3 **Maebashi** Japan
30 A2 **Mae Hong Son** Thai.
30 A2 **Mae Sariang** Thai.
30 A2 **Mae Suai** Thai.
91 C2 **Mafeteng** Lesotho
87 D3 **Mafia Island** Tanz.
91 C2 **Mafikeng** S. Africa
87 D3 **Mafinga** Tanz.
122 C3 **Mafra** Brazil
51 L3 **Magadan** Rus. Fed.
121 A6 **Magallanes, Estrecho de** *sea chan.* Chile
115 C4 **Magangue** Col.
112 A1 **Magdalena** Mex.
106 B2 **Magdalena** U.S.A.
112 A2 **Magdalena, Bahía** *b.* Mex.
69 E1 **Magdeburg** Ger.
65 C1 **Magherafelt** U.K.
55 E3 **Magnitogorsk** Rus. Fed.
110 B2 **Magnolia** U.S.A.
99 D1 **Magpie, Lac** *l.* Can.
82 A3 **Magṭa' Laḥjar** Maur.
119 D2 **Maguarinho, Cabo** *c.* Brazil
91 D2 **Magude** Moz.
30 A1 **Magwe** Myanmar
49 C2 **Mahābād** Iran
42 B2 **Mahajan** India
89 □D2 **Mahajanga** Madag.
29 C2 **Mahakam** *r.* Indon.
91 C1 **Mahalapye** Botswana
89 □D2 **Mahalevona** Madag.
43 C2 **Mahanadi** *r.* India
89 □D2 **Mahanoro** Madag.
31 B2 **Maha Sarakham** Thai.
89 □D2 **Mahavavy** *r.* Madag.
46 B2 **Mahd adh Dhahab** Saudi Arabia
118 C1 **Mahdia** Guyana
42 B2 **Mahesāna** India
42 B2 **Mahi** *r.* India
22 C1 **Mahia Peninsula** N.Z.
57 D3 **Mahilyow** Belarus
75 D2 **Mahón** Spain
42 B2 **Mahuva** India
78 C2 **Mahya Dağı** *mt.* Turkey
97 D2 **Maidstone** Can.
67 D3 **Maidstone** U.K.
83 D3 **Maiduguri** Nigeria
43 C2 **Mailani** India
69 D2 **Main** *r.* Ger.
86 B3 **Mai-Ndombe, Lac** *l.* Dem. Rep. Congo
69 E3 **Main-Donau-Kanal** *canal* Ger.
109 F1 **Maine** *state* U.S.A.
30 A1 **Maingkwan** Myanmar
64 C1 **Mainland** *i. Scotland* U.K.
64 □ **Mainland** *i. Scotland* U.K.
89 □D2 **Maintirano** Madag.
69 D2 **Mainz** Ger.
118 B1 **Maiquetía** Venez.
21 D2 **Maitland** *N.S.W.* Austr.
20 A2 **Maitland** *S.A.* Austr.
114 B3 **Maíz, Islas del** *is* Nic.
35 C3 **Maizuru** Japan
77 C2 **Maja Jezercë** *mt.* Albania
29 C2 **Majene** Indon.
Majorca *i.* Spain *see* Mallorca
91 C2 **Majwemasweu** S. Africa
86 B3 **Makabana** Congo
26 B3 **Makale** Indon.
45 E2 **Makanchi** Kazakh.
77 C2 **Makarska** Croatia
Makassar Indon. *see* Ujung Pandang
44 B2 **Makat** Kazakh.
91 D2 **Makatini Flats** *lowland* S. Africa
82 A4 **Makeni** Sierra Leone
88 B3 **Makgadikgadi** *salt pan* Botswana
55 D4 **Makhachkala** Rus. Fed.
44 B2 **Makhambet** Kazakh.
87 D3 **Makindu** Kenya
45 D1 **Makinsk** Kazakh.
59 D2 **Makiyivka** Ukr.
Makkah Saudi Arabia *see* Mecca
99 E1 **Makkovik** Can.
71 E2 **Makó** Hungary
86 B2 **Makokou** Gabon
87 D3 **Makongolosi** Tanz.
90 B2 **Makopong** Botswana
47 D2 **Makran** *reg.* Iran/Pak.
57 E2 **Maksatikha** Rus. Fed.
49 C2 **Mākū** Iran
30 A1 **Makum** India
35 B4 **Makurazaki** Japan
83 C4 **Makurdi** Nigeria
60 G2 **Malå** Sweden
114 B4 **Mala, Punta** *pt* Panama
86 A2 **Malabo** Equat. Guinea
28 A1 **Malacca, Strait of** Indon./Malaysia
102 D2 **Malad City** U.S.A.
56 C3 **Maladzyechna** Belarus
74 C2 **Málaga** Spain
85 B4 **Malakal** Sudan
29 C2 **Malang** Indon.
88 A1 **Malanje** Angola
61 G4 **Mälaren** *l.* Sweden
121 B4 **Malargüe** Arg.
48 B2 **Malatya** Turkey
89 C2 **Malawi** *country* Africa
Malawi, Lake Africa *see* Nyasa, Lake
57 D2 **Malaya Vishera** Rus. Fed.
32 B2 **Malaybalay** Phil.
49 C2 **Malāyer** Iran
28 B1 **Malaysia** *country* Asia
49 C2 **Malazgirt** Turkey
71 D1 **Malbork** Pol.
69 F1 **Malchin** Ger.
68 A2 **Maldegem** Belgium
24 C5 **Maldives** *country* Indian Ocean
121 C4 **Maldonado** Uru.
79 B3 **Maleas, Akra** *pt* Greece
71 D2 **Malé Karpaty** *hills* Slovakia
102 C2 **Malheur Lake** U.S.A.
82 B3 **Mali** *country* Africa
82 A3 **Mali** Guinea
26 C3 **Malili** Indon.
87 E3 **Malindi** Kenya
65 C1 **Malin Head** *hd* Rep. of Ireland
65 B1 **Malin More** Rep. of Ireland
79 C2 **Malkara** Turkey
56 C3 **Mal'kavichy** Belarus
78 C2 **Malko Tŭrnovo** Bulg.
21 C3 **Mallacoota** Austr.
21 C3 **Mallacoota Inlet** *b.* Austr.
64 B2 **Mallaig** U.K.
97 E1 **Mallery Lake** Can.
75 D2 **Mallorca** *i.* Spain
65 B2 **Mallow** Rep. of Ireland
60 H2 **Malmberget** Sweden
68 C2 **Malmédy** Belgium
90 A3 **Malmesbury** S. Africa
61 F4 **Malmö** Sweden
30 B1 **Malong** China
86 C4 **Malonga** Dem. Rep. Congo
61 E3 **Måløy** Norway
57 E2 **Maloyaroslavets** Rus. Fed.
57 E2 **Maloye Borisovo** Rus. Fed.
53 E6 **Malta** *country* Europe
56 C2 **Malta** Latvia
102 E1 **Malta** U.S.A.
90 A1 **Maltahöhe** Namibia
66 C1 **Malton** U.K.
27 C3 **Maluku** *is* Indon.
61 F3 **Malung** Sweden
91 C2 **Maluti Mountains** Lesotho
41 B3 **Malvan** India
110 B2 **Malvern** U.S.A.
58 B1 **Malyn** Ukr.
51 L2 **Malyy Anyuy** *r.* Rus. Fed.
49 C1 **Malyy Kavkaz** *mts* Asia
51 K2 **Malyy Lyakhovskiy, Ostrov** *i.* Rus. Fed.
91 C2 **Mamafubedu** S. Africa

87 C2 **Mambasa** Dem. Rep. Congo
86 B2 **Mambéré** *r.* C.A.R.
91 C2 **Mamelodi** S. Africa
118 B3 **Mamoré** *r.* Bol./Brazil
82 A3 **Mamou** Guinea
29 C2 **Mamuju** Indon.
82 B4 **Man** Côte d'Ivoire
118 B2 **Manacapuru** Brazil
75 D2 **Manacor** Spain
27 C2 **Manado** Indon.
114 B3 **Managua** Nic.
89 □D3 **Manakara** Madag.
46 B3 **Manākhah** Yemen
47 C2 **Manama** Bahrain
27 D3 **Manam Island** P.N.G.
89 □D3 **Mananara** *r.* Madag.
89 □D2 **Mananara Avaratra** Madag.
89 □D3 **Mananjary** Madag.
82 A3 **Manantali, Lac de** *l.* Mali
45 E2 **Manas Hu** *l.* China
27 C3 **Manatuto** East Timor
118 B2 **Manaus** Brazil
48 B2 **Manavgat** Turkey
66 B2 **Manchester** U.K.
109 E2 **Manchester** U.S.A.
89 □D3 **Mandabe** Madag.
61 E4 **Mandal** Norway
27 D3 **Mandala, Puncak** *mt.* Indon.
30 A1 **Mandalay** Myanmar
37 D1 **Mandalgovĭ** Mongolia
104 C1 **Mandan** U.S.A.
86 B1 **Mandara Mountains** Cameroon/Nigeria
76 A3 **Mandas** *Sardinia* Italy
68 C2 **Manderscheid** Ger.
42 B1 **Mandi** India
82 B3 **Mandiana** Guinea
42 B1 **Mandi Burewala** Pak.
43 C2 **Mandla** India
89 □D2 **Mandritsara** Madag.
42 B2 **Mandsaur** India
18 A3 **Mandurah** Austr.
41 B3 **Mandya** India
76 B1 **Manerbio** Italy
58 B1 **Manevychi** Ukr.
77 C2 **Manfredonia** Italy
82 B3 **Manga** Burkina
86 B3 **Mangai** Dem. Rep. Congo
22 C1 **Mangakino** N.Z.
78 C2 **Mangalia** Romania
41 B3 **Mangalore** India
91 C2 **Mangaung** S. Africa
29 B2 **Manggar** Indon.
44 B2 **Mangistau** Kazakh.
36 C2 **Mangnai** China
89 C2 **Mangochi** Malawi
27 C3 **Mangole** *i.* Indon.
122 B3 **Mangueirinha** Brazil
37 E1 **Mangui** China
105 D3 **Manhattan** U.S.A.
89 C3 **Manhica** Moz.
123 D2 **Manhuaçu** Brazil
89 □D2 **Mania** *r.* Madag.
76 B1 **Maniago** Italy
118 B2 **Manicoré** Brazil
99 D2 **Manicouagan** *r.* Can.
99 D1 **Manicougan, Réservoir** *resr* Can.
47 B2 **Manīfah** Saudi Arabia
32 B1 **Manila** Phil.
21 D2 **Manilla** Austr.
79 C3 **Manisa** Turkey
108 B2 **Manistee** U.S.A.
97 E2 **Manitoba** *prov.* Can.
97 E2 **Manitoba, Lake** Can.
108 B1 **Manitou Islands** U.S.A.
98 B2 **Manitoulin Island** Can.
108 B2 **Manitowoc** U.S.A.
98 C2 **Maniwaki** Can.
118 A1 **Manizales** Col.
105 E2 **Mankato** U.S.A.
82 B4 **Mankono** Côte d'Ivoire
97 D3 **Mankota** Can.
41 C4 **Mankulam** Sri Lanka
42 B2 **Manmad** India
41 B4 **Mannar, Gulf of** India/Sri Lanka
69 D3 **Mannheim** Ger.
96 C2 **Manning** Can.
20 A2 **Mannum** Austr.
97 C2 **Mannville** Can.
27 C3 **Manokwari** Indon.
87 C3 **Manono** Dem. Rep. Congo
31 A2 **Manoron** Myanmar
73 D3 **Manosque** France
99 C1 **Manouane, Lac** *l.* Can.
75 D1 **Manresa** Spain
89 B2 **Mansa** Zambia
95 F2 **Mansel Island** Can.
60 I2 **Mansel'kya** *ridge* Fin./Rus. Fed.
21 C3 **Mansfield** Austr.
66 C2 **Mansfield** U.K.
110 B2 **Mansfield** *LA* U.S.A.
108 C2 **Mansfield** *OH* U.S.A.
109 D2 **Mansfield** *PA* U.S.A.
118 A2 **Manta** Ecuador
111 E1 **Manteo** U.S.A.
72 C2 **Mantes-la-Jolie** France
123 C2 **Mantiqueira, Serra da** *mts* Brazil
76 B1 **Mantova** Italy
Mantua Italy *see* **Mantova**
119 C2 **Manuelzinho** Brazil
27 C3 **Manui** *i.* Indon.
22 B1 **Manukau** N.Z.
110 B2 **Many** U.S.A.
55 D4 **Manych-Gudilo, Ozero** *l.* Rus. Fed.
87 D3 **Manyoni** Tanz.
74 C2 **Manzanares** Spain
112 B3 **Manzanillo** Mex.
37 D1 **Manzhouli** China
91 D2 **Manzini** Swaziland
83 D3 **Mao** Chad
27 D3 **Maoke, Pegunungan** *mts* Indon.
91 C2 **Maokeng** S. Africa
33 A1 **Maokui Shan** *mt.* China
39 B3 **Maoming** China
89 C3 **Mapai** Moz.
43 C1 **Mapam Yumco** *l.* China
113 C3 **Mapastepec** Mex.
112 B2 **Mapimí** Mex.
112 B2 **Mapimí, Bolsón de** *des.* Mex.
32 A2 **Mapin** *i.* Phil.
89 C3 **Mapinhane** Moz.
97 D3 **Maple Creek** Can.
89 C3 **Maputo** Moz.
91 D2 **Maputo** *r.* Moz./S. Africa
43 C2 **Maquan He** *r.* China
88 A1 **Maquela do Zombo** Angola
121 B5 **Maquinchao** Arg.
105 E2 **Maquoketa** U.S.A.
118 B2 **Maraã** Brazil
119 D2 **Maraba** Brazil
119 C1 **Maracá, Ilha de** *i.* Brazil
118 A1 **Maracaibo** Venez.
118 A1 **Maracaibo, Lago de** *l.* Venez.
122 A2 **Maracaju, Serra de** *hills* Brazil
118 B1 **Maracay** Venez.
83 D2 **Marādah** Libya
83 C3 **Maradi** Niger
49 C2 **Marāgheh** Iran
119 D2 **Marajó, Baía de** *est.* Brazil
119 C2 **Marajó, Ilha de** *i.* Brazil
47 C2 **Marākī** Iran
87 D2 **Maralal** Kenya
18 C3 **Maralinga** Austr.
49 C2 **Marand** Iran
118 A2 **Marañón** *r.* Peru
98 B2 **Marathon** Can.
111 D4 **Marathon** U.S.A.
74 C2 **Marbella** Spain
18 A2 **Marble Bar** Austr.
91 C1 **Marble Hall** S. Africa
91 D3 **Marburg** S. Africa
69 D2 **Marburg an der Lahn** Ger.
67 D2 **March** U.K.
68 B2 **Marche-en-Famenne** Belgium
74 B2 **Marchena** Spain
120 B4 **Mar Chiquita, Lago** *l.* Arg.
109 E2 **Marcy, Mount** U.S.A.
42 B1 **Mardan** Pak.
121 C4 **Mar del Plata** Arg.
49 C2 **Mardın** Turkey
64 B2 **Maree, Loch** *l.* U.K.
76 B3 **Marettimo, Isola** *i. Sicily* Italy
57 D2 **Marevo** Rus. Fed.
106 C2 **Marfa** U.S.A.
18 A3 **Margaret River** Austr.
115 D3 **Margarita, Isla de** *i.* Venez.
91 D3 **Margate** S. Africa
67 D3 **Margate** U.K.
87 C2 **Margherita Peak** Dem. Rep. Congo/Uganda
44 C3 **Margo, Dasht-i** *des.* Afgh.
59 C2 **Marhanets'** Ukr.
30 A1 **Mari** Myanmar
120 B3 **María Elena** Chile
124 C5 **Mariana Trench** *sea feature* N. Pacific Ocean
110 B2 **Marianna** *AR* U.S.A.
111 C2 **Marianna** *FL* U.S.A.
70 C2 **Mariánské Lázně** Czech Rep.
112 B2 **Marías, Islas** *is* Mex.
46 B3 **Ma'rib** Yemen
77 C1 **Maribor** Slovenia
85 A4 **Maridi** *watercourse* Sudan
23 M2 **Marie Byrd Land** *reg.* Antarctica
115 D3 **Marie-Galante** *i.* Guadeloupe
61 G3 **Mariehamn** Fin.
90 A1 **Mariental** Namibia
61 F4 **Mariestad** Sweden
111 D2 **Marietta** *GA* U.S.A.
108 C3 **Marietta** *OH* U.S.A.
73 D3 **Marignane** France
51 K3 **Marii, Mys** *pt* Rus. Fed.
56 B3 **Marijampolė** Lith.
122 C2 **Marília** Brazil
74 B1 **Marín** Spain
77 C3 **Marina di Gioiosa Ionica** Italy
56 C3 **Mar"ina Horka** Belarus
108 B1 **Marinette** U.S.A.
122 B2 **Maringá** Brazil
74 B2 **Marinha Grande** Port.
108 B2 **Marion** *IN* U.S.A.
108 C2 **Marion** *OH* U.S.A.
111 E2 **Marion** *SC* U.S.A.
108 C3 **Marion** *VA* U.S.A.
111 D2 **Marion, Lake** U.S.A.
20 A3 **Marion Bay** Austr.
120 B3 **Mariscal Estigarribia** Para.
78 C2 **Maritsa** *r.* Bulg.
59 D2 **Mariupol'** Ukr.
85 C4 **Marka** Somalia
91 C1 **Marken** S. Africa
68 B1 **Markermeer** *l.* Neth.
51 I2 **Markha** *r.* Rus. Fed.
59 D2 **Markivka** Ukr.
110 B2 **Marksville** U.S.A.
69 D3 **Marktheidenfeld** Ger.
69 F2 **Marktredwitz** Ger.
68 C2 **Marl** Ger.
68 A3 **Marle** France

107 D2 **Marlin** U.S.A.
72 C3 **Marmande** France
79 C2 **Marmara Denizi** *g.* Turkey
79 C3 **Marmaris** Turkey
72 C2 **Marne-la-Vallée** France
89 □D2 **Maroantsetra** Madag.
69 E2 **Maroldsweisach** Ger.
89 □D2 **Maromokotro** *mt.* Madag.
89 C2 **Marondera** Zimbabwe
119 C1 **Maroni** *r.* Fr. Guiana
86 B1 **Maroua** Cameroon
89 □D2 **Marovoay** Madag.
108 B1 **Marquette** U.S.A.
85 A3 **Marra, Jebel** *mt.* Sudan
91 D2 **Marracuene** Moz.
82 B1 **Marrakech** Morocco
85 A3 **Marra Plateau** Sudan
20 A1 **Marree** Austr.
54 F2 **Marresale** Rus. Fed.
89 C2 **Marromeu** Moz.
89 C2 **Marrupa** Moz.
84 B2 **Marsa Alam** Egypt
83 D1 **Marsa al Burayqah** Libya
87 D2 **Marsabit** Kenya
76 B3 **Marsala** *Sicily* Italy
84 A1 **Marsa Matrûh** Egypt
69 D2 **Marsberg** Ger.
76 B2 **Marsciano** Italy
21 C2 **Marsden** Austr.
73 D3 **Marseille** France
105 D2 **Marshall** *MN* U.S.A.
105 E3 **Marshall** *MO* U.S.A.
107 E2 **Marshall** *TX* U.S.A.
17 E2 **Marshall Islands** *country* N. Pacific Ocean
105 E2 **Marshalltown** U.S.A.
108 A2 **Marshfield** U.S.A.
111 E3 **Marsh Harbour** Bahamas
110 B3 **Marsh Island** U.S.A.
56 A2 **Märsta** Sweden
31 A2 **Martaban, Gulf of** Myanmar
29 C2 **Martapura** *Kalimantan Selatan* Indon.
28 B2 **Martapura** *Sumatera* Indon.
109 E2 **Martha's Vineyard** *i.* U.S.A.
73 D2 **Martigny** Switz.
71 D2 **Martin** Slovakia
104 C2 **Martin** U.S.A.
113 C2 **Martínez** Mex.
115 D3 **Martinique** *terr.* West Indies
109 D3 **Martinsburg** U.S.A.
108 D3 **Martinsville** U.S.A.
22 C2 **Marton** N.Z.
75 D1 **Martorell** Spain
74 C2 **Martos** Spain
44 B1 **Martuk** Kazakh.
73 C3 **Marvejols** France
44 C3 **Mary** Turkm.
19 E2 **Maryborough** Austr.
90 B2 **Marydale** S. Africa
109 D3 **Maryland** *state* U.S.A.
105 D3 **Marysville** U.S.A.
105 E2 **Maryville** *MO* U.S.A.
111 D1 **Maryville** *TN* U.S.A.
87 D3 **Masai Steppe** *plain* Tanz.
87 D3 **Masaka** Uganda
33 B2 **Masan** S. Korea
87 D4 **Masasi** Tanz.
32 B1 **Masbate** Phil.
32 B1 **Masbate** *i.* Phil.
91 C2 **Maseru** Lesotho
44 B3 **Mashhad** Iran
91 C3 **Masibambane** S. Africa
47 C3 **Masīlah, Wādī al** *watercourse* Yemen
91 C2 **Masilo** S. Africa
87 D2 **Masindi** Uganda
47 C2 **Masīrah** *i.* Oman
47 C3 **Maşīrah, Gulf of** *b.* Oman
49 C2 **Masjed Soleymān** Iran
65 B2 **Mask, Lough** *l.* Rep. of Ireland
89 □E2 **Masoala, Tanjona** *c.* Madag.
105 E2 **Mason City** U.S.A.
Masqaţ Oman *see* **Muscat**
76 B2 **Massa** Italy
109 E2 **Massachusetts** *state* U.S.A.
109 E2 **Massachusetts Bay** U.S.A.
89 C3 **Massangena** Moz.
88 A1 **Massango** Angola
84 B3 **Massawa** Eritrea
109 E2 **Massena** U.S.A.
96 A2 **Masset** Can.
73 C2 **Massif Central** *mts* France
108 C2 **Massillon** U.S.A.
82 B3 **Massina** Mali
89 C3 **Massinga** Moz.
91 D1 **Massingir** Moz.
46 A2 **Mastābah** Saudi Arabia
22 C2 **Masterton** N.Z.
42 A2 **Mastung** Pak.
46 A2 **Mastūrah** Saudi Arabia
56 B3 **Masty** Belarus
35 B4 **Masuda** Japan
Masuku Gabon *see* **Franceville**
89 C3 **Masvingo** Zimbabwe
86 B3 **Matadi** Dem. Rep. Congo
114 B3 **Matagalpa** Nic.
98 C2 **Matagami** Can.
98 C2 **Matagami, Lac** *l.* Can.
107 D3 **Matagorda Island** U.S.A.
88 A2 **Matala** Angola
82 A3 **Matam** Senegal
112 B2 **Matamoros** *Coahuila* Mex.
113 C2 **Matamoros** *Tamaulipas* Mex.
87 D3 **Matandu** *r.* Tanz.
99 D2 **Matane** Can.
114 B2 **Matanzas** Cuba
41 C4 **Matara** Sri Lanka
29 C2 **Mataram** Indon.
18 C1 **Mataranka** Austr.
75 D1 **Mataró** Spain
91 C3 **Matatiele** S. Africa
22 A3 **Mataura** N.Z.
22 C1 **Matawai** N.Z.
120 B2 **Mategua** Bol.
113 B2 **Matehuala** Mex.
77 C2 **Matera** Italy
107 D3 **Mathis** U.S.A.
43 B2 **Mathura** India
32 B2 **Mati** Phil.
66 C2 **Matlock** U.K.
118 C3 **Mato Grosso** Brazil
122 B1 **Mato Grosso, Planalto do** *plat.* Brazil
91 D2 **Matola** Moz.
Matou China *see* **Pingguo**
35 B3 **Matsue** Japan
34 D2 **Matsumae** Japan
35 C4 **Matsusaka** Japan
39 C3 **Matsu Tao** *i.* Taiwan
35 B4 **Matsuyama** Japan
98 B1 **Mattagami** *r.* Can.
73 D2 **Matterhorn** *mt.* Italy/Switz.
102 C2 **Matterhorn** *mt.* U.S.A.
108 B3 **Mattoon** U.S.A.
118 B1 **Maturín** Venez.
91 C2 **Matwabeng** S. Africa
68 A2 **Maubeuge** France
72 C3 **Maubourguet** France
108 C2 **Maumee** *r.* U.S.A.
88 B2 **Maun** Botswana
30 A1 **Maungdaw** Myanmar
18 C2 **Maurice, Lake** *salt flat* Austr.
82 A3 **Mauritania** *country* Africa
81 H6 **Mauritius** *country* Indian Ocean
46 B2 **Māwān, Khashm** *hill* Saudi Arabia
86 B3 **Mawanga** Dem. Rep. Congo
39 B3 **Mawei** China
30 A1 **Mawkmai** Myanmar
30 A1 **Mawlaik** Myanmar
46 B2 **Mawqaq** Saudi Arabia
46 B3 **Mawza** Yemen
76 A3 **Maxia, Punta** *mt. Sardinia* Italy
51 J2 **Maya** *r.* Rus. Fed.
115 C2 **Mayaguana** *i.* Bahamas
64 B3 **Maybole** U.K.
72 B2 **Mayenne** France
72 B2 **Mayenne** *r.* France
96 C2 **Mayerthorpe** Can.
108 B3 **Mayfield** U.S.A.
55 D4 **Maykop** Rus. Fed.
94 B2 **Mayo** Can.
86 B3 **Mayoko** Congo
89 D2 **Mayotte** *terr.* Africa
51 J3 **Mayskiy** Rus. Fed.
108 C3 **Maysville** U.S.A.
86 B3 **Mayumba** Gabon
105 D1 **Mayville** U.S.A.
119 C2 **Mazagão** Brazil
72 C3 **Mazamet** France
45 D3 **Mazar** China
76 B3 **Mazara del Vallo** *Sicily* Italy
45 C3 **Mazār-e Sharīf** Afgh.
112 A2 **Mazatán** Mex.
114 A3 **Mazatenango** Guat.
112 B2 **Mazatlán** Mex.
56 B2 **Mažeikiai** Lith.
56 B2 **Mazirbe** Latvia
89 B3 **Mazunga** Zimbabwe
56 C3 **Mazyr** Belarus
91 D2 **Mbabane** Swaziland
86 B2 **Mbaïki** C.A.R.
89 C1 **Mbala** Zambia
87 D2 **Mbale** Uganda
86 B2 **Mbalmayo** Cameroon
86 B3 **Mbandaka** Dem. Rep. Congo
86 A2 **Mbanga** Cameroon
88 A1 **M'banza Congo** Angola
87 D3 **Mbeya** Tanz.
87 D4 **Mbinga** Tanz.
86 B2 **Mbomo** Congo
86 B2 **Mbouda** Cameroon
82 A3 **Mbour** Senegal
82 A3 **Mbout** Maur.
86 C3 **Mbuji-Mayi** Dem. Rep. Congo
87 D3 **Mbuyuni** Tanz.
107 D2 **McAlester** U.S.A.
107 D3 **McAllen** U.S.A.
96 B2 **McBride** Can.
102 C2 **McCall** U.S.A.
94 E2 **McClintock Channel** Can.
94 D2 **McClure Strait** Can.
110 B2 **McComb** U.S.A.
104 C2 **McConaughy, Lake** U.S.A.
104 C2 **McCook** U.S.A.
102 C2 **McDermitt** U.S.A.
102 D1 **McDonald Peak** U.S.A.
102 D1 **McGuire, Mount** U.S.A.
96 C2 **McLennan** Can.
96 B2 **McLeod Lake** Can.
102 B1 **McMinnville** *OR* U.S.A.
110 C1 **McMinnville** *TN* U.S.A.
105 D3 **McPherson** U.S.A.
91 C3 **Mdantsane** S. Africa
103 D3 **Mead, Lake** *resr* U.S.A.
97 D2 **Meadow Lake** Can.
108 C2 **Meadville** U.S.A.
34 D2 **Meaken-dake** *vol.* Japan
74 B1 **Mealhada** Port.
99 E1 **Mealy Mountains** Can.
96 C2 **Meander River** Can.
46 A2 **Mecca** Saudi Arabia
68 B2 **Mechelen** Belgium
68 B2 **Mechelen** Neth.

68 C2 **Mechernich** Ger.
68 C2 **Meckenheim** Ger.
74 B1 **Meda** Port.
28 A1 **Medan** Indon.
121 B5 **Medanosa, Punta** *pt* Arg.
41 C4 **Medawachchiya** Sri Lanka
75 D2 **Médéa** Alg.
118 A1 **Medellín** Col.
83 D1 **Medenine** Tunisia
102 B2 **Medford** U.S.A.
78 C2 **Medgidia** Romania
78 B1 **Mediaş** Romania
104 B2 **Medicine Bow Mountains** U.S.A.
104 B2 **Medicine Bow Peak** U.S.A.
97 C2 **Medicine Hat** Can.
105 D3 **Medicine Lodge** U.S.A.
123 D1 **Medina** Brazil
46 A2 **Medina** Saudi Arabia
74 C1 **Medinaceli** Spain
74 C1 **Medina del Campo** Spain
74 B1 **Medina de Rioseco** Spain
52 D6 **Mediterranean Sea**
97 C2 **Medley** Can.
55 E3 **Mednogorsk** Rus. Fed.
51 L2 **Medvezh'i, Ostrova** *is* Rus. Fed.
54 C2 **Medvezh'yegorsk** Rus. Fed.
18 A2 **Meekatharra** Austr.
104 B2 **Meeker** U.S.A.
43 B2 **Meerut** India
68 A2 **Meetkerke** Belgium
79 B3 **Megalopoli** Greece
99 C2 **Mégantic** Can.
43 C2 **Meghāsani** *mt.* India
48 A2 **Megisti** *i.* Greece
60 I1 **Mehamn** Norway
18 A2 **Meharry, Mount** Austr.
47 C2 **Mehrān** *watercourse* Iran
122 C1 **Meia Ponte** *r.* Brazil
86 B2 **Meiganga** Cameroon
33 B1 **Meihekou** China
Meijiang China *see* **Ningdu**
30 A1 **Meiktila** Myanmar
69 E2 **Meiningen** Ger.
70 C1 **Meißen** Ger.
39 B3 **Meizhou** China
120 B3 **Mejicana** *mt.* Arg.
120 A3 **Mejillones** Chile
85 B3 **Mek'elē** Eth.
82 C2 **Mekerrhane, Sebkha** *salt pan* Alg.
82 B1 **Meknès** Morocco
31 B2 **Mekong** *r.* Asia
31 B3 **Mekong, Mouths of the** Vietnam
28 B1 **Melaka** Malaysia
21 B3 **Melbourne** Austr.
111 D3 **Melbourne** U.S.A.
76 A2 **Mele, Capo** *c.* Italy
99 C1 **Mélèzes, Rivière aux** *r.* Can.
83 D3 **Mélfi** Chad
77 C2 **Melfi** Italy
97 D2 **Melfort** Can.
60 F3 **Melhus** Norway
74 B1 **Melide** Spain
82 B1 **Melilla** N. Africa
59 D2 **Melitopol'** Ukr.
69 D1 **Melle** Ger.
61 F4 **Mellerud** Sweden
69 E2 **Mellrichstadt** Ger.
68 D1 **Mellum** *i.* Ger.
120 C4 **Melo** Uru.
83 C1 **Melrhir, Chott** *salt l.* Alg.
67 C2 **Melton Mowbray** U.K.
72 C2 **Melun** France
97 D2 **Melville** Can.
19 D1 **Melville, Cape** Austr.
99 E1 **Melville, Lake** Can.
18 C1 **Melville Island** Austr.
94 D1 **Melville Island** Can.
95 F2 **Melville Peninsula** Can.
70 C2 **Memmingen** Ger.
29 B1 **Mempawah** Indon.
48 B3 **Memphis** Egypt
110 B1 **Memphis** *TN* U.S.A.
107 C2 **Memphis** *TX* U.S.A.
59 C1 **Mena** Ukr.
110 B2 **Mena** U.S.A.
82 C3 **Ménaka** Mali
Mènam Khong *r.* Laos/Thai. *see* **Mekong**
73 C3 **Mende** France
84 B3 **Mendefera** Eritrea
113 C2 **Méndez** Mex.
87 D2 **Mendī** Eth.
27 D3 **Mendi** P.N.G.
67 B3 **Mendip Hills** U.K.
121 B4 **Mendoza** Arg.
79 C3 **Menemen** Turkey
28 B2 **Menggala** Indon.
30 B1 **Mengzi** China
99 D1 **Menihek** Can.
20 B2 **Menindee** Austr.
20 B2 **Menindee Lake** Austr.
20 A3 **Meningie** Austr.
72 C2 **Mennecy** France
108 B1 **Menominee** U.S.A.
88 A2 **Menongue** Angola
75 D1 **Menorca** *i.* Spain
28 A2 **Mentawai, Kepulauan** *is* Indon.
28 B2 **Mentok** Indon.
18 B2 **Menzies** Austr.
68 C1 **Meppel** Neth.
68 C1 **Meppen** Ger.
91 D1 **Mepuze** Moz.
91 C2 **Meqheleng** S. Africa
76 B1 **Merano** Italy
27 D3 **Merauke** Indon.
20 B2 **Merbein** Austr.
103 B3 **Merced** U.S.A.
120 C3 **Mercedes** Arg.
95 G2 **Mercy, Cape** Can.
107 C1 **Meredith, Lake** U.S.A.
59 D2 **Merefa** Ukr.
84 A3 **Merga Oasis** Sudan
31 A2 **Mergui** Myanmar
31 A2 **Mergui Archipelago** *is* Myanmar
78 C2 **Meriç** *r.* Greece/Turkey
113 D2 **Mérida** Mex.
74 B2 **Mérida** Spain
118 A1 **Mérida** Venez.
110 C2 **Meridian** U.S.A.
21 C3 **Merimbula** Austr.
84 B3 **Merowe** Sudan
18 A3 **Merredin** Austr.
64 B3 **Merrick** *hill* U.K.
108 B1 **Merrill** U.S.A.
108 B2 **Merrillville** U.S.A.
96 B2 **Merritt** Can.
21 C2 **Merrygoen** Austr.
84 C3 **Mersa Fatma** Eritrea
68 C3 **Mersch** Lux.
69 E2 **Merseburg (Saale)** Ger.
66 B2 **Mersey** *r.* U.K.
28 B1 **Mersing** Malaysia
67 D3 **Mers-les-Bains** France
42 B2 **Merta** India
67 B3 **Merthyr Tydfil** U.K.
74 B2 **Mértola** Port.
87 D3 **Meru** *vol.* Tanz.
68 C3 **Merzig** Ger.
106 A2 **Mesa** U.S.A.
77 C2 **Mesagne** Italy
69 D2 **Meschede** Ger.
57 E3 **Meshchovsk** Rus. Fed.
59 E2 **Meshkovskaya** Rus. Fed.
79 B2 **Mesimeri** Greece
79 B3 **Mesolongi** Greece
89 D2 **Messalo** *r.* Moz.
77 C3 **Messina** *Sicily* Italy
91 D1 **Messina** S. Africa
77 C3 **Messina, Stretta di** *str.* Italy
79 B3 **Messiniakos Kolpos** *b.* Greece
78 B2 **Mesta** *r.* Bulg.
118 B1 **Meta** *r.* Col./Venez.
95 G2 **Meta Incognita Peninsula** Can.
120 B3 **Metán** Arg.
79 B3 **Methoni** Greece
77 C2 **Metković** Croatia
28 B2 **Metro** Indon.
85 B4 **Metu** Eth.
73 D2 **Metz** France
68 B2 **Meuse** *r.* Belgium/France
107 D2 **Mexia** U.S.A.
112 A1 **Mexicali** Mex.
112 A1 **Mexico** *country* Central America
113 C3 **México** Mex.
105 E3 **Mexico** U.S.A.
93 E5 **Mexico, Gulf of** Mex./U.S.A.
69 F1 **Meyenburg** Ger.
44 C3 **Meymaneh** Afgh.
54 D2 **Mezen'** Rus. Fed.
54 D2 **Mezen'** *r.* Rus. Fed.
54 E1 **Mezhdusharskiy, Ostrov** *i.* Rus. Fed.
71 E2 **Mezőtúr** Hungary
112 B2 **Mezquitic** Mex.
56 C2 **Mežvidi** Latvia
89 C2 **Mfuwe** Zambia
91 D2 **Mhlume** Swaziland
42 B2 **Mhow** India
113 C3 **Miahuatlán** Mex.
74 B2 **Miajadas** Spain
111 D3 **Miami** *FL* U.S.A.
107 E1 **Miami** *OK* U.S.A.
111 D3 **Miami Beach** U.S.A.
49 C2 **Miandowāb** Iran
89 □D2 **Miandrivazo** Madag.
49 C2 **Mīāneh** Iran
42 B1 **Mianwali** Pak.
38 A2 **Mianyang** China
89 □D2 **Miarinarivo** Madag.
55 F3 **Miass** Rus. Fed.
96 C2 **Mica Creek** Can.
71 E2 **Michalovce** Slovakia
108 B1 **Michigan** *state* U.S.A.
108 B2 **Michigan, Lake** U.S.A.
108 B2 **Michigan City** U.S.A.
98 B2 **Michipicoten Island** Can.
98 B2 **Michipicoten River** Can.
57 F3 **Michurinsk** Rus. Fed.
68 A2 **Middelburg** Neth.
91 C3 **Middelburg** *E. Cape* S. Africa
91 C2 **Middelburg** *Mpumalanga* S. Africa
68 B2 **Middelharnis** Neth.
102 B2 **Middle Alkali Lake** U.S.A.
41 D3 **Middle Andaman** *i.* India
104 D2 **Middle Loup** *r.* U.S.A.
108 C3 **Middlesboro** U.S.A.
66 C1 **Middlesbrough** U.K.
109 E2 **Middletown** *NY* U.S.A.
108 C3 **Middletown** *OH* U.S.A.
46 B3 **Mīdī** Yemen
98 C2 **Midland** Can.
108 C2 **Midland** *MI* U.S.A.
107 C2 **Midland** *TX* U.S.A.
65 B3 **Midleton** Rep. of Ireland
62 B1 **Miðvágur** Faroe Is
77 D2 **Midzhur** *mt.* Bulg./Yugo.
71 E1 **Mielec** Pol.
78 C1 **Miercurea-Ciuc** Romania
74 B1 **Mieres** Spain
69 E1 **Mieste** Ger.

113 C3 **Miguel Alemán, Presa** *resr* Mex.
112 B2 **Miguel Auza** Mex.
112 B2 **Miguel Hidalgo, Presa** *resr* Mex.
31 A2 **Migyaunglaung** Myanmar
57 E3 **Mikhaylov** Rus. Fed.
Mikhaylovgrad Bulg. *see* **Montana**
34 B2 **Mikhaylovka** Rus. Fed.
45 D1 **Mikhaylovskiy** Rus. Fed.
61 I3 **Mikkeli** Fin.
54 E2 **Mikun'** Rus. Fed.
35 C3 **Mikuni-sanmyaku** *mts* Japan
76 A1 **Milan** Italy
89 C2 **Milange** Moz.
Milano Italy *see* **Milan**
79 C3 **Milas** Turkey
105 D1 **Milbank** U.S.A.
67 D2 **Mildenhall** U.K.
20 B2 **Mildura** Austr.
30 B1 **Mile** China
104 B1 **Miles City** U.S.A.
109 D3 **Milford** *DE* U.S.A.
103 D3 **Milford** *UT* U.S.A.
67 A3 **Milford Haven** U.K.
22 A2 **Milford Sound** N.Z.
75 D2 **Miliana** Alg.
84 B3 **Milk, Wadi el** *watercourse* Sudan
51 L3 **Mil'kovo** Rus. Fed.
97 C3 **Milk River** Can.
73 C3 **Millau** France
111 D2 **Milledgeville** U.S.A.
105 E1 **Mille Lacs** *lakes* U.S.A.
98 A2 **Mille Lacs, Lac des** *l.* Can.
105 D2 **Miller** U.S.A.
59 E2 **Millerovo** Rus. Fed.
20 A2 **Millers Creek** Austr.
20 B3 **Millicent** Austr.
109 F1 **Millinocket** U.S.A.
21 D1 **Millmerran** Austr.
66 B1 **Millom** U.K.
96 C1 **Mills Lake** *l.* Can.
79 B3 **Milos** *i.* Greece
57 E3 **Miloslavskoye** Rus. Fed.
59 E2 **Milove** Ukr.
20 B1 **Milparinka** Austr.
22 A3 **Milton** N.Z.
67 C2 **Milton Keynes** U.K.
108 B2 **Milwaukee** U.S.A.
72 B3 **Mimizan** France
86 B3 **Mimongo** Gabon
47 C2 **Mīnāb** Iran
26 C2 **Minahasa, Semenanjung** *pen.* Indon.
47 C2 **Mina Jebel Ali** U.A.E.
28 B1 **Minas** Indon.
121 C4 **Minas** Uru.
123 D1 **Minas Novas** Brazil
113 C3 **Minatitlán** Mex.
30 A1 **Minbu** Myanmar
32 B2 **Mindanao** *i.* Phil.
69 D1 **Minden** Ger.
110 B2 **Minden** *LA* U.S.A.
105 D2 **Minden** *NE* U.S.A.
32 B1 **Mindoro** *i.* Phil.
32 A1 **Mindoro Strait** Phil.
86 B3 **Mindouli** Congo
67 B3 **Minehead** U.K.
122 B1 **Mineiros** Brazil
107 D2 **Mineral Wells** U.S.A.
45 E3 **Minfeng** China
87 C4 **Minga** Dem. Rep. Congo
49 C1 **Mingäçevir** Azer.
99 D1 **Mingan** Can.
20 B2 **Mingary** Austr.
38 B2 **Mingguang** China
30 A1 **Mingin** Myanmar
75 C2 **Minglanilla** Spain
87 D4 **Mingoyo** Tanz.
37 E1 **Mingshui** China
39 B3 **Mingxi** China
41 B4 **Minicoy** *i.* India
18 A2 **Minilya** Austr.
99 D1 **Minipi Lake** Can.
98 A1 **Miniss Lake** Can.
83 C4 **Minna** Nigeria
105 E2 **Minneapolis** U.S.A.
97 E2 **Minnedosa** Can.
105 E2 **Minnesota** *r.* U.S.A.
105 E1 **Minnesota** *state* U.S.A.
74 B1 **Miño** *r.* Port./Spain
Minorca *i.* Spain *see* **Menorca**
104 C1 **Minot** U.S.A.
56 C3 **Minsk** Belarus
71 E1 **Mińsk Mazowiecki** Pol.
99 D2 **Minto** Can.
98 C1 **Minto, Lac** *l.* Can.
38 A2 **Minxian** China
123 D1 **Mirabela** Brazil
123 D1 **Miralta** Brazil
79 C3 **Mirampelou, Kolpos** *b.* Greece
120 C3 **Miranda** Brazil
120 C2 **Miranda** *r.* Brazil
74 C1 **Miranda de Ebro** Spain
74 B1 **Mirandela** Port.
122 B2 **Mirandópolis** Brazil
47 C3 **Mirbāṭ** Oman
29 C1 **Miri** *Sarawak* Malaysia
121 C4 **Mirim, Lagoa** *l.* Brazil
47 D2 **Mīrjāveh** Iran
51 I2 **Mirnyy** Rus. Fed.
69 F1 **Mirow** Ger.
42 A2 **Mirpur Khas** Pak.
79 B3 **Mirtoö Pelagos** *sea* Greece
33 B2 **Miryang** S. Korea
43 C2 **Mirzapur** India
34 B1 **Mishan** China
114 B3 **Miskitos, Cayos** *is* Nic.
71 E2 **Miskolc** Hungary
27 C3 **Misoöl** *i.* Indon.
83 D1 **Miṣrātah** Libya
98 B1 **Missinaibi** *r.* Can.
98 B2 **Missinaibi Lake** Can.
96 B3 **Mission** Can.
98 B1 **Missisa Lake** Can.
110 C3 **Mississippi** *r.* U.S.A.
110 C2 **Mississippi** *state* U.S.A.
110 C3 **Mississippi Delta** U.S.A.
102 D1 **Missoula** U.S.A.
105 E3 **Missouri** *r.* U.S.A.
105 E3 **Missouri** *state* U.S.A.
99 C2 **Mistassibi** *r.* Can.
98 C1 **Mistassini, Lac** *l.* Can.
99 D1 **Mistastin Lake** Can.
71 D2 **Mistelbach** Austria
99 D1 **Mistinibi, Lac** *l.* Can.
98 C1 **Mistissini** Can.
19 D2 **Mitchell** Austr.
19 D1 **Mitchell** *r.* Austr.
105 D2 **Mitchell** U.S.A.
65 B2 **Mitchelstown** Rep. of Ireland
42 A2 **Mithi** Pak.
35 D3 **Mito** Japan
87 D3 **Mitole** Tanz.
21 D2 **Mittagong** Austr.
69 E2 **Mittelhausen** Ger.
69 D1 **Mittellandkanal** *canal* Ger.
69 F3 **Mitterteich** Ger.
118 A1 **Mitú** Col.
87 C4 **Mitumba, Chaîne des** *mts* Dem. Rep. Congo
87 C3 **Mitumba, Monts** *mts* Dem. Rep. Congo
86 B2 **Mitzic** Gabon
46 B2 **Miyah, Wādī al** *watercourse* Saudi Arabia
35 C4 **Miyake-jima** *i.* Japan
34 D3 **Miyako** Japan
35 B4 **Miyakonojō** Japan
44 B2 **Miyaly** Kazakh.
Miyang China *see* **Mile**
35 B4 **Miyazaki** Japan
83 D1 **Mizdah** Libya
65 B3 **Mizen Head** *hd* Rep. of Ireland
58 A2 **Mizhhir"ya** Ukr.
61 G4 **Mjölby** Sweden
61 F3 **Mjøsa** *l.* Norway
87 D3 **Mkomazi** Tanz.
71 C1 **Mladá Boleslav** Czech Rep.
77 D2 **Mladenovac** Yugo.
71 E1 **Mława** Pol.
91 C3 **Mlungisi** S. Africa
58 B1 **Mlyniv** Ukr.
91 C2 **Mmabatho** S. Africa
91 C2 **Mmathethe** Botswana
103 E3 **Moab** U.S.A.
91 D2 **Moamba** Moz.
87 C3 **Moba** Dem. Rep. Congo
86 C2 **Mobayi-Mbongo** Dem. Rep. Congo
105 E3 **Moberly** U.S.A.
110 C2 **Mobile** U.S.A.
110 C2 **Mobile Bay** U.S.A.
104 C1 **Mobridge** U.S.A.
89 D2 **Moçambique** Moz.
30 B1 **Môc Châu** Vietnam
91 C1 **Mochudi** Botswana
89 D2 **Mocimboa da Praia** Moz.
118 A1 **Mocoa** Col.
122 C2 **Mococa** Brazil
112 B2 **Mocorito** Mex.
112 B1 **Moctezuma** *Chihuahua* Mex.
113 B2 **Moctezuma** *San Luis Potosí* Mex.
112 B2 **Moctezuma** *Sonora* Mex.
89 C2 **Mocuba** Moz.
73 D2 **Modane** France
90 B2 **Modder** *r.* S. Africa
76 B2 **Modena** Italy
103 B3 **Modesto** U.S.A.
21 C3 **Moe** Austr.
68 C2 **Moers** Ger.
64 C3 **Moffat** U.K.
Mogadishu Somalia *see* **Muqdisho**
91 C1 **Mogalakwena** *r.* S. Africa
30 A1 **Mogaung** Myanmar
Mogilev Belarus *see* **Mahilyow**
122 C2 **Mogi-Mirim** Brazil
51 I3 **Mogocha** Rus. Fed.
30 A1 **Mogok** Myanmar
71 D2 **Mohács** Hungary
91 C3 **Mohale's Hoek** Lesotho
75 D2 **Mohammadia** Alg.
109 E2 **Mohawk** *r.* U.S.A.
87 D3 **Mohoro** Tanz.
58 B2 **Mohyliv Podil's'kyy** Ukr.
78 C1 **Moinești** Romania
60 F2 **Mo i Rana** Norway
72 C3 **Moissac** France
103 C3 **Mojave** U.S.A.
103 C3 **Mojave Desert** U.S.A.
30 B1 **Mojiang** China
123 C2 **Moji das Cruzes** Brazil
122 C2 **Moji-Guaçu** *r.* Brazil
22 B1 **Mokau** N.Z.
91 C2 **Mokhotlong** Lesotho
33 B3 **Mokp'o** S. Korea
113 C2 **Molango** Mex.
Moldavia *country* Europe *see* **Moldova**
60 E3 **Molde** Norway
58 B2 **Moldova** *country* Europe

78 B1 **Moldoveanu, Vârful** *mt.* Romania
58 B2 **Moldovei Centrale, Podişul** *plat.* Moldova
91 C1 **Molepolole** Botswana
56 C2 **Molėtai** Lith.
77 C2 **Molfetta** Italy
75 C1 **Molina de Aragón** Spain
118 A3 **Mollendo** Peru
21 C2 **Molong** Austr.
90 B2 **Molopo** *watercourse* Botswana/S. Africa
86 B2 **Moloundou** Cameroon
Moluccas *is* Indon. *see* Maluku
27 C3 **Molucca Sea** Indon.
20 B2 **Momba** Austr.
87 D3 **Mombasa** Kenya
122 B1 **Mombuca, Serra da** *hills* Brazil
61 F4 **Møn** *i.* Denmark
73 D3 **Monaco** *country* Europe
64 B2 **Monadhliath Mountains** U.K.
65 C1 **Monaghan** Rep. of Ireland
57 D3 **Monastyrshchina** Rus. Fed.
58 B2 **Monastyryshche** Ukr.
34 D2 **Monbetsu** Japan
76 A1 **Moncalieri** Italy
54 C2 **Monchegorsk** Rus. Fed.
68 C2 **Mönchengladbach** Ger.
112 B2 **Monclova** Mex.
99 D2 **Moncton** Can.
74 B1 **Mondego** *r.* Port.
91 D2 **Mondlo** S. Africa
76 A2 **Mondovì** Italy
79 B3 **Monemvasia** Greece
34 D1 **Moneron, Ostrov** *i.* Rus. Fed.
105 E3 **Monett** U.S.A.
76 B1 **Monfalcone** Italy
74 B1 **Monforte** Spain
30 B1 **Mông Cai** Vietnam
30 B1 **Mong Lin** Myanmar
36 C1 **Mongolia** *country* Asia
30 A1 **Mong Pawk** Myanmar
30 A1 **Mong Ping** Myanmar
88 B2 **Mongu** Zambia
103 C3 **Monitor Range** *mts* U.S.A.
82 C4 **Mono** *r.* Togo
103 C3 **Mono Lake** U.S.A.
77 C2 **Monopoli** Italy
75 C1 **Monreal del Campo** Spain
110 B2 **Monroe** *LA* U.S.A.
108 B2 **Monroe** *WI* U.S.A.
110 C2 **Monroeville** U.S.A.
82 A4 **Monrovia** Liberia
68 A2 **Mons** Belgium
90 B3 **Montagu** S. Africa
77 C3 **Montalto** *mt.* Italy
78 B2 **Montana** Bulg.
102 E1 **Montana** *state* U.S.A.
72 C2 **Montargis** France
72 C3 **Montauban** France
109 E2 **Montauk Point** U.S.A.
91 C2 **Mont-aux-Sources** *mt.* Lesotho
73 C2 **Montbard** France
73 D2 **Mont Blanc** *mt.* France/Italy
73 C2 **Montbrison** France
68 B3 **Montcornet** France
72 B3 **Mont-de-Marsan** France
72 C2 **Montdidier** France
119 C2 **Monte Alegre** Brazil
73 D3 **Monte-Carlo** Monaco
120 C4 **Monte Caseros** Arg.
114 C3 **Montego Bay** Jamaica
73 C3 **Montélimar** France
77 C2 **Montella** Italy
113 C2 **Montemorelos** Mex.
72 B2 **Montendre** France
77 C2 **Montenegro** *aut. rep.* Yugo.
89 C2 **Montepuez** Moz.
76 B2 **Montepulciano** Italy
103 B3 **Monterey** U.S.A.
103 B3 **Monterey Bay** U.S.A.
118 A1 **Montería** Col.
120 B2 **Montero** Bol.
113 B2 **Monterrey** Mex.
77 C2 **Montesano sulla Marcellana** Italy
77 C2 **Monte Sant'Angelo** Italy
119 E3 **Monte Santo** Brazil
76 A2 **Monte Santu, Capo di** *c.* *Sardinia* Italy
123 D1 **Montes Claros** Brazil
121 C4 **Montevideo** Uru.
105 D2 **Montevideo** U.S.A.
104 B3 **Monte Vista** U.S.A.
110 C2 **Montgomery** U.S.A.
73 D2 **Monthey** Switz.
110 B2 **Monticello** *AR* U.S.A.
103 E3 **Monticello** *UT* U.S.A.
72 C2 **Montignac** France
74 C2 **Montilla** Spain
99 D2 **Mont-Joli** Can.
98 C2 **Mont-Laurier** Can.
72 C2 **Montluçon** France
99 C2 **Montmagny** Can.
72 C2 **Montmorillon** France
19 E2 **Monto** Austr.
102 D2 **Montpelier** *ID* U.S.A.
109 E2 **Montpelier** *VT* U.S.A.
73 C3 **Montpellier** France
98 C2 **Montréal** Can.
97 D2 **Montreal Lake** Can.
97 D2 **Montreal Lake** *l.* Can.
67 D3 **Montreuil** France
73 D2 **Montreux** Switz.
64 C2 **Montrose** U.K.
104 B3 **Montrose** U.S.A.
115 D3 **Montserrat** *terr.* West Indies
30 A1 **Monywa** Myanmar
76 A1 **Monza** Italy
75 D1 **Monzón** Spain
91 C1 **Mookane** Botswana
21 D1 **Moonie** Austr.
21 C1 **Moonie** *r.* Austr.
20 A2 **Moonta** Austr.
18 A2 **Moore, Lake** *salt flat* Austr.
105 D1 **Moorhead** U.S.A.
21 C3 **Mooroopna** Austr.
90 A3 **Moorreesburg** S. Africa
98 B1 **Moose** *r.* Can.
98 B1 **Moose Factory** Can.
109 F1 **Moosehead Lake** U.S.A.
97 D2 **Moose Jaw** Can.
105 E1 **Moose Lake** U.S.A.
97 D2 **Moosomin** Can.
98 B1 **Moosonee** Can.
20 B2 **Mootwingee** Austr.
91 C1 **Mopane** S. Africa
82 B3 **Mopti** Mali
118 A3 **Moquegua** Peru
61 F3 **Mora** Sweden
105 E1 **Mora** U.S.A.
43 B2 **Moradabad** India
89 □D2 **Moramanga** Madag.
71 D2 **Morava** *r.* Europe
64 B2 **Moray Firth** *b.* U.K.
68 C3 **Morbach** Ger.
42 B2 **Morbi** India
72 B3 **Morcenx** France
37 E1 **Mordaga** China
97 E3 **Morden** Can.
59 E1 **Mordovo** Rus. Fed.
66 B1 **Morecambe** U.K.
66 B1 **Morecambe Bay** U.K.
21 C1 **Moree** Austr.
27 D3 **Morehead** P.N.G.
108 C3 **Morehead** U.S.A.
111 E2 **Morehead City** U.S.A.
113 B3 **Morelia** Mex.
75 C1 **Morella** Spain
74 B2 **Morena, Sierra** *mts* Spain
78 C2 **Moreni** Romania
96 A2 **Moresby, Mount** Can.
96 A2 **Moresby Island** Can.
110 B3 **Morgan City** U.S.A.
111 D1 **Morganton** U.S.A.
108 D3 **Morgantown** U.S.A.
73 D2 **Morges** Switz.
34 D2 **Mori** Japan
96 B2 **Morice Lake** Can.
34 D3 **Morioka** Japan
21 D2 **Morisset** Austr.
72 B2 **Morlaix** France
19 C1 **Mornington Island** Austr.
27 D3 **Morobe** P.N.G.
82 B1 **Morocco** *country* Africa
87 D3 **Morogoro** Tanz.
32 B2 **Moro Gulf** Phil.
90 B2 **Morokweng** S. Africa
89 □D3 **Morombe** Madag.
36 C1 **Mörön** Mongolia
89 □D3 **Morondava** Madag.
89 D2 **Moroni** Comoros
27 C2 **Morotai** *i.* Indon.
87 D2 **Moroto** Uganda
66 C1 **Morpeth** U.K.
122 C1 **Morrinhos** Brazil
97 E3 **Morris** Can.
105 D1 **Morris** U.S.A.
111 D1 **Morristown** U.S.A.
55 D3 **Morshanka** Rus. Fed.
122 B1 **Mortes, Rio das** *r.* Brazil
20 B3 **Mortlake** Austr.
21 D3 **Moruya** Austr.
64 B2 **Morvern** *reg.* U.K.
21 C3 **Morwell** Austr.
70 B2 **Mosbach** Ger.
57 E2 **Moscow** Rus. Fed.
102 C1 **Moscow** U.S.A.
68 C2 **Mosel** *r.* Ger.
73 D2 **Moselle** *r.* France
102 C1 **Moses Lake** U.S.A.
60 □A3 **Mosfellsbær** Iceland
22 B3 **Mosgiel** N.Z.
57 D2 **Moshenskoye** Rus. Fed.
87 D3 **Moshi** Tanz.
60 F2 **Mosjøen** Norway
Moskva Rus. Fed. *see* Moscow
71 D2 **Mosonmagyaróvár** Hungary
61 F4 **Moss** Norway
90 B3 **Mossel Bay** S. Africa
90 B3 **Mossel Bay** *b.* S. Africa
86 B3 **Mossendjo** Congo
20 B2 **Mossgiel** Austr.
19 D1 **Mossman** Austr.
119 E2 **Mossoró** Brazil
21 D2 **Moss Vale** Austr.
70 C1 **Most** Czech Rep.
82 C1 **Mostaganem** Alg.
77 C2 **Mostar** Bos.-Herz.
120 C4 **Mostardas** Brazil
61 G4 **Motala** Sweden
64 C3 **Motherwell** U.K.
75 C2 **Motilla del Palancar** Spain
90 B1 **Motokwe** Botswana
74 C2 **Motril** Spain
78 B2 **Motru** Romania
113 D2 **Motul** Mex.
79 C3 **Moudros** Greece
86 B3 **Mouila** Gabon
20 B3 **Moulamein** Austr.
73 C2 **Moulins** France
31 A2 **Moulmein** Myanmar
111 D2 **Moultrie** U.S.A.
111 E2 **Moultrie, Lake** U.S.A.
108 B3 **Mound City** U.S.A.

83 D4 **Moundou** Chad
105 E3 **Mountain Grove** U.S.A.
110 B1 **Mountain Home** *AR* U.S.A.
102 C2 **Mountain Home** *ID* U.S.A.
111 D1 **Mount Airy** U.S.A.
20 A3 **Mount Barker** Austr.
21 C3 **Mount Beauty** Austr.
89 C2 **Mount Darwin** Zimbabwe
109 F2 **Mount Desert Island** U.S.A.
91 C3 **Mount Fletcher** S. Africa
91 C3 **Mount Frere** S. Africa
20 B3 **Mount Gambier** Austr.
27 D3 **Mount Hagen** P.N.G.
21 C2 **Mount Hope** Austr.
19 C2 **Mount Isa** Austr.
18 A2 **Mount Magnet** Austr.
20 B2 **Mount Manara** Austr.
22 C1 **Mount Maunganui** N.Z.
105 E2 **Mount Pleasant** *IA* U.S.A.
108 C2 **Mount Pleasant** *MI* U.S.A.
107 E2 **Mount Pleasant** *TX* U.S.A.
67 A3 **Mount's Bay** U.K.
102 B2 **Mount Shasta** U.S.A.
108 B3 **Mount Vernon** *IL* U.S.A.
108 C2 **Mount Vernon** *OH* U.S.A.
102 B1 **Mount Vernon** *WA* U.S.A.
19 D2 **Moura** Austr.
83 E3 **Mourdi, Dépression du** *depr.* Chad
65 C1 **Mourne Mountains** U.K.
68 A2 **Mouscron** Belgium
83 D3 **Moussoro** Chad
26 C2 **Moutong** Indon.
83 C2 **Mouydir, Monts du** *plat.* Alg.
68 B3 **Mouzon** France
65 B1 **Moy** *r.* Rep. of Ireland
85 B4 **Moyale** Eth.
91 C3 **Moyeni** Lesotho
45 D2 **Moyynty** Kazakh.
89 C3 **Mozambique** *country* Africa
81 G5 **Mozambique Channel** Africa
57 E2 **Mozhaysk** Rus. Fed.
87 D3 **Mpanda** Tanz.
89 C2 **Mpika** Zambia
89 C1 **Mporokoso** Zambia
91 C2 **Mpumalanga** *prov.* S. Africa
56 C2 **Mshinskaya** Rus. Fed.
75 D2 **M'Sila** Alg.
57 D2 **Msta** *r.* Rus. Fed.
57 D2 **Mstinskiy Most** Rus. Fed.
57 D3 **Mstsislaw** Belarus
57 E3 **Mtsensk** Rus. Fed.
87 E4 **Mtwara** Tanz.
86 B3 **Muanda** Dem. Rep. Congo
30 B2 **Muang Khammouan** Laos
31 B2 **Muang Không** Laos
30 B1 **Muang Ngoy** Laos
30 B2 **Muang Pakbeng** Laos
30 B2 **Muang Pakxan** Laos
30 B1 **Muang Sing** Laos
30 B2 **Muang Vangviang** Laos
30 B2 **Muang Xaignabouri** Laos
28 B1 **Muar** Malaysia
28 B2 **Muarabungo** Indon.
28 B2 **Muaradua** Indon.
29 C2 **Muaralaung** Indon.
28 A2 **Muarasiberut** Indon.
28 B2 **Muaratembesi** Indon.
29 C2 **Muarateweh** Indon.
87 D2 **Mubende** Uganda
83 D3 **Mubi** Nigeria
88 B2 **Muconda** Angola
123 E1 **Mucuri** Brazil
123 E1 **Mucuri** *r.* Brazil
34 A2 **Mudanjiang** China
34 A1 **Mudan Jiang** *r.* China
79 C2 **Mudanya** Turkey
69 E1 **Müden (Örtze)** Ger.
21 C2 **Mudgee** Austr.
31 A2 **Mudon** Myanmar
48 B1 **Mudurnu** Turkey
89 C2 **Mueda** Moz.
89 B2 **Mufulira** Zambia
88 B2 **Mufumbwe** Zambia
79 C3 **Muğla** Turkey
84 B2 **Muhammad Qol** Sudan
69 F2 **Mühlberg** Ger.
69 E2 **Mühlhausen (Thüringen)** Ger.
31 B3 **Mui Ca Mau** *c.* Vietnam
65 C2 **Muine Bheag** Rep. of Ireland
89 C2 **Muite** Moz.
33 B2 **Muju** S. Korea
58 A2 **Mukacheve** Ukr.
29 C1 **Mukah** *Sarawak* Malaysia
31 B2 **Mukdahan** Thai.
18 A3 **Mukinbudin** Austr.
28 B2 **Mukomuko** Indon.
89 C2 **Mulanje, Mount** Malawi
69 F2 **Mulde** *r.* Ger.
112 A2 **Mulegé** Mex.
107 C2 **Muleshoe** U.S.A.
74 C2 **Mulhacén** *mt.* Spain
68 C2 **Mülheim an der Ruhr** Ger.
73 D2 **Mulhouse** France
34 B2 **Muling** China
34 B1 **Muling** *r.* China
64 B2 **Mull** *i.* U.K.
21 C2 **Mullaley** Austr.
104 C2 **Mullen** U.S.A.
29 C1 **Muller, Pegunungan** *mts* Indon.
18 A2 **Mullewa** Austr.
65 C2 **Mullingar** Rep. of Ireland
64 B3 **Mull of Galloway** *c.* U.K.
64 B3 **Mull of Kintyre** *hd* U.K.
64 A3 **Mull of Oa** *hd* U.K.
88 B2 **Mulobezi** Zambia
42 B1 **Multan** Pak.
41 B3 **Mumbai** India
88 B2 **Mumbwa** Zambia
113 D2 **Muna** Mex.
69 E2 **Münchberg** Ger.
München Ger. *see* **Munich**
108 B2 **Muncie** U.S.A.
18 B3 **Mundrabilla** Austr.
87 C2 **Mungbere** Dem. Rep. Congo
43 C2 **Munger** India
20 A1 **Mungeranie** Austr.
21 C1 **Mungindi** Austr.
70 C2 **Munich** Ger.
123 D2 **Muniz Freire** Brazil
69 E1 **Münster** *Niedersachsen* Ger.
68 C2 **Münster** *Nordrhein-Westfalen* Ger.
65 B2 **Munster** *reg.* Rep. of Ireland
68 C2 **Münsterland** *reg.* Ger.
30 B1 **Mường Nhie** Vietnam
60 H2 **Muonio** Fin.
60 H2 **Muonioälven** *r.* Fin./Sweden
85 C4 **Muqdisho** Somalia
71 D2 **Mur** *r.* Austria
87 C3 **Muramvya** Burundi
87 D3 **Muranga** Kenya
49 B2 **Murat** *r.* Turkey
79 C2 **Muratlı** Turkey
18 A2 **Murchison** *watercourse* Austr.
75 C2 **Murcia** Spain
79 C2 **Mürefte** Turkey
78 B1 **Mureș** *r.* Romania
72 C3 **Muret** France
110 C1 **Murfreesboro** U.S.A.
45 C3 **Murghab** *r.* Afgh.
45 D3 **Murghob** Tajik.
123 D2 **Muriaé** Brazil
88 B1 **Muriege** Angola
69 F1 **Müritz** *l.* Ger.
54 C2 **Murmansk** Rus. Fed.
55 D3 **Murom** Rus. Fed.
34 D2 **Muroran** Japan
74 B1 **Muros** Spain
35 B4 **Muroto** Japan
35 B4 **Muroto-zaki** *pt* Japan
111 D1 **Murphy** U.S.A.
21 C1 **Murra Murra** Austr.
20 A3 **Murray** *r.* Austr.
96 B2 **Murray** *r.* Can.
108 B3 **Murray** U.S.A.
111 D2 **Murray, Lake** U.S.A.
20 A3 **Murray Bridge** Austr.
90 B3 **Murraysburg** S. Africa
20 B3 **Murrayville** Austr.
20 B2 **Murrumbidgee** *r.* Austr.
89 C2 **Murrupula** Moz.
21 D2 **Murrurundi** Austr.
77 C1 **Murska Sobota** Slovenia
22 C1 **Murupara** N.Z.
43 C2 **Murwara** India
21 D1 **Murwillumbah** Austr.
83 D2 **Murzūq** Libya
49 C2 **Muş** Turkey
78 B2 **Musala** *mt.* Bulg.
33 B1 **Musan** N. Korea
46 B3 **Musaymir** Yemen
47 C2 **Muscat** Oman
105 E2 **Muscatine** U.S.A.
18 C2 **Musgrave Ranges** *mts* Austr.
86 B3 **Mushie** Dem. Rep. Congo
28 B2 **Musi** *r.* Indon.
108 B2 **Muskegon** U.S.A.
107 D1 **Muskogee** U.S.A.
96 B2 **Muskwa** *r.* Can.
42 A1 **Muslimbagh** Pak.
46 A3 **Musmar** Sudan
87 D3 **Musoma** Tanz.
64 C3 **Musselburgh** U.K.
56 B2 **Mustjala** Estonia
21 D2 **Muswellbrook** Austr.
84 A2 **Mut** Egypt
89 C2 **Mutare** Zimbabwe
34 D2 **Mutsu** Japan
89 C2 **Mutuali** Moz.
60 I2 **Muurola** Fin.
38 A2 **Mu Us Shamo** *des.* China
88 A1 **Muxaluando** Angola
54 C2 **Muyezerskiy** Rus. Fed.
87 D3 **Muyinga** Burundi
44 B2 **Muynak** Uzbek.
42 B1 **Muzaffargarh** Pak.
43 C2 **Muzaffarpur** India
112 B2 **Múzquiz** Mex.
43 C1 **Muztag** *mt.* China
87 C3 **Mwanza** Dem. Rep. Congo
87 D3 **Mwanza** Tanz.
86 C3 **Mweka** Dem. Rep. Congo
89 B2 **Mwenda** Zambia
86 C3 **Mwene-Ditu** Dem. Rep. Congo
89 C3 **Mwenezi** Zimbabwe
87 C3 **Mweru, Lake** Dem. Rep. Congo/Zambia
86 C3 **Mwimba** Dem. Rep. Congo
88 B2 **Mwinilunga** Zambia
56 C3 **Myadzyel** Belarus
30 A2 **Myanaung** Myanmar
30 A1 **Myanmar** *country* Asia
31 A2 **Myaungmya** Myanmar
30 A1 **Myingyan** Myanmar
30 A1 **Myitkyina** Myanmar
59 C2 **Mykolayiv** Ukr.
79 C3 **Mykonos** Greece
79 C3 **Mykonos** *i.* Greece
54 E2 **Myla** Rus. Fed.
43 D2 **Mymensingh** Bangl.
30 A1 **Myohaung** Myanmar
33 B1 **Myŏnggan** N. Korea
56 C2 **Myory** Belarus
60 □B3 **Mýrdalsjökull** *ice cap* Iceland

59 C2 **Myrhorod** Ukr.
58 C2 **Myronivka** Ukr.
111 E2 **Myrtle Beach** U.S.A.
21 C3 **Myrtleford** Austr.
102 B2 **Myrtle Point** U.S.A.
57 E2 **Myshkin** Rus. Fed.
71 C1 **Myślibórz** Pol.
41 B3 **Mysore** India
51 N2 **Mys Shmidta** Rus. Fed.
31 B2 **My Tho** Vietnam
79 C3 **Mytilini** Greece
57 E3 **Mytishchi** Rus. Fed.
91 C3 **Mzamomhle** S. Africa
89 C2 **Mzimba** Malawi
89 C2 **Mzuzu** Malawi

N

65 C2 **Naas** Rep. of Ireland
90 A2 **Nababeep** S. Africa
55 E3 **Naberezhnyye Chelny** Rus. Fed.
27 D3 **Nabire** Indon.
48 B2 **Nāblus** West Bank
91 C1 **Naboomspruit** S. Africa
89 D2 **Nacala** Moz.
31 A2 **Nachuge** India
107 E2 **Nacogdoches** U.S.A.
112 B1 **Nacozari de García** Mex.
42 B2 **Nadiad** India
58 A2 **Nadvirna** Ukr.
54 C2 **Nadvoitsy** Rus. Fed.
54 G2 **Nadym** Rus. Fed.
61 F4 **Næstved** Denmark
79 B3 **Nafpaktos** Greece
79 B3 **Nafplio** Greece
83 D1 **Nafūsah, Jabal** *hills* Libya
46 B2 **Nafy** Saudi Arabia
32 B1 **Naga** Phil.
98 B1 **Nagagami** *r.* Can.
35 C3 **Nagano** Japan
35 C3 **Nagaoka** Japan
43 D2 **Nagaon** India
42 B1 **Nagar** India
42 B2 **Nagar Parkar** Pak.
35 A4 **Nagasaki** Japan
35 B4 **Nagato** Japan
42 B2 **Nagaur** India
41 B4 **Nagercoil** India
42 A2 **Nagha Kalat** Pak.
43 B2 **Nagina** India
35 C3 **Nagoya** Japan
43 B2 **Nagpur** India
36 C2 **Nagqu** China
71 D2 **Nagyatád** Hungary
71 D2 **Nagykanizsa** Hungary
96 B1 **Nahanni Butte** Can.
44 A3 **Nahāvand** Iran
69 E1 **Nahrendorf** Ger.
121 A5 **Nahuel Huapí, Lago** *l.* Arg.
99 D1 **Nain** Can.
49 D2 **Nā'īn** Iran
64 C2 **Nairn** U.K.
87 D3 **Nairobi** Kenya
87 D3 **Naivasha** Kenya
49 D2 **Najafābād** Iran
46 B2 **Najd** *reg.* Saudi Arabia
74 C1 **Nájera** Spain
33 C1 **Najin** N. Korea
46 B3 **Najrān** Saudi Arabia
Nakambé *watercourse* Burkina/Ghana *see* **White Volta**
35 C3 **Nakatsugawa** Japan
46 A3 **Nakfa** Eritrea
34 B2 **Nakhodka** Rus. Fed.
31 B2 **Nakhon Pathom** Thai.
31 B2 **Nakhon Ratchasima** Thai.
31 B2 **Nakhon Sawan** Thai.
31 A3 **Nakhon Si Thammarat** Thai.
98 B1 **Nakina** Can.
89 C1 **Nakonde** Zambia
61 F4 **Nakskov** Denmark
87 D3 **Nakuru** Kenya
96 C2 **Nakusp** Can.
43 D2 **Nalbari** India
55 D4 **Nal'chik** Rus. Fed.
83 D1 **Nālūt** Libya
91 C2 **Namahadi** S. Africa
45 D2 **Namangan** Uzbek.
90 A2 **Namaqualand** *reg.* Namibia
90 A2 **Namaqualand** *reg.* S. Africa
19 E2 **Nambour** Austr.
21 D2 **Nambucca Heads** Austr.
33 B2 **Namch'ŏn** N. Korea
31 A2 **Nam Chon Reservoir** Thai.
43 D1 **Nam Co** *salt l.* China
30 B1 **Nam Đinh** Vietnam
88 A3 **Namib Desert** Namibia
88 A2 **Namibe** Angola
88 A3 **Namibia** *country* Africa
40 D2 **Namjagbarwa Feng** *mt.* China
27 C3 **Namlea** Indon.
21 C2 **Namoi** *r.* Austr.
102 C2 **Nampa** U.S.A.
82 B3 **Nampala** Mali
33 B2 **Namp'o** N. Korea
89 C2 **Nampula** Moz.
30 A1 **Namrup** India
30 A1 **Namsang** Myanmar
60 F3 **Namsos** Norway
31 A2 **Nam Tok** Thai.
51 J2 **Namtsy** Rus. Fed.
30 A1 **Namtu** Myanmar
68 B2 **Namur** Belgium
88 B2 **Namwala** Zambia
33 B2 **Namwŏn** S. Korea
30 A1 **Namya Ra** Myanmar
30 B2 **Nan** Thai.
96 B3 **Nanaimo** Can.
39 B3 **Nan'an** China
90 A1 **Nananib Plateau** Namibia
35 C3 **Nanao** Japan
39 B3 **Nanchang** *Jiangxi* China
39 B3 **Nanchang** *Jiangxi* China
38 A2 **Nanchong** China
31 A3 **Nancowry** *i.* India
73 D2 **Nancy** France
43 C1 **Nanda Devi** *mt.* India
39 A3 **Nandan** China
41 B3 **Nānded** India
42 B2 **Nandurbar** India
41 B3 **Nandyal** India
39 B3 **Nanfeng** China
86 B2 **Nanga Eboko** Cameroon
29 C2 **Nangahpinoh** Indon.
45 D3 **Nanga Parbat** *mt.* Jammu and Kashmir
29 C2 **Nangatayap** Indon.
38 B2 **Nangong** China
87 D3 **Nangulangwa** Tanz.
38 C2 **Nanhui** China
38 B2 **Nanjing** China
Nanking China *see* **Nanjing**
88 A2 **Nankova** Angola
39 B3 **Nan Ling** *mts* China
39 A3 **Nanning** China
95 H2 **Nanortalik** Greenland
39 A3 **Nanpan Jiang** *r.* China
43 C2 **Nanpara** India
39 B3 **Nanping** China
37 E3 **Nansei-shotō** *is* Japan
72 B2 **Nantes** France
38 C2 **Nantong** China
109 F2 **Nantucket Island** U.S.A.
123 D1 **Nanuque** Brazil
32 B2 **Nanusa, Kepulauan** *is* Indon.
39 B3 **Nanxiong** China
38 B2 **Nanyang** China
38 B2 **Nanzhang** China
99 C1 **Naococane, Lac** *l.* Can.
42 A2 **Naokot** Pak.
103 B3 **Napa** U.S.A.
94 D2 **Napaktulik Lake** Can.
95 H2 **Napasoq** Greenland
22 C1 **Napier** N.Z.
76 B2 **Naples** Italy
111 D3 **Naples** U.S.A.
118 A2 **Napo** *r.* Ecuador
Napoli Italy *see* **Naples**
82 B3 **Nara** Mali
61 I4 **Narach** Belarus
20 B3 **Naracoorte** Austr.
113 C2 **Naranjos** Mex.
31 B3 **Narathiwat** Thai.
73 C3 **Narbonne** France
31 A2 **Narcondam Island** India
95 G1 **Nares Strait** Can./Greenland
90 A1 **Narib** Namibia
55 D4 **Narimanov** Rus. Fed.
35 D3 **Narita** Japan
42 B2 **Narmada** *r.* India
42 B2 **Narnaul** India
76 B2 **Narni** Italy
58 B1 **Narodychi** Ukr.
57 E2 **Naro-Fominsk** Rus. Fed.
21 D3 **Narooma** Austr.
56 C3 **Narowlya** Belarus
21 C2 **Narrabri** Austr.
21 C2 **Narrandera** Austr.
21 C2 **Narromine** Austr.
56 C2 **Narva** Estonia
56 C2 **Narva Bay** Estonia/Rus. Fed.
60 G2 **Narvik** Norway
56 C2 **Narvskoye Vodokhranilishche** *resr* Estonia/Rus. Fed.
54 E2 **Nar'yan-Mar** Rus. Fed.
45 D2 **Naryn** Kyrg.
42 B2 **Nashik** India
109 E2 **Nashua** U.S.A.
110 C1 **Nashville** U.S.A.
85 B4 **Nasir** Sudan
96 B2 **Nass** *r.* Can.
114 C2 **Nassau** Bahamas
84 B2 **Nasser, Lake** *resr* Egypt
61 F4 **Nässjö** Sweden
98 C1 **Nastapoca** *r.* Can.
98 C1 **Nastapoka Islands** Can.
88 B3 **Nata** Botswana
119 E2 **Natal** Brazil
99 D1 **Natashquan** Can.
99 D1 **Natashquan** *r.* Can.
110 B2 **Natchez** U.S.A.
110 B2 **Natchitoches** U.S.A.
21 C3 **Nathalia** Austr.
75 D1 **Nati, Punta** *pt* Spain
82 C3 **Natitingou** Benin
119 D3 **Natividade** Brazil
35 D3 **Natori** Japan
29 B1 **Natuna, Kepulauan** *is* Indon.
29 B1 **Natuna Besar** *i.* Indon.
88 A3 **Nauchas** Namibia
69 F1 **Nauen** Ger.
56 B2 **Naujoji Akmenė** Lith.
69 E2 **Naumburg (Saale)** Ger.
17 E3 **Nauru** *country* S. Pacific Ocean
113 C2 **Nautla** Mex.
56 C3 **Navahrudak** Belarus
74 B2 **Navalmoral de la Mata** Spain
74 B2 **Navalvillar de Pela** Spain
65 C2 **Navan** Rep. of Ireland
56 C2 **Navapolatsk** Belarus

51 M2 **Navarin, Mys** *c.* Rus. Fed.
121 B6 **Navarino, Isla** *i.* Chile
64 B1 **Naver** *r.* U.K.
57 D3 **Navlya** Rus. Fed.
78 C2 **Năvodari** Romania
45 C2 **Navoi** Uzbek.
112 B2 **Navojoa** Mex.
112 B2 **Navolato** Mex.
42 A2 **Nawabshah** Pak.
30 A1 **Nawnghkio** Myanmar
30 A1 **Nawngleng** Myanmar
49 C2 **Naxçıvan** Azer.
79 C3 **Naxos** *i.* Greece
112 B2 **Nayar** Mex.
34 D2 **Nayoro** Japan
48 B2 **Nazareth** Israel
112 B2 **Nazas** Mex.
112 B2 **Nazas** *r.* Mex.
118 A3 **Nazca** Peru
79 C3 **Nazilli** Turkey
85 B4 **Nazrēt** Eth.
47 C2 **Nazwá** Oman
89 B1 **Nchelenge** Zambia
90 B1 **Ncojane** Botswana
88 A1 **N'dalatando** Angola
86 C2 **Ndélé** C.A.R.
86 B3 **Ndendé** Gabon
83 D3 **Ndjamena** Chad
89 B2 **Ndola** Zambia
65 C1 **Neagh, Lough** *l.* U.K.
18 C2 **Neale, Lake** *salt flat* Austr.
79 B2 **Nea Roda** Greece
67 B3 **Neath** U.K.
21 C1 **Nebine Creek** *r.* Austr.
44 B3 **Nebitdag** Turkm.
118 B1 **Neblina, Pico da** *mt.* Brazil
57 D2 **Nebolchi** Rus. Fed.
104 C2 **Nebraska** *state* U.S.A.
105 D2 **Nebraska City** U.S.A.
76 B3 **Nebrodi, Monti** *mts Sicily* Italy
121 C4 **Necochea** Arg.
99 C1 **Nedlouc, Lac** *l.* Can.
103 D4 **Needles** U.S.A.
97 E2 **Neepawa** Can.
55 E3 **Neftekamsk** Rus. Fed.
50 F2 **Nefteyugansk** Rus. Fed.
88 A1 **Negage** Angola
85 B4 **Negēlē** Eth.
118 A2 **Negra, Punta** *pt* Peru
31 A2 **Negrais, Cape** Myanmar
121 B5 **Negro** *r.* Arg.
118 C2 **Negro** *r.* S. America
120 C4 **Negro** *r.* Uru.
74 B2 **Negro, Cabo** *c.* Morocco
32 B2 **Negros** *i.* Phil.
37 E1 **Nehe** China
38 A3 **Neijiang** China
97 D2 **Neilburg** Can.
38 A1 **Nei Mongol Zizhiqu** *aut. reg.* China
118 A1 **Neiva** Col.
97 E2 **Nejanilini Lake** Can.
85 B4 **Nek'emtē** Eth.
57 F2 **Nekrasovskoye** Rus. Fed.
57 D2 **Nelidovo** Rus. Fed.
41 B3 **Nellore** India
96 C3 **Nelson** Can.
97 E2 **Nelson** *r.* Can.
22 B2 **Nelson** N.Z.
20 B3 **Nelson, Cape** Austr.
21 D2 **Nelson Bay** Austr.
97 E2 **Nelson House** Can.
102 E1 **Nelson Reservoir** U.S.A.
91 D2 **Nelspruit** S. Africa
82 B3 **Néma** Maur.
56 B2 **Neman** Rus. Fed.
72 C2 **Nemours** France
34 D2 **Nemuro** Japan
58 B2 **Nemyriv** Ukr.
65 B2 **Nenagh** Rep. of Ireland
67 D2 **Nene** *r.* U.K.
37 E1 **Nenjiang** China
105 E3 **Neosho** U.S.A.
43 C2 **Nepal** *country* Asia
43 C2 **Nepalganj** Nepal
103 D3 **Nephi** U.S.A.
65 B1 **Nephin** *hill* Rep. of Ireland
65 B1 **Nephin Beg Range** *hills* Rep. of Ireland
99 D2 **Nepisiguit** *r.* Can.
87 C2 **Nepoko** *r.* Dem. Rep. Congo
72 C3 **Nérac** France
21 D1 **Nerang** Austr.
37 D1 **Nerchinsk** Rus. Fed.
57 F2 **Nerekhta** Rus. Fed.
77 C2 **Neretva** *r.* Bos.-Herz./Croatia
88 B2 **Neriquinha** Angola
56 B3 **Neris** *r.* Lith.
57 E2 **Nerl'** *r.* Rus. Fed.
54 F2 **Nerokhi** Rus. Fed.
122 C1 **Nerópolis** Brazil
51 J3 **Neryungri** Rus. Fed.
60 C2 **Neskaupstaður** Iceland
64 B2 **Ness, Loch** *l.* U.K.
104 D3 **Ness City** U.S.A.
79 B2 **Nestos** *r.* Greece
68 B1 **Netherlands** *country* Europe
115 D3 **Netherlands Antilles** *terr.* West Indies
95 G2 **Nettilling Lake** Can.
69 F1 **Neubrandenburg** Ger.
73 D2 **Neuchâtel** Switz.
68 C2 **Neuerburg** Ger.
68 B3 **Neufchâteau** Belgium
73 D2 **Neufchâteau** France
72 C2 **Neufchâtel-en-Bray** France
69 D2 **Neuhof** Ger.
70 B1 **Neumünster** Ger.
70 B2 **Neunkirchen** Ger.
121 B4 **Neuquén** Arg.
121 B4 **Neuquén** *r.* Arg.
69 F1 **Neuruppin** Ger.
68 C2 **Neuss** Ger.
69 D1 **Neustadt am Rübenberge** Ger.
69 E3 **Neustadt an der Aisch** Ger.
69 F1 **Neustrelitz** Ger.
68 C2 **Neuwied** Ger.
105 E3 **Nevada** U.S.A.
103 C3 **Nevada** *state* U.S.A.
74 C2 **Nevada, Sierra** *mts* Spain
103 B2 **Nevada, Sierra** *mts* U.S.A.
56 C2 **Nevel'** Rus. Fed.
73 C2 **Nevers** France
21 C2 **Nevertire** Austr.
77 C2 **Nevesinje** Bos.-Herz.
55 D4 **Nevinnomyssk** Rus. Fed.
96 B2 **New Aiyansh** Can.
108 B3 **New Albany** U.S.A.
119 C1 **New Amsterdam** Guyana
109 E2 **Newark** *NJ* U.S.A.
108 C2 **Newark** *OH* U.S.A.
66 C2 **Newark-on-Trent** U.K.
109 E2 **New Bedford** U.S.A.
111 E1 **New Bern** U.S.A.
111 D2 **Newberry** U.S.A.
107 E2 **New Boston** U.S.A.
107 D3 **New Braunfels** U.S.A.
65 C2 **Newbridge** Rep. of Ireland
99 D2 **New Brunswick** *prov.* Can.
67 C3 **Newbury** U.K.
17 E4 **New Caledonia** *terr.* S. Pacific Ocean
21 D2 **Newcastle** Austr.
99 D2 **Newcastle** Can.
91 C2 **Newcastle** S. Africa
65 D1 **Newcastle** U.K.
108 C2 **New Castle** U.S.A.
104 C2 **Newcastle** U.S.A.
66 B2 **Newcastle-under-Lyme** U.K.
66 C1 **Newcastle upon Tyne** U.K.
65 B2 **Newcastle West** Rep. of Ireland
42 B2 **New Delhi** India
96 C3 **New Denver** Can.
21 D2 **New England Range** *mts* Austr.
99 E2 **Newfoundland** *i.* Can.
99 E1 **Newfoundland** *prov.* Can.
99 D2 **New Glasgow** Can.
27 D3 **New Guinea** *i.* Indon./P.N.G.
109 E2 **New Hampshire** *state* U.S.A.
109 E2 **New Haven** U.S.A.
96 B2 **New Hazelton** Can.
New Hebrides *country* S. Pacific Ocean *see* **Vanuatu**
110 B2 **New Iberia** U.S.A.
109 E3 **New Jersey** *state* U.S.A.
98 C2 **New Liskeard** Can.
18 A2 **Newman** Austr.
106 B2 **New Mexico** *state* U.S.A.
110 B3 **New Orleans** U.S.A.
108 C2 **New Philadelphia** U.S.A.
22 B1 **New Plymouth** N.Z.
67 C3 **Newport** *England* U.K.
67 B3 **Newport** *Wales* U.K.
110 B1 **Newport** *AR* U.S.A.
102 B2 **Newport** *OR* U.S.A.
109 E2 **Newport** *RI* U.S.A.
109 E2 **Newport** *VT* U.S.A.
102 C1 **Newport** *WA* U.S.A.
109 D3 **Newport News** U.S.A.
111 E3 **New Providence** *i.* Bahamas
67 A3 **Newquay** U.K.
110 B2 **New Roads** U.S.A.
65 C2 **New Ross** Rep. of Ireland
65 C1 **Newry** U.K.
New Siberia Islands Rus. Fed. *see* Novosibirskiye Ostrova
20 B2 **New South Wales** *state* Austr.
105 E2 **Newton** *IA* U.S.A.
105 D3 **Newton** *KS* U.S.A.
67 B3 **Newton Abbot** U.K.
64 B3 **Newton Stewart** U.K.
65 B2 **Newtown** Rep. of Ireland
67 B2 **Newtown** U.K.
104 C1 **New Town** U.S.A.
65 D1 **Newtownabbey** U.K.
65 D1 **Newtownards** U.K.
64 C3 **Newtown St Boswells** U.K.
65 C1 **Newtownstewart** U.K.
105 E2 **New Ulm** U.S.A.
109 E2 **New York** U.S.A.
109 D2 **New York** *state* U.S.A.
22 B2 **New Zealand** *country* Oceania
47 C2 **Neyrīz** Iran
44 B3 **Neyshābūr** Iran
113 C3 **Nezahualcóyotl, Presa** *resr* Mex.
29 B1 **Ngabang** Indon.
43 C2 **Ngamring** China
43 C1 **Ngangla Ringco** *salt l.* China
45 E3 **Nganglong Kangri** *mt.* China
43 C1 **Nganglong Kangri** *mts* China
43 C1 **Ngangzê Co** *salt l.* China
30 A2 **Ngao** Thai.
86 B2 **Ngaoundéré** Cameroon
22 C1 **Ngaruawahia** N.Z.
30 A2 **Ngathainggyaung** Myanmar
86 B3 **Ngo** Congo
31 B2 **Ngoc Linh** *mt.* Vietnam
83 D4 **Ngol Bembo** Nigeria
36 C2 **Ngoring Hu** *l.* China
83 D3 **Ngourti** Niger
83 D3 **Nguigmi** Niger

27 D2 **Ngulu** *atoll* Micronesia
83 D3 **Nguru** Nigeria
91 D2 **Ngwelezana** S. Africa
89 C2 **Nhamalabué** Moz.
31 B2 **Nha Trang** Vietnam
20 B3 **Nhill** Austr.
91 D2 **Nhlangano** Swaziland
19 C1 **Nhulunbuy** Austr.
109 D2 **Niagara Falls** Can.
82 C3 **Niamey** Niger
87 C2 **Niangara** Dem. Rep. Congo
82 B3 **Niangay, Lac** *l.* Mali
28 A1 **Nias** *i.* Indon.
114 B3 **Nicaragua** *country* Central America
114 B3 **Nicaragua, Lago de** *l.* Nic.
73 D3 **Nice** France
41 D4 **Nicobar Islands** India
48 B2 **Nicosia** Cyprus
114 B4 **Nicoya, Golfo de** *b.* Costa Rica
56 B2 **Nida** Lith.
71 E1 **Nidzica** Pol.
70 B1 **Niebüll** Ger.
69 D2 **Niederaula** Ger.
86 B2 **Niefang** Equat. Guinea
69 F1 **Niemegk** Ger.
69 D1 **Nienburg (Weser)** Ger.
68 B1 **Nieuwe-Niedorp** Neth.
119 C1 **Nieuw Nickerie** Suriname
90 A3 **Nieuwoudtville** S. Africa
68 A2 **Nieuwpoort** Belgium
48 B2 **Niğde** Turkey
83 D3 **Niger** *country* Africa
83 C4 **Niger** *r.* Africa
83 C4 **Niger, Mouths of the** Nigeria
83 C4 **Nigeria** *country* Africa
98 B2 **Nighthawk Lake** Can.
79 B2 **Nigrita** Greece
35 C3 **Niigata** Japan
35 B4 **Niihama** Japan
35 C4 **Nii-jima** *i.* Japan
35 C3 **Niitsu** Japan
68 B2 **Nijmegen** Neth.
68 C1 **Nijverdal** Neth.
60 J2 **Nikel'** Rus. Fed.
51 M3 **Nikol'skoye** Rus. Fed.
59 C2 **Nikopol'** Ukr.
48 B1 **Niksar** Turkey
47 D2 **Nīkshahr** Iran
77 C2 **Nikšić** Yugo.
84 B1 **Nile** *r.* Africa
108 B2 **Niles** U.S.A.
42 B2 **Nimach** India
73 C3 **Nîmes** France
21 C3 **Nimmitabel** Austr.
85 B4 **Nimule** Sudan
21 C1 **Nindigully** Austr.
41 B4 **Nine Degree Channel** India
21 C3 **Ninety Mile Beach** Austr.
22 B1 **Ninety Mile Beach** N.Z.
38 C3 **Ningbo** China
39 B3 **Ningde** China
39 B3 **Ningdu** China
38 B2 **Ningguo** China
39 C3 **Ninghai** China
Ningjiang China *see* Songyuan
36 C2 **Ningjing Shan** *mts* China
38 A2 **Ningxia** *aut. reg.* China
38 B2 **Ningyang** China
30 B1 **Ninh Binh** Vietnam
31 B2 **Ninh Hoa** Vietnam
34 D2 **Ninohe** Japan
105 D2 **Niobrara** *r.* U.S.A.
82 B3 **Niono** Mali
82 B3 **Nioro** Mali
72 B2 **Niort** France
97 D2 **Nipawin** Can.
98 B2 **Nipigon** Can.
98 B2 **Nipigon, Lake** Can.
99 D1 **Nipishish Lake** Can.
98 C2 **Nipissing, Lake** Can.
103 C3 **Nipton** U.S.A.
119 D3 **Niquelândia** Brazil
41 B3 **Nirmal** India
77 D2 **Niš** Yugo.
77 D2 **Nišava** *r.* Yugo.
76 B3 **Niscemi** *Sicily* Italy
35 B4 **Nishino-omote** Japan
123 D2 **Niterói** Brazil
64 C3 **Nith** *r.* U.K.
71 D2 **Nitra** Slovakia
17 G4 **Niue** *terr.* S. Pacific Ocean
60 H3 **Nivala** Fin.
68 B2 **Nivelles** Belgium
41 B3 **Nizamabad** India
55 E3 **Nizhnekamsk** Rus. Fed.
51 H3 **Nizhneudinsk** Rus. Fed.
50 G2 **Nizhnevartovsk** Rus. Fed.
57 F3 **Nizhniy Kislyay** Rus. Fed.
55 D3 **Nizhniy Lomov** Rus. Fed.
55 D3 **Nizhniy Novgorod** Rus. Fed.
54 E2 **Nizhniy Odes** Rus. Fed.
54 E3 **Nizhniy Tagil** Rus. Fed.
51 G2 **Nizhnyaya Tunguska** *r.* Rus. Fed.
59 C1 **Nizhyn** Ukr.
87 D3 **Njinjo** Tanz.
87 D3 **Njombe** Tanz.
89 C2 **Nkhotakota** Malawi
86 A2 **Nkongsamba** Cameroon
91 C3 **Nkululeko** S. Africa
91 C3 **Nkwenkwezi** S. Africa
35 B4 **Nobeoka** Japan
20 B1 **Noccundra** Austr.
112 A1 **Nogales** Mex.
106 A2 **Nogales** U.S.A.
72 C2 **Nogent-le-Rotrou** France
57 E2 **Noginsk** Rus. Fed.
42 B2 **Nohar** India
68 C3 **Nohfelden** Ger.
72 B2 **Noirmoutier, Île de** *i.* France
72 B2 **Noirmoutier-en-l'Île** France
35 C4 **Nojima-zaki** *c.* Japan
42 B2 **Nokha** India
61 H3 **Nokia** Fin.
42 A2 **Nok Kundi** Pak.
86 B2 **Nola** C.A.R.
54 D3 **Nolinsk** Rus. Fed.
91 C3 **Nomonde** S. Africa
91 D2 **Nondweni** S. Africa
30 B2 **Nong Khai** Thai.
20 A2 **Nonning** Austr.
112 B2 **Nonoava** Mex.
33 B2 **Nonsan** S. Korea
31 B2 **Nonthaburi** Thai.
90 B3 **Nonzwakazi** S. Africa
45 C3 **Norak** Tajik.
98 C2 **Noranda** Can.
50 C1 **Nordaustlandet** *i.* Svalbard
96 C2 **Nordegg** Can.
68 C1 **Norden** Ger.
51 H1 **Nordenshel'da, Arkhipelag** *is* Rus. Fed.
68 C1 **Norderney** Ger.
68 C1 **Norderney** *i.* Ger.
69 E1 **Norderstedt** Ger.
61 E3 **Nordfjordeid** Norway
69 E2 **Nordhausen** Ger.
69 D1 **Nordholz** Ger.
68 C1 **Nordhorn** Ger.
60 I1 **Nordkapp** *c.* Norway
60 F3 **Nordli** Norway
70 C2 **Nördlingen** Ger.
60 G3 **Nordmaling** Sweden
62 B1 **Norðoyar** *i.* Faroe Is
65 C2 **Nore** *r.* Rep. of Ireland
105 D2 **Norfolk** *NE* U.S.A.
109 D3 **Norfolk** *VA* U.S.A.
17 E4 **Norfolk Island** *terr.* S. Pacific Ocean
61 E3 **Norheimsund** Norway
50 G2 **Noril'sk** Rus. Fed.
107 D1 **Norman** U.S.A.
Normandes, Îles *is* Europe *see* **Channel Islands**
19 D1 **Normanton** Austr.
61 G4 **Norrköping** Sweden
61 G4 **Norrtälje** Sweden
18 B3 **Norseman** Austr.
60 G3 **Norsjö** Sweden
66 C1 **Northallerton** U.K.
18 A2 **Northampton** Austr.
67 C2 **Northampton** U.K.
41 D3 **North Andaman** *i.* India
97 D2 **North Battleford** Can.
98 C2 **North Bay** Can.
98 C1 **North Belcher Islands** Can.
64 C2 **North Berwick** U.K.
North Cape Norway *see* Nordkapp
22 B1 **North Cape** *c.* N.Z.
98 A1 **North Caribou Lake** Can.
111 E1 **North Carolina** *state* U.S.A.
98 B2 **North Channel** *lake channel* Can.
64 A3 **North Channel** U.K.
96 B3 **North Cowichan** Can.
104 C1 **North Dakota** *state* U.S.A.
67 C3 **North Downs** *hills* U.K.
111 E3 **Northeast Providence Channel** Bahamas
69 D2 **Northeim** Ger.
91 C1 **Northern** *prov.* S. Africa
90 A2 **Northern Cape** *prov.* S. Africa
97 E2 **Northern Indian Lake** Can.
65 C1 **Northern Ireland** *prov.* U.K.
16 D2 **Northern Mariana Islands** *terr.* N. Pacific Ocean
18 C1 **Northern Territory** *admin. div.* Austr.
64 C2 **North Esk** *r.* U.K.
105 E2 **Northfield** U.S.A.
67 D3 **North Foreland** *c.* U.K.
22 B1 **North Island** N.Z.
97 E2 **North Knife Lake** Can.
30 A1 **North Lakhimpur** India
104 C2 **North Platte** U.S.A.
104 C2 **North Platte** *r.* U.S.A.
64 C1 **North Ronaldsay** *i.* U.K.
62 D2 **North Sea** *sea* Europe
98 A1 **North Spirit Lake** *l.* Can.
21 D1 **North Stradbroke Island** Austr.
22 B1 **North Taranaki Bight** *b.* N.Z.
98 C1 **North Twin Island** Can.
66 B1 **North Tyne** *r.* U.K.
64 A2 **North Uist** *i.* U.K.
99 D2 **Northumberland Strait** Can.
90 C2 **North West** *prov.* S. Africa
18 A2 **North West Cape** Austr.
111 E3 **Northwest Providence Channel** Bahamas
99 E1 **North West River** Can.
96 B1 **Northwest Territories** *admin. div.* Can.
66 C1 **North York Moors** *moorland* U.K.
108 C3 **Norton** U.S.A.
89 C2 **Norton** Zimbabwe
108 C2 **Norwalk** U.S.A.
61 F3 **Norway** *country* Europe
97 E2 **Norway House** Can.
128 A3 **Norwegian Sea** N. Atlantic Ocean
67 D2 **Norwich** U.K.
109 E2 **Norwich** *CT* U.S.A.
109 D2 **Norwich** *NY* U.S.A.

34 D2 **Noshiro** Japan
59 C1 **Nosivka** Ukr.
90 B2 **Nosop** *watercourse* Africa
54 E2 **Nosovaya** Rus. Fed.
47 C2 **Noşratābād** Iran
Nossob *watercourse* Africa *see* Nosop
89 □D2 **Nosy Bé** *i.* Madag.
89 □E2 **Nosy Boraha** *i.* Madag.
71 D1 **Noteć** *r.* Pol.
61 E4 **Notodden** Norway
35 C3 **Noto-hantō** *pen.* Japan
99 D2 **Notre Dame, Monts** *mts* Can.
99 E2 **Notre Dame Bay** Can.
98 C1 **Nottaway** *r.* Can.
67 C2 **Nottingham** U.K.
82 A2 **Nouâdhibou** Maur.
82 A3 **Nouakchott** Maur.
82 A3 **Nouâmghâr** Maur.
82 B3 **Nouna** Burkina
122 B2 **Nova Esperança** Brazil
123 D2 **Nova Friburgo** Brazil
77 C1 **Nova Gradiška** Croatia
122 C2 **Nova Granada** Brazil
123 D2 **Nova Iguaçu** Brazil
59 C2 **Nova Kakhovka** Ukr.
123 D1 **Nova Lima** Brazil
122 B2 **Nova Londrina** Brazil
59 C2 **Nova Odesa** Ukr.
118 B1 **Nova Paraiso** Brazil
99 D2 **Nova Scotia** *prov.* Can.
123 D1 **Nova Venécia** Brazil
51 K1 **Novaya Sibir', Ostrov** *i.* Rus. Fed.
54 E1 **Novaya Zemlya** *is* Rus. Fed.
71 D2 **Nové Zámky** Slovakia
59 C1 **Novhorod-Sivers'kyy** Ukr.
78 B2 **Novi Iskŭr** Bulg.
76 A2 **Novi Ligure** Italy
77 D2 **Novi Pazar** Yugo.
77 C1 **Novi Sad** Yugo.
55 D3 **Novoanninskiy** Rus. Fed.
118 B2 **Novo Aripuanã** Brazil
59 D2 **Novoazovs'k** Ukr.
59 E2 **Novocherkassk** Rus. Fed.
54 D2 **Novodvinsk** Rus. Fed.
120 C3 **Novo Hamburgo** Brazil
122 C2 **Novo Horizonte** Brazil
58 B1 **Novohrad-Volyns'kyy** Ukr.
71 D2 **Novo Mesto** Slovenia
59 D3 **Novomikhaylovskiy** Rus. Fed.
57 E3 **Novomoskovsk** Rus. Fed.
59 D2 **Novomoskovs'k** Ukr.
59 C2 **Novomyrhorod** Ukr.
59 C2 **Novooleksiyivka** Ukr.
59 D2 **Novopskov** Ukr.
59 D3 **Novorossiysk** Rus. Fed.
56 C2 **Novorzhev** Rus. Fed.
55 E3 **Novosergiyevka** Rus. Fed.
59 D2 **Novoshakhtinsk** Rus. Fed.
50 G3 **Novosibirsk** Rus. Fed.
51 K1 **Novosibirskiye Ostrova** *is* Rus. Fed.
57 E3 **Novosil'** Rus. Fed.
57 D2 **Novosokol'niki** Rus. Fed.
59 C2 **Novotroyits'ke** Ukr.
59 C2 **Novoukrayinka** Ukr.
58 A1 **Novovolyns'k** Ukr.
57 D3 **Novozybkov** Rus. Fed.
71 D2 **Nový Jičín** Czech Rep.
54 E2 **Novyy Bor** Rus. Fed.
59 D1 **Novyy Oskol** Rus. Fed.
54 G2 **Novyy Port** Rus. Fed.
50 G2 **Novyy Urengoy** Rus. Fed.
37 E1 **Novyy Urgal** Rus. Fed.
71 D1 **Nowogard** Pol.
21 D2 **Nowra** Austr.
49 D2 **Now Shahr** Iran
42 B1 **Nowshera** Pak.
71 E2 **Nowy Sącz** Pol.
71 E2 **Nowy Targ** Pol.
55 G2 **Noyabr'sk** Rus. Fed.
73 C2 **Noyon** France
89 C2 **Nsanje** Malawi
86 B3 **Ntandembele** Dem. Rep. Congo
87 D3 **Ntungamo** Uganda
47 C2 **Nu'aym** *reg.* Oman
84 B2 **Nubian Desert** Sudan
118 A3 **Nudo Coropuna** *mt.* Peru
97 E1 **Nueltin Lake** Can.
121 A5 **Nueva Lubecka** Arg.
113 B2 **Nueva Rosita** Mex.
112 B1 **Nuevo Casas Grandes** Mex.
112 B2 **Nuevo Ideal** Mex.
113 C2 **Nuevo Laredo** Mex.
85 C4 **Nugaal** *watercourse* Somalia
73 C2 **Nuits-St-Georges** France
30 A1 **Nu Jiang** *r.* China
44 B2 **Nukus** Uzbek.
18 B2 **Nullagine** Austr.
18 B3 **Nullarbor Plain** Austr.
83 D4 **Numan** Nigeria
35 C3 **Numazu** Japan
61 E3 **Numedal** *val.* Norway
27 C3 **Numfoor** *i.* Indon.
21 C3 **Numurkah** Austr.
95 J3 **Nunap Isua** *c.* Greenland
97 E1 **Nunavut** *admin. div.* Can.
67 C2 **Nuneaton** U.K.
74 B1 **Nuñomoral** Spain
76 A2 **Nuoro** *Sardinia* Italy
46 B2 **Nuqrah** Saudi Arabia
20 A2 **Nuriootpa** Austr.
60 I3 **Nurmes** Fin.
69 E3 **Nürnberg** Ger.
30 A1 **Nu Shan** *mts* China
42 A2 **Nushki** Pak.
95 H2 **Nuuk** Greenland
95 H2 **Nuussuaq** Greenland
95 H2 **Nuussuaq** *pen.* Greenland
46 A2 **Nuweiba el Muzeina** Egypt
90 B3 **Nuweveldberge** *mts* S. Africa
54 F2 **Nyagan'** Rus. Fed.
43 D1 **Nyainqêntanglha Feng** *mt.* China
43 D2 **Nyainqêntanglha Shan** *mts* China
85 A3 **Nyala** Sudan
43 C2 **Nyalam** China
54 D2 **Nyandoma** Rus. Fed.
86 B3 **Nyanga** *r.* Gabon
89 C2 **Nyanga** Zimbabwe
89 C1 **Nyasa, Lake** Africa
61 F4 **Nyborg** Denmark
60 I1 **Nyborg** Norway
61 G4 **Nybro** Sweden
87 D3 **Nyeri** Kenya
43 C1 **Nyima** China
36 C3 **Nyingchi** China
71 E2 **Nyíregyháza** Hungary
61 F4 **Nykøbing** Denmark
61 G4 **Nyköping** Sweden
91 C1 **Nylstroom** S. Africa
21 C2 **Nymagee** Austr.
61 G4 **Nynäshamn** Sweden
21 C2 **Nyngan** Austr.
56 B3 **Nyoman** *r.* Belarus/Lith.
73 D3 **Nyons** France
54 E2 **Nyrob** Rus. Fed.
102 C2 **Nyssa** U.S.A.
87 C3 **Nyunzu** Dem. Rep. Congo
59 C2 **Nyzhn'ohirs'kyy** Ukr.
87 D3 **Nzega** Tanz.
82 B4 **Nzérékoré** Guinea
88 A1 **N'zeto** Angola

O

104 C2 **Oahe, Lake** U.S.A.
20 B2 **Oakbank** Austr.
110 B2 **Oakdale** U.S.A.
21 D1 **Oakey** Austr.
102 B1 **Oak Harbor** U.S.A.
108 C3 **Oak Hill** U.S.A.
103 B3 **Oakland** U.S.A.
18 B2 **Oakover** *r.* Austr.
102 B2 **Oakridge** U.S.A.
111 D1 **Oak Ridge** U.S.A.
22 B3 **Oamaru** N.Z.
32 B1 **Oas** Phil.
113 C3 **Oaxaca** Mex.
54 F2 **Ob'** *r.* Rus. Fed.
86 B2 **Obala** Cameroon
64 B2 **Oban** U.K.
74 B1 **O Barco** Spain
21 C2 **Oberon** Austr.
69 F3 **Oberpfälzer Wald** *mts* Ger.
69 F3 **Oberviechtach** Ger.
27 C3 **Obi** *i.* Indon.
119 C2 **Óbidos** Brazil
34 D2 **Obihiro** Japan
37 E1 **Obluch'ye** Rus. Fed.
57 E2 **Obninsk** Rus. Fed.
87 C2 **Obo** C.A.R.
85 C3 **Obock** Djibouti
86 B3 **Obouya** Congo
57 E3 **Oboyan'** Rus. Fed.
112 B2 **Obregón, Presa** *resr* Mex.
77 D2 **Obrenovac** Yugo.
55 E3 **Obshchiy Syrt** *hills* Rus. Fed.
54 G2 **Obskaya Guba** *sea chan.* Rus. Fed.
82 B4 **Obuasi** Ghana
58 C1 **Obukhiv** Ukr.
54 D2 **Ob"yachevo** Rus. Fed.
111 D3 **Ocala** U.S.A.
112 B2 **Ocampo** Mex.
74 C2 **Ocaña** Spain
118 A1 **Occidental, Cordillera** *mts* Col.
118 A3 **Occidental, Cordillera** *mts* Peru
109 D3 **Ocean City** U.S.A.
96 B2 **Ocean Falls** Can.
103 C4 **Oceanside** U.S.A.
59 C2 **Ochakiv** Ukr.
54 E3 **Ocher** Rus. Fed.
69 E3 **Ochsenfurt** Ger.
111 D2 **Oconee** *r.* U.S.A.
113 C3 **Ocosingo** Mex.
84 B2 **Oda, Jebel** *mt.* Sudan
34 D2 **Ōdate** Japan
35 C3 **Odawara** Japan
61 E3 **Odda** Norway
74 B2 **Odemira** Port.
79 C3 **Ödemiş** Turkey
61 F4 **Odense** Denmark
69 D3 **Odenwald** *reg.* Ger.
70 C1 **Oderbucht** *b.* Ger.
58 C2 **Odesa** Ukr.
107 C2 **Odessa** U.S.A.
82 B4 **Odienné** Côte d'Ivoire
71 D2 **Odra** *r.* Ger./Pol.
119 D2 **Oeiras** Brazil
68 D2 **Oelde** Ger.
104 C2 **Oelrichs** U.S.A.
69 F2 **Oelsnitz** Ger.
68 B1 **Oenkerk** Neth.
77 C2 **Ofanto** *r.* Italy
69 D2 **Offenbach am Main** Ger.
70 B2 **Offenburg** Ger.
85 C4 **Ogadēn** *reg.* Eth.
34 C3 **Oga-hantō** *pen.* Japan
35 C3 **Ōgaki** Japan

104 C2 **Ogallala** U.S.A.
83 C4 **Ogbomoso** Nigeria
102 D2 **Ogden** U.S.A.
109 D2 **Ogdensburg** U.S.A.
94 B2 **Ogilvie** *r.* Can.
94 B2 **Ogilvie Mountains** Can.
111 D2 **Oglethorpe, Mount** U.S.A.
98 B1 **Ogoki** *r.* Can.
98 B1 **Ogoki Reservoir** Can.
56 B2 **Ogre** Latvia
77 C1 **Ogulin** Croatia
108 B3 **Ohio** *r.* U.S.A.
108 C2 **Ohio** *state* U.S.A.
69 E2 **Ohrdruf** Ger.
77 D2 **Ohrid** Macedonia
119 C1 **Oiapoque** Brazil
109 D2 **Oil City** U.S.A.
68 A3 **Oise** *r.* France
35 B4 **Ōita** Japan
112 B2 **Ojinaga** Mex.
120 B3 **Ojos del Salado, Nevado** *mt.* Arg./Chile
57 F2 **Oka** *r.* Rus. Fed.
88 A3 **Okahandja** Namibia
88 A3 **Okakarara** Namibia
96 C3 **Okanagan Falls** Can.
96 C3 **Okanagan Lake** Can.
102 C1 **Okanogan** U.S.A.
102 C1 **Okanogan** *r.* U.S.A.
42 B1 **Okara** Pak.
88 B2 **Okavango** *r.* Botswana/Namibia
88 B2 **Okavango Delta** *swamp* Botswana
35 C3 **Okaya** Japan
35 B4 **Okayama** Japan
35 C4 **Okazaki** Japan
111 D3 **Okeechobee, Lake** U.S.A.
111 D2 **Okefenokee Swamp** U.S.A.
67 A3 **Okehampton** U.K.
42 A2 **Okha** India
51 K3 **Okha** Rus. Fed.
43 C2 **Okhaldhunga** Nepal
51 K3 **Okhotka** *r.* Rus. Fed.
51 K3 **Okhotsk** Rus. Fed.
124 C2 **Okhotsk, Sea of** Japan/Rus. Fed.
59 C1 **Okhtyrka** Ukr.
37 E3 **Okinawa** *i.* Japan
35 B3 **Oki-shotō** *is* Japan
107 D1 **Oklahoma** *state* U.S.A.
107 D1 **Oklahoma City** U.S.A.
107 D1 **Okmulgee** U.S.A.
46 A2 **Oko, Wadi** *watercourse* Sudan
86 B3 **Okondja** Gabon
96 C2 **Okotoks** Can.
57 D3 **Okovskiy Les** *for.* Rus. Fed.
86 B3 **Okoyo** Congo
60 H1 **Øksfjord** Norway
30 A2 **Oktwin** Myanmar
54 D2 **Oktyabr'skiy** *Arkhangel'skaya Oblast'* Rus. Fed.
51 L3 **Oktyabr'skiy** *Kamchatskaya Oblast'* Rus. Fed.
55 E3 **Oktyabr'skiy** *Respublika Bashkortostan* Rus. Fed.
54 F2 **Oktyabr'skoye** Rus. Fed.
51 H1 **Oktyabr'skoy Revolyutsii, Ostrov** *i.* Rus. Fed.
57 D2 **Okulovka** Rus. Fed.
34 C2 **Okushiri-tō** *i.* Japan
60 □A3 **Ólafsvík** Iceland
61 G4 **Öland** *i.* Sweden
20 B2 **Olary** Austr.
121 B4 **Olavarría** Arg.
71 D1 **Oława** Pol.
76 A2 **Olbia** *Sardinia* Italy
94 B2 **Old Crow** Can.
69 D1 **Oldenburg** Ger.
70 C1 **Oldenburg in Holstein** Ger.
68 C1 **Oldenzaal** Neth.
65 B3 **Old Head of Kinsale** *hd* Rep. of Ireland
96 C2 **Olds** Can.
97 D2 **Old Wives Lake** Can.
109 D2 **Olean** U.S.A.
71 E1 **Olecko** Pol.
51 J2 **Olekminsk** Rus. Fed.
59 C2 **Oleksandriya** Ukr.
54 C2 **Olenegorsk** Rus. Fed.
51 I2 **Olenek** Rus. Fed.
51 I2 **Olenek** *r.* Rus. Fed.
57 D2 **Olenino** Rus. Fed.
58 B1 **Olevs'k** Ukr.
74 B2 **Olhão** Port.
90 A2 **Olifants** *watercourse* Namibia
91 D1 **Olifants** S. Africa
91 D1 **Olifants** *r. Northern* S. Africa
90 A3 **Olifants** *r. W. Cape* S. Africa
90 B2 **Olifantshoek** S. Africa
122 C2 **Olímpia** Brazil
119 E2 **Olinda** Brazil
91 C1 **Oliphants Drift** S. Africa
75 C2 **Oliva** Spain
123 D2 **Oliveira** Brazil
74 B2 **Olivenza** Spain
120 B3 **Ollagüe** Chile
118 A2 **Olmos** Peru
108 B3 **Olney** U.S.A.
71 D2 **Olomouc** Czech Rep.
32 B1 **Olongapo** Phil.
72 B3 **Oloron-Ste-Marie** France
75 D1 **Olot** Spain
37 D1 **Olovyannaya** Rus. Fed.
68 C2 **Olpe** Ger.
71 E1 **Olsztyn** Pol.
78 B2 **Olt** *r.* Romania
49 C1 **Oltu** Turkey
79 B3 **Olympia** Greece
102 B1 **Olympia** U.S.A.
79 B2 **Olympos** *mt.* Greece
102 B1 **Olympus, Mount** U.S.A.
51 M3 **Olyutorskiy, Mys** *c.* Rus. Fed.
65 C1 **Omagh** U.K.
105 D2 **Omaha** U.S.A.
47 C3 **Oman** *country* Asia
47 C2 **Oman, Gulf of** Asia
88 A3 **Omaruru** Namibia
88 B2 **Omatako** *watercourse* Namibia
84 B3 **Omdurman** Sudan
21 C3 **Omeo** Austr.
113 C3 **Ometepec** Mex.
46 A3 **Om Hajēr** Eritrea
96 B2 **Omineca Mountains** Can.
35 C3 **Ōmiya** Japan
68 C1 **Ommen** Neth.
51 L2 **Omolon** *r.* Rus. Fed.
68 B3 **Omont** France
50 F3 **Omsk** Rus. Fed.
51 L2 **Omsukchan** Rus. Fed.
78 C1 **Omu, Vârful** *mt.* Romania
35 A4 **Ōmura** Japan
109 D3 **Onancock** U.S.A.
108 C1 **Onaping Lake** Can.
99 C2 **Onatchiway, Lac** *l.* Can.
88 A2 **Oncócua** Angola
90 B3 **Onderstedorings** S. Africa
88 A2 **Ondjiva** Angola
54 C2 **Onega** Rus. Fed.
54 C2 **Onega** *r.* Rus. Fed.
54 C2 **Onega, Lake** *l.* Rus. Fed.
96 B2 **One Hundred Mile House** Can.
109 D2 **Oneida Lake** U.S.A.
105 D2 **O'Neill** U.S.A.
109 D2 **Oneonta** U.S.A.
78 C1 **Oneşti** Romania
Onezhskoye, Ozero *l.* Rus. Fed. *see* **Onega, Lake**
90 B2 **Ongers** *watercourse* S. Africa
33 B2 **Ongjin** N. Korea
41 C3 **Ongole** India
89 □D3 **Onilahy** *r.* Madag.
83 C4 **Onitsha** Nigeria
90 A2 **Onseepkans** S. Africa
18 A2 **Onslow** Austr.
111 E2 **Onslow Bay** U.S.A.
98 A1 **Ontario** *prov.* Can.
102 C2 **Ontario** U.S.A.
109 D2 **Ontario, Lake** Can./U.S.A.
19 C2 **Oodnadatta** Austr.
Oostende Belgium *see* Ostend
68 B2 **Oosterhout** Neth.
68 A2 **Oosterschelde** *est.* Neth.
68 B1 **Oost-Vlieland** Neth.
96 B2 **Ootsa Lake** Can.
96 B2 **Ootsa Lake** *l.* Can.
86 C3 **Opala** Dem. Rep. Congo
98 C1 **Opataca, Lac** *l.* Can.
71 D2 **Opava** Czech Rep.
111 C2 **Opelika** U.S.A.
110 B2 **Opelousas** U.S.A.
98 C1 **Opinaca, Réservoir** *resr* Can.
99 D1 **Opiscotéo, Lac** *l.* Can.
56 C2 **Opochka** Rus. Fed.
112 A2 **Opodepe** Mex.
71 D1 **Opole** Pol.
44 B2 **Opornyy** Kazakh.
74 B1 **Oporto** Port.
22 C1 **Opotiki** N.Z.
61 E3 **Oppdal** Norway
22 B1 **Opunake** N.Z.
88 A2 **Opuwo** Namibia
78 B1 **Oradea** Romania
77 D2 **Orahovac** Yugo.
82 B1 **Oran** Alg.
120 B3 **Orán** Arg.
33 B1 **Ŏrang** N. Korea
21 C2 **Orange** Austr.
73 C3 **Orange** France
90 A2 **Orange** *r.* Namibia/S. Africa
107 E2 **Orange** U.S.A.
111 D2 **Orangeburg** U.S.A.
108 C2 **Orangeville** Can.
69 F1 **Oranienburg** Ger.
90 A2 **Oranjemund** Namibia
88 B3 **Orapa** Botswana
78 B1 **Orăştie** Romania
76 B2 **Orbetello** Italy
21 C3 **Orbost** Austr.
18 B1 **Ord, Mount** Austr.
74 B1 **Ordes** Spain
48 B1 **Ordu** Turkey
Ordzhonikidze Rus. Fed. *see* Vladikavkaz
59 C2 **Ordzhonikidze** Ukr.
61 G4 **Örebro** Sweden
102 B2 **Oregon** *state* U.S.A.
102 B1 **Oregon City** U.S.A.
55 C3 **Orekhovo-Zuyevo** Rus. Fed.
57 E3 **Orel** Rus. Fed.
51 K3 **Orel', Ozero** *l.* Rus. Fed.
103 D2 **Orem** U.S.A.
79 C3 **Ören** Turkey
55 E3 **Orenburg** Rus. Fed.
22 A3 **Orepuki** N.Z.
61 F4 **Öresund** *str.* Denmark/Sweden
67 D2 **Orford Ness** *hd* U.K.
79 C3 **Orhaneli** Turkey
79 C2 **Orhangazi** Turkey
51 I3 **Orhon Gol** *r.* Mongolia
120 B2 **Oriental, Cordillera** *mts* Bol.
118 A1 **Oriental, Cordillera** *mts* Col.
118 A3 **Oriental, Cordillera** *mts* Peru

75 C2 **Orihuela** Spain
59 D2 **Orikhiv** Ukr.
98 C2 **Orillia** Can.
118 B1 **Orinoco** *r.* Col./Venez.
118 B1 **Orinoco Delta** Venez.
61 H4 **Orissaare** Estonia
76 A3 **Oristano** *Sardinia* Italy
61 I3 **Orivesi** *l.* Fin.
119 C2 **Oriximiná** Brazil
113 C3 **Orizaba** Mex.
113 C3 **Orizaba, Pico de** *vol.* Mex.
60 E3 **Orkanger** Norway
61 F4 **Örkelljunga** Sweden
60 E3 **Orkla** *r.* Norway
64 C1 **Orkney Islands** *is* U.K.
122 C2 **Orlândia** Brazil
111 D3 **Orlando** U.S.A.
72 C2 **Orléans** France
42 A2 **Ormara** Pak.
32 B1 **Ormoc** Phil.
66 B2 **Ormskirk** U.K.
60 G3 **Örnsköldsvik** Sweden
82 B3 **Orodara** Burkina
102 C1 **Orofino** U.S.A.
Oroqen Zizhiqi China *see* **Alihe**
32 B2 **Oroquieta** Phil.
76 A2 **Orosei** *Sardinia* Italy
76 A2 **Orosei, Golfo di** *b. Sardinia* Italy
71 E2 **Orosháza** Hungary
103 B3 **Oroville** U.S.A.
20 A2 **Orroroo** Austr.
57 D3 **Orsha** Belarus
55 E3 **Orsk** Rus. Fed.
61 E3 **Ørsta** Norway
74 B1 **Ortegal, Cabo** *c.* Spain
72 B3 **Orthez** France
74 B1 **Ortigueira** Spain
105 D1 **Ortonville** U.S.A.
51 J2 **Orulgan, Khrebet** *mts* Rus. Fed.
49 C2 **Orūmīyeh, Daryācheh-ye** *salt l.* Iran
120 B2 **Oruro** Bol.
76 B2 **Orvieto** Italy
114 B4 **Osa, Península de** *pen.* Costa Rica
105 E3 **Osage** *r.* U.S.A.
35 C4 **Ōsaka** Japan
69 E1 **Oschersleben (Bode)** Ger.
76 A2 **Oschiri** *Sardinia* Italy
57 E3 **Osetr** *r.* Rus. Fed.
45 D2 **Osh** Kyrg.
88 A2 **Oshakati** Namibia
98 C2 **Oshawa** Can.
34 C2 **Ō-shima** *i.* Japan
35 C4 **Ō-shima** *i.* Japan
108 B2 **Oshkosh** U.S.A.
49 C2 **Oshnovīyeh** Iran
83 C4 **Oshogbo** Nigeria
86 B3 **Oshwe** Dem. Rep. Congo
77 C1 **Osijek** Croatia
96 B2 **Osilinka** *r.* Can.
76 B2 **Osimo** Italy
91 D2 **Osizweni** S. Africa
105 E2 **Oskaloosa** U.S.A.
61 G4 **Oskarshamn** Sweden
57 E3 **Oskol** *r.* Rus. Fed.
61 F4 **Oslo** Norway
61 F4 **Oslofjorden** *sea chan.* Norway
48 B1 **Osmancık** Turkey
79 C2 **Osmaneli** Turkey
48 B2 **Osmaniye** Turkey
56 C2 **Os'mino** Rus. Fed.
68 D1 **Osnabrück** Ger.
121 A5 **Osorno** Chile
74 C1 **Osorno** Spain
96 C3 **Osoyoos** Can.
68 B2 **Oss** Neth.
19 D4 **Ossa, Mount** Austr.
51 L3 **Ossora** Rus. Fed.
57 D2 **Ostashkov** Rus. Fed.
69 D1 **Oste** *r.* Ger.
68 A2 **Ostend** Belgium
69 E1 **Osterburg (Altmark)** Ger.
61 F3 **Österdalälven** *l.* Sweden
69 D1 **Osterholz-Scharmbeck** Ger.
69 E2 **Osterode am Harz** Ger.
60 F3 **Östersund** Sweden
Ostfriesische Inseln *is* Ger. *see* **East Frisian Islands**
68 C1 **Ostfriesland** *reg.* Ger.
61 G3 **Östhammar** Sweden
71 D2 **Ostrava** Czech Rep.
71 D1 **Ostróda** Pol.
57 E3 **Ostrogozhsk** Rus. Fed.
71 E1 **Ostrołęka** Pol.
69 F2 **Ostrov** Czech Rep.
56 C2 **Ostrov** Rus. Fed.
71 E1 **Ostrowiec Świętokrzyski** Pol.
71 E1 **Ostrów Mazowiecka** Pol.
71 D1 **Ostrów Wielkopolski** Pol.
78 B2 **Osŭm** *r.* Bulg.
35 B4 **Ōsumi-kaikyō** *sea chan.* Japan
35 B4 **Ōsumi-shotō** *is* Japan
74 B2 **Osuna** Spain
109 D2 **Oswego** U.S.A.
67 B2 **Oswestry** U.K.
35 C3 **Ōta** Japan
22 B3 **Otago Peninsula** N.Z.
22 C2 **Otaki** N.Z.
34 D2 **Otaru** Japan
88 A2 **Otavi** Namibia
102 C1 **Othello** U.S.A.
88 A3 **Otjiwarongo** Namibia
85 B3 **Otoro, Jebel** *mt.* Sudan
61 E4 **Otra** *r.* Norway
77 C2 **Otranto, Strait of** Albania/Italy
35 C3 **Ōtsu** Japan
61 E3 **Otta** Norway
98 C2 **Ottawa** Can.
98 C2 **Ottawa** *r.* Can.
108 B2 **Ottawa** *IL* U.S.A.
105 D3 **Ottawa** *KS* U.S.A.
98 B1 **Otter Rapids** Can.
68 B2 **Ottignies** Belgium
105 E2 **Ottumwa** U.S.A.
118 A2 **Otuzco** Peru
20 B3 **Otway, Cape** Austr.
110 B2 **Ouachita** *r.* U.S.A.
110 B2 **Ouachita, Lake** U.S.A.
110 B2 **Ouachita Mountains** U.S.A.
86 C2 **Ouadda** C.A.R.
83 D3 **Ouaddaï** *reg.* Chad
82 B3 **Ouagadougou** Burkina
82 B3 **Ouahigouya** Burkina
82 B3 **Oualâta** Maur.
86 C2 **Ouanda-Djailé** C.A.R.
82 B2 **Ouarâne** *reg.* Maur.
83 C1 **Ouargla** Alg.
82 B1 **Ouarzazate** Morocco
68 A2 **Oudenaarde** Belgium
90 B3 **Oudtshoorn** S. Africa
75 C2 **Oued Tlélat** Alg.
72 A2 **Ouessant, Île d'** *i.* France
86 B2 **Ouésso** Congo
82 B1 **Oujda** Morocco
75 D2 **Ouled Farès** Alg.
60 I2 **Oulu** Fin.
60 I3 **Oulujärvi** *l.* Fin.
76 A1 **Oulx** Italy
83 E3 **Oum-Chalouba** Chad
83 D3 **Oum-Hadjer** Chad
83 E3 **Ounianga Kébir** Chad
68 B2 **Oupeye** Belgium
68 C3 **Our, Vallée de l'** *val.* Ger./Lux.
74 B1 **Ourense** Spain
122 C2 **Ourinhos** Brazil
123 D2 **Ouro Preto** Brazil
68 B2 **Ourthe** *r.* Belgium
66 C2 **Ouse** *r.* U.K.
Outaouais, Rivière des *r.* Can. *see* **Ottawa**
99 D2 **Outardes** *r.* Can.
99 D1 **Outardes Quatre, Réservoir** *resr* Can.
64 A2 **Outer Hebrides** *is* U.K.
88 A3 **Outjo** Namibia
97 D2 **Outlook** Can.
60 I3 **Outokumpu** Fin.
20 B3 **Ouyen** Austr.
74 B1 **Ovar** Port.
60 H2 **Överkalix** Sweden
103 D3 **Overton** U.S.A.
60 H2 **Övertorneå** Sweden
74 B1 **Oviedo** Spain
61 E3 **Øvre Årdal** Norway
61 F3 **Øvre Rendal** Norway
58 B1 **Ovruch** Ukr.
86 B3 **Owando** Congo
35 C4 **Owase** Japan
105 E2 **Owatonna** U.S.A.
108 B3 **Owensboro** U.S.A.
103 C3 **Owens Lake** U.S.A.
98 B2 **Owen Sound** Can.
83 C4 **Owerri** Nigeria
108 C2 **Owosso** U.S.A.
102 C2 **Owyhee** U.S.A.
102 C2 **Owyhee** *r.* U.S.A.
22 B2 **Oxford** N.Z.
67 C3 **Oxford** U.K.
110 C2 **Oxford** U.S.A.
97 E2 **Oxford Lake** Can.
20 B2 **Oxley** Austr.
Ox Mountains Rep. of Ireland *see* Slieve Gamph
103 C4 **Oxnard** U.S.A.
35 C3 **Oyama** Japan
86 B2 **Oyem** Gabon
97 C2 **Oyen** Can.
73 D2 **Oyonnax** France
32 B2 **Ozamiz** Phil.
110 C2 **Ozark** U.S.A.
105 E3 **Ozark Plateau** U.S.A.
105 E3 **Ozarks, Lake of the** U.S.A.
51 L3 **Ozernovskiy** Rus. Fed.
56 B3 **Ozersk** Rus. Fed.
57 E3 **Ozery** Rus. Fed.
55 D3 **Ozinki** Rus. Fed.

P

95 H2 **Paamiut** Greenland
90 A3 **Paarl** S. Africa
71 D1 **Pabianice** Pol.
43 C2 **Pabna** Bangl.
42 A2 **Pab Range** *mts* Pak.
118 B1 **Pacaraima, Serra** *mts* Brazil
77 C3 **Pachino** *Sicily* Italy
113 C2 **Pachuca** Mex.
124 **Pacific Ocean**
71 D1 **Paczków** Pol.
28 B2 **Padang** Indon.
28 B2 **Padangpanjang** Indon.
28 A1 **Padangsidimpuan** Indon.
69 D2 **Paderborn** Ger.
Padova Italy *see* **Padua**
107 D3 **Padre Island** U.S.A.
20 B3 **Padthaway** Austr.
76 B1 **Padua** Italy

108 B3 **Paducah** *KY* U.S.A.
107 C2 **Paducah** *TX* U.S.A.
33 B1 **Paegam** N. Korea
33 B1 **Paekdu-san** *mt.* China/N. Korea
33 A2 **Paengnyŏng-do** *i.* S. Korea
22 C1 **Paeroa** N.Z.
48 B2 **Pafos** Cyprus
77 C2 **Pag** Croatia
32 B2 **Pagadian** Phil.
28 B2 **Pagai Selatan** *i.* Indon.
28 B2 **Pagai Utara** *i.* Indon.
27 D1 **Pagan** *i.* N. Mariana Is
29 C2 **Pagatan** Indon.
106 A1 **Page** U.S.A.
56 B2 **Pagėgiai** Lith.
104 B3 **Pagosa Springs** U.S.A.
56 C2 **Paide** Estonia
61 I3 **Päijänne** *l.* Fin.
43 C2 **Paikü Co** *l.* China
106 A1 **Painted Desert** U.S.A.
64 B3 **Paisley** U.K.
60 H2 **Pajala** Sweden
118 C1 **Pakaraima Mountains** Guyana
42 A2 **Pakistan** *country* Asia
30 A1 **Pakokku** Myanmar
56 B2 **Pakruojis** Lith.
71 D2 **Paks** Hungary
98 A1 **Pakwash Lake** Can.
31 B2 **Pakxé** Laos
83 D4 **Pala** Chad
28 B2 **Palabuhanratu, Teluk** *b.* Indon.
79 C3 **Palaikastro** Greece
79 B3 **Palaiochora** Greece
90 B1 **Palamakoloi** Botswana
75 D1 **Palamós** Spain
51 L3 **Palana** Rus. Fed.
32 B1 **Palanan** Phil.
29 C2 **Palangkaraya** Indon.
42 B2 **Palanpur** India
88 B3 **Palapye** Botswana
51 L2 **Palatka** Rus. Fed.
111 D3 **Palatka** U.S.A.
27 C2 **Palau** *country* N. Pacific Ocean
31 A2 **Palaw** Myanmar
32 A2 **Palawan** *i.* Phil.
61 H4 **Paldiski** Estonia
28 B2 **Palembang** Indon.
74 C1 **Palencia** Spain
113 C3 **Palenque** Mex.
76 B3 **Palermo** *Sicily* Italy
107 D2 **Palestine** U.S.A.
30 A1 **Paletwa** Myanmar
42 B2 **Pali** India
77 C2 **Palinuro, Capo** *c.* Italy
79 B3 **Paliouri, Akra** *pt* Greece
68 B3 **Paliseul** Belgium
60 I3 **Paljakka** *hill* Fin.
56 C2 **Palkino** Rus. Fed.
41 B4 **Palk Strait** India/Sri Lanka
22 C2 **Palliser, Cape** N.Z.
74 B2 **Palma del Río** Spain
75 D2 **Palma de Mallorca** Spain
122 B3 **Palmas** Brazil
82 B4 **Palmas, Cape** Liberia
122 C3 **Palmeira** Brazil
119 D2 **Palmeirais** Brazil
23 P2 **Palmer Land** *reg.* Antarctica
22 C2 **Palmerston North** N.Z.
77 C3 **Palmi** Italy
113 C2 **Palmillas** Mex.
118 A1 **Palmira** Col.
103 C4 **Palm Springs** U.S.A.
17 G3 **Palmyra Atoll** N. Pacific Ocean
85 B3 **Paloich** Sudan
113 C3 **Palomares** Mex.
75 C2 **Palos, Cabo de** *c.* Spain
60 I3 **Paltamo** Fin.
26 B3 **Palu** Indon.
51 M2 **Palyavaam** *r.* Rus. Fed.
72 C3 **Pamiers** France
45 D3 **Pamir** *mts* Asia
111 E1 **Pamlico Sound** *sea chan.* U.S.A.
120 B2 **Pampa Grande** Bol.
121 B4 **Pampas** *reg.* Arg.
118 A1 **Pamplona** Col.
75 C1 **Pamplona** Spain
79 D2 **Pamukova** Turkey
28 B2 **Panaitan** *i.* Indon.
41 B3 **Panaji** India
114 B4 **Panama** *country* Central America
114 C4 **Panamá** Panama
114 C4 **Panamá, Golfo de** *g.* Panama
114 C4 **Panama Canal** Panama
110 C2 **Panama City** U.S.A.
103 C3 **Panamint Range** *mts* U.S.A.
29 B1 **Panarik** Indon.
32 B1 **Panay** *i.* Phil.
77 D2 **Pančevo** Yugo.
32 B1 **Pandan** Phil.
43 C2 **Pandaria** India
41 B3 **Pandharpur** India
56 B2 **Panevėžys** Lith.
29 C2 **Pangkalanbuun** Indon.
28 A1 **Pangkalansusu** Indon.
28 B2 **Pangkalpinang** Indon.
95 G2 **Pangnirtung** Can.
54 G2 **Pangody** Rus. Fed.
57 F3 **Panino** Rus. Fed.
42 B2 **Panipat** India
42 A2 **Panjgur** Pak.
33 C1 **Pan Ling** *mts* China
43 C2 **Panna** India
18 A2 **Pannawonica** Austr.
122 B2 **Panorama** Brazil
33 B1 **Panshi** China
113 C2 **Pánuco** Mex.
39 A3 **Panxian** China
30 B1 **Panzhihua** China
77 C3 **Paola** Italy
31 B2 **Paôy Pêt** Cambodia
71 D2 **Pápa** Hungary
22 B1 **Papakura** N.Z.
113 C2 **Papantla** Mex.
68 C1 **Papenburg** Ger.
27 D3 **Papua, Gulf of** P.N.G.
27 D3 **Papua New Guinea** *country* Oceania
30 A2 **Papun** Myanmar
18 A2 **Paraburdoo** Austr.
122 C1 **Paracatu** Brazil
123 C1 **Paracatu** *r.* Brazil
20 A2 **Parachilna** Austr.
77 D2 **Paraćin** Yugo.
123 D1 **Pará de Minas** Brazil
103 B3 **Paradise** U.S.A.
110 B1 **Paragould** U.S.A.
119 C3 **Paraguai** *r.* Brazil
120 C3 **Paraguay** *r.* Arg./Para.
120 C3 **Paraguay** *country* S. America
123 D2 **Paraíba do Sul** *r.* Brazil
122 B1 **Paraíso** Brazil
113 C3 **Paraíso** Mex.
82 C4 **Parakou** Benin
119 C1 **Paramaribo** Suriname
51 L3 **Paramushir, Ostrov** *i.* Rus. Fed.
120 B4 **Paraná** Arg.
122 A3 **Paraná** *r.* S. America
122 C3 **Paranaguá** Brazil
122 B1 **Paranaíba** Brazil
122 B2 **Paranaíba** *r.* Brazil
122 B2 **Paranapanema** *r.* Brazil
122 B2 **Paranapiacaba, Serra** *mts* Brazil
122 B2 **Paranavaí** Brazil
22 B2 **Paraparaumu** N.Z.
123 D2 **Parati** Brazil
122 B1 **Paraúna** Brazil
73 C2 **Paray-le-Monial** France
41 B3 **Parbhani** India
69 E1 **Parchim** Ger.
123 E1 **Pardo** *r. Bahia* Brazil
122 B2 **Pardo** *r. Mato Grosso do Sul* Brazil
122 C2 **Pardo** *r. São Paulo* Brazil
71 D1 **Pardubice** Czech Rep.
118 B3 **Parecis, Serra dos** *hills* Brazil
98 C2 **Parent, Lac** *l.* Can.
26 B3 **Parepare** Indon.
57 D2 **Parfino** Rus. Fed.
79 B3 **Parga** Greece
77 C3 **Parghelia** Italy
115 D3 **Paria, Gulf of** Trin. and Tob./Venez.
118 B1 **Parima, Serra** *mts* Brazil
119 C2 **Parintins** Brazil
72 C2 **Paris** France
110 C1 **Paris** *TN* U.S.A.
107 D2 **Paris** *TX* U.S.A.
61 H3 **Parkano** Fin.
106 A2 **Parker** U.S.A.
108 C3 **Parkersburg** U.S.A.
21 C2 **Parkes** Austr.
108 A1 **Park Falls** U.S.A.
105 D1 **Park Rapids** U.S.A.
76 B2 **Parma** Italy
119 D2 **Parnaíba** Brazil
119 D2 **Parnaíba** *r.* Brazil
79 B3 **Parnassos** *mt.* Greece
22 B2 **Parnassus** N.Z.
79 B3 **Parnon** *mts* Greece
56 B2 **Pärnu** Estonia
33 B2 **P'aro-ho** *l.* S. Korea
20 B2 **Paroo** *watercourse* Austr.
44 C3 **Paropamisus** *mts* Afgh.
79 C3 **Paros** *i.* Greece
112 B2 **Parras** Mex.
94 C2 **Parry, Cape** Can.
94 D1 **Parry Islands** Can.
98 B2 **Parry Sound** Can.
105 D3 **Parsons** U.S.A.
69 D2 **Partenstein** Ger.
72 B2 **Parthenay** France
65 B2 **Partry Mountains** Rep. of Ireland
119 C2 **Paru** *r.* Brazil
103 C4 **Pasadena** U.S.A.
30 A2 **Pasawng** Myanmar
110 C2 **Pascagoula** U.S.A.
78 C1 **Paşcani** Romania
102 C1 **Pasco** U.S.A.
70 C1 **Pasewalk** Ger.
97 D2 **Pasfield Lake** Can.
28 B1 **Pasir Putih** Malaysia
42 A2 **Pasni** Pak.
121 A5 **Paso Río Mayo** Arg.
103 B3 **Paso Robles** U.S.A.
123 D2 **Passa Tempo** Brazil
70 C2 **Passau** Ger.
120 C3 **Passo Fundo** Brazil
123 C2 **Passos** Brazil
56 C2 **Pastavy** Belarus
118 A2 **Pastaza** *r.* Peru
118 A1 **Pasto** Col.
29 C2 **Pasuruan** Indon.
56 B2 **Pasvalys** Lith.
121 A6 **Patagonia** *reg.* Arg.
43 C2 **Patan** Nepal
22 B1 **Patea** N.Z.
109 E2 **Paterson** U.S.A.

Pathein Myanmar *see* Bassein
104 B2 **Pathfinder Reservoir** U.S.A.
29 C2 **Pati** Indon.
30 A1 **Pātkai Bum** *mts* India/Myanmar
79 C3 **Patmos** *i.* Greece
43 C2 **Patna** India
49 C2 **Patnos** Turkey
122 B3 **Pato Branco** Brazil
120 C4 **Patos, Lagoa dos** *l.* Brazil
123 C1 **Patos de Minas** Brazil
120 B4 **Patquía** Arg.
79 B3 **Patra** Greece
43 C2 **Patratu** India
122 C1 **Patrocínio** Brazil
31 B3 **Pattani** Thai.
31 B2 **Pattaya** Thai.
96 B2 **Pattullo, Mount** Can.
97 D2 **Patuanak** Can.
112 B3 **Pátzcuaro** Mex.
72 B3 **Pau** France
72 B2 **Pauillac** France
30 A1 **Pauk** Myanmar
119 D2 **Paulistana** Brazil
119 E2 **Paulo Afonso** Brazil
91 D2 **Paulpietersburg** S. Africa
107 D2 **Pauls Valley** U.S.A.
30 A2 **Paungde** Myanmar
123 D1 **Pavão** Brazil
76 A1 **Pavia** Italy
56 B2 **Pāvilosta** Latvia
78 C2 **Pavlikeni** Bulg.
45 D1 **Pavlodar** Kazakh.
59 D2 **Pavlohrad** Ukr.
59 E1 **Pavlovsk** Rus. Fed.
59 D2 **Pavlovskaya** Rus. Fed.
28 B2 **Payakumbuh** Indon.
102 C2 **Payette** U.S.A.
102 C2 **Payette** *r.* U.S.A.
54 F2 **Pay-Khoy, Khrebet** *hills* Rus. Fed.
120 C4 **Paysandú** Uru.
49 C1 **Pazar** Turkey
78 B2 **Pazardzhik** Bulg.
79 C3 **Pazarköy** Turkey
76 B1 **Pazin** Croatia
31 A2 **Pe** Myanmar
96 C2 **Peace** *r.* Can.
96 C2 **Peace River** Can.
103 E3 **Peale, Mount** U.S.A.
110 C2 **Pearl** *r.* U.S.A.
107 D3 **Pearsall** U.S.A.
89 C2 **Pebane** Moz.
77 D2 **Peć** Yugo.
123 D1 **Peçanha** Brazil
54 E2 **Pechora** Rus. Fed.
54 E2 **Pechora** *r.* Rus. Fed.
56 C2 **Pechory** Rus. Fed.
107 C2 **Pecos** U.S.A.
107 C3 **Pecos** *r.* U.S.A.
71 D2 **Pécs** Hungary
123 D1 **Pedra Azul** Brazil
122 C2 **Pedregulho** Brazil
119 D2 **Pedreiras** Brazil
41 C4 **Pedro, Point** Sri Lanka
119 D2 **Pedro Afonso** Brazil
122 B1 **Pedro Gomes** Brazil
120 C3 **Pedro Juan Caballero** Para.
64 C3 **Peebles** U.K.
111 E2 **Pee Dee** *r.* U.S.A.
94 C2 **Peel** *r.* Can.
66 A1 **Peel** Isle of Man
96 C2 **Peerless Lake** Can.
22 B2 **Pegasus Bay** N.Z.
69 E3 **Pegnitz** Ger.
30 A2 **Pegu** Myanmar
30 A2 **Pegu Yoma** *mts* Myanmar
121 B4 **Pehuajó** Arg.
69 E1 **Peine** Ger.
56 C2 **Peipus, Lake** Estonia/Rus. Fed.
79 B3 **Peiraias** Greece
122 B2 **Peixe** *r.* Brazil
29 B2 **Pekalongan** Indon.
28 B1 **Pekan** Malaysia
28 B1 **Pekanbaru** Indon.
Peking China *see* **Beijing**
98 B2 **Pelee Island** Can.
27 C3 **Peleng** *i.* Indon.
60 I2 **Pelkosenniemi** Fin.
90 A2 **Pella** S. Africa
27 D3 **Pelleluhu Islands** P.N.G.
60 H2 **Pello** Fin.
96 A1 **Pelly** *r.* Can.
95 F2 **Pelly Bay** Can.
96 A1 **Pelly Mountains** Can.
120 C4 **Pelotas** Brazil
120 C3 **Pelotas, Rio das** *r.* Brazil
109 F1 **Pemadumcook Lake** U.S.A.
29 B1 **Pemangkat** Indon.
28 A1 **Pematangsiantar** Indon.
89 D2 **Pemba** Moz.
88 B2 **Pemba** Zambia
87 D3 **Pemba Island** Tanz.
96 B2 **Pemberton** Can.
105 D1 **Pembina** *r.* U.S.A.
98 C2 **Pembroke** Can.
67 A3 **Pembroke** U.K.
74 C1 **Peña Cerredo** *mt.* Spain
113 C2 **Peña Nevada, Cerro** *mt.* Mex.
122 B2 **Penápolis** Brazil
74 B1 **Peñaranda de Bracamonte** Spain
75 C1 **Peñarroya** *mt.* Spain
74 B2 **Peñarroya-Pueblonuevo** Spain
74 B1 **Peñas, Cabo de** *c.* Spain
121 A5 **Penas, Golfo de** *g.* Chile
74 B1 **Peña Ubiña** *mt.* Spain
102 C1 **Pendleton** U.S.A.
96 B2 **Pendleton Bay** Can.
102 C1 **Pend Oreille Lake** U.S.A.
42 B3 **Penganga** *r.* India
91 D1 **Penge** S. Africa
38 C2 **Penglai** China
39 A3 **Pengshui** China
74 B2 **Peniche** Port.
66 B1 **Penicuik** U.K.
76 B2 **Penne** Italy
66 B1 **Pennines** *hills* U.K.
109 D2 **Pennsylvania** *state* U.S.A.
95 G2 **Penny Icecap** Can.
109 F2 **Penobscot** *r.* U.S.A.
20 B3 **Penola** Austr.
18 C3 **Penong** Austr.
66 B1 **Penrith** U.K.
110 C2 **Pensacola** U.S.A.
23 Q1 **Pensacola Mountains** Antarctica
29 C1 **Pensiangan** *Sabah* Malaysia
96 C3 **Penticton** Can.
64 C1 **Pentland Firth** *sea chan.* U.K.
67 B2 **Penygadair** *hill* U.K.
55 D3 **Penza** Rus. Fed.
67 A3 **Penzance** U.K.
108 B2 **Peoria** U.S.A.
75 C1 **Perales del Alfambra** Spain
19 E2 **Percy Isles** Austr.
54 F2 **Peregrebnoye** Rus. Fed.
118 A1 **Pereira** Col.
122 B2 **Pereira Barreto** Brazil
58 A2 **Peremyshlyany** Ukr.
57 E2 **Pereslavl'-Zalesskiy** Rus. Fed.
59 C1 **Pereyaslav-Khmel'nyts'kyy** Ukr.
121 B4 **Pergamino** Arg.
60 H3 **Perhonjoki** *r.* Fin.
99 C1 **Péribonca, Lac** *l.* Can.
120 B3 **Perico** Arg.
112 B2 **Pericos** Mex.
72 C2 **Périgueux** France
115 C4 **Perija, Sierra de** *mts* Venez.
121 A5 **Perito Moreno** Arg.
69 E1 **Perleberg** Ger.
54 E3 **Perm'** Rus. Fed.
20 A2 **Pernatty Lagoon** *salt flat* Austr.
73 C2 **Péronne** France
113 C3 **Perote** Mex.
72 C3 **Perpignan** France
67 A3 **Perranporth** U.K.
111 D2 **Perry** *FL* U.S.A.
111 D2 **Perry** *GA* U.S.A.
105 E2 **Perry** *IA* U.S.A.
107 C1 **Perryton** U.S.A.
105 F3 **Perryville** U.S.A.
67 B2 **Pershore** U.K.
59 D2 **Pershotravens'k** Ukr.
18 A3 **Perth** Austr.
64 C2 **Perth** U.K.
73 D3 **Pertuis** France
76 A2 **Pertusato, Capo** *c. Corsica* France
118 A2 **Peru** *country* S. America
125 H4 **Peru-Chile Trench** *sea feature* S. Pacific Ocean
76 B2 **Perugia** Italy
122 C2 **Peruíbe** Brazil
68 A2 **Péruwelz** Belgium
58 C2 **Pervomays'k** Ukr.
59 C2 **Pervomays'ke** Ukr.
57 F3 **Pervomayskiy** Rus. Fed.
59 D2 **Pervomays'kyy** Ukr.
76 B2 **Pesaro** Italy
76 B2 **Pescara** Italy
76 B2 **Pescara** *r.* Italy
42 B1 **Peshawar** Pak.
77 D2 **Peshkopi** Albania
77 C1 **Pesnica** Slovenia
57 E2 **Pestovo** Rus. Fed.
112 B3 **Petatlán** Mex.
108 B2 **Petenwell Lake** U.S.A.
20 A2 **Peterborough** Austr.
98 C2 **Peterborough** Can.
67 C2 **Peterborough** U.K.
64 D2 **Peterhead** U.K.
18 B2 **Petermann Ranges** *mts* Austr.
97 D2 **Peter Pond Lake** Can.
96 A2 **Petersburg** *AK* U.S.A.
109 D3 **Petersburg** *VA* U.S.A.
69 D1 **Petershagen** Ger.
99 D1 **Petit Lac Manicouagan** *l.* Can.
99 E1 **Petit Mécatina** *r.* Can.
113 D2 **Peto** Mex.
108 C1 **Petoskey** U.S.A.
48 B2 **Petra** Jordan
34 B2 **Petra Velikogo, Zaliv** *b.* Rus. Fed.
79 B2 **Petrich** Bulg.
119 D2 **Petrolina** Brazil
45 C1 **Petropavlovsk** Kazakh.
51 L3 **Petropavlovsk-Kamchatskiy** Rus. Fed.
78 B1 **Petroșani** Romania
57 F3 **Petrovskoye** Rus. Fed.
54 C2 **Petrozavodsk** Rus. Fed.
91 C2 **Petrusburg** S. Africa
91 C2 **Petrus Steyn** S. Africa
90 B3 **Petrusville** S. Africa
57 E2 **Petushki** Rus. Fed.
28 A1 **Peureula** Indon.
51 M2 **Pevek** Rus. Fed.
70 B2 **Pforzheim** Ger.
70 C2 **Pfunds** Austria

69 D3 **Pfungstadt** Ger.
91 C2 **Phahameng** S. Africa
91 D1 **Phalaborwa** S. Africa
42 B2 **Phalodi** India
31 A3 **Phangnga** Thai.
31 B2 **Phan Rang** Vietnam
31 B2 **Phan Thiêt** Vietnam
31 B3 **Phatthalung** Thai.
30 A2 **Phayao** Thai.
97 D2 **Phelps Lake** Can.
111 C2 **Phenix City** U.S.A.
31 A2 **Phet Buri** Thai.
31 B2 **Phichit** Thai.
109 D3 **Philadelphia** U.S.A.
104 C2 **Philip** U.S.A.
68 B2 **Philippeville** Belgium
68 A2 **Philippine** Neth.
32 B1 **Philippines** *country* Asia
32 B1 **Philippine Sea** N. Pacific Ocean
94 B2 **Philip Smith Mountains** U.S.A.
90 B3 **Philipstown** S. Africa
21 C3 **Phillip Island** Austr.
105 D3 **Phillipsburg** U.S.A.
91 C2 **Phiritona** S. Africa
31 B2 **Phitsanulok** Thai.
Phnom Penh Cambodia *see* Phnum Pénh
31 B2 **Phnum Pénh** Cambodia
106 A2 **Phoenix** U.S.A.
31 B2 **Phon** Thai.
30 B1 **Phôngsali** Laos
30 B2 **Phrae** Thai.
31 A3 **Phuket** Thai.
31 B2 **Phumĭ Sâmraông** Cambodia
91 C2 **Phuthaditjhaba** S. Africa
76 A1 **Piacenza** Italy
76 B2 **Pianosa, Isola** *i.* Italy
78 C1 **Piatra Neamț** Romania
119 D2 **Piauí** *r.* Brazil
76 B1 **Piave** *r.* Italy
85 B4 **Pibor** *r.* Sudan
85 B4 **Pibor Post** Sudan
110 C2 **Picayune** U.S.A.
120 B3 **Pichanal** Arg.
112 A2 **Pichilingue** Mex.
66 C1 **Pickering** U.K.
98 A1 **Pickle Lake** Can.
119 D2 **Picos** Brazil
121 B5 **Pico Truncado** Arg.
21 D2 **Picton** Austr.
113 C3 **Piedras Negras** Guat.
113 B2 **Piedras Negras** Mex.
61 I3 **Pieksämäki** Fin.
60 I3 **Pielinen** *l.* Fin.
104 C2 **Pierre** U.S.A.
73 C3 **Pierrelatte** France
91 D2 **Pietermaritzburg** S. Africa
91 C1 **Pietersburg** S. Africa
78 B1 **Pietrosa** *mt.* Romania
96 C2 **Pigeon Lake** Can.
121 B4 **Pigüé** Arg.
60 I3 **Pihtipudas** Fin.
113 C3 **Pijijiapan** Mex.
104 C3 **Pikes Peak** U.S.A.
90 A3 **Piketberg** S. Africa
108 C3 **Pikeville** U.S.A.
71 D1 **Piła** Pol.
121 C4 **Pilar** Arg.
120 C3 **Pilar** Para.
21 C2 **Pilliga** Austr.
122 C1 **Pilões, Serra dos** *mts* Brazil
118 B3 **Pimenta Bueno** Brazil
121 C4 **Pinamar** Arg.
48 B2 **Pınarbaşı** Turkey
114 B2 **Pinar del Río** Cuba
71 E1 **Pińczów** Pol.
119 D2 **Pindaré** *r.* Brazil
79 B2 **Pindos** *mts* Greece
110 B2 **Pine Bluff** U.S.A.
18 C1 **Pine Creek** Austr.
104 B2 **Pinedale** U.S.A.
54 D2 **Pinega** Rus. Fed.
97 D2 **Pinehouse Lake** Can.
79 B3 **Pineios** *r.* Greece
96 C1 **Pine Point** Can.
104 C2 **Pine Ridge** U.S.A.
91 D2 **Pinetown** S. Africa
38 B2 **Pingdingshan** China
39 A3 **Pingguo** China
39 B3 **Pingjiang** China
38 A2 **Pingliang** China
39 C3 **P'ingtung** Taiwan
39 A3 **Pingxiang** *Guangxi* China
39 B3 **Pingxiang** *Jiangxi* China
38 B2 **Pingyin** China
119 D2 **Pinheiro** Brazil
69 D1 **Pinneberg** Ger.
113 C3 **Pinotepa Nacional** Mex.
56 C3 **Pinsk** Belarus
103 D3 **Pioche** U.S.A.
87 C3 **Piodi** Dem. Rep. Congo
71 E1 **Pionki** Pol.
71 D1 **Piotrków Trybunalski** Pol.
105 D2 **Pipestone** U.S.A.
99 C2 **Pipmuacan, Réservoir** *resr* Can.
122 B2 **Piquiri** *r.* Brazil
122 C1 **Piracanjuba** Brazil
122 C2 **Piracicaba** Brazil
123 D1 **Piracicaba** *r.* Brazil
122 C2 **Piraçununga** Brazil
122 C2 **Piraí do Sul** Brazil
122 C2 **Pirajuí** Brazil
122 B1 **Piranhas** Brazil
119 E2 **Piranhas** *r.* Brazil
123 D1 **Pirapora** Brazil
122 C1 **Pires do Rio** Brazil
119 D2 **Piripiri** Brazil
27 C3 **Piru** Indon.
76 B2 **Pisa** Italy
120 A2 **Pisagua** Chile
118 A3 **Pisco** Peru
70 C2 **Písek** Czech Rep.
47 D2 **Pīshīn** Iran
113 D2 **Pisté** Mex.
77 C2 **Pisticci** Italy
76 B2 **Pistoia** Italy
74 C1 **Pisuerga** *r.* Spain
102 B2 **Pit** *r.* U.S.A.
122 B2 **Pitanga** Brazil
123 D1 **Pitangui** Brazil
17 H6 **Pitcairn Islands** *terr.* S. Pacific Ocean
60 H2 **Piteå** Sweden
60 H2 **Piteälven** *r.* Sweden
78 B2 **Pitești** Romania
43 C2 **Pithoragarh** India
64 C2 **Pitlochry** U.K.
96 B2 **Pitt Island** Can.
105 E3 **Pittsburg** U.S.A.
109 D2 **Pittsburgh** U.S.A.
109 E2 **Pittsfield** U.S.A.
21 D1 **Pittsworth** Austr.
123 C2 **Piumhí** Brazil
118 A2 **Piura** Peru
58 C2 **Pivdennyy Buh** *r.* Ukr.
99 E2 **Placentia** Can.
103 B3 **Placerville** U.S.A.
114 C2 **Placetas** Cuba
107 C2 **Plainview** U.S.A.
122 C1 **Planaltina** Brazil
110 B2 **Plaquemine** U.S.A.
74 B1 **Plasencia** Spain
115 C4 **Plato** Col.
105 D2 **Platte** *r.* U.S.A.
109 E2 **Plattsburgh** U.S.A.
69 F1 **Plau** Ger.
69 F2 **Plauen** Ger.
69 F1 **Plauer See** *l.* Ger.
57 E3 **Plavsk** Rus. Fed.
31 B2 **Plây Cu** Vietnam
97 E2 **Playgreen Lake** Can.
121 B4 **Plaza Huincul** Arg.
107 D3 **Pleasanton** U.S.A.
22 B2 **Pleasant Point** N.Z.
72 C2 **Pleaux** France
98 B1 **Pledger Lake** Can.
22 C1 **Plenty, Bay of** *g.* N.Z.
104 C1 **Plentywood** U.S.A.
54 D2 **Plesetsk** Rus. Fed.
99 C1 **Pletipi, Lac** *l.* Can.
68 C2 **Plettenberg** Ger.
90 B3 **Plettenberg Bay** S. Africa
78 B2 **Pleven** Bulg.
77 C2 **Pljevlja** Yugo.
77 C2 **Ploče** Croatia
71 D1 **Płock** Pol.
77 C2 **Pločno** *mt.* Bos.-Herz.
72 B2 **Ploemeur** France
78 C2 **Ploiești** Romania
57 F2 **Ploskoye** Rus. Fed.
72 B2 **Plouzané** France
78 B2 **Plovdiv** Bulg.
56 B2 **Plungė** Lith.
56 C3 **Plyeshchanitsy** Belarus
115 D3 **Plymouth** Montserrat
67 A3 **Plymouth** U.K.
108 B2 **Plymouth** U.S.A.
56 C2 **Plyussa** Rus. Fed.
70 C2 **Plzeň** Czech Rep.
82 B3 **Pô** Burkina
76 B1 **Po** *r.* Italy
45 E2 **Pobedy, Pik** *mt.* China/Kyrg.
110 B1 **Pocahontas** U.S.A.
102 D2 **Pocatello** U.S.A.
58 B1 **Pochayiv** Ukr.
57 D3 **Pochep** Rus. Fed.
57 D3 **Pochinok** Rus. Fed.
113 C3 **Pochutla** Mex.
109 D3 **Pocomoke City** U.S.A.
123 C2 **Poços de Caldas** Brazil
57 D2 **Poddor'ye** Rus. Fed.
59 D1 **Podgorenskiy** Rus. Fed.
77 C2 **Podgorica** Yugo.
50 G3 **Podgornoye** Rus. Fed.
51 H2 **Podkamennaya Tunguska** *r.* Rus. Fed.
57 E2 **Podol'sk** Rus. Fed.
77 D2 **Podujevo** Yugo.
90 A2 **Pofadder** S. Africa
57 D3 **Pogar** Rus. Fed.
77 D2 **Pogradec** Albania
33 B2 **P'ohang** S. Korea
78 B2 **Poiana Mare** Romania
86 C3 **Poie** Dem. Rep. Congo
86 B3 **Pointe-Noire** Congo
94 D2 **Point Lake** Can.
108 C3 **Point Pleasant** U.S.A.
72 C2 **Poitiers** France
42 B2 **Pokaran** India
43 C2 **Pokhara** Nepal
51 J2 **Pokrovsk** Rus. Fed.
59 D2 **Pokrovskoye** Rus. Fed.
106 A1 **Polacca** U.S.A.
74 B1 **Pola de Lena** Spain
71 D1 **Poland** *country* Europe
56 C2 **Polatsk** Belarus
45 C3 **Pol-e Khomrī** Afgh.
29 C2 **Polewali** Indon.
86 B2 **Poli** Cameroon
70 C1 **Police** Pol.
77 C2 **Policoro** Italy
73 D2 **Poligny** France
32 B1 **Polillo Islands** Phil.
58 B1 **Polis'ke** Ukr.
71 D1 **Polkowice** Pol.
77 C3 **Pollino, Monte** *mt.* Italy
59 D2 **Polohy** Ukr.

58 B1 **Polonne** Ukr.
102 D1 **Polson** U.S.A.
59 C2 **Poltava** Ukr.
34 B2 **Poltavka** Rus. Fed.
59 D2 **Poltavskaya** Rus. Fed.
56 C2 **Põlva** Estonia
79 B2 **Polygyros** Greece
74 B2 **Pombal** Port.
76 B2 **Pomezia** Italy
60 I2 **Pomokaira** *reg.* Fin.
78 C2 **Pomorie** Bulg.
70 C1 **Pomorska, Zatoka** *b.* Pol.
123 D1 **Pompéu** Brazil
107 D1 **Ponca City** U.S.A.
115 D3 **Ponce** Puerto Rico
41 B3 **Pondicherry** India
95 F2 **Pond Inlet** Can.
74 B1 **Ponferrada** Spain
85 A4 **Pongo** *watercourse* Sudan
91 D2 **Pongola** *r.* S. Africa
96 C2 **Ponoka** Can.
122 B3 **Ponta Grossa** Brazil
122 C1 **Pontalina** Brazil
73 D2 **Pont-à-Mousson** France
73 D2 **Pontarlier** France
70 B2 **Pontcharra** France
110 B2 **Pontchartrain, Lake** U.S.A.
74 B2 **Ponte de Sor** Port.
97 D3 **Ponteix** Can.
118 C3 **Pontes-e-Lacerda** Brazil
74 B1 **Pontevedra** Spain
108 B2 **Pontiac** *IL* U.S.A.
108 C2 **Pontiac** *MI* U.S.A.
29 B2 **Pontianak** Indon.
72 B2 **Pontivy** France
119 C1 **Pontoetoe** Suriname
97 E2 **Ponton** Can.
67 B3 **Pontypool** U.K.
76 B2 **Ponziane, Isole** *is* Italy
67 C3 **Poole** U.K.
Poona India *see* **Pune**
20 B2 **Pooncarie** Austr.
120 B2 **Poopó, Lago de** *l.* Bol.
118 A1 **Popayán** Col.
51 I2 **Popigay** *r.* Rus. Fed.
20 B2 **Popiltah** Austr.
97 E2 **Poplar** *r.* Can.
105 E3 **Poplar Bluff** U.S.A.
86 B3 **Popokabaka** Dem. Rep. Congo
78 C2 **Popovo** Bulg.
71 E2 **Poprad** Slovakia
119 D3 **Porangatu** Brazil
42 A2 **Porbandar** India
94 B2 **Porcupine** *r.* Can./U.S.A.
76 B1 **Poreč** Croatia
61 H3 **Pori** Fin.
22 B2 **Porirua** N.Z.
56 C2 **Porkhov** Rus. Fed.
72 B2 **Pornic** France
51 K3 **Poronaysk** Rus. Fed.
79 B3 **Poros** Greece
61 E4 **Porsgrunn** Norway
65 C1 **Portadown** U.K.
65 D1 **Portaferry** U.K.
108 B2 **Portage** U.S.A.
97 E3 **Portage la Prairie** Can.
96 B3 **Port Alberni** Can.
74 B2 **Portalegre** Port.
107 C2 **Portales** U.S.A.
96 A2 **Port Alexander** U.S.A.
96 B2 **Port Alice** Can.
102 B1 **Port Angeles** U.S.A.
19 D4 **Port Arthur** Austr.
107 E3 **Port Arthur** U.S.A.
64 A3 **Port Askaig** U.K.
20 A2 **Port Augusta** Austr.
115 C3 **Port-au-Prince** Haiti
99 E1 **Port aux Choix** Can.
41 D3 **Port Blair** India
20 B3 **Port Campbell** Austr.
22 B3 **Port Chalmers** N.Z.
111 D3 **Port Charlotte** U.S.A.
115 C3 **Port-de-Paix** Haiti
96 A2 **Port Edward** Can.
123 D1 **Porteirinha** Brazil
119 C2 **Portel** Brazil
98 B2 **Port Elgin** Can.
91 C3 **Port Elizabeth** S. Africa
64 A3 **Port Ellen** U.K.
66 A1 **Port Erin** Isle of Man
90 A3 **Porterville** S. Africa
103 C3 **Porterville** U.S.A.
20 B3 **Port Fairy** Austr.
22 C1 **Port Fitzroy** N.Z.
86 A3 **Port-Gentil** Gabon
83 C4 **Port Harcourt** Nigeria
96 B2 **Port Hardy** Can.
Port Harrison Can. *see* Inukjuak
99 D2 **Port Hawkesbury** Can.
18 A2 **Port Hedland** Austr.
67 A2 **Porthmadog** U.K.
99 E1 **Port Hope Simpson** Can.
108 C2 **Port Huron** U.S.A.
74 B2 **Portimão** Port.
21 C2 **Portland** *N.S.W.* Austr.
20 B3 **Portland** *Vic.* Austr.
109 E2 **Portland** *ME* U.S.A.
102 B1 **Portland** *OR* U.S.A.
96 A2 **Portland Canal** *inlet* Can.
65 C2 **Portlaoise** Rep. of Ireland
107 D3 **Port Lavaca** U.S.A.
20 A2 **Port Lincoln** Austr.
82 A4 **Port Loko** Sierra Leone
21 D2 **Port Macquarie** Austr.
96 B2 **Port McNeill** Can.
99 D2 **Port-Menier** Can.
27 D3 **Port Moresby** P.N.G.
64 A1 **Port Nis** U.K.
90 A2 **Port Nolloth** S. Africa
Port-Nouveau-Québec Can. *see* **Kangiqsualujjuaq**
Porto Port. *see* **Oporto**
118 B2 **Porto Acre** Brazil
120 C4 **Porto Alegre** Brazil
119 C3 **Porto Artur** Brazil
119 C3 **Porto dos Gaúchos Óbidos** Brazil
118 C3 **Porto Esperidião** Brazil
76 B2 **Portoferraio** Italy
119 D2 **Porto Franco** Brazil
115 D3 **Port of Spain** Trin. and Tob.
76 B1 **Portogruaro** Italy
76 B2 **Portomaggiore** Italy
122 B2 **Porto Mendes** Brazil
120 C3 **Porto Murtinho** Brazil
119 D3 **Porto Nacional** Brazil
82 C4 **Porto-Novo** Benin
122 B2 **Porto Primavera, Represa** *resr* Brazil
119 C2 **Porto Santana** Brazil
123 E1 **Porto Seguro** Brazil
76 B2 **Porto Tolle** Italy
76 A2 **Porto Torres** *Sardinia* Italy
73 D3 **Porto-Vecchio** *Corsica* France
118 B2 **Porto Velho** Brazil
118 A2 **Portoviejo** Ecuador
20 B3 **Port Phillip Bay** Austr.
20 A2 **Port Pirie** Austr.
64 A2 **Portree** U.K.
96 B3 **Port Renfrew** Can.
65 C1 **Portrush** U.K.
84 B1 **Port Said** Egypt
111 C3 **Port St Joe** U.S.A.
91 C3 **Port St Johns** S. Africa
91 D3 **Port Shepstone** S. Africa
67 C3 **Portsmouth** U.K.
109 E2 **Portsmouth** *NH* U.S.A.
108 C3 **Portsmouth** *OH* U.S.A.
109 D3 **Portsmouth** *VA* U.S.A.
65 C1 **Portstewart** U.K.
84 B3 **Port Sudan** Sudan
110 C3 **Port Sulphur** U.S.A.
67 B3 **Port Talbot** U.K.
74 B2 **Portugal** *country* Europe
65 B2 **Portumna** Rep. of Ireland
73 C3 **Port-Vendres** France
61 I3 **Porvoo** Fin.
33 B2 **Poryŏng** S. Korea
120 C3 **Posadas** Arg.
57 E2 **Poshekhon'ye** Rus. Fed.
60 I2 **Posio** Fin.
26 C3 **Poso** Indon.
119 D3 **Posse** Brazil
69 E2 **Pößneck** Ger.
107 C2 **Post** U.S.A.
98 C1 **Poste-de-la-Baleine** Can.
90 B2 **Postmasburg** S. Africa
77 C2 **Posušje** Bos.-Herz.
107 E1 **Poteau** U.S.A.
77 C2 **Potenza** Italy
91 C1 **Potgietersrus** S. Africa
119 D2 **Poti** *r.* Brazil
49 C1 **P'ot'i** Georgia
83 D3 **Potiskum** Nigeria
109 D3 **Potomac, South Branch** *r.* U.S.A.
120 B2 **Potosí** Bol.
32 B1 **Pototan** Phil.
106 C3 **Potrero del Llano** Mex.
69 F1 **Potsdam** Ger.
109 E2 **Potsdam** U.S.A.
109 D2 **Pottstown** U.S.A.
109 D2 **Pottsville** U.S.A.
99 E2 **Pouch Cove** Can.
109 E2 **Poughkeepsie** U.S.A.
66 B2 **Poulton-le-Fylde** U.K.
123 C2 **Pouso Alegre** Brazil
31 B2 **Poŭthĭsăt** Cambodia
71 D2 **Považská Bystrica** Slovakia
77 C2 **Povlen** *mt.* Yugo.
74 B1 **Póvoa de Varzim** Port.
104 B2 **Powell** U.S.A.
103 D3 **Powell, Lake** *resr* U.S.A.
96 B3 **Powell River** Can.
122 B1 **Poxoréu** Brazil
Poyang China *see* **Boyang**
39 B3 **Poyang Hu** *l.* China
77 D2 **Požarevac** Yugo.
113 C2 **Poza Rica** Mex.
77 C1 **Požega** Croatia
77 D2 **Požega** Yugo.
47 D2 **Pozm Tīāb** Iran
71 D1 **Poznań** Pol.
74 C2 **Pozoblanco** Spain
76 B2 **Pozzuoli** Italy
28 B2 **Prabumulih** Indon.
70 C2 **Prachatice** Czech Rep.
31 A2 **Prachuap Khiri Khan** Thai.
123 E1 **Prado** Brazil
70 C1 **Prague** Czech Rep.
Praha Czech Rep. *see* Prague
107 C2 **Prairie Dog Town Fork** *r.* U.S.A.
108 A2 **Prairie du Chien** U.S.A.
28 A1 **Prapat** Indon.
122 C1 **Prata** Brazil
73 E3 **Prato** Italy
105 D3 **Pratt** U.S.A.
29 C2 **Praya** Indon.
31 B2 **Preăh Vihear** Cambodia
57 F2 **Prechistoye** Rus. Fed.
97 D2 **Preeceville** Can.
56 C2 **Preiļi** Latvia
21 C2 **Premer** Austr.
73 C2 **Prémery** France
69 F1 **Premnitz** Ger.
70 C1 **Prenzlau** Ger.

31 A2 **Preparis Island** Cocos Is
31 A2 **Preparis North Channel** Cocos Is
31 A2 **Preparis South Channel** Cocos Is
71 D2 **Přerov** Czech Rep.
106 A2 **Prescott** U.S.A.
77 D2 **Preševo** Yugo.
120 B3 **Presidencia Roque Sáenz Peña** Arg.
119 D2 **Presidente Dutra** Brazil
122 B2 **Presidente Epitácio** Brazil
122 B2 **Presidente Prudente** Brazil
106 C3 **Presidio** U.S.A.
71 E2 **Prešov** Slovakia
109 F1 **Presque Isle** U.S.A.
69 F2 **Pressel** Ger.
66 B2 **Preston** U.K.
102 D2 **Preston** U.S.A.
64 B3 **Prestwick** U.K.
123 C1 **Preto** *r.* Brazil
91 C2 **Pretoria** S. Africa
79 B3 **Preveza** Greece
31 B2 **Prey Vêng** Cambodia
77 C2 **Priboj** Yugo.
103 D3 **Price** U.S.A.
56 B3 **Prienai** Lith.
90 B2 **Prieska** S. Africa
71 D2 **Prievidza** Slovakia
77 C2 **Prijedor** Bos.-Herz.
44 A2 **Prikaspiyskaya Nizmennost'** *lowland* Kazakh./Rus. Fed.
77 D2 **Prilep** Macedonia
59 D2 **Primorsko-Akhtarsk** Rus. Fed.
97 D2 **Primrose Lake** Can.
97 D2 **Prince Albert** Can.
90 B3 **Prince Albert** S. Africa
94 D2 **Prince Albert Peninsula** Can.
90 B3 **Prince Albert Road** S. Africa
94 C2 **Prince Alfred, Cape** Can.
90 A3 **Prince Alfred Hamlet** S. Africa
95 F2 **Prince Charles Island** Can.
23 E2 **Prince Charles Mountains** Antarctica
99 D2 **Prince Edward Island** *prov.* Can.
96 B2 **Prince George** Can.
94 E2 **Prince of Wales Island** Can.
96 A2 **Prince of Wales Island** U.S.A.
94 D2 **Prince of Wales Strait** Can.
94 D1 **Prince Patrick Island** Can.
94 E2 **Prince Regent Inlet** *sea chan.* Can.
96 A2 **Prince Rupert** Can.
19 D1 **Princess Charlotte Bay** Austr.
96 B2 **Princess Royal Island** Can.
96 B3 **Princeton** Can.
102 B2 **Prineville** U.S.A.
58 B1 **Pripet** *r.* Belarus/Ukr.
58 A1 **Pripet Marshes** Belarus/Ukr.
77 D2 **Priština** Yugo.
69 F1 **Pritzwalk** Ger.
77 C2 **Privlaka** Croatia
57 F2 **Privolzhsk** Rus. Fed.
77 D2 **Prizren** Yugo.
119 C1 **Professor van Blommestein Meer** *resr* Suriname
113 D2 **Progreso** Mex.
49 C1 **Prokhladnyy** Rus. Fed.
77 D2 **Prokuplje** Yugo.
94 C3 **Prophet** *r.* Can.
96 B2 **Prophet River** Can.
19 D2 **Proserpine** Austr.
57 E3 **Protvino** Rus. Fed.
78 C2 **Provadiya** Bulg.
109 E2 **Providence** U.S.A.
114 B3 **Providencia, Isla de** *i.* Caribbean Sea
51 N2 **Provideniya** Rus. Fed.
103 D2 **Provo** U.S.A.
97 C2 **Provost** Can.
122 B3 **Prudentópolis** Brazil
94 B2 **Prudhoe Bay** U.S.A.
68 C2 **Prüm** Ger.
73 D3 **Prunelli-di-Fiumorbo** *Corsica* France
71 E1 **Pruszków** Pol.
78 C1 **Prut** *r.* Europe
59 C1 **Pryluky** Ukr.
59 D2 **Prymors'k** Ukr.
107 D1 **Pryor** U.S.A.
Prypyats' *r.* Belarus *see* **Pripet**
71 E2 **Przemyśl** Pol.
79 C3 **Psara** *i.* Greece
49 C1 **Psebay** Rus. Fed.
59 D3 **Pshekha** *r.* Rus. Fed.
56 C2 **Pskov** Rus. Fed.
56 C2 **Pskov, Lake** Estonia/Rus. Fed.
79 B2 **Ptolemaïda** Greece
77 C1 **Ptuj** Slovenia
118 A2 **Pucallpa** Peru
39 B3 **Pucheng** China
33 B2 **Puch'ŏn** S. Korea
71 D1 **Puck** Pol.
60 I2 **Pudasjärvi** Fin.
54 C2 **Pudozh** Rus. Fed.
Puduchcheri India *see* Pondicherry
113 C3 **Puebla** Mex.
104 C3 **Pueblo** U.S.A.
121 B4 **Puelén** Arg.
74 C2 **Puente-Genil** Spain
121 A5 **Puerto Aisén** Chile
120 B2 **Puerto Alegre** Bol.
113 C3 **Puerto Ángel** Mex.
114 B4 **Puerto Armuelles** Panama
118 B1 **Puerto Ayacucho** Venez.
114 B3 **Puerto Cabezas** Nic.
118 B1 **Puerto Carreño** Col.
121 A5 **Puerto Cisnes** Chile
112 A2 **Puerto Cortés** Mex.
113 C3 **Puerto Escondido** Mex.
120 B2 **Puerto Frey** Bol.
118 B1 **Puerto Inírida** Col.
120 C2 **Puerto Isabel** Bol.
118 A2 **Puerto Leguizamo** Col.
112 A2 **Puerto Libertad** Mex.
74 C2 **Puertollano** Spain
121 B5 **Puerto Madryn** Arg.
118 B3 **Puerto Maldonado** Peru
121 A5 **Puerto Montt** Chile
121 A6 **Puerto Natáles** Chile
118 B1 **Puerto Nuevo** Col.
112 A1 **Puerto Peñasco** Mex.
120 C3 **Puerto Pinasco** Para.
115 C3 **Puerto Plata** Dom. Rep.
118 A2 **Puerto Portillo** Peru
32 A2 **Puerto Princesa** Phil.
115 D3 **Puerto Rico** *terr.* West Indies
126 B3 **Puerto Rico Trench** *sea feature* Caribbean Sea
121 B6 **Puerto Santa Cruz** Arg.
112 B2 **Puerto Vallarta** Mex.
55 D3 **Pugachev** Rus. Fed.
42 B2 **Pugal** India
22 B2 **Pukaki, Lake** N.Z.
97 D2 **Pukatawagan** Can.
33 B1 **Pukchin** N. Korea
33 B1 **Pukch'ŏng** N. Korea
22 B1 **Pukekohe** N.Z.
33 B1 **Puksubaek-san** *mt.* N. Korea
76 B2 **Pula** Croatia
76 A3 **Pula** *Sardinia* Italy
120 B3 **Pulacayo** Bol.
60 I3 **Pulkkila** Fin.
102 C1 **Pullman** U.S.A.
32 B1 **Pulog, Mount** Phil.
118 A2 **Puná, Isla** *i.* Ecuador
22 B2 **Punakaiki** N.Z.
91 D1 **Punda Maria** S. Africa
41 B3 **Pune** India
33 B1 **P'ungsan** N. Korea
89 C2 **Púnguè** *r.* Moz.
121 B4 **Punta Alta** Arg.
121 A6 **Punta Arenas** Chile
76 A2 **Punta Balestrieri** *mt.* *Sardinia* Italy
114 B3 **Puntarenas** Costa Rica
118 A1 **Punto Fijo** Venez.
60 I3 **Puolanka** Fin.
38 B3 **Puqi** China
50 G2 **Pur** *r.* Rus. Fed.
43 C3 **Puri** India
68 B1 **Purmerend** Neth.
43 C2 **Pūrnia** India
43 C2 **Puruliya** India
118 B2 **Purus** *r.* Peru
29 B2 **Purwakarta** Indon.
33 B1 **Puryŏng** N. Korea
33 B2 **Pusan** S. Korea
56 C2 **Pushkinskiye Gory** Rus. Fed.
56 C2 **Pustoshka** Rus. Fed.
30 A1 **Putao** Myanmar
39 B3 **Putian** China
Puting China *see* **De'an**
29 C2 **Puting, Tanjung** *pt* Indon.
69 F1 **Putlitz** Ger.
26 A2 **Putrajaya** Malaysia
70 C1 **Puttgarden** Ger.
118 A2 **Putumayo** *r.* Col.
29 C1 **Putusibau** Indon.
59 C1 **Putyvl'** Ukr.
95 F2 **Puvurnituq** Can.
38 B2 **Puyang** China
72 C3 **Puylaurens** France
67 A2 **Pwllheli** U.K.
31 A2 **Pyapon** Myanmar
50 G2 **Pyasina** *r.* Rus. Fed.
55 D4 **Pyatigorsk** Rus. Fed.
59 C2 **P"yatykhatky** Ukr.
30 A2 **Pyè** Myanmar
56 C3 **Pyetrykaw** Belarus
61 H3 **Pyhäjärvi** *l.* Fin.
60 H3 **Pyhäjoki** *r.* Fin.
60 I3 **Pyhäsalmi** Fin.
30 A2 **Pyinmana** Myanmar
33 B2 **Pyŏksŏng** N. Korea
33 B2 **P'yŏnggang** N. Korea
33 B2 **P'yŏngyang** N. Korea
103 C2 **Pyramid Lake** U.S.A.
75 D1 **Pyrenees** *mts* Europe
79 B3 **Pyrgos** Greece
59 C1 **Pyryatyn** Ukr.
71 C1 **Pyrzyce** Pol.
56 C2 **Pytalovo** Rus. Fed.
30 A2 **Pyu** Myanmar
79 B3 **Pyxaria** *mt.* Greece

Q

95 G1 **Qaanaaq** Greenland
91 C3 **Qacha's Nek** Lesotho
37 D2 **Qagan Nur** China
36 C2 **Qaidam Pendi** *basin* China
42 A1 **Qalāt** Afgh.
46 B2 **Qal'at Bīshah** Saudi Arabia
97 E1 **Qamanirjuaq Lake** *l.* Can.
Qamanittuaq Can. *see* Baker Lake
46 B3 **Qam Hadīl** Saudi Arabia
46 B2 **Qaryat al Ulyā** Saudi Arabia

95 H2 **Qasigiannguit** Greenland
47 D2 **Qaṣr-e-Qand** Iran
49 C2 **Qaṣr-e Shīrīn** Iran
95 H2 **Qassimiut** Greenland
46 B3 **Qa'tabah** Yemen
47 C2 **Qatar** *country* Asia
84 A2 **Qattâra Depression** Egypt
49 C1 **Qazax** Azer.
49 C1 **Qazımämmäd** Azer.
49 C2 **Qazvīn** Iran
84 B2 **Qena** Egypt
95 H2 **Qeqertarsuaq** *i.* Greenland
95 H2 **Qeqertarsuatsiaat** Greenland
95 H2 **Qeqertarsuup Tunua** *b.* Greenland
47 C2 **Qeshm** Iran
38 A3 **Qianjiang** *Chongqing* China
38 B2 **Qianjiang** *Hubei* China
33 A1 **Qian Shan** *mts* China
38 C2 **Qidong** China
36 B2 **Qiemo** China
39 A3 **Qijiang** China
36 C2 **Qijiaojing** China
95 G2 **Qikiqtarjuaq** Can.
42 A2 **Qila Ladgasht** Pak.
36 C2 **Qilian Shan** *mts* China
95 I2 **Qillak** *i.* Greenland
38 B3 **Qimen** China
95 G1 **Qimusseriarsuaq** *b.* Greenland
38 C2 **Qingdao** China
36 C2 **Qinghai Hu** *salt l.* China
36 C2 **Qinghai Nanshan** *mts* China
39 B3 **Qingyuan** *Guangdong* China
33 A1 **Qingyuan** *Liaoning* China
36 B2 **Qing Zang Gaoyuan** *plat.* China
38 B2 **Qingzhou** China
38 B2 **Qinhuangdao** China
38 A2 **Qin Ling** *mts* China
Qinting China *see* **Lianhua**
38 B2 **Qinyang** China
39 A3 **Qinzhou** China
39 B4 **Qionghai** China
36 C2 **Qionglai Shan** *mts* China
39 B4 **Qiongshan** China
37 E1 **Qiqihar** China
47 C2 **Qīr** Iran
34 B1 **Qitaihe** China
38 B2 **Qixian** China
Qogir Feng *mt.* China/Jammu and Kashmir *see* **K2**
49 D2 **Qom** Iran
49 D2 **Qomishēh** Iran
Qomolangma Feng *mt.* China/Nepal *see* **Everest, Mount**
Qoqek China *see* **Tacheng**
68 C1 **Quakenbrück** Ger.
31 B2 **Quang Ngai** Vietnam
31 B2 **Quang Tri** Vietnam
39 B3 **Quanzhou** *Fujian* China
39 B3 **Quanzhou** *Guangxi* China
76 A3 **Quartu Sant'Elena** *Sardinia* Italy
106 A2 **Quartzsite** U.S.A.
49 C1 **Quba** Azer.
21 C3 **Queanbeyan** Austr.
99 C2 **Québec** Can.
98 C1 **Québec** *prov.* Can.
69 E2 **Quedlinburg** Ger.
96 A2 **Queen Charlotte** Can.
96 A2 **Queen Charlotte Islands** Can.
96 B2 **Queen Charlotte Sound** *sea chan.* Can.
96 B2 **Queen Charlotte Strait** Can.
94 D1 **Queen Elizabeth Islands** Can.
94 E2 **Queen Maud Gulf** Can.
23 B2 **Queen Maud Land** *reg.* Antarctica
23 N1 **Queen Maud Mountains** Antarctica
20 B1 **Queensland** *state* Austr.
19 D4 **Queenstown** Austr.
22 A3 **Queenstown** N.Z.
91 C3 **Queenstown** S. Africa
89 C2 **Quelimane** Moz.
106 B2 **Quemado** U.S.A.
122 B2 **Querência do Norte** Brazil
113 B2 **Querétaro** Mex.
69 E2 **Querfurt** Ger.
96 B2 **Quesnel** Can.
96 B2 **Quesnel Lake** Can.
42 A1 **Quetta** Pak.
32 A2 **Quezon** Phil.
32 B1 **Quezon City** Phil.
88 A2 **Quibala** Angola
118 A1 **Quibdó** Col.
72 B2 **Quiberon** France
72 C3 **Quillan** France
121 C4 **Quilmes** Arg.
41 B4 **Quilon** India
19 D2 **Quilpie** Austr.
121 A4 **Quilpué** Chile
88 A1 **Quimbele** Angola
120 B3 **Quimilí** Arg.
72 B2 **Quimper** France
72 B2 **Quimperlé** France
105 E3 **Quincy** *IL* U.S.A.
109 E2 **Quincy** *MA* U.S.A.
31 B2 **Qui Nhơn** Vietnam
75 C1 **Quinto** Spain
89 D2 **Quionga** Moz.
21 D2 **Quirindi** Austr.
89 C3 **Quissico** Moz.
88 A2 **Quitapa** Angola
118 A2 **Quito** Ecuador
119 E2 **Quixadá** Brazil
39 A3 **Qujing** China
20 A2 **Quorn** Austr.
47 C2 **Qurayat** Oman
45 C3 **Qŭrghonteppa** Tajik.
84 B2 **Quseir** Egypt
39 B3 **Quzhou** China

R

71 D2 **Raab** *r.* Austria
60 H3 **Raahe** Fin.
68 C1 **Raalte** Neth.
29 C2 **Raas** *i.* Indon.
29 C2 **Raba** Indon.
82 B1 **Rabat** Morocco
46 A2 **Rābigh** Saudi Arabia
58 B2 **Râbniţa** Moldova
99 E2 **Race, Cape** Can.
110 B3 **Raceland** U.S.A.
31 B3 **Rach Gia** Vietnam
108 B2 **Racine** U.S.A.
46 B3 **Radā'** Yemen
78 C1 **Rădăuţi** Romania
108 B3 **Radcliff** U.S.A.
42 B2 **Radhanpur** India
98 C1 **Radisson** Can.
71 E1 **Radom** Pol.
71 D1 **Radomsko** Pol.
58 B1 **Radomyshl'** Ukr.
77 D2 **Radoviš** Macedonia
56 B2 **Radviliškis** Lith.
46 A2 **Raḑwá, Jabal** *mt.* Saudi Arabia
58 B1 **Radyvyliv** Ukr.
43 C2 **Rae Bareli** India
96 C1 **Rae-Edzo** Can.
96 C1 **Rae Lakes** Can.
68 C2 **Raeren** Belgium
22 C1 **Raetihi** N.Z.
120 B4 **Rafaela** Arg.
86 C2 **Rafaï** C.A.R.
46 B2 **Rafḩā'** Saudi Arabia
47 C1 **Rafsanjān** Iran
32 B2 **Ragang, Mount** *vol.* Phil.
77 B3 **Ragusa** *Sicily* Italy
27 C3 **Raha** Indon.
56 D3 **Rahachow** Belarus
42 B2 **Rahimyar Khan** Pak.
41 B3 **Raichur** India
43 C2 **Raigarh** India
96 C2 **Rainbow Lake** Can.
102 B1 **Rainier, Mount** *vol.* U.S.A.
98 A2 **Rainy Lake** Can.
97 E3 **Rainy River** Can.
43 C2 **Raipur** India
61 H3 **Raisio** Fin.
41 C3 **Rajahmundry** India
29 C1 **Rajang** *r. Sarawak* Malaysia
42 B2 **Rajanpur** Pak.
41 B4 **Rajapalaiyam** India
42 B2 **Rajasthan Canal** India
42 B2 **Rajgarh** India
42 B2 **Rajkot** India
42 B2 **Rajpur** India
43 C2 **Rajshahi** Bangl.
22 B2 **Rakaia** *r.* N.Z.
58 A2 **Rakhiv** Ukr.
59 D1 **Rakitnoye** Rus. Fed.
56 C2 **Rakke** Estonia
56 C2 **Rakvere** Estonia
111 E1 **Raleigh** U.S.A.
42 B2 **Ramgarh** India
49 C2 **Rāmhormoz** Iran
46 B3 **Ramlat Dahm** *des.* Saudi Arabia/Yemen
Ramlat Rabyānah *des.* Libya *see* **Rebiana Sand Sea**
78 C1 **Râmnicu Sărat** Romania
78 B1 **Râmnicu Vâlcea** Romania
57 E3 **Ramon'** Rus. Fed.
91 C1 **Ramotswa** Botswana
43 B2 **Rampur** India
30 A2 **Ramree Island** Myanmar
66 A1 **Ramsey** Isle of Man
67 D3 **Ramsgate** U.K.
43 C2 **Ranaghat** India
29 C1 **Ranau** *Sabah* Malaysia
121 A4 **Rancagua** Chile
43 C2 **Ranchi** India
61 F4 **Randers** Denmark
22 B2 **Rangiora** N.Z.
22 C1 **Rangitaiki** *r.* N.Z.
Rangoon Myanmar *see* **Yangôn**
43 C2 **Rangpur** Bangl.
97 E1 **Rankin Inlet** Can.
21 C2 **Rankin's Springs** Austr.
64 B2 **Rannoch Moor** *moorland* U.K.
42 B2 **Rann of Kachchh** *marsh* India
31 A3 **Ranong** Thai.
27 C3 **Ransiki** Indon.
29 C2 **Rantaupanjang** Indon.
28 A1 **Rantauprapat** Indon.
60 I2 **Ranua** Fin.
46 B2 **Ranyah, Wādī** *watercourse* Saudi Arabia
104 C2 **Rapid City** U.S.A.
56 B2 **Rapla** Estonia
42 A2 **Rapur** India
121 B5 **Rasa, Punta** *pt* Arg.
47 C2 **Ra's al Khaymah** U.A.E.
85 B3 **Ras Dashen** *mt.* Eth.
56 B2 **Raseiniai** Lith.
49 C2 **Rasht** Iran

42 A2 **Ras Koh** *mt.* Pak.
56 C2 **Rasony** Belarus
55 D3 **Rasskazovo** Rus. Fed.
47 C2 **Ras Tannūrah** Saudi Arabia
69 D1 **Rastede** Ger.
61 F3 **Rätan** Sweden
42 B2 **Ratangarh** India
31 A2 **Rat Buri** Thai.
30 A1 **Rathedaung** Myanmar
69 F1 **Rathenow** Ger.
65 C1 **Rathlin Island** U.K.
42 B2 **Ratlam** India
41 B3 **Ratnagiri** India
41 C4 **Ratnapura** Sri Lanka
58 A1 **Ratne** Ukr.
106 C1 **Raton** U.S.A.
64 D2 **Rattray Head** *hd* U.K.
69 E1 **Ratzeburg** Ger.
60 □B2 **Raufarhöfn** Iceland
22 C1 **Raukumara Range** *mts* N.Z.
61 H3 **Rauma** Fin.
29 C2 **Raung, Gunung** *vol.* Indon.
43 C2 **Raurkela** India
102 D1 **Ravalli** U.S.A.
49 C2 **Ravānsar** Iran
76 B2 **Ravenna** Italy
70 B2 **Ravensburg** Ger.
42 B1 **Ravi** *r.* Pak.
42 B1 **Rawalpindi** Pak.
71 D1 **Rawicz** Pol.
18 B3 **Rawlinna** Austr.
104 B2 **Rawlins** U.S.A.
121 B5 **Rawson** Arg.
41 C3 **Rāyagarha** India
37 E1 **Raychikhinsk** Rus. Fed.
46 B3 **Raydah** Yemen
55 E3 **Rayevskiy** Rus. Fed.
102 B1 **Raymond** U.S.A.
21 D2 **Raymond Terrace** Austr.
107 D3 **Raymondville** U.S.A.
113 C2 **Rayón** Mex.
31 B2 **Rayong** Thai.
46 A2 **Rayyis** Saudi Arabia
72 B2 **Raz, Pointe du** *pt* France
49 C2 **Razāzah, Buḩayrat ar** *l.* Iraq
78 C2 **Razgrad** Bulg.
78 C2 **Razim, Lacul** *lag.* Romania
78 B2 **Razlog** Bulg.
72 B2 **Ré, Île de** *i.* France
67 C3 **Reading** U.K.
109 D2 **Reading** U.S.A.
83 E2 **Rebiana Sand Sea** *des.* Libya
34 D1 **Rebun-tō** *i.* Japan
18 B3 **Recherche, Archipelago of the** *is* Austr.
57 D3 **Rechytsa** Belarus
119 E2 **Recife** Brazil
91 C3 **Recife, Cape** S. Africa
68 C2 **Recklinghausen** Ger.
120 C3 **Reconquista** Arg.
110 B2 **Red** *r.* U.S.A.
99 E1 **Red Bay** Can.
103 B2 **Red Bluff** U.S.A.
66 C1 **Redcar** U.K.
97 C2 **Redcliff** Can.
20 B2 **Red Cliffs** Austr.
96 C2 **Red Deer** Can.
94 D3 **Red Deer** *r.* Can.
97 D2 **Red Deer Lake** Can.
102 B2 **Redding** U.S.A.
67 C2 **Redditch** U.K.
105 D2 **Redfield** U.S.A.
98 A1 **Red Lake** Can.
105 E1 **Red Lakes** U.S.A.
102 E1 **Red Lodge** U.S.A.
102 B2 **Redmond** U.S.A.
105 D2 **Red Oak** U.S.A.
74 B2 **Redondo** Port.
127 B2 **Red Sea** Africa/Asia
96 B1 **Redstone** *r.* Can.
105 E2 **Red Wing** U.S.A.
105 D2 **Redwood Falls** U.S.A.
65 C2 **Ree, Lough** *l.* Rep. of Ireland
102 B2 **Reedsport** U.S.A.
22 B2 **Reefton** N.Z.
70 C2 **Regen** Ger.
123 E1 **Regência** Brazil
70 C2 **Regensburg** Ger.
82 C2 **Reggane** Alg.
77 C3 **Reggio di Calabria** Italy
76 B2 **Reggio nell'Emilia** Italy
78 B1 **Reghin** Romania
97 D2 **Regina** Can.
90 A1 **Rehoboth** Namibia
69 F2 **Reichenbach** Ger.
111 E1 **Reidsville** U.S.A.
67 C3 **Reigate** U.K.
73 C2 **Reims** France
69 E1 **Reinbek** Ger.
97 D2 **Reindeer** *r.* Can.
97 E2 **Reindeer Island** Can.
97 D2 **Reindeer Lake** Can.
60 F2 **Reine** Norway
68 C3 **Reinsfeld** Ger.
91 C2 **Reitz** S. Africa
90 B2 **Reivilo** S. Africa
97 D1 **Reliance** Can.
75 D2 **Relizane** Alg.
47 C2 **Remeshk** Iran
73 D2 **Remiremont** France
68 C2 **Remscheid** Ger.
70 B1 **Rendsburg** Ger.
109 D1 **Renfrew** Can.
28 B2 **Rengat** Indon.
58 B2 **Reni** Ukr.
20 B2 **Renmark** Austr.
68 D2 **Rennerod** Ger.
72 B2 **Rennes** France
97 D1 **Rennie Lake** Can.
76 B2 **Reno** *r.* Italy
103 C3 **Reno** U.S.A.
38 A3 **Renshou** China
108 B2 **Rensselaer** U.S.A.
43 C2 **Renukut** India
22 B2 **Renwick** N.Z.
26 C3 **Reo** Indon.
104 D3 **Republican** *r.* U.S.A.
95 F2 **Repulse Bay** Can.
118 A2 **Requena** Peru
75 C2 **Requena** Spain
122 B2 **Reserva** Brazil
120 C3 **Resistencia** Arg.
78 B1 **Reşiţa** Romania
94 E2 **Resolute Bay** Can.
95 G2 **Resolution Island** Can.
73 C2 **Rethel** France
79 B3 **Rethymno** Greece
81 H6 **Réunion** *terr.* Indian Ocean
75 D1 **Reus** Spain
70 B2 **Reutlingen** Ger.
96 C2 **Revelstoke** Can.
112 A3 **Revillagigedo, Islas** *is* Mex.
96 A2 **Revillagigedo Island** U.S.A.
43 C2 **Rewa** India
102 D2 **Rexburg** U.S.A.
60 □A3 **Reykjanestá** *pt* Iceland
60 □A3 **Reykjavík** Iceland
113 C2 **Reynosa** Mex.
56 C2 **Rēzekne** Latvia
Rhein *r.* Ger. *see* **Rhine**
68 C1 **Rheine** Ger.
69 F1 **Rheinsberg** Ger.
Rhin *r.* France *see* **Rhine**
68 C2 **Rhine** *r.* Europe
108 B1 **Rhinelander** U.S.A.
69 F1 **Rhinluch** *marsh* Ger.
69 F1 **Rhinow** Ger.
109 E2 **Rhode Island** *state* U.S.A.
79 C3 **Rhodes** Greece
79 C3 **Rhodes** *i.* Greece
78 B2 **Rhodope Mountains** Bulg./Greece
73 C3 **Rhône** *r.* France/Switz.
Rhuthun U.K. *see* **Ruthin**
66 B2 **Rhyl** U.K.
122 C1 **Rianópolis** Brazil
28 B1 **Riau, Kepulauan** *is* Indon.
74 B1 **Ribadeo** Spain
74 B1 **Ribadesella** Spain
122 B2 **Ribas do Rio Pardo** Brazil
89 C2 **Ribáuè** Moz.
66 B2 **Ribble** *r.* U.K.
122 C2 **Ribeira** *r.* Brazil
122 C2 **Ribeirão Preto** Brazil
72 C2 **Ribérac** France
120 B2 **Riberalta** Bol.
70 C1 **Ribnitz-Damgarten** Ger.
108 A1 **Rice Lake** U.S.A.
91 D2 **Richards Bay** S. Africa
94 B2 **Richardson Mountains** Can.
103 D3 **Richfield** U.S.A.
102 C1 **Richland** U.S.A.
108 A2 **Richland Center** U.S.A.
21 D2 **Richmond** *N.S.W.* Austr.
19 D2 **Richmond** *Qld* Austr.
22 B2 **Richmond** N.Z.
90 B3 **Richmond** S. Africa
66 C1 **Richmond** U.K.
108 C3 **Richmond** *IN* U.S.A.
108 C3 **Richmond** *KY* U.S.A.
109 D3 **Richmond** *VA* U.S.A.
98 C2 **Rideau Lakes** Can.
103 C3 **Ridgecrest** U.S.A.
70 C1 **Riesa** Ger.
68 D1 **Rieste** Ger.
91 B2 **Riet** *r.* S. Africa
69 D2 **Rietberg** Ger.
76 B2 **Rieti** Italy
104 B3 **Rifle** U.S.A.
56 B2 **Rīga** Latvia
56 B2 **Riga, Gulf of** Estonia/Latvia
47 C2 **Rīgān** Iran
102 D2 **Rigby** U.S.A.
99 E1 **Rigolet** Can.
61 H3 **Riihimäki** Fin.
76 B1 **Rijeka** Croatia
102 C2 **Riley** U.S.A.
46 B2 **Rimah, Wādī al** *watercourse* Saudi Arabia
71 E2 **Rimavská Sobota** Slovakia
96 C2 **Rimbey** Can.
76 B2 **Rimini** Italy
99 D2 **Rimouski** Can.
61 F3 **Ringebu** Norway
61 E4 **Ringkøbing** Denmark
60 G2 **Ringvassøy** *i.* Norway
69 D1 **Rinteln** Ger.
118 A2 **Riobamba** Ecuador
118 B3 **Rio Branco** Brazil
122 C3 **Rio Branco do Sul** Brazil
122 B2 **Rio Brilhante** Brazil
122 C2 **Rio Claro** Brazil
121 B4 **Río Colorado** Arg.
121 B4 **Río Cuarto** Arg.
123 D2 **Rio de Janeiro** Brazil
121 B6 **Río Gallegos** Arg.
121 B6 **Río Grande** Arg.
120 C4 **Rio Grande** Brazil
112 B2 **Río Grande** Mex.
107 D3 **Rio Grande** *r.* Mex./U.S.A.
107 D3 **Rio Grande City** U.S.A.
118 A1 **Ríohacha** Col.
118 A2 **Rioja** Peru
113 D2 **Río Lagartos** Mex.
73 C2 **Riom** France
120 B2 **Río Mulatos** Bol.
122 C3 **Rio Negro** Brazil
123 D1 **Rió Pardo de Minas** Brazil
106 B1 **Rio Rancho** U.S.A.

118 A2 **Río Tigre** Ecuador
122 B1 **Rio Verde** Brazil
113 C2 **Río Verde** Mex.
122 B1 **Rio Verde de Mato Grosso** Brazil
58 C1 **Ripky** Ukr.
66 C2 **Ripley** U.K.
75 D1 **Ripoll** Spain
66 C1 **Ripon** U.K.
34 D1 **Rishiri-tō** *i.* Japan
61 E4 **Risør** Norway
90 B2 **Ritchie** S. Africa
31 A2 **Ritchie's Archipelago** *is* India
102 C1 **Ritzville** U.S.A.
120 B3 **Rivadavia** Arg.
76 B1 **Riva del Garda** Italy
114 B3 **Rivas** Nic.
120 C4 **Rivera** Uru.
97 D2 **Riverhurst** Can.
20 B2 **Riverina** *reg.* Austr.
90 B3 **Riversdale** S. Africa
103 C4 **Riverside** U.S.A.
104 B2 **Riverton** U.S.A.
99 D2 **Riverview** Can.
72 C3 **Rivesaltes** France
99 D2 **Rivière-du-Loup** Can.
58 B1 **Rivne** Ukr.
22 B2 **Riwaka** N.Z.
46 B2 **Riyadh** Saudi Arabia
49 C1 **Rize** Turkey
38 B2 **Rizhao** China
73 C2 **Roanne** France
109 D3 **Roanoke** U.S.A.
111 E1 **Roanoke** *r.* U.S.A.
111 E1 **Roanoke Rapids** U.S.A.
20 A3 **Robe** Austr.
69 F1 **Röbel** Ger.
60 H3 **Robertsfors** Sweden
90 A3 **Robertson** S. Africa
82 A4 **Robertsport** Liberia
99 C2 **Roberval** Can.
18 A2 **Robinson Range** *hills* Austr.
20 B2 **Robinvale** Austr.
97 D2 **Roblin** Can.
96 C2 **Robson, Mount** Can.
76 B3 **Rocca Busambra** *mt. Sicily* Italy
121 C4 **Rocha** Uru.
66 B2 **Rochdale** U.K.
122 B1 **Rochedo** Brazil
68 B2 **Rochefort** Belgium
72 B2 **Rochefort** France
105 E2 **Rochester** *MN* U.S.A.
109 E2 **Rochester** *NH* U.S.A.
109 D2 **Rochester** *NY* U.S.A.
108 B2 **Rockford** U.S.A.
19 E2 **Rockhampton** Austr.
111 D2 **Rock Hill** U.S.A.
18 A3 **Rockingham** Austr.
108 A2 **Rock Island** U.S.A.
104 B1 **Rock Springs** *MT* U.S.A.
107 C3 **Rocksprings** U.S.A.
104 B2 **Rock Springs** *WY* U.S.A.
104 C3 **Rocky Ford** U.S.A.
111 E1 **Rocky Mount** U.S.A.
96 C2 **Rocky Mountain House** Can.
92 D3 **Rocky Mountains** U.S.A.
68 B3 **Rocroi** France
70 C1 **Rødbyhavn** Denmark
99 E1 **Roddickton** Can.
72 C3 **Rodez** France
57 F2 **Rodniki** Rus. Fed.
Rodos Greece *see* **Rhodes**
Rodos *i.* Greece *see* **Rhodes**
18 A2 **Roebourne** Austr.
18 B1 **Roebuck Bay** Austr.
91 C1 **Roedtan** S. Africa
68 B2 **Roermond** Neth.
68 A2 **Roeselare** Belgium
95 F2 **Roes Welcome Sound** *sea chan.* Can.
110 B1 **Rogers** U.S.A.
90 B3 **Roggeveldberge** *esc.* S. Africa
60 G2 **Rognan** Norway
102 B2 **Rogue** *r.* U.S.A.
56 B2 **Roja** Latvia
28 B1 **Rokan** *r.* Indon.
56 C2 **Rokiškis** Lith.
58 B1 **Rokytne** Ukr.
122 B2 **Rolândia** Brazil
105 E3 **Rolla** U.S.A.
19 D2 **Roma** Austr.
27 C3 **Roma** *i.* Indon.
Roma Italy *see* **Rome**
91 C2 **Roma** Lesotho
107 D3 **Roma** U.S.A.
111 E2 **Romain, Cape** U.S.A.
78 C1 **Roman** Romania
78 B1 **Romania** *country* Europe
59 C3 **Roman-Kosh** *mt.* Ukr.
37 D1 **Romanovka** Rus. Fed.
73 D2 **Rombas** France
32 B1 **Romblon** Phil.
76 B2 **Rome** Italy
111 C2 **Rome** *GA* U.S.A.
109 D2 **Rome** *NY* U.S.A.
67 D3 **Romford** U.K.
73 C2 **Romilly-sur-Seine** France
59 C1 **Romny** Ukr.
72 C2 **Romorantin-Lanthenay** France
67 C3 **Romsey** U.K.
64 □ **Ronas Hill** *hill* U.K.
119 C3 **Roncador, Serra do** *hills* Brazil
74 B2 **Ronda** Spain
122 B2 **Rondon** Brazil
122 B1 **Rondonópolis** Brazil
42 B1 **Rondu** Jammu and Kashmir
39 A3 **Rong'an** China
39 A3 **Rongjiang** China
30 A1 **Rongklang Range** *mts* Myanmar
Rongmei China *see* **Hefeng**
61 F4 **Rønne** Denmark
23 P2 **Ronne Ice Shelf** Antarctica
69 D1 **Ronnenberg** Ger.
68 A2 **Ronse** Belgium
68 B1 **Roordahuizum** Neth.
43 B2 **Roorkee** India
68 B2 **Roosendaal** Neth.
103 E2 **Roosevelt** U.S.A.
96 B2 **Roosevelt, Mount** Can.
23 K2 **Roosevelt Island** Antarctica
72 B3 **Roquefort** France
118 B1 **Roraima, Mount** Guyana
61 F3 **Røros** Norway
121 B4 **Rosario** Arg.
112 A1 **Rosario** *Baja California Norte* Mex.
112 B2 **Rosario** *Sinaloa* Mex.
112 B2 **Rosario** *Sonora* Mex.
119 C3 **Rosário Oeste** Brazil
112 A2 **Rosarito** Mex.
77 C3 **Rosarno** Italy
72 B2 **Roscoff** France
65 B2 **Roscommon** Rep. of Ireland
65 C2 **Roscrea** Rep. of Ireland
115 D3 **Roseau** Dominica
105 D1 **Roseau** U.S.A.
102 B2 **Roseburg** U.S.A.
107 D3 **Rosenberg** U.S.A.
69 D1 **Rosengarten** Ger.
70 C2 **Rosenheim** Ger.
97 D2 **Rosetown** Can.
90 A2 **Rosh Pinah** Namibia
78 C2 **Roșiori de Vede** Romania
61 F4 **Roskilde** Denmark
57 D3 **Roslavl'** Rus. Fed.
77 C3 **Rossano** Italy
65 B1 **Rossan Point** Rep. of Ireland
23 K1 **Ross Ice Shelf** Antarctica
99 D2 **Rossignol, Lake** Can.
96 C3 **Rossland** Can.
65 C2 **Rosslare** Rep. of Ireland
82 A3 **Rosso** Maur.
73 D3 **Rosso, Capo** *c. Corsica* France
67 B3 **Ross-on-Wye** U.K.
59 D1 **Rossosh'** Rus. Fed.
96 A1 **Ross River** Can.
23 K2 **Ross Sea** Antarctica
60 F2 **Røssvatnet** *l.* Norway
47 C2 **Rostāq** Iran
97 D2 **Rosthern** Can.
70 C1 **Rostock** Ger.
57 E2 **Rostov** Rus. Fed.
59 D2 **Rostov-na-Donu** Rus. Fed.
60 H2 **Rosvik** Sweden
106 C2 **Roswell** U.S.A.
27 D2 **Rota** *i.* N. Mariana Is
27 C3 **Rote** *i.* Indon.
69 D1 **Rotenburg (Wümme)** Ger.
70 C2 **Roth** Ger.
66 C1 **Rothbury** U.K.
66 C2 **Rotherham** U.K.
64 B3 **Rothesay** U.K.
21 C2 **Roto** Austr.
73 D3 **Rotondo, Monte** *mt. Corsica* France
22 C1 **Rotorua** N.Z.
22 C1 **Rotorua, Lake** N.Z.
69 E2 **Rottenbach** Ger.
70 C2 **Rottenmann** Austria
68 B2 **Rotterdam** Neth.
70 B2 **Rottweil** Ger.
73 C1 **Roubaix** France
72 C2 **Rouen** France
Roulers Belgium *see* Roeselare
21 D2 **Round Mountain** *mt.* Austr.
99 E2 **Round Pond** *l.* Can.
107 D2 **Round Rock** U.S.A.
102 E1 **Roundup** U.S.A.
64 C1 **Rousay** *i.* U.K.
98 C2 **Rouyn** Can.
60 I2 **Rovaniemi** Fin.
59 D2 **Roven'ki** Rus. Fed.
59 D2 **Roven'ky** Ukr.
76 B1 **Rovereto** Italy
31 B2 **Rôviĕng Tbong** Cambodia
76 B1 **Rovigo** Italy
76 B1 **Rovinj** Croatia
21 C1 **Rowena** Austr.
32 B2 **Roxas** *Mindanao* Phil.
32 B1 **Roxas** *Mindoro* Phil.
32 A1 **Roxas** *Palawan* Phil.
32 B1 **Roxas** *Panay* Phil.
20 A2 **Roxby Downs** Austr.
106 C1 **Roy** U.S.A.
108 B1 **Royale, Isle** *i.* U.S.A.
72 B2 **Royan** France
67 C2 **Royston** U.K.
58 C2 **Rozdil'na** Ukr.
59 C2 **Rozdol'ne** Ukr.
68 B3 **Rozoy-sur-Serre** France
55 D3 **Rtishchevo** Rus. Fed.
22 C1 **Ruapehu, Mount** *vol.* N.Z.
22 A3 **Ruapuke Island** N.Z.
46 B3 **Rub' al Khālī** *des.* Saudi Arabia
59 D2 **Rubizhne** Ukr.
45 E1 **Rubtsovsk** Rus. Fed.
103 C2 **Ruby Mountains** U.S.A.
34 C2 **Rudnaya Pristan'** Rus. Fed.
57 D3 **Rudnya** Rus. Fed.
44 C1 **Rudnyy** Kazakh.

50 E1 **Rudol'fa, Ostrov** *i.* Rus. Fed.
69 E2 **Rudolstadt** Ger.
87 D3 **Rufiji** *r.* Tanz.
121 B4 **Rufino** Arg.
89 B2 **Rufunsa** Zambia
38 C2 **Rugao** China
67 C2 **Rugby** U.K.
104 C1 **Rugby** U.S.A.
70 C1 **Rügen** *i.* Ger.
69 E2 **Ruhla** Ger.
56 B2 **Ruhnu** *i.* Estonia
68 C2 **Ruhr** *r.* Ger.
39 C3 **Rui'an** China
106 B2 **Ruidoso** U.S.A.
112 B2 **Ruiz** Mex.
87 D3 **Rukwa, Lake** Tanz.
64 A2 **Rum** *i.* U.K.
77 C1 **Ruma** Yugo.
46 B2 **Rumāh** Saudi Arabia
85 A4 **Rumbek** Sudan
73 D2 **Rumilly** France
18 C1 **Rum Jungle** Austr.
22 B2 **Runanga** N.Z.
66 B2 **Runcorn** U.K.
88 A2 **Rundu** Namibia
36 B2 **Ruoqiang** China
98 C1 **Rupert** *r.* Can.
98 C1 **Rupert Bay** Can.
89 C2 **Rusape** Zimbabwe
78 C2 **Ruse** Bulg.
45 D3 **Rushon** Tajik.
104 C2 **Rushville** U.S.A.
21 C3 **Rushworth** Austr.
97 D2 **Russell** Can.
22 B1 **Russell** N.Z.
110 C2 **Russellville** *AL* U.S.A.
110 B1 **Russellville** *AR* U.S.A.
108 B3 **Russellville** *KY* U.S.A.
69 D2 **Rüsselsheim** Ger.
50 D3 **Russian Federation** *country* Asia/Europe
91 C2 **Rustenburg** S. Africa
110 B2 **Ruston** U.S.A.
66 B2 **Ruthin** U.K.
109 E2 **Rutland** U.S.A.
43 B1 **Rutog** China
87 E4 **Ruvuma** *r.* Moz./Tanz.
47 C2 **Ruweis** U.A.E.
45 C1 **Ruzayevka** Kazakh.
55 D3 **Ruzayevka** Rus. Fed.
87 C3 **Rwanda** *country* Africa
57 E3 **Ryazan'** Rus. Fed.
57 F3 **Ryazhsk** Rus. Fed.
54 C2 **Rybachiy, Poluostrov** *pen.* Rus. Fed.
57 E2 **Rybinsk** Rus. Fed.
57 E2 **Rybinskoye Vodokhranilishche** *resr* Rus. Fed.
71 D1 **Rybnik** Pol.
57 E3 **Rybnoye** Rus. Fed.
67 D3 **Rye** U.K.
57 D3 **Ryl'sk** Rus. Fed.
35 C3 **Ryōtsu** Japan
Ryukyu Islands Japan *see* Nansei-shotō
71 E1 **Rzeszów** Pol.
57 D2 **Rzhev** Rus. Fed.

S

47 C2 **Sa'ādatābād** Iran
69 E2 **Saale** *r.* Ger.
69 E2 **Saalfeld** Ger.
70 B2 **Saarbrücken** Ger.
56 B2 **Saaremaa** *i.* Estonia
60 I2 **Saarenkylä** Fin.
61 I3 **Saarijärvi** Fin.
70 B2 **Saarlouis** Ger.
48 B2 **Sab' Ābār** Syria
75 D1 **Sabadell** Spain
29 C1 **Sabah** *state* Malaysia
29 C2 **Sabalana** *i.* Indon.
123 D1 **Sabará** Brazil
76 B2 **Sabaudia** Italy
83 D2 **Sabhā** Libya
91 D2 **Sabie** *r.* Moz./S. Africa
113 B2 **Sabinas** Mex.
113 B2 **Sabinas Hidalgo** Mex.
99 D2 **Sable, Cape** Can.
111 D3 **Sable, Cape** U.S.A.
99 E2 **Sable Island** Can.
74 B1 **Sabugal** Port.
46 B3 **Şabyā** Saudi Arabia
44 B3 **Sabzevār** Iran
88 A2 **Sachanga** Angola
98 A1 **Sachigo Lake** *l.* Can.
33 B3 **Sach'on** S. Korea
94 C2 **Sachs Harbour** Can.
103 B3 **Sacramento** U.S.A.
106 B2 **Sacramento Mountains** U.S.A.
103 B2 **Sacramento Valley** U.S.A.
91 C3 **Sada** S. Africa
75 C1 **Sádaba** Spain
46 B3 **Şa'dah** Yemen
30 A1 **Sadiya** India
35 C3 **Sadoga-shima** *i.* Japan
75 D2 **Sa Dragonera** *i.* Spain
61 F4 **Säffle** Sweden
106 B2 **Safford** U.S.A.
67 D2 **Saffron Walden** U.K.
82 B1 **Safi** Morocco
54 D2 **Safonovo** *Arkhangel'skaya Oblast'* Rus. Fed.
57 D2 **Safonovo** *Smolenskaya Oblast'* Rus. Fed.
46 B2 **Safrā' as Sark** *esc.* Saudi Arabia
43 C2 **Saga** China
35 B4 **Saga** Japan
30 A1 **Sagaing** Myanmar
35 C3 **Sagamihara** Japan
43 B2 **Sagar** India
108 C2 **Saginaw** U.S.A.
108 C2 **Saginaw Bay** U.S.A.
74 B2 **Sagres** Port.
114 B2 **Sagua la Grande** Cuba
109 F1 **Saguenay** *r.* Can.
75 C2 **Sagunto** Spain
44 B2 **Sagyndyk, Mys** *pt* Kazakh.
74 B1 **Sahagún** Spain
82 C3 **Sahara** *des.* Africa
Şaḩara el Gharbîya *des.* Egypt *see* **Western Desert**
Şaḩara el Sharqîya *des.* Egypt *see* **Eastern Desert**
Saharan Atlas *mts* Alg. *see* **Atlas Saharien**
43 B2 **Saharanpur** India
43 C2 **Saharsa** India
112 B2 **Sahuayo** Mex.
46 B2 **Şāḩūq** *reg.* Saudi Arabia
48 B2 **Saïda** Lebanon
43 C2 **Saidpur** Bangl.
Saigon Vietnam *see* **Hô Chi Minh**
30 A1 **Saiha** India
37 D2 **Saihan Tal** China
35 B4 **Saiki** Japan
61 I3 **Saimaa** *l.* Fin.
112 B2 **Sain Alto** Mex.
64 C3 **St Abb's Head** *hd* U.K.
67 C3 **St Albans** U.K.
67 B3 **St Alban's Head** *hd* U.K.
64 C2 **St Andrews** U.K.
68 B2 **St Anthonis** Neth.
99 E1 **St Anthony** Can.
102 D2 **St Anthony** U.S.A.
20 B3 **St Arnaud** Austr.
99 E1 **St-Augustin** Can.
99 E1 **St-Augustin** *r.* Can.
111 D3 **St Augustine** U.S.A.
67 A3 **St Austell** U.K.
72 C2 **St-Avertin** France
66 B1 **St Bees Head** *hd* U.K.
73 D3 **St-Bonnet-en-Champsaur** France
67 A3 **St Bride's Bay** U.K.
72 B2 **St-Brieuc** France
98 C2 **St Catharines** Can.
67 C3 **St Catherine's Point** U.K.
105 E3 **St Charles** U.S.A.
108 C2 **St Clair, Lake** Can./U.S.A.
73 D2 **St-Claude** France
105 E1 **St Cloud** U.S.A.
108 A1 **St Croix** *r.* U.S.A.
115 D3 **St Croix Island** *i.* Virgin Is (U.S.A.)
73 C2 **St-Dizier** France
99 D2 **Ste-Anne-des-Monts** Can.
73 D2 **St-Égrève** France
94 B2 **St Elias Mountains** Can.
99 D1 **Ste Marguerite** *r.* Can.
97 E2 **Ste Rose du Lac** Can.
72 B2 **Saintes** France
73 C2 **St-Étienne** France
76 A2 **St-Florent** *Corsica* France
73 C2 **St-Flour** France
104 C3 **St Francis** U.S.A.
73 D2 **St Gallen** Switz.
72 C3 **St-Gaudens** France
21 C1 **St George** Austr.
103 D3 **St George** U.S.A.
111 D3 **St George Island** U.S.A.
99 C2 **St-Georges** Can.
115 D3 **St George's** Grenada
65 C3 **St George's Channel** Rep. of Ireland/U.K.
80 C5 **St Helena** *terr.* S. Atlantic Ocean
90 A3 **St Helena Bay** S. Africa
90 A3 **St Helena Bay** *b.* S. Africa
66 B2 **St Helens** U.K.
102 B1 **St Helens, Mount** *vol.* U.S.A.
63 C4 **St Helier** Channel Is
68 B2 **St-Hubert** Belgium
109 E1 **St-Hyacinthe** Can.
108 C1 **St Ignace** U.S.A.
98 B2 **St Ignace Island** Can.
67 A3 **St Ives** U.K.
96 A2 **St James, Cape** Can.
99 C2 **St-Jean, Lac** *l.* Can.
72 B2 **St-Jean-d'Angély** France
72 B2 **St-Jean-de-Monts** France
98 C2 **St-Jean-sur-Richelieu** Can.
109 E1 **St-Jérôme** Can.
102 C1 **St Joe** *r.* U.S.A.
99 D2 **St John** Can.
109 F1 **St John** *r.* U.S.A.
115 D3 **St John's** Antigua
99 E2 **St John's** Can.
106 B2 **St Johns** U.S.A.
109 E2 **St Johnsbury** U.S.A.
105 E3 **St Joseph** U.S.A.
98 A1 **St Joseph, Lake** Can.
98 B2 **St Joseph Island** Can.
72 C2 **St-Junien** France
62 B2 **St Kilda** *i.* U.K.
115 D3 **St Kitts and Nevis** *country* West Indies
68 A2 **St-Laureins** Belgium
St-Laurent, Golfe du Can. *see* **St Lawrence, Gulf of**
119 C1 **St-Laurent-du-Maroni** Fr. Guiana
99 D2 **St Lawrence** *inlet* Can.

99 D2 **St Lawrence, Gulf of** Can.
51 N2 **St Lawrence Island** U.S.A.
72 B2 **St-Lô** France
82 A3 **St Louis** Senegal
105 E3 **St Louis** U.S.A.
105 E1 **St Louis** *r.* U.S.A.
115 D3 **St Lucia** *country* West Indies
91 D2 **St Lucia Estuary** S. Africa
115 D3 **St Maarten** *i.* West Indies
72 B2 **St-Malo** France
72 B2 **St-Malo, Golfe de** *g.* France
90 A3 **St Martin, Cape** S. Africa
109 D2 **St Marys** U.S.A.
98 C2 **St Maurice** *r.* Can.
73 D2 **St Moritz** Switz.
72 B2 **St-Nazaire** France
68 B2 **St-Niklaas** Belgium
72 C1 **St-Omer** France
97 C2 **St Paul** Can.
105 E2 **St Paul** U.S.A.
105 E2 **St Peter** U.S.A.
63 C4 **St Peter Port** Channel Is
57 D2 **St Petersburg** Rus. Fed.
111 D3 **St Petersburg** U.S.A.
99 E2 **St-Pierre** St Pierre and Miquelon
99 E2 **St Pierre and Miquelon** *terr.* N. America
72 B2 **St-Pierre-d'Oléron** France
73 C2 **St-Pourçain-sur-Sioule** France
99 D2 **St Quentin** Can.
73 C2 **St-Quentin** France
99 D2 **St Siméon** Can.
97 E2 **St Theresa Point** Can.
98 B2 **St Thomas** Can.
73 D3 **St-Tropez** France
73 D3 **St-Tropez, Cap de** *c.* France
70 C2 **St Veit an der Glan** Austria
20 A3 **St Vincent, Gulf** Austr.
115 D3 **St Vincent and the Grenadines** *i.* West Indies
68 C2 **St-Vith** Belgium
97 D2 **St Walburg** Can.
68 C3 **St Wendel** Ger.
72 C2 **St-Yrieix-la-Perche** France
27 D1 **Saipan** *i.* N. Mariana Is
120 B2 **Sajama, Nevado** *mt.* Bol.
90 B2 **Sak** *watercourse* S. Africa
35 C4 **Sakai** Japan
35 B4 **Sakaide** Japan
46 B2 **Sakākah** Saudi Arabia
104 C1 **Sakakawea, Lake** U.S.A.
79 D2 **Sakarya** Turkey
79 D2 **Sakarya** *r.* Turkey
34 C3 **Sakata** Japan
33 B1 **Sakchu** N. Korea
51 K3 **Sakhalin** *i.* Rus. Fed.
91 C2 **Sakhile** S. Africa
49 C1 **Şäki** Azer.
56 B3 **Šakiai** Lith.
37 E3 **Sakishima-shotō** *is* Japan
30 B2 **Sakon Nakhon** Thai.
90 B3 **Sakrivier** S. Africa
35 D3 **Sakura** Japan
59 C2 **Saky** Ukr.
61 G4 **Sala** Sweden
98 C2 **Salaberry-de-Valleyfield** Can.
56 B2 **Salacgrīva** Latvia
77 C2 **Sala Consilina** Italy
103 C4 **Salada, Laguna** *salt l.* Mex.
120 B4 **Salado** *r.* Arg.
113 C2 **Salado** *r.* Mex.
82 B4 **Salaga** Ghana
90 B1 **Salajwe** Botswana
83 D3 **Salal** Chad
46 A2 **Salâla** Sudan
47 C3 **Şalālah** Oman
113 B2 **Salamanca** Mex.
74 B1 **Salamanca** Spain
74 B1 **Salas** Spain
27 C3 **Salawati** *i.* Indon.
26 C3 **Salayar** *i.* Indon.
116 B4 **Sala y Gómez, Isla** *i.* S. Pacific Ocean
72 C2 **Salbris** France
56 C3 **Šalčininkai** Lith.
74 C1 **Saldaña** Spain
90 A3 **Saldanha** S. Africa
56 B2 **Saldus** Latvia
21 C3 **Sale** Austr.
54 F2 **Salekhard** Rus. Fed.
41 B3 **Salem** India
105 E3 **Salem** *MO* U.S.A.
102 B2 **Salem** *OR* U.S.A.
64 B2 **Salen** U.K.
77 B2 **Salerno** Italy
66 B2 **Salford** U.K.
119 E2 **Salgado** *r.* Brazil
71 D2 **Salgótarján** Hungary
119 E2 **Salgueiro** Brazil
104 B3 **Salida** U.S.A.
79 C3 **Salihli** Turkey
56 C3 **Salihorsk** Belarus
89 C2 **Salima** Malawi
89 C2 **Salimo** Moz.
105 D3 **Salina** U.S.A.
76 B3 **Salina, Isola** *i.* Italy
113 C3 **Salina Cruz** Mex.
123 D1 **Salinas** Brazil
112 B2 **Salinas** Mex.
103 B3 **Salinas** U.S.A.
119 D2 **Salinópolis** Brazil
67 C3 **Salisbury** U.K.
109 D3 **Salisbury** *MD* U.S.A.
111 D1 **Salisbury** *NC* U.S.A.
67 B3 **Salisbury Plain** U.K.
119 D2 **Salitre** *r.* Brazil
60 I2 **Salla** Fin.
95 F2 **Salluit** Can.
43 C2 **Sallyana** Nepal
49 C2 **Salmās** Iran
102 D1 **Salmon** U.S.A.
102 C1 **Salmon** *r.* U.S.A.
96 C2 **Salmon Arm** Can.
102 C2 **Salmon River Mountains** U.S.A.
68 C3 **Salmtal** Ger.
61 H3 **Salo** Fin.
55 D4 **Sal'sk** Rus. Fed.
90 B3 **Salt** *watercourse* S. Africa
106 A2 **Salt** *r.* U.S.A.
120 B3 **Salta** Arg.
113 B2 **Saltillo** Mex.
102 D2 **Salt Lake City** U.S.A.
122 C2 **Salto** Brazil
120 C4 **Salto** Uru.
123 E1 **Salto da Divisa** Brazil
122 B2 **Salto del Guaíra** Para.
103 C4 **Salton Sea** *salt l.* U.S.A.
111 D2 **Saluda** U.S.A.
76 A2 **Saluzzo** Italy
119 E3 **Salvador** Brazil
47 C2 **Salwah** Saudi Arabia
30 A2 **Salween** *r.* Myanmar
49 C2 **Salyan** Azer.
70 C2 **Salzburg** Austria
69 E1 **Salzgitter** Ger.
69 D2 **Salzkotten** Ger.
69 E1 **Salzwedel** Ger.
112 B1 **Samalayuca** Mex.
34 D2 **Samani** Japan
32 B1 **Samar** *i.* Phil.
55 E3 **Samara** Rus. Fed.
29 C2 **Samarinda** Indon.
45 C3 **Samarkand** Uzbek.
49 C2 **Sāmarrā'** Iraq
49 C1 **Şamaxı** Azer.
87 C3 **Šamba** Dem. Rep. Congo
29 C1 **Sambaliung** *mts* Indon.
43 C2 **Sambalpur** India
29 C2 **Sambar, Tanjung** *pt* Indon.
29 B1 **Sambas** Indon.
89 □E2 **Sambava** Madag.
58 A2 **Sambir** Ukr.
121 C4 **Samborombón, Bahía** *b.* Arg.
33 B2 **Samch'ŏk** S. Korea
49 C2 **Samdi Dag** *mt.* Turkey
87 D3 **Same** Tanz.
46 B2 **Samīrah** Saudi Arabia
33 B1 **Samjiyŏn** N. Korea
17 G4 **Samoa** *country* S. Pacific Ocean
77 C1 **Samobor** Croatia
79 C3 **Samos** *i.* Greece
79 C2 **Samothraki** Greece
79 C2 **Samothraki** *i.* Greece
29 C2 **Sampit** Indon.
87 C3 **Sampwe** Dem. Rep. Congo
107 E2 **Sam Rayburn Reservoir** U.S.A.
48 B1 **Samsun** Turkey
49 C1 **Samtredia** Georgia
31 B3 **Samui, Ko** *i.* Thai.
31 B2 **Samut Songkhram** Thai.
82 B3 **San** Mali
46 B3 **Şan'ā'** Yemen
86 A2 **Sanaga** *r.* Cameroon
49 C2 **Sanandaj** Iran
114 B3 **San Andrés, Isla de** *i.* Caribbean Sea
106 B2 **San Andres Mountains** U.S.A.
113 C3 **San Andrés Tuxtla** Mex.
107 C2 **San Angelo** U.S.A.
107 D3 **San Antonio** U.S.A.
103 C4 **San Antonio, Mount** U.S.A.
75 D2 **San Antonio Abad** Spain
120 B3 **San Antonio de los Cobres** Arg.
121 B5 **San Antonio Oeste** Arg.
76 B2 **San Benedetto del Tronto** Italy
112 A3 **San Benedicto, Isla** *i.* Mex.
103 C4 **San Bernardino** U.S.A.
103 C4 **San Bernardino Mountains** U.S.A.
111 C3 **San Blas, Cape** U.S.A.
120 B2 **San Borja** Bol.
112 B2 **San Buenaventura** Mex.
32 B1 **San Carlos** Phil.
115 D4 **San Carlos** Venez.
121 A5 **San Carlos de Bariloche** Arg.
115 C4 **San Carlos del Zulia** Venez.
103 C4 **San Clemente Island** U.S.A.
72 C2 **Sancoins** France
118 A1 **San Cristóbal** Venez.
113 C3 **San Cristóbal de las Casas** Mex.
114 C2 **Sancti Spíritus** Cuba
91 D1 **Sand** *r.* S. Africa
29 C1 **Sandakan** *Sabah* Malaysia
61 E3 **Sandane** Norway
79 B2 **Sandanski** Bulg.
64 C1 **Sanday** *i.* U.K.
107 C2 **Sanderson** U.S.A.
118 B3 **Sandia** Peru
103 C4 **San Diego** U.S.A.
48 B2 **Sandıklı** Turkey
61 E4 **Sandnes** Norway
60 F2 **Sandnessjøen** Norway
86 C3 **Sandoa** Dem. Rep. Congo
71 E1 **Sandomierz** Pol.
57 E2 **Sandovo** Rus. Fed.
30 A2 **Sandoway** Myanmar
62 B1 **Sandoy** *i.* Faroe Is
102 C1 **Sandpoint** U.S.A.
39 B3 **Sandu** China
62 B1 **Sandur** Faroe Is

108 C2 **Sandusky** U.S.A.
90 A3 **Sandveld** *mts* S. Africa
61 F4 **Sandvika** Norway
61 G3 **Sandviken** Sweden
99 E1 **Sandwich Bay** Can.
97 D2 **Sandy Bay** Can.
19 E2 **Sandy Cape** Austr.
98 A1 **Sandy Lake** Can.
98 A1 **Sandy Lake** *l.* Can.
112 A1 **San Felipe** *Baja California Norte* Mex.
113 B2 **San Felipe** *Guanajuato* Mex.
118 B1 **San Felipe** Venez.
112 A2 **San Fernando** *Baja California Norte* Mex.
113 C2 **San Fernando** *Tamaulipas* Mex.
32 B1 **San Fernando** *Luzon* Phil.
32 B1 **San Fernando** *Luzon* Phil.
74 B2 **San Fernando** Spain
115 D3 **San Fernando** Trin. and Tob.
118 B1 **San Fernando de Apure** Venez.
111 D3 **Sanford** *FL* U.S.A.
109 E2 **Sanford** *ME* U.S.A.
120 B4 **San Francisco** Arg.
103 B3 **San Francisco** U.S.A.
75 D2 **San Francisco Javier** Spain
42 B3 **Sangamner** India
51 J2 **Sangar** Rus. Fed.
76 A3 **San Gavino Monreale** *Sardinia* Italy
69 E2 **Sangerhausen** Ger.
29 C1 **Sanggau** Indon.
86 B3 **Sangha** *r.* Congo
77 C3 **San Giovanni in Fiore** Italy
32 B2 **Sangir** *i.* Indon.
27 C2 **Sangir, Kepulauan** *is* Indon.
33 B2 **Sangju** S. Korea
29 C1 **Sangkulirang** Indon.
41 B3 **Sangli** India
86 B2 **Sangmélima** Cameroon
89 C3 **Sango** Zimbabwe
73 D2 **San Gottardo, Passo del** *pass* Switz.
104 B3 **Sangre de Cristo Range** *mts* U.S.A.
43 C2 **Sangsang** China
112 A2 **San Hipólito, Punta** *pt* Mex.
112 A2 **San Ignacio** Mex.
98 C1 **Sanikiluaq** Can.
39 A3 **Sanjiang** China
103 B3 **San Joaquin** *r.* U.S.A.
121 B5 **San Jorge, Golfo de** *g.* Arg.
114 B4 **San José** Costa Rica
32 B1 **San Jose** *Luzon* Phil.
32 B1 **San Jose** *Mindoro* Phil.
103 B3 **San Jose** U.S.A.
112 A2 **San José, Isla** *i.* Mex.
112 B2 **San José de Bavicora** Mex.
32 B1 **San Jose de Buenavista** Phil.
112 A2 **San José de Comondú** Mex.
112 B2 **San José del Cabo** Mex.
118 A1 **San José del Guaviare** Col.
120 B4 **San Juan** Arg.
114 B3 **San Juan** *r.* Costa Rica/Nic.
115 D3 **San Juan** Puerto Rico
103 D3 **San Juan** *r.* U.S.A.
75 D2 **San Juan Bautista** Spain
113 C3 **San Juan Bautista Tuxtepec** Mex.
102 B1 **San Juan Islands** U.S.A.
112 B2 **San Juanito** Mex.
104 B3 **San Juan Mountains** U.S.A.
121 B5 **San Julián** Arg.
43 C2 **Sankh** *r.* India
Sankt-Peterburg Rus. Fed. *see* **St Petersburg**
48 B2 **Şanlıurfa** Turkey
106 B3 **San Lorenzo** Mex.
74 B2 **Sanlúcar de Barrameda** Spain
112 B2 **San Lucas** Mex.
121 B4 **San Luis** Arg.
113 B2 **San Luis de la Paz** Mex.
106 A2 **San Luisito** Mex.
103 B3 **San Luis Obispo** U.S.A.
113 B2 **San Luis Potosí** Mex.
112 A1 **San Luis Río Colorado** Mex.
107 D3 **San Marcos** U.S.A.
76 B2 **San Marino** *country* Europe
76 B2 **San Marino** San Marino
112 B2 **San Martin de Bolaños** Mex.
121 A5 **San Martín de los Andes** Arg.
121 B5 **San Matías, Golfo** *g.* Arg.
38 B2 **Sanmenxia** China
114 B3 **San Miguel** El Salvador
120 B3 **San Miguel de Tucumán** Arg.
113 C3 **San Miguel Sola de Vega** Mex.
39 B3 **Sanming** China
121 B4 **San Nicolás de los Arroyos** Arg.
103 C4 **San Nicolas Island** U.S.A.
91 C2 **Sannieshof** S. Africa
71 E2 **Sanok** Pol.
32 B1 **San Pablo** Phil.
112 B2 **San Pablo Balleza** Mex.
120 B3 **San Pedro** Arg.
120 B2 **San Pedro** Bol.
82 B4 **San-Pédro** Côte d'Ivoire
112 A2 **San Pedro** Mex.
106 A2 **San Pedro** *watercourse* U.S.A.
74 B2 **San Pedro, Sierra de** *mts* Spain
112 B2 **San Pedro de las Colonias** Mex.
114 B3 **San Pedro Sula** Hond.
76 A3 **San Pietro, Isola di** *i.* *Sardinia* Italy
112 A1 **San Quintín, Cabo** *c.* Mex.
121 B4 **San Rafael** Arg.
76 A2 **San Remo** Italy
114 B3 **San Salvador** El Salvador
120 B3 **San Salvador de Jujuy** Arg.
77 C2 **San Severo** Italy
77 C2 **Sanski Most** Bos.-Herz.
120 B2 **Santa Ana** Bol.
114 B3 **Santa Ana** El Salvador
112 A1 **Santa Ana** Mex.
103 C4 **Santa Ana** U.S.A.
112 B2 **Santa Bárbara** Mex.
103 C4 **Santa Barbara** U.S.A.
122 B2 **Santa Bárbara, Serra de** *hills* Brazil
120 B3 **Santa Catalina** Chile
118 B2 **Santa Clara** Col.
114 C2 **Santa Clara** Cuba
103 C4 **Santa Clarita** U.S.A.
77 C3 **Santa Croce, Capo** *c.* *Sicily* Italy
121 B6 **Santa Cruz** *r.* Arg.
120 B2 **Santa Cruz** Bol.
32 B1 **Santa Cruz** Phil.
103 B3 **Santa Cruz** U.S.A.
113 C3 **Santa Cruz Barillas** Guat.
123 E1 **Santa Cruz Cabrália** Brazil
75 C2 **Santa Cruz de Moya** Spain
82 A2 **Santa Cruz de Tenerife** Canary Is
103 C4 **Santa Cruz Island** U.S.A.
120 B4 **Santa Fé** Arg.
106 B1 **Santa Fe** U.S.A.
122 B1 **Santa Helena de Goiás** Brazil
121 B4 **Santa Isabel** Arg.
122 B1 **Santa Luisa, Serra de** *hills* Brazil
120 C3 **Santa Maria** Brazil
112 B1 **Santa María** *r.* Mex.
103 B4 **Santa Maria** U.S.A.
91 D2 **Santa Maria, Cabo de** *c.* Moz.
74 B2 **Santa Maria, Cabo de** *c.* Port.
119 D2 **Santa Maria das Barreiras** Brazil
77 C3 **Santa Maria di Leuca, Capo** *c.* Italy
118 A1 **Santa Marta** Col.
103 C4 **Santa Monica** U.S.A.
119 D3 **Santana** Brazil
74 C1 **Santander** Spain
76 A3 **Sant'Antioco** *Sardinia* Italy
76 A3 **Sant'Antioco, Isola di** *i.* *Sardinia* Italy
119 C2 **Santarém** Brazil
74 B2 **Santarém** Port.
122 B1 **Santa Rita do Araguaia** Brazil
121 B4 **Santa Rosa** Arg.
120 C3 **Santa Rosa** Brazil
103 B3 **Santa Rosa** *CA* U.S.A.
106 C2 **Santa Rosa** *NM* U.S.A.
114 B3 **Santa Rosa de Copán** Hond.
103 B4 **Santa Rosa Island** U.S.A.
112 A2 **Santa Rosalía** Mex.
120 C3 **Santiago** Brazil
121 A4 **Santiago** Chile
115 C3 **Santiago** Dom. Rep.
112 B2 **Santiago** Mex.
114 B4 **Santiago** Panama
32 B1 **Santiago** Phil.
74 B1 **Santiago de Compostela** Spain
112 B2 **Santiago Ixcuintla** Mex.
112 B2 **Santiago Papasquiaro** Mex.
75 D1 **Sant Jordi, Golf de** *g.* Spain
123 D2 **Santo Amaro de Campos** Brazil
123 C2 **Santo André** Brazil
120 C3 **Santo Angelo** Brazil
122 B2 **Santo Antônio da Platina** Brazil
119 E3 **Santo Antônio de Jesus** Brazil
118 B2 **Santo Antônio do Içá** Brazil
123 C2 **Santo Antônio do Monte** Brazil
115 D3 **Santo Domingo** Dom. Rep.
106 B1 **Santo Domingo Pueblo** U.S.A.
79 C3 **Santorini** *i.* Greece
123 C2 **Santos** Brazil
120 C3 **Santo Tomé** Arg.
121 A5 **San Valentín, Cerro** *mt.* Chile
114 B3 **San Vicente** El Salvador
112 A1 **San Vicente** Mex.
118 A3 **San Vicente de Cañete** Peru
76 B2 **San Vincenzo** Italy
76 B3 **San Vito, Capo** *c.* *Sicily* Italy
39 A4 **Sanya** China
123 C2 **São Bernardo do Campo** Brazil
120 C3 **São Borja** Brazil
122 C2 **São Carlos** Brazil
119 C3 **São Félix** *Mato Grosso* Brazil
119 C2 **São Félix** *Pará* Brazil
123 D2 **São Fidélis** Brazil
123 D1 **São Francisco** Brazil
119 E3 **São Francisco** *r.* Brazil
122 C3 **São Francisco, Ilha de** *i.* Brazil

122 C3 **São Francisco do Sul** Brazil
120 C4 **São Gabriel** Brazil
123 D2 **São Gonçalo** Brazil
123 C1 **São Gotardo** Brazil
123 D2 **São João da Barra** Brazil
123 C2 **São João da Boa Vista** Brazil
74 B1 **São João da Madeira** Port.
123 D1 **São João do Paraíso** Brazil
123 D2 **São João Nepomuceno** Brazil
122 C2 **São Joaquim da Barra** Brazil
122 C2 **São José do Rio Preto** Brazil
123 C2 **São José dos Campos** Brazil
122 C3 **São José dos Pinhais** Brazil
123 C2 **São Lourenço** Brazil
119 D2 **São Luís** Brazil
122 C2 **São Manuel** Brazil
122 C1 **São Marcos** *r.* Brazil
119 D2 **São Marcos, Baía de** *b.* Brazil
123 E1 **São Mateus** Brazil
123 C2 **São Paulo** Brazil
119 D2 **São Raimundo Nonato** Brazil
123 C1 **São Romão** Brazil
123 C2 **São Sebastião, Ilha do** *i.* Brazil
122 C2 **São Sebastião do Paraíso** Brazil
122 B1 **São Simão** Brazil
122 B1 **São Simão, Barragem de** *resr* Brazil
27 C2 **Sao-Siu** Indon.
123 D2 **São Tomé, Cabo de** *c.* Brazil
80 D3 **São Tomé and Príncipe** *country* Africa
123 C2 **São Vicente** Brazil
74 B2 **São Vicente, Cabo de** *c.* Port.
27 C3 **Saparua** Indon.
57 F3 **Sapozhok** Rus. Fed.
34 D2 **Sapporo** Japan
77 C2 **Sapri** Italy
49 C2 **Saqqez** Iran
49 C2 **Sarāb** Iran
31 B2 **Sara Buri** Thai.
77 C2 **Sarajevo** Bos.-Herz.
44 B1 **Saraktash** Rus. Fed.
109 E2 **Saranac Lake** U.S.A.
77 D3 **Sarandë** Albania
32 B2 **Sarangani Islands** Phil.
55 D3 **Saransk** Rus. Fed.
55 E3 **Sarapul** Rus. Fed.
111 D3 **Sarasota** U.S.A.
58 B2 **Sarata** Ukr.
104 B2 **Saratoga** U.S.A.
109 E2 **Saratoga Springs** U.S.A.
29 C1 **Saratok** *Sarawak* Malaysia
55 D3 **Saratov** Rus. Fed.
47 D2 **Saravan** Iran
31 B2 **Saravan** Laos
29 C1 **Sarawak** *state* Malaysia
78 C2 **Saray** Turkey
79 C3 **Sarayköy** Turkey
47 D2 **Sarbāz** Iran
42 B2 **Sardarshahr** India
Sardegna *i.* Italy *see* Sardinia
76 A2 **Sardinia** *i.* Italy
60 G2 **Sarektjåkkå** *mt.* Sweden
45 C3 **Sar-e Pol** Afgh.
126 B3 **Sargasso Sea** Atlantic Ocean
42 B1 **Sargodha** Pak.
83 D4 **Sarh** Chad
47 D2 **Sarhad** *reg.* Iran
49 D2 **Sārī** Iran
79 C3 **Sarıgöl** Turkey
49 C1 **Sarıkamış** Turkey
29 C1 **Sarikei** *Sarawak* Malaysia
19 D2 **Sarina** Austr.
83 D2 **Sarīr Tibesti** *des.* Libya
33 B2 **Sariwŏn** N. Korea
79 C2 **Sarıyer** Turkey
45 D2 **Sarkand** Kazakh.
79 C2 **Şarköy** Turkey
27 D3 **Sarmi** Indon.
108 C2 **Sarnia** Can.
58 B1 **Sarny** Ukr.
28 B2 **Sarolangun** Indon.
79 B3 **Saronikos Kolpos** *g.* Greece
79 C2 **Saros Körfezi** *b.* Turkey
55 D3 **Sarova** Rus. Fed.
73 D2 **Sarrebourg** France
74 B1 **Sarria** Spain
75 C1 **Sarrión** Spain
73 D3 **Sartène** *Corsica* France
71 D2 **Sárvár** Hungary
44 B2 **Sarykamyshkoye Ozero** *salt l.* Turkm./Uzbek.
45 D2 **Saryozek** Kazakh.
45 D2 **Saryshagan** Kazakh.
50 F3 **Sarysu** *watercourse* Kazakh.
45 D3 **Sary-Tash** Kyrg.
43 C2 **Sasaram** India
35 A4 **Sasebo** Japan
97 D2 **Saskatchewan** *prov.* Can.
97 D2 **Saskatchewan** *r.* Can.
97 D2 **Saskatoon** Can.
51 I2 **Saskylakh** Rus. Fed.
91 C2 **Sasolburg** S. Africa
55 D3 **Sasovo** Rus. Fed.
82 B4 **Sassandra** Côte d'Ivoire
76 A2 **Sassari** *Sardinia* Italy
70 C1 **Sassnitz** Ger.
82 A3 **Satadougou** Mali
91 D1 **Satara** S. Africa
43 C2 **Satna** India
42 B2 **Satpura Range** *mts* India
31 B2 **Sattahip** Thai.
78 B1 **Satu Mare** Romania
31 B3 **Satun** Thai.
112 B2 **Saucillo** Mex.
61 E4 **Sauda** Norway
60 □B2 **Sauðárkrókur** Iceland
46 B2 **Saudi Arabia** *country* Asia
73 C3 **Saugues** France
73 C2 **Saulieu** France
98 B2 **Sault Ste Marie** Can.
108 C1 **Sault Ste Marie** U.S.A.
45 C1 **Saumalkol'** Kazakh.
27 C3 **Saumlakki** Indon.
72 B2 **Saumur** France
88 B1 **Saurimo** Angola
77 D2 **Sava** *r.* Europe
59 E1 **Savala** *r.* Rus. Fed.
111 D2 **Savannah** *GA* U.S.A.
110 C1 **Savannah** *TN* U.S.A.
111 D2 **Savannah** *r.* U.S.A.
31 B2 **Savannakhét** Laos
98 A1 **Savant Lake** *l.* Can.
79 C3 **Savaştepe** Turkey
57 F2 **Savino** Rus. Fed.
54 D2 **Savinskiy** Rus. Fed.
76 A2 **Savona** Italy
61 I3 **Savonlinna** Fin.
60 I2 **Savukoski** Fin.
42 B2 **Sawai Madhopur** India
30 A2 **Sawankhalok** Thai.
104 B3 **Sawatch Range** *mts* U.S.A.
21 D2 **Sawtell** Austr.
27 C3 **Sawu Sea** Indon.
47 C3 **Sayḥūt** Yemen
37 D2 **Saynshand** Mongolia
109 D2 **Sayre** U.S.A.
112 B3 **Sayula** *Jalisco* Mex.
113 C3 **Sayula** *Veracruz* Mex.
96 B2 **Sayward** Can.
57 E2 **Sazonovo** Rus. Fed.
82 B2 **Sbaa** Alg.
66 B1 **Scafell Pike** *hill* U.K.
77 C3 **Scalea** Italy
64 C1 **Scapa Flow** *inlet* U.K.
98 C2 **Scarborough** Can.
115 D3 **Scarborough** Trin. and Tob.
66 C1 **Scarborough** U.K.
32 A1 **Scarborough Shoal** *sea feature* Phil.
64 A2 **Scarinish** U.K.
Scarpanto *i.* Greece *see* Karpathos
73 D2 **Schaffhausen** Switz.
68 B1 **Schagen** Neth.
70 C2 **Schärding** Austria
68 A2 **Scharendijke** Neth.
69 D1 **Scharhörn** *sea feature* Ger.
69 D1 **Scheeßel** Ger.
99 D1 **Schefferville** Can.
103 D3 **Schell Creek Range** *mts* U.S.A.
109 E2 **Schenectady** U.S.A.
69 E3 **Scheßlitz** Ger.
68 C1 **Schiermonnikoog** *i.* Neth.
76 B1 **Schio** Italy
69 F2 **Schkeuditz** Ger.
69 E1 **Schladen** Ger.
69 E2 **Schleiz** Ger.
70 B1 **Schleswig** Ger.
69 D2 **Schloß Holte-Stukenbrock** Ger.
69 D2 **Schlüchtern** Ger.
69 E3 **Schlüsselfeld** Ger.
69 D2 **Schmallenberg** Ger.
69 D1 **Schneverdingen** Ger.
69 E1 **Schönebeck (Elbe)** Ger.
69 E1 **Schöningen** Ger.
68 B2 **Schoonhoven** Neth.
27 D3 **Schouten Islands** P.N.G.
70 C2 **Schwandorf** Ger.
29 C2 **Schwaner, Pegunungan** *mts* Indon.
69 E1 **Schwarzenbek** Ger.
69 F2 **Schwarzenberg** Ger.
90 A2 **Schwarzrand** *mts* Namibia
70 C2 **Schwaz** Austria
70 C1 **Schwedt an der Oder** Ger.
69 E2 **Schweinfurt** Ger.
69 E1 **Schwerin** Ger.
69 E1 **Schweriner See** *l.* Ger.
73 D2 **Schwyz** Switz.
76 B3 **Sciacca** *Sicily* Italy
63 B4 **Scilly, Isles of** U.K.
108 C3 **Scioto** *r.* U.S.A.
104 B1 **Scobey** U.S.A.
21 D2 **Scone** Austr.
23 R3 **Scotia Ridge** *sea feature* S. Atlantic Ocean
126 C8 **Scotia Sea** S. Atlantic Ocean
64 C2 **Scotland** *admin. div.* U.K.
96 B2 **Scott, Cape** Can.
91 D3 **Scottburgh** S. Africa
104 C3 **Scott City** U.S.A.
104 C2 **Scottsbluff** U.S.A.
110 C2 **Scottsboro** U.S.A.
64 B1 **Scourie** U.K.
109 D2 **Scranton** U.S.A.
66 C2 **Scunthorpe** U.K.
73 E2 **Scuol** Switz.
97 E2 **Seal** *r.* Can.
90 B3 **Seal, Cape** S. Africa
20 B3 **Sea Lake** Austr.
107 D3 **Sealy** U.S.A.
110 B1 **Searcy** U.S.A.
66 B1 **Seascale** U.K.
102 B1 **Seattle** U.S.A.
109 E2 **Sebago Lake** U.S.A.
112 A2 **Sebastián Vizcaíno, Bahía** *b.* Mex.
78 B1 **Sebeş** Romania

28 B2 **Sebesi** *i.* Indon.
56 C2 **Sebezh** Rus. Fed.
48 B1 **Şebinkarahisar** Turkey
111 D3 **Sebring** U.S.A.
96 B3 **Sechelt** Can.
118 A2 **Sechura** Peru
41 B3 **Secunderabad** India
105 E3 **Sedalia** U.S.A.
73 C2 **Sedan** France
22 B2 **Seddon** N.Z.
106 A2 **Sedona** U.S.A.
69 E2 **Seeburg** Ger.
69 E1 **Seehausen (Altmark)** Ger.
90 A2 **Seeheim** Namibia
72 C2 **Sées** France
69 E2 **Seesen** Ger.
69 E1 **Seevetal** Ger.
82 A4 **Sefadu** Sierra Leone
91 C1 **Sefare** Botswana
28 B1 **Segamat** Malaysia
54 C2 **Segezha** Rus. Fed.
82 B3 **Ségou** Mali
74 C1 **Segovia** Spain
83 D2 **Séguédine** Niger
82 B4 **Séguéla** Côte d'Ivoire
107 D3 **Seguin** U.S.A.
75 C2 **Segura** *r.* Spain
74 C2 **Segura, Sierra de** *mts* Spain
88 B3 **Sehithwa** Botswana
61 H3 **Seinäjoki** Fin.
72 C2 **Seine** *r.* France
72 B2 **Seine, Baie de** *b.* France
71 E1 **Sejny** Pol.
28 B2 **Sekayu** Indon.
82 B4 **Sekondi** Ghana
27 C3 **Selaru** *i.* Indon.
29 C2 **Selatan, Tanjung** *pt* Indon.
94 A2 **Selawik** U.S.A.
66 C2 **Selby** U.K.
88 B3 **Selebi-Phikwe** Botswana
73 D2 **Sélestat** France
60 □A3 **Selfoss** Iceland
82 A3 **Sélibabi** Maur.
106 A1 **Seligman** U.S.A.
84 A2 **Selîma Oasis** Sudan
79 C3 **Selimiye** Turkey
82 B3 **Sélingué, Lac de** *l.* Mali
57 D2 **Selizharovo** Rus. Fed.
61 E4 **Seljord** Norway
97 E2 **Selkirk** Can.
64 C3 **Selkirk** U.K.
96 C2 **Selkirk Mountains** Can.
110 C2 **Selma** *AL* U.S.A.
103 C3 **Selma** *CA* U.S.A.
57 D3 **Sel'tso** Rus. Fed.
118 A2 **Selvas** *reg.* Brazil
102 C1 **Selway** *r.* U.S.A.
97 D1 **Selwyn Lake** Can.
96 A1 **Selwyn Mountains** Can.
19 C2 **Selwyn Range** *hills* Austr.
28 B2 **Semangka, Teluk** *b.* Indon.
29 C2 **Semarang** Indon.
29 B1 **Sematan** *Sarawak* Malaysia
86 B2 **Sembé** Congo
49 C2 **Şemdinli** Turkey
59 C1 **Semenivka** Ukr.
54 D3 **Semenov** Rus. Fed.
29 C2 **Semeru, Gunung** *vol.* Indon.
57 E3 **Semiluki** Rus. Fed.
104 B2 **Seminoe Reservoir** U.S.A.
107 C2 **Seminole** U.S.A.
111 D2 **Seminole, Lake** U.S.A.
45 E1 **Semipalatinsk** Kazakh.
29 C1 **Semitau** Indon.
49 D2 **Semnān** Iran
29 C1 **Semporna** *Sabah* Malaysia
73 C2 **Semur-en-Auxois** France
118 B2 **Sena Madureira** Brazil
88 B2 **Senanga** Zambia
35 B4 **Sendai** *Kagoshima* Japan
35 D3 **Sendai** *Miyagi* Japan
82 A3 **Senegal** *country* Africa
82 A3 **Sénégal** *r.* Maur./Senegal
70 C1 **Senftenberg** Ger.
87 D3 **Sengerema** Tanz.
119 D3 **Senhor do Bonfim** Brazil
76 B2 **Senigallia** Italy
77 B2 **Senj** Croatia
60 G2 **Senja** *i.* Norway
90 B2 **Senlac** S. Africa
72 C2 **Senlis** France
31 B2 **Senmonorom** Cambodia
85 B3 **Sennar** Sudan
98 C2 **Senneterre** Can.
91 C3 **Senqu** *r.* Lesotho
73 C2 **Sens** France
77 D1 **Senta** Yugo.
96 B2 **Sentinel Peak** Can.
43 B2 **Seoni** India
33 B2 **Seoul** S. Korea
123 D2 **Sepetiba, Baía de** *b.* Brazil
27 D3 **Sepik** *r.* P.N.G.
29 C1 **Sepinang** Indon.
99 D1 **Sept-Îles** Can.
27 C3 **Seram** *i.* Indon.
27 C3 **Seram Sea** Indon.
28 B2 **Serang** Indon.
77 D2 **Serbia** *aut. rep.* Yugo.
57 E3 **Serebryanyye Prudy** Rus. Fed.
28 B1 **Seremban** Malaysia
87 D3 **Serengeti Plain** Tanz.
55 D3 **Sergach** Rus. Fed.
57 E2 **Sergiyev Posad** Rus. Fed.
29 C1 **Seria** Brunei
29 C1 **Serian** *Sarawak* Malaysia
79 B3 **Serifos** *i.* Greece
48 B2 **Serik** Turkey
27 C3 **Sermata, Kepulauan** *is* Indon.
54 F3 **Serov** Rus. Fed.
88 B3 **Serowe** Botswana
74 B2 **Serpa** Port.
57 E3 **Serpukhov** Rus. Fed.
122 B1 **Serranópolis** Brazil
68 A3 **Serre** *r.* France
79 B2 **Serres** Greece
119 E3 **Serrinha** Brazil
123 D1 **Sêrro** Brazil
122 C2 **Sertãozinho** Brazil
29 C2 **Seruyan** *r.* Indon.
36 C2 **Sêrxu** China
88 A2 **Sesfontein** Namibia
76 B2 **Sessa Aurunca** Italy
75 D2 **Ses Salines, Cap de** *c.* Spain
76 A2 **Sestri Levante** Italy
73 C3 **Sète** France
123 D1 **Sete Lagoas** Brazil
60 G2 **Setermoen** Norway
61 E4 **Setesdal** *val.* Norway
83 C1 **Sétif** Alg.
35 B4 **Seto-naikai** *sea* Japan
82 B1 **Settat** Morocco
66 B1 **Settle** U.K.
74 B2 **Setúbal** Port.
74 B2 **Setúbal, Baía de** *b.* Port.
98 A1 **Seul, Lac** *l.* Can.
44 A2 **Sevan, Lake** Armenia
Sevana Lich *l.* Armenia *see* Sevan, Lake
59 C3 **Sevastopol'** Ukr.
73 C3 **Sévérac-le-Château** France
98 B1 **Severn** *r.* Can.
90 B2 **Severn** S. Africa
67 B3 **Severn** *r.* U.K.
54 D2 **Severnaya Dvina** *r.* Rus. Fed.
51 H1 **Severnaya Zemlya** *is* Rus. Fed.
54 D2 **Severnyy** *Nenetskiy Avtonomnyy Okrug* Rus. Fed.
54 F2 **Severnyy** *Respublika Komi* Rus. Fed.
54 C2 **Severodvinsk** Rus. Fed.
54 C2 **Severomorsk** Rus. Fed.
59 D3 **Severskaya** Rus. Fed.
103 D3 **Sevier** *r.* U.S.A.
103 D3 **Sevier Lake** U.S.A.
Sevilla Spain *see* **Seville**
74 B2 **Seville** Spain
96 A2 **Sewell Inlet** Can.
96 C2 **Sexsmith** Can.
112 B2 **Sextín** *r.* Mex.
54 G1 **Seyakha** Rus. Fed.
81 H4 **Seychelles** *country* Indian Ocean
60 □C2 **Seyðisfjörður** Iceland
59 C1 **Seym** *r.* Rus. Fed./Ukr.
51 L2 **Seymchan** Rus. Fed.
21 C3 **Seymour** Austr.
108 B3 **Seymour** *IN* U.S.A.
107 D2 **Seymour** *TX* U.S.A.
73 C2 **Sézanne** France
79 B3 **Sfakia** Greece
78 C1 **Sfântu Gheorghe** Romania
83 D1 **Sfax** Tunisia
's-Gravenhage Neth. *see* **The Hague**
64 A2 **Sgurr Alasdair** *hill* U.K.
38 A2 **Shaanxi** *prov.* China
59 D2 **Shabel'sk** Rus. Fed.
45 D3 **Shache** China
23 R1 **Shackleton Range** *mts* Antarctica
67 B3 **Shaftesbury** U.K.
43 C2 **Shahdol** India
45 C3 **Shah Fuladi** *mt.* Afgh.
43 B2 **Shahjahanpur** India
49 D2 **Shahr-e Kord** Iran
57 E2 **Shakhovskaya** Rus. Fed.
45 C3 **Shakhrisabz** Uzbek.
59 E2 **Shakhty** Rus. Fed.
54 D3 **Shakhun'ya** Rus. Fed.
34 D2 **Shakotan-misaki** *c.* Japan
36 C2 **Shaluli Shan** *mts* China
97 E2 **Shamattawa** Can.
107 C1 **Shamrock** U.S.A.
36 C2 **Shandan** China
38 B2 **Shandong** *prov.* China
38 C2 **Shandong Bandao** *pen.* China
89 B2 **Shangani** *r.* Zimbabwe
38 C2 **Shanghai** China
38 C2 **Shanghai** *mun.* China
39 B3 **Shanghang** China
38 B2 **Shangqiu** China
39 B3 **Shangrao** China
38 B2 **Shangshui** China
38 C2 **Shangyu** China
37 E1 **Shangzhi** China
38 A2 **Shangzhou** China
65 B2 **Shannon** *r.* Rep. of Ireland
65 B2 **Shannon, Mouth of the** Rep. of Ireland
39 B3 **Shantou** China
38 B2 **Shanxi** *prov.* China
39 B3 **Shaoguan** China
39 B3 **Shaowu** China
38 C2 **Shaoxing** China
39 B3 **Shaoyang** China
84 C2 **Shaqrā'** Saudi Arabia
58 B2 **Sharhorod** Ukr.
47 C2 **Sharjah** U.A.E.
56 C2 **Sharkawshchyna** Belarus
18 A2 **Shark Bay** Austr.
46 A2 **Sharm el Sheikh** Egypt
108 C2 **Sharon** U.S.A.
54 D3 **Shar'ya** Rus. Fed.
89 B3 **Shashe** *r.* Botswana/Zimbabwe
85 B4 **Shashemenē** Eth.

102 B2 **Shasta, Mount** *vol.* U.S.A.
102 B2 **Shasta Lake** U.S.A.
57 E2 **Shatura** Rus. Fed.
97 D3 **Shaunavon** Can.
108 B2 **Shawano** U.S.A.
99 C2 **Shawinigan** Can.
107 D1 **Shawnee** U.S.A.
18 B2 **Shay Gap** Austr.
57 E3 **Shchekino** Rus. Fed.
57 E2 **Shchelkovo** Rus. Fed.
57 E3 **Shchigry** Rus. Fed.
59 C1 **Shchors** Ukr.
56 B3 **Shchuchyn** Belarus
59 D1 **Shebekino** Rus. Fed.
45 C3 **Sheberghān** Afgh.
108 B2 **Sheboygan** U.S.A.
59 D3 **Shebsh** *r.* Rus. Fed.
65 C2 **Sheelin, Lough** *l.* Rep. of Ireland
66 C2 **Sheffield** U.K.
57 E2 **Sheksna** Rus. Fed.
57 E2 **Sheksninskoye Vodokhranilishche** *resr* Rus. Fed.
51 M2 **Shelagskiy, Mys** *pt* Rus. Fed.
99 D2 **Shelburne** Can.
102 D1 **Shelby** U.S.A.
108 B3 **Shelbyville** *IN* U.S.A.
110 C1 **Shelbyville** *TN* U.S.A.
51 L2 **Shelikhova, Zaliv** *g.* Rus. Fed.
97 D2 **Shellbrook** Can.
102 B1 **Shelton** U.S.A.
105 D2 **Shenandoah** U.S.A.
109 D3 **Shenandoah** *r.* U.S.A.
109 D3 **Shenandoah Mountains** U.S.A.
86 A2 **Shendam** Nigeria
54 D2 **Shenkursk** Rus. Fed.
33 A1 **Shenyang** China
39 B3 **Shenzhen** China
58 B1 **Shepetivka** Ukr.
21 C3 **Shepparton** Austr.
67 D3 **Sheppey, Isle of** *i.* U.K.
99 D2 **Sherbrooke** *N.S.* Can.
99 C2 **Sherbrooke** *Que.* Can.
84 B3 **Shereiq** Sudan
104 B2 **Sheridan** U.S.A.
107 D2 **Sherman** U.S.A.
68 B2 **'s-Hertogenbosch** Neth.
64 □ **Shetland Islands** *is* U.K.
44 B2 **Shetpe** Kazakh.
Shevchenko Kazakh. *see* **Aktau**
105 D1 **Sheyenne** *r.* U.S.A.
47 B3 **Shibām** Yemen
34 D2 **Shibetsu** Japan
39 B3 **Shicheng** China
64 B2 **Shiel, Loch** *l.* U.K.
45 E2 **Shihezi** China
Shijiao China *see* **Fogang**
38 B2 **Shijiazhuang** China
42 A2 **Shikarpur** Pak.
35 B4 **Shikoku** *i.* Japan
34 D2 **Shikotsu-ko** *l.* Japan
54 D2 **Shilega** Rus. Fed.
43 C2 **Shiliguri** India
43 D2 **Shillong** India
57 F3 **Shilovo** Rus. Fed.
85 C3 **Shimbiris** *mt.* Somalia
35 C3 **Shimizu** Japan
41 B3 **Shimoga** India
35 B4 **Shimonoseki** Japan
64 B1 **Shin, Loch** *l.* U.K.
35 C4 **Shingū** Japan
91 D1 **Shingwedzi** S. Africa
91 D1 **Shingwedzi** *r.* S. Africa
87 D3 **Shinyanga** Tanz.
35 C4 **Shiono-misaki** *c.* Japan
106 B1 **Shiprock** U.S.A.
39 A3 **Shiqian** China
38 A2 **Shiquan** China
Shiquanhe China *see* **Gar**
Shiquan He *r.* China/Pak. *see* **Indus**
35 C3 **Shirane-san** *mt.* Japan
49 D3 **Shīrāz** Iran
34 D2 **Shiretoko-misaki** *c.* Japan
34 D2 **Shiriya-zaki** *c.* Japan
42 B2 **Shiv** India
43 B2 **Shivpuri** India
38 B2 **Shiyan** China
38 A2 **Shizuishan** China
35 C4 **Shizuoka** Japan
57 D3 **Shklow** Belarus
77 C2 **Shkodër** Albania
51 H1 **Shmidta, Ostrov** *i.* Rus. Fed.
103 C3 **Shoshone** U.S.A.
103 C3 **Shoshone Mountains** *mts* U.S.A.
91 C1 **Shoshong** Botswana
59 C1 **Shostka** Ukr.
38 B2 **Shouxian** China
106 A2 **Show Low** U.S.A.
59 C2 **Shpola** Ukr.
110 B2 **Shreveport** U.S.A.
67 B2 **Shrewsbury** U.K.
30 A1 **Shuangjiang** China
55 E4 **Shubarkuduk** Kazakh.
84 B1 **Shubrâ el Kheima** Egypt
57 D2 **Shugozero** Rus. Fed.
88 B2 **Shumba** Zimbabwe
78 C2 **Shumen** Bulg.
56 C2 **Shumilina** Belarus
57 D3 **Shumyachi** Rus. Fed.
94 A2 **Shungnak** U.S.A.
46 B3 **Shuqrah** Yemen
57 F2 **Shushkodom** Rus. Fed.
49 C2 **Shushtar** Iran
96 C2 **Shuswap Lake** Can.
57 F2 **Shuya** Rus. Fed.
57 F2 **Shuyskoye** Rus. Fed.
30 A1 **Shwebo** Myanmar
30 A1 **Shwedwin** Myanmar
30 A2 **Shwegyin** Myanmar
45 C2 **Shymkent** Kazakh.
59 C2 **Shyroke** Ukr.
27 C3 **Sia** Indon.
42 A2 **Siahan Range** *mts* Pak.
42 B1 **Sialkot** Pak.
32 B2 **Siargao** *i.* Phil.
56 B2 **Šiauliai** Lith.
77 C2 **Šibenik** Croatia
28 A2 **Siberut** *i.* Indon.
42 A2 **Sibi** Pak.
78 B1 **Sibiu** Romania
28 A1 **Sibolga** Indon.
30 A1 **Sibsagar** India
29 C1 **Sibu** *Sarawak* Malaysia
86 B2 **Sibut** C.A.R.
32 B1 **Sibuyan** *i.* Phil.
32 B1 **Sibuyan Sea** Phil.
96 C2 **Sicamous** Can.
31 A3 **Sichon** Thai.
38 A2 **Sichuan** *prov.* China
38 A3 **Sichuan Pendi** *basin* China
73 D3 **Sicié, Cap** *c.* France
Sicilia *i.* Italy *see* **Sicily**
76 B3 **Sicilian Channel** Italy/Tunisia
76 B3 **Sicily** *i.* Italy
118 A3 **Sicuani** Peru
42 B2 **Siddhapur** India
79 C3 **Sideros, Akra** *pt* Greece
75 D2 **Sidi Aïssa** Alg.
75 D2 **Sidi Ali** Alg.
82 B1 **Sidi Bel Abbès** Alg.
82 A2 **Sidi Ifni** Morocco
28 A1 **Sidikalang** Indon.
64 C2 **Sidlaw Hills** U.K.
67 B3 **Sidmouth** U.K.
102 B1 **Sidney** Can.
104 C1 **Sidney** *MT* U.S.A.
104 C2 **Sidney** *NE* U.S.A.
108 C2 **Sidney** *OH* U.S.A.
111 D2 **Sidney Lanier, Lake** U.S.A.
122 B2 **Sidrolândia** Brazil
71 E1 **Siedlce** Pol.
68 C2 **Sieg** *r.* Ger.
68 D2 **Siegen** Ger.
31 B2 **Siĕmréab** Cambodia
76 B2 **Siena** Italy
71 D1 **Sieradz** Pol.
121 B5 **Sierra Grande** Arg.
82 A4 **Sierra Leone** *country* Africa
112 B2 **Sierra Mojada** Mex.
106 A2 **Sierra Vista** U.S.A.
73 D2 **Sierre** Switz.
79 B3 **Sifnos** *i.* Greece
75 C2 **Sig** Alg.
95 H2 **Sigguup Nunaa** *pen.* Greenland
71 E2 **Sighetu Marmației** Romania
78 B1 **Sighișoara** Romania
28 A1 **Sigli** Indon.
60 □B2 **Siglufjörður** Iceland
70 B2 **Sigmaringen** Ger.
68 B3 **Signy-l'Abbaye** France
74 C1 **Sigüenza** Spain
82 B3 **Siguiri** Guinea
56 B2 **Sigulda** Latvia
31 B2 **Sihanoukville** Cambodia
60 I3 **Siilinjärvi** Fin.
49 C2 **Siirt** Turkey
28 B2 **Sijunjung** Indon.
42 B2 **Sikar** India
82 B3 **Sikasso** Mali
105 F3 **Sikeston** U.S.A.
34 B2 **Sikhote-Alin'** *mts* Rus. Fed.
79 C3 **Sikinos** *i.* Greece
112 B2 **Silao** Mex.
43 D2 **Silchar** India
45 D1 **Siletiteniz, Ozero** *salt l.* Kazakh.
43 C2 **Silgarhi** Nepal
48 B2 **Silifke** Turkey
43 C1 **Siling Co** *salt l.* China
78 C2 **Silistra** Bulg.
48 A1 **Silivri** Turkey
61 F3 **Siljan** *l.* Sweden
61 E4 **Silkeborg** Denmark
56 C2 **Sillamäe** Estonia
110 B1 **Siloam Springs** U.S.A.
91 D2 **Silobela** S. Africa
56 B2 **Šilutė** Lith.
49 C2 **Silvan** Turkey
106 B2 **Silver City** U.S.A.
104 B3 **Silverton** U.S.A.
30 B1 **Simao** China
98 C2 **Simàrd, Lac** *l.* Can.
79 C3 **Simav** Turkey
79 C3 **Simav Dağları** *mts* Turkey
86 C2 **Simba** Dem. Rep. Congo
109 D2 **Simcoe, Lake** Can.
28 A1 **Simeuluë** *i.* Indon.
59 C3 **Simferopol'** Ukr.
78 B1 **Șimleu Silvaniei** Romania
68 C3 **Simmern (Hunsrück)** Ger.
97 D2 **Simonhouse** Can.
19 C2 **Simpson Desert** Austr.
61 F4 **Simrishamn** Sweden
28 A1 **Sinabang** Indon.
84 B2 **Sinai** *pen.* Egypt
39 A3 **Sinan** China
33 B2 **Sinanju** N. Korea
30 A1 **Sinbo** Myanmar
118 A1 **Sincelejo** Col.
79 C3 **Sındırgı** Turkey
54 E2 **Sindor** Rus. Fed.
79 C2 **Sinekçi** Turkey
74 B2 **Sines** Port.

74 B2 **Sines, Cabo de** *c.* Port.
43 C2 **Singahi** India
28 B1 **Singapore** *country* Asia
29 C2 **Singaraja** Indon.
87 D3 **Singida** Tanz.
30 A1 **Singkaling Hkamti** Myanmar
29 B1 **Singkawang** Indon.
28 A1 **Singkil** Indon.
21 D2 **Singleton** Austr.
30 A1 **Singu** Myanmar
76 A2 **Siniscola** *Sardinia* Italy
77 C2 **Sinj** Croatia
84 B3 **Sinkat** Sudan
Sinkiang *aut. reg.* China *see* **Xinjiang Uygur Zizhiqu**
48 B1 **Sinop** Turkey
33 B1 **Sinp'o** N. Korea
29 C1 **Sintang** Indon.
107 D3 **Sinton** U.S.A.
33 A1 **Sinŭiju** N. Korea
71 D2 **Siófok** Hungary
73 D2 **Sion** Switz.
105 D2 **Sioux Center** U.S.A.
105 D2 **Sioux City** U.S.A.
105 D2 **Sioux Falls** U.S.A.
98 A1 **Sioux Lookout** Can.
33 A1 **Siping** China
97 E2 **Sipiwesk Lake** Can.
28 A2 **Sipura** *i.* Indon.
61 E4 **Sira** *r.* Norway
Siracusa Italy *see* **Syracuse**
19 C1 **Sir Edward Pellew Group** *is* Austr.
84 B1 **Sirhān, Wādī an** *watercourse* Saudi Arabia
47 C2 **Sīrīk** Iran
30 B2 **Siri Kit Dam** Thai.
96 B1 **Sir James MacBrien, Mount** Can.
47 C2 **Sīrjān** Iran
49 C2 **Şırnak** Turkey
42 B2 **Sirohi** India
42 B2 **Sirsa** India
83 D1 **Sirte** Libya
83 D1 **Sirte, Gulf of** Libya
56 B2 **Širvintos** Lith.
113 C2 **Sisal** Mex.
90 B2 **Sishen** S. Africa
49 C2 **Sisian** Armenia
97 D2 **Sisipuk Lake** Can.
31 B2 **Sisŏphŏn** Cambodia
73 D3 **Sisteron** France
43 C2 **Sitapur** India
91 D2 **Siteki** Swaziland
96 A2 **Sitka** U.S.A.
30 A2 **Sittang** *r.* Myanmar
68 B2 **Sittard** Neth.
30 A1 **Sittwe** Myanmar
29 C2 **Situbondo** Indon.
48 B2 **Sivas** Turkey
79 C3 **Sivaslı** Turkey
48 B2 **Siverek** Turkey
48 B2 **Sivrihisar** Turkey
84 A2 **Siwa** Egypt
42 B1 **Siwalik Range** *mts* India/Nepal
73 D3 **Six-Fours-les-Plages** France
38 B2 **Sixian** China
91 C2 **Siyabuswa** S. Africa
61 F4 **Sjælland** *i.* Denmark
77 D2 **Sjenica** Yugo.
60 G2 **Sjøvegan** Norway
59 C2 **Skadovs'k** Ukr.
61 F4 **Skagen** Denmark
61 E4 **Skagerrak** *str.* Denmark/Norway
102 B1 **Skagit** *r.* U.S.A.
96 A2 **Skagway** U.S.A.
42 B1 **Skardu** Jammu and Kashmir
71 E1 **Skarżysko-Kamienna** Pol.
71 D2 **Skawina** Pol.
96 B2 **Skeena** *r.* Can.
96 B2 **Skeena Mountains** Can.
66 D2 **Skegness** U.K.
60 H3 **Skellefteå** Sweden
60 H3 **Skellefteälven** *r.* Sweden
65 C2 **Skerries** Rep. of Ireland
79 B3 **Skiathos** *i.* Greece
65 B3 **Skibbereen** Rep. of Ireland
60 □B2 **Skíðadals-jökull** *glacier* Iceland
66 B1 **Skiddaw** *hill* U.K.
61 E4 **Skien** Norway
71 E1 **Skierniewice** Pol.
83 C1 **Skikda** Alg.
20 B3 **Skipton** Austr.
66 B2 **Skipton** U.K.
61 E4 **Skive** Denmark
60 H1 **Skjervøy** Norway
79 B3 **Skopelos** *i.* Greece
57 E3 **Skopin** Rus. Fed.
77 D2 **Skopje** Macedonia
61 F4 **Skövde** Sweden
109 F2 **Skowhegan** U.S.A.
56 B2 **Skrunda** Latvia
96 A1 **Skukum, Mount** Can.
91 D1 **Skukuza** S. Africa
56 B2 **Skuodas** Lith.
64 A2 **Skye** *i.* U.K.
79 B3 **Skyros** Greece
79 B3 **Skyros** *i.* Greece
61 F4 **Slagelse** Denmark
65 C2 **Slaney** *r.* Rep. of Ireland
56 C2 **Slantsy** Rus. Fed.
77 C1 **Slatina** Croatia
78 B2 **Slatina** Romania
97 C1 **Slave** *r.* Can.
82 C4 **Slave Coast** Africa
96 C2 **Slave Lake** Can.
50 G3 **Slavgorod** Rus. Fed.
77 C1 **Slavonski Brod** Croatia
58 B1 **Slavuta** Ukr.
58 C1 **Slavutych** Ukr.
59 D2 **Slavyansk-na-Kubani** Rus. Fed.
57 D3 **Slawharad** Belarus
71 D1 **Sławno** Pol.
65 A2 **Slea Head** *hd* Rep. of Ireland
98 C1 **Sleeper Islands** Can.
65 D1 **Slieve Donard** *hill* U.K.
65 B1 **Slieve Gamph** *hills* Rep. of Ireland
65 B1 **Sligo** Rep. of Ireland
65 B1 **Sligo Bay** Rep. of Ireland
61 G4 **Slite** Sweden
78 C2 **Sliven** Bulg.
78 C2 **Slobozia** Romania
96 C3 **Slocan** Can.
56 C3 **Slonim** Belarus
68 B1 **Sloten** Neth.
67 C3 **Slough** U.K.
71 D2 **Slovakia** *country* Europe
76 B1 **Slovenia** *country* Europe
59 D2 **Slov"yans'k** Ukr.
58 B1 **Sluch** *r.* Ukr.
71 D1 **Słupsk** Pol.
56 C3 **Slutsk** Belarus
65 A2 **Slyne Head** *hd* Rep. of Ireland
99 D1 **Smallwood Reservoir** Can.
56 C3 **Smalyavichy** Belarus
56 C3 **Smarhon'** Belarus
97 D2 **Smeaton** Can.
77 D2 **Smederevo** Yugo.
77 D2 **Smederevska Palanka** Yugo.
59 C2 **Smila** Ukr.
56 C2 **Smiltene** Latvia
96 B2 **Smithers** Can.
111 E1 **Smithfield** U.S.A.
109 D3 **Smith Mountain Lake** U.S.A.
98 C2 **Smiths Falls** Can.
105 D3 **Smoky Hills** U.S.A.
60 E3 **Smøla** *i.* Norway
57 D3 **Smolensk** Rus. Fed.
78 B2 **Smolyan** Bulg.
98 B2 **Smooth Rock Falls** Can.
Smyrna Turkey *see* **İzmir**
60 □B3 **Snæfell** *mt.* Iceland
66 A1 **Snaefell** *hill* Isle of Man
100 B2 **Snake** *r.* U.S.A.
102 D2 **Snake River Plain** U.S.A.
96 C1 **Snare Lake** Can.
68 B1 **Sneek** Neth.
65 B3 **Sneem** Rep. of Ireland
90 B3 **Sneeuberge** *mts* S. Africa
76 B1 **Snežnik** *mt.* Slovenia
59 C2 **Snihurivka** Ukr.
61 E3 **Snøhetta** *mt.* Norway
97 D1 **Snowbird Lake** *l.* Can.
66 A2 **Snowdon** *mt.* U.K.
Snowdrift Can. *see* **Łutselk'e**
97 C1 **Snowdrift** *r.* Can.
106 A2 **Snowflake** U.S.A.
97 D2 **Snow Lake** Can.
20 A2 **Snowtown** Austr.
21 C3 **Snowy** *r.* Austr.
21 C3 **Snowy Mountains** Austr.
107 C2 **Snyder** U.S.A.
89 □D2 **Soalala** Madag.
85 B4 **Sobat** *r.* Sudan
68 C3 **Sobernheim** Ger.
57 F2 **Sobinka** Rus. Fed.
119 D3 **Sobradinho, Barragem de** *resr* Brazil
119 D2 **Sobral** Brazil
55 C4 **Sochi** Rus. Fed.
33 B2 **Sŏch'ŏn** S. Korea
118 A1 **Socorro** Col.
106 B2 **Socorro** U.S.A.
112 A3 **Socorro, Isla** *i.* Mex.
24 B4 **Socotra** *i.* Yemen
31 B3 **Soc Trăng** Vietnam
74 C2 **Socuéllamos** Spain
60 I2 **Sodankylä** Fin.
102 D2 **Soda Springs** U.S.A.
61 G3 **Söderhamn** Sweden
61 G4 **Södertälje** Sweden
84 A3 **Sodiri** Sudan
85 B4 **Sodo** Eth.
61 G3 **Södra Kvarken** *str.* Fin./Sweden
68 D2 **Soest** Ger.
78 B2 **Sofia** Bulg.
Sofiya Bulg. *see* **Sofia**
36 C2 **Sog** China
61 E3 **Sognefjorden** *inlet* Norway
79 D2 **Söğüt** Turkey
84 B2 **Sohâg** Egypt
68 B2 **Soignies** Belgium
73 C2 **Soissons** France
58 A1 **Sokal'** Ukr.
79 C3 **Söke** Turkey
49 C1 **Sokhumi** Georgia
82 C4 **Sokodé** Togo
57 F2 **Sokol** Rus. Fed.
69 F2 **Sokolov** Czech Rep.
83 C3 **Sokoto** Nigeria
83 C3 **Sokoto** *r.* Nigeria
58 B2 **Sokyryany** Ukr.
41 B3 **Solāpur** India
103 B3 **Soledad** U.S.A.
57 F2 **Soligalich** Rus. Fed.
67 C2 **Solihull** U.K.
54 E3 **Solikamsk** Rus. Fed.
55 E3 **Sol'-Iletsk** Rus. Fed.
68 C2 **Solingen** Ger.
60 G3 **Sollefteå** Sweden
61 G4 **Sollentuna** Sweden
69 D2 **Solling** *hills* Ger.

57 E2 **Solnechnogorsk** Rus. Fed.
28 B2 **Solok** Indon.
17 E3 **Solomon Islands** *country* S. Pacific Ocean
73 D2 **Solothurn** Switz.
69 D1 **Soltau** Ger.
57 D2 **Sol'tsy** Rus. Fed.
64 C3 **Solway Firth** *est.* U.K.
88 B2 **Solwezi** Zambia
79 C3 **Soma** Turkey
85 C4 **Somalia** *country* Africa
88 B1 **Sombo** Angola
77 C1 **Sombor** Yugo.
112 B2 **Sombrerete** Mex.
108 C3 **Somerset** U.S.A.
91 C3 **Somerset East** S. Africa
94 E2 **Somerset Island** Can.
90 A3 **Somerset West** S. Africa
69 E2 **Sömmerda** Ger.
43 C2 **Son** *r.* India
61 E4 **Sønderborg** Denmark
69 E2 **Sondershausen** Ger.
Søndre Strømfjord *inlet* Greenland *see* Kangerlussuaq
30 B1 **Sông Da, Hô** *resr* Vietnam
87 D4 **Songea** Tanz.
33 B1 **Sŏnggan** N. Korea
33 B1 **Songhua Hu** *resr* China
Sŏngjin N. Korea *see* Kimch'aek
31 B3 **Songkhla** Thai.
37 E1 **Songling** China
33 B2 **Sŏngnam** S. Korea
33 B2 **Songnim** N. Korea
88 A1 **Songo** Angola
89 C2 **Songo** Moz.
37 E1 **Songyuan** China
Sonid Youqi China *see* Saihan Tal
57 E2 **Sonkovo** Rus. Fed.
30 B1 **Son La** Vietnam
42 A2 **Sonmiani** Pak.
69 E2 **Sonneberg** Ger.
112 A2 **Sonora** *r.* Mex.
103 B3 **Sonora** *CA* U.S.A.
107 C2 **Sonora** *TX* U.S.A.
106 A2 **Sonoyta** Mex.
85 A4 **Sopo** *watercourse* Sudan
71 D2 **Sopron** Hungary
76 B2 **Sora** Italy
98 C2 **Sorel** Can.
19 D4 **Sorell** Austr.
74 C1 **Soria** Spain
44 B2 **Sor Mertvyy Kultuk** *dry lake* Kazakh.
58 B2 **Soroca** Moldova
122 C2 **Sorocaba** Brazil
55 E3 **Sorochinsk** Rus. Fed.
27 D2 **Sorol** *atoll* Micronesia
27 C3 **Sorong** Indon.
87 D2 **Soroti** Uganda
60 H1 **Sørøya** *i.* Norway
60 G2 **Sorsele** Sweden
32 B1 **Sorsogon** Phil.
54 C2 **Sortavala** Rus. Fed.
60 G2 **Sortland** Norway
33 B2 **Sŏsan** S. Korea
91 C2 **Soshanguve** S. Africa
57 E3 **Sosna** *r.* Rus. Fed.
121 B4 **Sosneado** *mt.* Arg.
54 E2 **Sosnogorsk** Rus. Fed.
54 D2 **Sosnovka** Rus. Fed.
56 C2 **Sosnovyy Bor** Rus. Fed.
71 D1 **Sosnowiec** Pol.
59 D2 **Sosyka** *r.* Rus. Fed.
113 C2 **Soto la Marina** Mex.
86 B2 **Souanké** Congo
72 C3 **Souillac** France
Sŏul S. Korea *see* **Seoul**
72 B2 **Soulac-sur-Mer** France
72 B3 **Soulom** France
48 B2 **Soûr** Lebanon
75 D2 **Sour el Ghozlane** Alg.
97 D3 **Souris** *Man.* Can.
99 D2 **Souris** *P.E.I.* Can.
97 E3 **Souris** *r.* Can.
119 E2 **Sousa** Brazil
83 D1 **Sousse** Tunisia
72 B3 **Soustons** France
90 B3 **South Africa, Republic of** *country* Africa
67 C3 **Southampton** U.K.
95 F2 **Southampton Island** Can.
41 D3 **South Andaman** *i.* India
20 A1 **South Australia** *state* Austr.
98 B2 **South Baymouth** Can.
108 B2 **South Bend** U.S.A.
111 D2 **South Carolina** *state* U.S.A.
124 B4 **South China Sea** N. Pacific Ocean
104 C2 **South Dakota** *state* U.S.A.
67 C3 **South Downs** *hills* U.K.
97 D2 **Southend** Can.
67 D3 **Southend-on-Sea** U.K.
22 B2 **Southern Alps** *mts* N.Z.
18 A3 **Southern Cross** Austr.
97 E2 **Southern Indian Lake** Can.
127 **Southern Ocean**
111 E1 **Southern Pines** U.S.A.
64 B3 **Southern Uplands** *hills* U.K.
117 F6 **South Georgia and South Sandwich Islands** *terr.* S. Atlantic Ocean
64 A2 **South Harris** *pen.* U.K.
97 E1 **South Henik Lake** *l.* Can.
22 B2 **South Island** N.Z.
103 B3 **South Lake Tahoe** U.S.A.
96 B1 **South Nahanni** *r.* Can.
23 Q4 **South Orkney Islands** S. Atlantic Ocean
104 C2 **South Platte** *r.* U.S.A.
66 B2 **Southport** U.K.
111 E2 **Southport** U.S.A.
64 C1 **South Ronaldsay** *i.* U.K.
91 D3 **South Sand Bluff** *pt* S. Africa
97 D2 **South Saskatchewan** *r.* Can.
97 E2 **South Seal** *r.* Can.
23 P3 **South Shetland Islands** Antarctica
66 C1 **South Shields** U.K.
22 B1 **South Taranaki Bight** *b.* N.Z.
98 C1 **South Twin Island** Can.
64 A2 **South Uist** *i.* U.K.
21 D2 **South West Rocks** Austr.
67 D2 **Southwold** U.K.
77 C3 **Soverato** Italy
56 B2 **Sovetsk** Rus. Fed.
54 F2 **Sovetskiy** Rus. Fed.
59 C2 **Sovyets'kyy** Ukr.
91 C2 **Soweto** S. Africa
34 D1 **Sōya-misaki** *c.* Japan
58 C1 **Sozh** *r.* Europe
78 C2 **Sozopol** Bulg.
74 C1 **Spain** *country* Europe
67 C2 **Spalding** U.K.
103 D2 **Spanish Fork** U.S.A.
65 B2 **Spanish Point** Rep. of Ireland
76 B3 **Sparagio, Monte** *mt. Sicily* Italy
103 C3 **Sparks** U.S.A.
108 A2 **Sparta** U.S.A.
111 D2 **Spartanburg** U.S.A.
79 B3 **Sparti** Greece
77 C3 **Spartivento, Capo** *c.* Italy
57 D3 **Spas-Demensk** Rus. Fed.
57 F2 **Spas-Klepiki** Rus. Fed.
34 B2 **Spassk-Dal'niy** Rus. Fed.
57 F3 **Spassk-Ryazanskiy** Rus. Fed.
79 B3 **Spatha, Akra** *pt* Greece
104 C2 **Spearfish** U.S.A.
107 C1 **Spearman** U.S.A.
105 D2 **Spencer** U.S.A.
20 A2 **Spencer Gulf** *est.* Austr.
66 C1 **Spennymoor** U.K.
22 B2 **Spenser Mountains** N.Z.
69 D3 **Spessart** *reg.* Ger.
64 C2 **Spey** *r.* U.K.
70 B2 **Speyer** Ger.
68 C1 **Spiekeroog** *i.* Ger.
68 B2 **Spijkenisse** Neth.
96 C2 **Spirit River** Can.
71 E2 **Spišská Nová Ves** Slovakia
50 C1 **Spitsbergen** *i.* Svalbard
70 C2 **Spittal an der Drau** Austria
77 C2 **Split** Croatia
97 E2 **Split Lake** Can.
97 E2 **Split Lake** *l.* Can.
102 C1 **Spokane** U.S.A.
108 A1 **Spooner** U.S.A.
70 C1 **Spree** *r.* Ger.
90 A2 **Springbok** S. Africa
99 E2 **Springdale** Can.
110 B1 **Springdale** U.S.A.
69 D1 **Springe** Ger.
106 C1 **Springer** U.S.A.
106 B2 **Springerville** U.S.A.
104 C3 **Springfield** *CO* U.S.A.
108 B3 **Springfield** *IL* U.S.A.
109 E2 **Springfield** *MA* U.S.A.
105 E3 **Springfield** *MO* U.S.A.
108 C3 **Springfield** *OH* U.S.A.
102 B2 **Springfield** *OR* U.S.A.
91 C3 **Springfontein** S. Africa
99 D2 **Springhill** Can.
111 D3 **Spring Hill** U.S.A.
22 B2 **Springs Junction** N.Z.
66 D2 **Spurn Head** *hd* U.K.
96 B3 **Squamish** Can.
77 C3 **Squillace, Golfo di** *g.* Italy
Srbija *aut. rep.* Yugo. *see* Serbia
78 C2 **Sredets** Bulg.
51 L3 **Sredinnyy Khrebet** *mts* Rus. Fed.
51 L2 **Srednekolymsk** Rus. Fed.
57 E3 **Sredne-Russkaya Vozvyshennost'** *hills* Rus. Fed.
51 I2 **Sredne-Sibirskoye Ploskogor'ye** *plat.* Rus. Fed.
78 B2 **Srednogorie** Bulg.
37 D1 **Sretensk** Rus. Fed.
29 C1 **Sri Aman** *Sarawak* Malaysia
41 B4 **Sri Jayewardenepura Kotte** Sri Lanka
41 C3 **Srikakulam** India
41 C4 **Sri Lanka** *country* Asia
42 B1 **Srinagar** Jammu and Kashmir
41 B3 **Srivardhan** India
69 D1 **Stade** Ger.
69 E1 **Stadensen** Ger.
68 C1 **Stadskanaal** Neth.
69 D2 **Stadtallendorf** Ger.
69 D1 **Stadthagen** Ger.
69 E2 **Staffelstein** Ger.
67 B2 **Stafford** U.K.
67 C3 **Staines** U.K.
59 D2 **Stakhanov** Ukr.
Stalingrad Rus. Fed. *see* Volgograd
71 E1 **Stalowa Wola** Pol.
67 C2 **Stamford** U.K.
109 E2 **Stamford** *CT* U.S.A.
107 D2 **Stamford** *TX* U.S.A.
90 A1 **Stampriet** Namibia

60 F2 **Stamsund** Norway
91 C2 **Standerton** S. Africa
108 C2 **Standish** U.S.A.
91 D2 **Stanger** S. Africa
121 C6 **Stanley** Falkland Is
104 C1 **Stanley** U.S.A.
51 I3 **Stanovoye Nagor'ye** *mts* Rus. Fed.
51 J3 **Stanovoy Khrebet** *mts* Rus. Fed.
21 D1 **Stanthorpe** Austr.
71 E1 **Starachowice** Pol.
Stara Planina *mts* Bulg./Yugo. *see* **Balkan Mountains**
57 D2 **Staraya Russa** Rus. Fed.
57 D2 **Staraya Toropa** Rus. Fed.
78 C2 **Stara Zagora** Bulg.
71 D1 **Stargard Szczeciński** Pol.
110 C2 **Starkville** U.S.A.
59 D2 **Starobil's'k** Ukr.
57 D3 **Starodub** Rus. Fed.
71 D1 **Starogard Gdański** Pol.
58 B2 **Starokostyantyniv** Ukr.
59 D2 **Starominskaya** Rus. Fed.
59 D2 **Staroshcherbinovskaya** Rus. Fed.
59 D2 **Starotitarovskaya** Rus. Fed.
57 F3 **Staroyur'yevo** Rus. Fed.
67 B3 **Start Point** U.K.
56 C3 **Staryya Darohi** Belarus
57 E3 **Staryy Oskol** Rus. Fed.
69 E2 **Staßfurt** Ger.
109 D2 **State College** U.S.A.
111 D2 **Statesboro** U.S.A.
111 D1 **Statesville** U.S.A.
61 E4 **Stavanger** Norway
55 D4 **Stavropol'** Rus. Fed.
55 D4 **Stavropol'skaya Vozvyshennost'** *hills* Rus. Fed.
20 B3 **Stawell** Austr.
91 C2 **Steadville** S. Africa
104 B2 **Steamboat Springs** U.S.A.
69 E2 **Stedten** Ger.
96 C2 **Steen River** Can.
102 C2 **Steens Mountain** U.S.A.
68 C1 **Steenwijk** Neth.
94 D2 **Stefansson Island** Can.
69 E3 **Steigerwald** *mts* Ger.
97 E3 **Steinbach** Can.
68 C1 **Steinfurt** Ger.
88 A3 **Steinhausen** Namibia
60 F3 **Steinkjer** Norway
90 A2 **Steinkopf** S. Africa
90 B2 **Stella** S. Africa
73 D2 **Stenay** France
69 E1 **Stendal** Ger.
Stepanakert Azer. *see* **Xankändi**
99 E2 **Stephenville** Can.
107 D2 **Stephenville** U.S.A.
90 B3 **Sterling** S. Africa
104 C2 **Sterling** *CO* U.S.A.
108 B2 **Sterling** *IL* U.S.A.
55 E3 **Sterlitamak** Rus. Fed.
69 E1 **Sternberg** Ger.
96 C2 **Stettler** Can.
108 C2 **Steubenville** U.S.A.
67 C3 **Stevenage** U.K.
97 E2 **Stevenson Lake** Can.
94 B2 **Stevens Village** U.S.A.
96 B2 **Stewart** Can.
94 B2 **Stewart** *r.* Can.
22 A3 **Stewart Island** N.Z.
95 F2 **Stewart Lake** Can.
70 C2 **Steyr** Austria
90 B3 **Steytlerville** S. Africa
96 A2 **Stikine** *r.* Can.
96 A2 **Stikine Plateau** Can.
90 B3 **Stilbaai** S. Africa
107 D1 **Stillwater** U.S.A.
77 D2 **Štip** Macedonia
64 C2 **Stirling** U.K.
60 F3 **Stjørdalshalsen** Norway
71 D2 **Stockerau** Austria
61 G4 **Stockholm** Sweden
66 B2 **Stockport** U.K.
103 B3 **Stockton** U.S.A.
66 C1 **Stockton-on-Tees** U.K.
31 B2 **Stœng Trêng** Cambodia
64 B1 **Stoer, Point of** U.K.
66 B2 **Stoke-on-Trent** U.K.
77 C2 **Stolac** Bos.-Herz.
68 C2 **Stolberg (Rheinland)** Ger.
56 C3 **Stolin** Belarus
69 F2 **Stollberg** Ger.
69 D1 **Stolzenau** Ger.
64 C2 **Stonehaven** U.K.
97 E2 **Stonewall** Can.
97 D2 **Stony Rapids** Can.
60 G2 **Storavan** *l.* Sweden
61 F4 **Store Bælt** *sea chan.* Denmark
60 F3 **Støren** Norway
60 F2 **Storforshei** Norway
94 D2 **Storkerson Peninsula** Can.
105 D2 **Storm Lake** U.S.A.
61 E3 **Stornosa** *mt.* Norway
64 A1 **Stornoway** U.K.
54 E2 **Storozhevsk** Rus. Fed.
58 B2 **Storozhynets'** Ukr.
60 F3 **Storsjön** *l.* Sweden
60 H2 **Storslett** Norway
60 G2 **Storuman** Sweden
67 C3 **Stour** *r. England* U.K.
67 C2 **Stour** *r. England* U.K.
98 A1 **Stout Lake** Can.
56 C3 **Stowbtsy** Belarus
65 C1 **Strabane** U.K.
70 C2 **Strakonice** Czech Rep.
70 C1 **Stralsund** Ger.
90 A3 **Strand** S. Africa
61 E3 **Stranda** Norway
65 D1 **Strangford Lough** *inlet* U.K.
64 B3 **Stranraer** U.K.
73 D2 **Strasbourg** France
108 C2 **Stratford** Can.
22 B1 **Stratford** N.Z.
107 C1 **Stratford** U.S.A.
67 C2 **Stratford-upon-Avon** U.K.
96 C2 **Strathmore** Can.
64 C2 **Strathspey** *val.* U.K.
70 C2 **Straubing** Ger.
19 C3 **Streaky Bay** Austr.
108 B2 **Streator** U.S.A.
78 B2 **Strehaia** Romania
79 B2 **Strimonas** *r.* Greece
121 B5 **Stroeder** Arg.
69 D1 **Ströhen** Ger.
77 C3 **Stromboli, Isola** *i.* Italy
64 C1 **Stromness** U.K.
60 G3 **Strömsund** Sweden
64 C1 **Stronsay** *i.* U.K.
21 D2 **Stroud** Austr.
67 B3 **Stroud** U.K.
68 C1 **Strücklingen (Saterland)** Ger.
77 D2 **Struga** Macedonia
56 C2 **Strugi-Krasnyye** Rus. Fed.
90 B3 **Struis Bay** S. Africa
77 D2 **Strumica** Macedonia
90 B2 **Strydenburg** S. Africa
58 A2 **Stryy** Ukr.
96 B2 **Stuart Lake** Can.
98 A1 **Stull Lake** Can.
57 E3 **Stupino** Rus. Fed.
108 B2 **Sturgeon Bay** U.S.A.
98 C2 **Sturgeon Falls** Can.
98 A2 **Sturgeon Lake** Can.
108 B2 **Sturgis** *MI* U.S.A.
104 C2 **Sturgis** *SD* U.S.A.
18 B1 **Sturt Creek** *watercourse* Austr.
20 B1 **Sturt Stony Desert** Austr.
91 C3 **Stutterheim** S. Africa
70 B2 **Stuttgart** Ger.
110 B2 **Stuttgart** U.S.A.
60 □A2 **Stykkishólmur** Iceland
58 B1 **Styr** *r.* Belarus/Ukr.
123 D1 **Suaçuí Grande** *r.* Brazil
84 B3 **Suakin** Sudan
46 A3 **Suara, Mount** Eritrea
77 C1 **Subotica** Yugo.
78 C1 **Suceava** Romania
65 B2 **Suck** *r.* Rep. of Ireland
120 B2 **Sucre** Bol.
122 B2 **Sucuriú** *r.* Brazil
57 E2 **Suda** Rus. Fed.
59 C3 **Sudak** Ukr.
85 B3 **Sudan** *country* Africa
98 B2 **Sudbury** Can.
85 A4 **Sudd** *swamp* Sudan
57 F2 **Sudislavl'** Rus. Fed.
57 F2 **Sudogda** Rus. Fed.
62 B1 **Suðuroy** *i.* Faroe Is
75 C2 **Sueca** Spain
84 B2 **Suez** Egypt
84 B2 **Suez, Gulf of** Egypt
48 B2 **Suez Canal** Egypt
109 D3 **Suffolk** U.S.A.
47 C2 **Şuḩār** Oman
37 D1 **Sühbaatar** Mongolia
69 E2 **Suhl** Ger.
77 C1 **Suhopolje** Croatia
38 B2 **Suide** China
34 B2 **Suifenhe** China
37 E1 **Suihua** China
38 A2 **Suining** China
38 B2 **Suiping** China
38 B2 **Suizhou** China
42 B2 **Sujangarh** India
42 B1 **Sujanpur** India
42 A2 **Sujawal** Pak.
28 B2 **Sukabumi** Indon.
29 B2 **Sukadana** Indon.
57 E3 **Sukhinichi** Rus. Fed.
57 F2 **Sukhona** *r.* Rus. Fed.
30 A2 **Sukhothai** Thai.
42 A2 **Sukkur** Pak.
57 E2 **Sukromny** Rus. Fed.
27 C3 **Sula, Kepulauan** *is* Indon.
42 A1 **Sulaiman Ranges** *mts* Pak.
26 C3 **Sulawesi** *i.* Indon.
118 A2 **Sullana** Peru
105 E3 **Sullivan** U.S.A.
107 D2 **Sulphur Springs** U.S.A.
32 B2 **Sulu Archipelago** *is* Phil.
32 A2 **Sulu Sea** N. Pacific Ocean
69 E3 **Sulzbach-Rosenberg** Ger.
47 C2 **Sumāil** Oman
28 A1 **Sumatera** *i.* Indon.
Sumatra *i.* Indon. *see* **Sumatera**
26 C3 **Sumba** *i.* Indon.
29 C2 **Sumbawa** *i.* Indon.
29 C2 **Sumbawabesar** Indon.
87 D3 **Sumbawanga** Tanz.
88 A2 **Sumbe** Angola
64 □ **Sumburgh** U.K.
64 □ **Sumburgh Head** *hd* U.K.
29 C2 **Sumenep** Indon.
35 D4 **Sumisu-jima** *i.* Japan
99 D2 **Summerside** Can.
108 C3 **Summersville** U.S.A.
96 B2 **Summit Lake** Can.
71 D2 **Šumperk** Czech Rep.
49 C1 **Sumqayıt** Azer.
111 D2 **Sumter** U.S.A.
59 C1 **Sumy** Ukr.

43 D2 **Sunamganj** Bangl.
33 B2 **Sunan** N. Korea
47 C2 **Şunaynah** Oman
20 B3 **Sunbury** Austr.
109 D2 **Sunbury** U.S.A.
33 B2 **Sunch'ŏn** N. Korea
33 B3 **Sunch'ŏn** S. Korea
91 C2 **Sun City** S. Africa
28 B2 **Sunda, Selat** *str.* Indon.
104 C2 **Sundance** U.S.A.
43 C2 **Sundarbans** *reg.* Bangl./India
66 C1 **Sunderland** U.K.
96 C2 **Sundre** Can.
61 G3 **Sundsvall** Sweden
91 D2 **Sundumbili** S. Africa
28 B2 **Sungailiat** Indon.
28 B2 **Sungaipenuh** Indon.
28 B1 **Sungei Petani** Malaysia
48 B1 **Sungurlu** Turkey
61 E3 **Sunndalsøra** Norway
102 C1 **Sunnyside** U.S.A.
103 B3 **Sunnyvale** U.S.A.
51 I2 **Suntar** Rus. Fed.
42 A2 **Suntsar** Pak.
82 B4 **Sunyani** Ghana
50 D2 **Suoyarvi** Rus. Fed.
106 A2 **Superior** *AZ* U.S.A.
105 D2 **Superior** *NE* U.S.A.
108 A1 **Superior** *WI* U.S.A.
108 B1 **Superior, Lake** Can./U.S.A.
57 D3 **Suponevo** Rus. Fed.
49 C2 **Sūq ash Shuyūkh** Iraq
38 B2 **Suqian** China
46 A2 **Sūq Suwayq** Saudi Arabia
47 C2 **Şūr** Oman
42 A2 **Surab** Pak.
29 C2 **Surabaya** Indon.
29 C2 **Surakarta** Indon.
42 B2 **Surat** India
42 B2 **Suratgarh** India
31 A3 **Surat Thani** Thai.
57 D3 **Surazh** Rus. Fed.
77 D2 **Surdulica** Yugo.
42 B2 **Surendranagar** India
50 F2 **Surgut** Rus. Fed.
32 B2 **Surigao** Phil.
31 B2 **Surin** Thai.
119 C1 **Suriname** *country* S. America
Surt Libya *see* **Sirte**
Surt, Khalīj *g.* Libya *see* Sirte, Gulf of
28 B2 **Surulangun** Indon.
57 F2 **Susanino** Rus. Fed.
103 B2 **Susanville** U.S.A.
48 B1 **Suşehri** Turkey
99 D2 **Sussex** Can.
69 D1 **Süstedt** Ger.
68 C1 **Sustrum** Ger.
51 K2 **Susuman** Rus. Fed.
79 C3 **Susurluk** Turkey
43 B1 **Sutak** Jammu and Kashmir
90 B3 **Sutherland** S. Africa
67 C2 **Sutton Coldfield** U.K.
34 D2 **Suttsu** Japan
57 E3 **Suvorov** Rus. Fed.
71 E1 **Suwałki** Pol.
31 B2 **Suwannaphum** Thai.
111 D3 **Suwannee** *r.* U.S.A.
Suweis, Qanâ el *canal* Egypt *see* **Suez Canal**
33 B2 **Suwŏn** S. Korea
47 C2 **Sūzā** Iran
57 F2 **Suzdal'** Rus. Fed.
38 B2 **Suzhou** *Anhui* China
38 C2 **Suzhou** *Jiangsu* China
35 C3 **Suzu** Japan
35 C3 **Suzu-misaki** *pt* Japan
50 B1 **Svalbard** *terr.* Arctic Ocean
59 D2 **Svatove** Ukr.
31 B2 **Svay Riĕng** Cambodia
61 F3 **Sveg** Sweden
56 C2 **Švenčionys** Lith.
61 F4 **Svendborg** Denmark
Sverdlovsk Rus. Fed. *see* Yekaterinburg
77 D2 **Sveti Nikole** Macedonia
56 B3 **Svetlogorsk** Rus. Fed.
56 B3 **Svetlyy** Rus. Fed.
61 I3 **Svetogorsk** Rus. Fed.
78 C2 **Svilengrad** Bulg.
78 B2 **Svinecea Mare, Vârful** *mt.* Romania
78 C2 **Svishtov** Bulg.
71 D2 **Svitavy** Czech Rep.
59 C2 **Svitlovods'k** Ukr.
37 E1 **Svobodnyy** Rus. Fed.
60 F2 **Svolvær** Norway
56 C3 **Svyetlahorsk** Belarus
111 D2 **Swainsboro** U.S.A.
88 A3 **Swakopmund** Namibia
20 B3 **Swan Hill** Austr.
96 C2 **Swan Hills** Can.
97 D2 **Swan Lake** *l.* Can.
97 D2 **Swan River** Can.
21 D2 **Swansea** Austr.
67 B3 **Swansea** U.K.
91 C2 **Swartruggens** S. Africa
Swatow China *see* Shantou
91 D2 **Swaziland** *country* Africa
61 G3 **Sweden** *country* Europe
107 C2 **Sweetwater** U.S.A.
104 B2 **Sweetwater** *r.* U.S.A.
90 B3 **Swellendam** S. Africa
71 D1 **Świdnica** Pol.
71 D1 **Świdwin** Pol.
71 D1 **Świebodzin** Pol.
71 D1 **Świecie** Pol.
97 D2 **Swift Current** Can.
65 C1 **Swilly, Lough** *inlet* Rep. of Ireland
67 C3 **Swindon** U.K.
70 C1 **Świnoujście** Pol.
73 D2 **Switzerland** *country* Europe
65 C2 **Swords** Rep. of Ireland
56 C3 **Syanno** Belarus
57 D2 **Sychevka** Rus. Fed.
21 D2 **Sydney** Austr.
99 D2 **Sydney** Can.
99 D2 **Sydney Mines** Can.
59 D2 **Syeverodonets'k** Ukr.
54 E2 **Syktyvkar** Rus. Fed.
110 C2 **Sylacauga** U.S.A.
43 D2 **Sylhet** Bangl.
70 B1 **Sylt** *i.* Ger.
79 C3 **Symi** *i.* Greece
59 D2 **Synel'nykove** Ukr.
77 C3 **Syracuse** *Sicily* Italy
104 C3 **Syracuse** *KS* U.S.A.
109 D2 **Syracuse** *NY* U.S.A.
45 C2 **Syrdar'ya** *r.* Asia
48 B2 **Syria** *country* Asia
31 A2 **Syriam** Myanmar
Syrian Desert Asia *see* Bādiyat ash Shām
79 B3 **Syros** *i.* Greece
55 D3 **Syzran'** Rus. Fed.
70 C1 **Szczecin** Pol.
71 D1 **Szczecinek** Pol.
71 E1 **Szczytno** Pol.
71 E2 **Szeged** Hungary
71 D2 **Székesfehérvár** Hungary
71 D2 **Szekszárd** Hungary
71 E2 **Szentes** Hungary
71 D2 **Szentgotthárd** Hungary
71 D2 **Szigetvár** Hungary
71 E2 **Szolnok** Hungary
71 D2 **Szombathely** Hungary

T

46 B2 **Tābah** Saudi Arabia
44 B3 **Ţabas** Iran
49 D3 **Tābask, Kūh-e** *mt.* Iran
118 B2 **Tabatinga** Brazil
82 B2 **Tabelbala** Alg.
97 C3 **Taber** Can.
70 C2 **Tábor** Czech Rep.
87 D3 **Tabora** Tanz.
82 B4 **Tabou** Côte d'Ivoire
49 C2 **Tabrīz** Iran
46 A2 **Tabūk** Saudi Arabia
45 E2 **Tacheng** China
70 C2 **Tachov** Czech Rep.
32 B1 **Tacloban** Phil.
118 A3 **Tacna** Peru
102 B1 **Tacoma** U.S.A.
120 C4 **Tacuarembó** Uru.
106 B3 **Tacupeto** Mex.
85 C3 **Tadjoura** Djibouti
48 B2 **Tadmur** Syria
97 E2 **Tadoule Lake** Can.
33 B2 **Taegu** S. Korea
33 B2 **Taejŏn** S. Korea
33 B3 **Taejŏng** S. Korea
33 B2 **T'aepaek** S. Korea
75 C1 **Tafalla** Spain
120 B3 **Tafi Viejo** Arg.
47 D2 **Taftān, Kūh-e** *mt.* Iran
59 D2 **Taganrog** Rus. Fed.
59 D2 **Taganrog, Gulf of** Rus. Fed./Ukr.
30 A1 **Tagaung** Myanmar
32 B1 **Tagaytay City** Phil.
32 B2 **Tagbilaran** Phil.
32 B1 **Tagudin** Phil.
32 B2 **Tagum** Phil.
74 B2 **Tagus** *r.* Port./Spain
28 B1 **Tahan, Gunung** *mt.* Malaysia
83 C2 **Tahat, Mont** *mt.* Alg.
37 E1 **Tahe** China
107 E1 **Tahlequah** U.S.A.
103 B3 **Tahoe, Lake** U.S.A.
103 B3 **Tahoe City** U.S.A.
94 D2 **Tahoe Lake** Can.
83 C3 **Tahoua** Niger
47 C2 **Tahrūd** Iran
96 B3 **Tahsis** Can.
38 B2 **Tai'an** China
Taibus Qi China *see* Baochang
39 C3 **T'aichung** Taiwan
22 C1 **Taihape** N.Z.
38 C2 **Tai Hu** *l.* China
20 A3 **Tailem Bend** Austr.
39 C3 **T'ainan** Taiwan
123 D1 **Taiobeiras** Brazil
39 C3 **T'aipei** Taiwan
28 B1 **Taiping** Malaysia
39 B3 **Taishan** China
121 A5 **Taitao, Península de** *pen.* Chile
39 C3 **T'aitung** Taiwan
60 I2 **Taivalkoski** Fin.
60 H2 **Taivaskero** *hill* Fin.
39 C3 **Taiwan** *country* Asia
39 B3 **Taiwan Strait** China/Taiwan
38 B2 **Taiyuan** China
38 B2 **Taizhou** *Jiangsu* China
39 C3 **Taizhou** *Zhejiang* China
46 B3 **Ta'izz** Yemen
45 D3 **Tajikistan** *country* Asia
43 B2 **Taj Mahal** India
Tajo *r.* Port. *see* **Tagus**
26 A1 **Tak** Thai.
22 B2 **Takaka** N.Z.
35 B4 **Takamatsu** Japan

35 C3 **Takaoka** Japan
22 B1 **Takapuna** N.Z.
35 C3 **Takasaki** Japan
90 B1 **Takatokwane** Botswana
35 C3 **Takayama** Japan
35 C3 **Takefu** Japan
28 A1 **Takengon** Indon.
31 B2 **Takêv** Cambodia
31 B2 **Ta Khmau** Cambodia
34 D2 **Takikawa** Japan
96 B2 **Takla Lake** Can.
96 B2 **Takla Landing** Can.
45 E3 **Taklimakan Shamo** *des.* China
96 A2 **Taku** *r.* Can./U.S.A.
31 A3 **Takua Pa** Thai.
83 C4 **Takum** Nigeria
56 C3 **Talachyn** Belarus
42 B1 **Talagang** Pak.
118 A2 **Talara** Peru
42 A2 **Talar-i-Band** *mts* Pak.
27 C2 **Talaud, Kepulauan** *is* Indon.
74 C2 **Talavera de la Reina** Spain
121 A4 **Talca** Chile
121 A4 **Talcahuano** Chile
57 E2 **Taldom** Rus. Fed.
45 D2 **Taldykorgan** Kazakh.
27 C3 **Taliabu** *i.* Indon.
32 B1 **Talisay** Phil.
29 C2 **Taliwang** Indon.
49 C2 **Tall 'Afar** Iraq
111 D2 **Tallahassee** U.S.A.
56 B2 **Tallinn** Estonia
110 B2 **Tallulah** U.S.A.
72 B2 **Talmont-St-Hilaire** France
58 C2 **Tal'ne** Ukr.
85 B3 **Talodi** Sudan
59 E1 **Talovaya** Rus. Fed.
94 E2 **Taloyoak** Can.
56 B2 **Talsi** Latvia
120 A3 **Taltal** Chile
97 C1 **Taltson** *r.* Can.
28 A1 **Talu** Indon.
21 C1 **Talwood** Austr.
82 B4 **Tamale** Ghana
83 C2 **Tamanrasset** Alg.
67 A3 **Tamar** *r.* U.K.
113 C2 **Tamazunchale** Mex.
82 A3 **Tambacounda** Senegal
29 B1 **Tambelan, Kepulauan** *is* Indon.
59 E1 **Tambov** Rus. Fed.
113 C2 **Tamiahua, Laguna de** *lag.* Mex.
111 D3 **Tampa** U.S.A.
111 D3 **Tampa Bay** U.S.A.
61 H3 **Tampere** Fin.
113 C2 **Tampico** Mex.
37 D1 **Tamsagbulag** Mongolia
70 C2 **Tamsweg** Austria
21 D2 **Tamworth** Austr.
67 C2 **Tamworth** U.K.
87 E3 **Tana** *r.* Kenya
35 C4 **Tanabe** Japan
60 I1 **Tana Bru** Norway
85 B3 **T'ana Hāyk'** *l.* Eth.
29 C2 **Tanahgrogot** Indon.
26 C3 **Tanahjampea** *i.* Indon.
18 C1 **Tanami Desert** Austr.
94 A2 **Tanana** U.S.A.
76 A1 **Tanaro** *r.* Italy
33 B1 **Tanch'ŏn** N. Korea
32 B2 **Tandag** Phil.
78 C2 **Țăndărei** Romania
121 C4 **Tandil** Arg.
42 A2 **Tando Adam** Pak.
42 A2 **Tando Muhammmad Khan** Pak.
82 B2 **Tanezrouft** *reg.* Alg./Mali
87 D3 **Tanga** Tanz.
87 C3 **Tanganyika, Lake** Africa
Tanger Morocco *see* **Tangier**
69 E1 **Tangermünde** Ger.
43 C1 **Tanggula Shan** *mts* China
82 B1 **Tangier** Morocco
43 C1 **Tangra Yumco** *salt l.* China
38 B2 **Tangshan** China
36 C2 **Taniantaweng Shan** *mts* China
27 C3 **Tanimbar, Kepulauan** *is* Indon.
32 B2 **Tanjay** Phil.
28 A1 **Tanjungbalai** Indon.
28 B2 **Tanjungkarang-Telukbetung** Indon.
29 B2 **Tanjungpandan** Indon.
28 B1 **Tanjungpinang** Indon.
29 C1 **Tanjungredeb** Indon.
29 C1 **Tanjungselor** Indon.
42 B1 **Tank** Pak.
83 C3 **Tanout** Niger
43 C2 **Tansen** Nepal
84 B1 **Tanta** Egypt
87 D3 **Tanzania** *country* Africa
37 E1 **Taonan** China
106 B1 **Taos** U.S.A.
82 B2 **Taoudenni** Mali
56 C2 **Tapa** Estonia
113 C3 **Tapachula** Mex.
119 C2 **Tapajós** *r.* Brazil
28 A1 **Tapaktuan** Indon.
113 C3 **Tapanatepec** Mex.
118 B2 **Tapauá** Brazil
82 B4 **Tapeta** Liberia
42 B2 **Tāpi** *r.* India
109 D3 **Tappahannock** U.S.A.
118 B2 **Tapurucuara** Brazil
122 B1 **Taquarí** Brazil
122 A1 **Taquarí** *r.* Brazil
122 B1 **Taquari, Serra do** *hills* Brazil
122 C2 **Taquaritinga** Brazil
83 D4 **Taraba** *r.* Nigeria
Țarābulus Libya *see* **Tripoli**
29 C1 **Tarakan** Indon.
56 A3 **Taran, Mys** *pt* Rus. Fed.
Taranaki, Mount *vol.* N.Z. *see* **Egmont, Mt**
74 C1 **Tarancón** Spain
77 C2 **Taranto** Italy
77 C2 **Taranto, Golfo di** *g.* Italy
118 A2 **Tarapoto** Peru
59 E2 **Tarasovskiy** Rus. Fed.
118 A2 **Tarauacá** Brazil
118 B2 **Tarauacá** *r.* Brazil
45 D2 **Taraz** Kazakh.
75 C1 **Tarazona** Spain
45 E2 **Tarbagatay, Khrebet** *mts* Kazakh.
64 A2 **Tarbert** *Scotland* U.K.
64 B3 **Tarbert** *Scotland* U.K.
72 C3 **Tarbes** France
64 B2 **Tarbet** U.K.
19 C3 **Tarcoola** Austr.
21 D2 **Taree** Austr.
78 C2 **Târgoviște** Romania
78 B1 **Târgu Jiu** Romania
78 B1 **Târgu Mureș** Romania
78 C1 **Târgu Neamț** Romania
47 C2 **Tarif** U.A.E.
120 B3 **Tarija** Bol.
47 B3 **Tarīm** Yemen
45 E2 **Tarim He** *r.* China
45 E3 **Tarim Pendi** *basin* China
27 D3 **Taritatu** *r.* Indon.
50 G2 **Tarko-Sale** Rus. Fed.
82 B4 **Tarkwa** Ghana
32 B1 **Tarlac** Phil.
60 G2 **Tärnaby** Sweden
45 C3 **Tarnak** *r.* Afgh.
78 B1 **Târnăveni** Romania
71 E1 **Tarnobrzeg** Pol.
71 E1 **Tarnów** Pol.
82 B1 **Taroudannt** Morocco
76 B2 **Tarquinia** Italy
75 D1 **Tarragona** Spain
75 D1 **Tàrrega** Spain
48 B2 **Tarsus** Turkey
120 B3 **Tartagal** Arg.
72 B3 **Tartas** France
56 C2 **Tartu** Estonia
48 B2 **Țarțūs** Syria
123 D1 **Tarumirim** Brazil
57 E3 **Tarusa** Rus. Fed.
76 B1 **Tarvisio** Italy
Tashi Chho Bhutan *see* Thimphu
49 D3 **Tashk, Daryācheh-ye** *l.* Iran
45 C2 **Tashkent** Uzbek.
99 D1 **Tasiujaq** Can.
45 E2 **Taskesken** Kazakh.
22 B2 **Tasman Bay** N.Z.
19 D4 **Tasmania** *state* Austr.
22 B2 **Tasman Mountains** N.Z.
124 D8 **Tasman Sea** S. Pacific Ocean
83 C2 **Tassili du Hoggar** *plat.* Alg.
83 C2 **Tassili n'Ajjer** *plat.* Alg.
71 D2 **Tatabánya** Hungary
58 B2 **Tatarbunary** Ukr.
51 K3 **Tatarskiy Proliv** *str.* Rus. Fed.
35 C4 **Tateyama** Japan
96 C1 **Tathlina Lake** Can.
46 B3 **Tathlīth** Saudi Arabia
46 B2 **Tathlīth, Wādī** *watercourse* Saudi Arabia
21 C3 **Tathra** Austr.
30 A1 **Tatkon** Myanmar
96 B2 **Tatla Lake** Can.
42 A2 **Tatta** Pak.
122 C2 **Tatuí** Brazil
107 C2 **Tatum** U.S.A.
49 C2 **Tatvan** Turkey
119 D2 **Taua** Brazil
123 C2 **Taubaté** Brazil
69 D3 **Tauberbischofsheim** Ger.
22 C1 **Taumarunui** N.Z.
30 A1 **Taunggyi** Myanmar
30 A2 **Taunqup** Myanmar
67 B3 **Taunton** U.K.
68 C2 **Taunus** *hills* Ger.
22 C1 **Taupo** N.Z.
22 C1 **Taupo, Lake** N.Z.
56 B2 **Tauragė** Lith.
22 C1 **Tauranga** N.Z.
79 C3 **Tavas** Turkey
54 F3 **Tavda** Rus. Fed.
74 B2 **Tavira** Port.
67 A3 **Tavistock** U.K.
31 A2 **Tavoy** Myanmar
79 C3 **Tavșanlı** Turkey
67 A3 **Taw** *r.* U.K.
108 C2 **Tawas City** U.S.A.
29 C1 **Tawau** *Sabah* Malaysia
32 A2 **Tawitawi** *i.* Phil.
113 C3 **Taxco** Mex.
45 D3 **Taxkorgan** China
64 C2 **Tay** *r.* U.K.
64 C2 **Tay, Firth of** *est.* U.K.
64 B2 **Tay, Loch** *l.* U.K.
96 B2 **Taylor** Can.
107 D2 **Taylor** U.S.A.
108 B3 **Taylorville** U.S.A.
46 A2 **Taymā'** Saudi Arabia
51 H2 **Taymura** *r.* Rus. Fed.
51 H2 **Taymyr, Ozero** *l.* Rus. Fed.
51 G2 **Taymyr, Poluostrov** *pen.* Rus. Fed.
31 B2 **Tây Ninh** Vietnam
32 A1 **Taytay** Phil.

50 G2 **Taz** *r.* Rus. Fed.
82 B1 **Taza** Morocco
97 D2 **Tazin Lake** Can.
54 G2 **Tazovskaya Guba** *sea chan.* Rus. Fed.
49 C1 **T'bilisi** Georgia
59 E2 **Tbilisskaya** Rus. Fed.
86 B3 **Tchibanga** Gabon
86 B2 **Tcholliré** Cameroon
71 D1 **Tczew** Pol.
112 B2 **Teacapán** Mex.
22 A3 **Te Anau** N.Z.
22 A3 **Te Anau, Lake** N.Z.
113 C3 **Teapa** Mex.
22 C1 **Te Awamutu** N.Z.
83 C1 **Tébessa** Alg.
28 B2 **Tebingtinggi** *Sumatera* Indon.
28 A1 **Tebingtinggi** *Sumatera* Indon.
82 B4 **Techiman** Ghana
112 B3 **Tecomán** Mex.
112 B2 **Tecoripa** Mex.
113 B3 **Técpan** Mex.
112 B2 **Tecuala** Mex.
78 C1 **Tecuci** Romania
44 C3 **Tedzhen** Turkm.
44 C3 **Tedzhen** *r.* Turkm.
36 C1 **Teeli** Rus. Fed.
66 C1 **Tees** *r.* U.K.
79 C3 **Tefenni** Turkey
29 B2 **Tegal** Indon.
114 B3 **Tegucigalpa** Hond.
83 C3 **Teguidda-n-Tessoumt** Niger
82 B4 **Téhini** Côte d'Ivoire
49 D2 **Tehrān** Iran
113 C3 **Tehuacán** Mex.
113 C3 **Tehuantepec, Golfo de** *g.* Mex.
113 C3 **Tehuantepec, Istmo de** *isth.* Mex.
67 A2 **Teifi** *r.* U.K.
Tejo *r.* Spain *see* **Tagus**
113 B3 **Tejupilco** Mex.
22 B2 **Tekapo, Lake** N.Z.
113 D2 **Tekax** Mex.
84 B3 **Tekezē Wenz** *r.* Eritrea/Eth.
79 C2 **Tekirdağ** Turkey
22 C1 **Te Kuiti** N.Z.
43 C2 **Tel** *r.* India
49 C1 **T'elavi** Georgia
48 B2 **Tel Aviv-Yafo** Israel
113 D2 **Telchac Puerto** Mex.
96 A2 **Telegraph Creek** Can.
122 B2 **Telêmaco Borba** Brazil
67 B2 **Telford** U.K.
28 A2 **Telo** Indon.
54 E2 **Tel'pos-Iz, Gora** *mt.* Rus. Fed.
56 B2 **Telšiai** Lith.
28 B1 **Teluk Anson** Malaysia
29 B2 **Telukbatang** Indon.
28 A1 **Telukdalam** Indon.
98 C2 **Temagami Lake** Can.
29 C2 **Temanggung** Indon.
91 C2 **Temba** S. Africa
28 B2 **Tembilahan** Indon.
88 A1 **Tembo Aluma** Angola
67 B2 **Teme** *r.* U.K.
28 B1 **Temerloh** Malaysia
45 D1 **Temirtau** Kazakh.
21 C2 **Temora** Austr.
107 D2 **Temple** U.S.A.
65 C2 **Templemore** Rep. of Ireland
113 C2 **Tempoal** Mex.
59 D2 **Temryuk** Rus. Fed.
121 A4 **Temuco** Chile
22 B2 **Temuka** N.Z.
113 C2 **Tenabo** Mex.
41 C3 **Tenali** India
31 A2 **Tenasserim** Myanmar
67 A3 **Tenby** U.K.
85 C3 **Tendaho** Eth.
73 D3 **Tende** France
76 A2 **Tende, Col de** *pass* France/Italy
41 D4 **Ten Degree Channel** India
35 D3 **Tendō** Japan
82 B3 **Ténenkou** Mali
83 D2 **Ténéré du Tafassâsset** *des.* Niger
82 A2 **Tenerife** *i.* Canary Is
75 D2 **Ténès** Alg.
29 C2 **Tengah, Kepulauan** *is* Indon.
30 A1 **Tengchong** China
29 C2 **Tenggarong** Indon.
38 A2 **Tengger Shamo** *des.* China
39 B3 **Tengxian** China
87 C4 **Tenke** Dem. Rep. Congo
82 B3 **Tenkodogo** Burkina
19 C1 **Tennant Creek** Austr.
110 C1 **Tennessee** *r.* U.S.A.
110 C1 **Tennessee** *state* U.S.A.
113 C3 **Tenosique** Mex.
21 D1 **Tenterfield** Austr.
122 B2 **Teodoro Sampaio** Brazil
123 D1 **Teófilo Otôni** Brazil
113 C3 **Teopisca** Mex.
112 B2 **Tepache** Mex.
22 B1 **Te Paki** N.Z.
112 B3 **Tepalcatepec** Mex.
112 B2 **Tepatitlán** Mex.
112 B2 **Tepehuanes** Mex.
77 D2 **Tepelenë** Albania
112 B2 **Tepic** Mex.
70 C1 **Teplice** Czech Rep.
57 E3 **Teploye** Rus. Fed.
112 B2 **Tequila** Mex.
76 B2 **Teramo** Italy
57 E3 **Terbuny** Rus. Fed.
58 B2 **Terebovlya** Ukr.
55 D4 **Terek** *r.* Rus. Fed.
119 D2 **Teresina** Brazil
123 D2 **Teresópolis** Brazil
31 A3 **Teressa Island** India
48 B1 **Terme** Turkey
45 C3 **Termez** Uzbek.
76 B3 **Termini Imerese** *Sicily* Italy
113 C3 **Términos, Laguna de** *lag.* Mex.
77 B2 **Termoli** Italy
27 C2 **Ternate** Indon.
68 A2 **Terneuzen** Neth.
76 B2 **Terni** Italy
58 B2 **Ternopil'** Ukr.
96 B2 **Terrace** Can.
98 B2 **Terrace Bay** Can.
90 B2 **Terra Firma** S. Africa
108 B3 **Terre Haute** U.S.A.
99 E2 **Terrenceville** Can.
68 B1 **Terschelling** *i.* Neth.
76 A3 **Tertenia** *Sardinia* Italy
75 C1 **Teruel** Spain
60 H2 **Tervola** Fin.
77 C2 **Tešanj** Bos.-Herz.
84 B3 **Teseney** Eritrea
34 D2 **Teshio-gawa** *r.* Japan
96 A1 **Teslin** Can.
96 A1 **Teslin Lake** Can.
122 B1 **Tesouro** Brazil
83 C3 **Tessaoua** Niger
89 C2 **Tete** Moz.
58 C1 **Teteriv** *r.* Ukr.
69 F1 **Teterow** Ger.
58 B2 **Tetiyiv** Ukr.
82 B1 **Tétouan** Morocco
77 D2 **Tetovo** Macedonia
120 B3 **Teuco** *r.* Arg.
112 B2 **Teul de González Ortega** Mex.
69 D1 **Teutoburger Wald** *hills* Ger.
76 B2 **Tevere** *r.* Italy
22 A3 **Teviot** N.Z.
19 E2 **Tewantin** Austr.
22 C2 **Te Wharau** N.Z.
107 E2 **Texarkana** U.S.A.
21 D1 **Texas** Austr.
107 D2 **Texas** *state* U.S.A.
107 E3 **Texas City** U.S.A.
68 B1 **Texel** *i.* Neth.
107 D2 **Texoma, Lake** U.S.A.
91 C2 **Teyateyaneng** Lesotho
57 F2 **Teykovo** Rus. Fed.
57 F2 **Teza** *r.* Rus. Fed.
43 D2 **Tezpur** India
30 A1 **Tezu** India
97 E1 **Tha-anne** *r.* Can.
91 C2 **Thabana-Ntlenyana** *mt.* Lesotho
91 C2 **Thaba Putsoa** *mt.* Lesotho
91 C1 **Thabazimbi** S. Africa
91 C2 **Thabong** S. Africa
31 A2 **Thagyettaw** Myanmar
30 B1 **Thai Binh** Vietnam
31 B2 **Thailand** *country* Asia
31 B2 **Thailand, Gulf of** Asia
30 B1 **Thai Nguyên** Vietnam
31 A3 **Thalang** Thai.
42 B1 **Thal Desert** Pak.
69 E2 **Thale (Harz)** Ger.
30 B2 **Tha Li** Thai.
91 C1 **Thamaga** Botswana
84 C3 **Thamar, Jabal** *mt.* Yemen
47 C3 **Thamarīt** Oman
98 B2 **Thames** *r.* Can.
22 C1 **Thames** N.Z.
67 D3 **Thames** *est.* U.K.
67 D3 **Thames** *r.* U.K.
31 A2 **Thanbyuzayat** Myanmar
30 B2 **Thanh Hoa** Vietnam
41 B3 **Thanjavur** India
42 A2 **Thano Bula Khan** Pak.
30 B1 **Than Uyen** Vietnam
42 A2 **Thar Desert** India/Pak.
20 B1 **Thargomindah** Austr.
30 A2 **Tharrawaddy** Myanmar
49 C2 **Tharthār, Buḩayrat ath** *l.* Iraq
79 B2 **Thasos** *i.* Greece
30 B1 **Thât Khê** Vietnam
30 A2 **Thaton** Myanmar
30 A1 **Thaungdut** Myanmar
30 A2 **Thayetmyo** Myanmar
30 A1 **Thazi** Myanmar
114 C2 **The Bahamas** *country* West Indies
66 B1 **The Cheviot** *hill* U.K.
102 B1 **The Dalles** U.S.A.
67 C2 **The Fens** *reg.* U.K.
82 A3 **The Gambia** *country* Africa
84 B2 **The Great Oasis** Egypt
47 C2 **The Gulf** Asia
68 B1 **The Hague** Neth.
97 E1 **Thelon** *r.* Can.
69 E2 **Themar** Ger.
64 A1 **The Minch** *sea chan.* U.K.
118 B2 **Theodore Roosevelt** *r.* Brazil
97 D2 **The Pas** Can.
79 B2 **Thermaïkos Kolpos** *g.* Greece
104 B2 **Thermopolis** U.S.A.
21 C3 **The Rock** Austr.
98 B2 **Thessalon** Can.
79 B2 **Thessaloniki** Greece
67 D2 **Thetford** U.K.
99 C2 **Thetford Mines** Can.
30 A1 **The Triangle** *mts* Myanmar
99 D1 **Thévenet, Lac** *l.* Can.
67 D2 **The Wash** *b.* U.K.
107 D2 **The Woodlands** U.S.A.
110 B3 **Thibodaux** U.S.A.
97 E2 **Thicket Portage** Can.

105 D1 **Thief River Falls** U.S.A.
73 C2 **Thiers** France
87 D3 **Thika** Kenya
41 B4 **Thiladhunmathee Atoll** Maldives
43 C2 **Thimphu** Bhutan
73 D2 **Thionville** France
Thira *i.* Greece *see* **Santorini**
66 C1 **Thirsk** U.K.
Thiruvananthapuram India *see* **Trivandrum**
61 E4 **Thisted** Denmark
97 E1 **Thlewiaza** *r.* Can.
30 A2 **Thoen** Thai.
91 D1 **Thohoyandou** S. Africa
69 E1 **Thomasburg** Ger.
65 C2 **Thomastown** Rep. of Ireland
111 D2 **Thomasville** U.S.A.
68 C2 **Thommen** Belgium
97 E2 **Thompson** Can.
96 E3 **Thompson** *r.* U.S.A.
102 C1 **Thompson Falls** U.S.A.
96 B2 **Thompson Sound** Can.
64 C3 **Thornhill** U.K.
23 C2 **Thorshavnheiane** *reg.* Antarctica
72 B2 **Thouars** France
96 C2 **Three Hills** Can.
31 A2 **Three Pagodas Pass** Myanmar/Thai.
82 B4 **Three Points, Cape** Ghana
31 B2 **Thu Dâu Môt** Vietnam
68 B2 **Thuin** Belgium
89 B3 **Thuli** Zimbabwe
98 B2 **Thunder Bay** Can.
31 A3 **Thung Song** Thai.
69 E2 **Thüringer Becken** *reg.* Ger.
69 E2 **Thüringer Wald** *mts* Ger.
65 C2 **Thurles** Rep. of Ireland
64 C1 **Thurso** U.K.
64 C1 **Thurso** *r.* U.K.
119 D2 **Tianguá** Brazil
38 B2 **Tianjin** China
38 B2 **Tianjin** *mun.* China
39 A3 **Tianlin** China
38 B2 **Tianmen** China
38 A2 **Tianshui** China
122 B2 **Tibagi** *r.* Brazil
86 B2 **Tibati** Cameroon
Tiberias, Lake *l.* Israel *see* **Galilee, Sea of**
83 D2 **Tibesti** *mts* Chad
Tibet *aut. reg.* China *see* **Xizang Zizhiqu**
Tibet, Plateau of China *see* **Qing Zang Gaoyuan**
20 B1 **Tibooburra** Austr.
112 A2 **Tiburón, Isla** *i.* Mex.
82 B3 **Tîchît** Maur.
82 A2 **Tichla** Western Sahara
113 D2 **Ticul** Mex.
82 A3 **Tidjikja** Maur.
68 B2 **Tiel** Neth.
33 A1 **Tieling** China
68 A2 **Tielt** Belgium
68 B2 **Tienen** Belgium
36 B2 **Tien Shan** *mts* China/Kyrg.
Tientsin China *see* **Tianjin**
61 G3 **Tierp** Sweden
113 C3 **Tierra Blanca** Mex.
113 C3 **Tierra Colorada** Mex.
121 B6 **Tierra del Fuego, Isla Grande de** *i.* Arg./Chile
122 C2 **Tietê** Brazil
Tiflis Georgia *see* **T'bilisi**
111 D2 **Tifton** U.S.A.
58 B2 **Tighina** Moldova
99 D2 **Tignish** Can.
118 A2 **Tigre** *r.* Ecuador/Peru
49 C2 **Tigris** *r.* Asia
113 C2 **Tihuatlán** Mex.
112 A1 **Tijuana** Mex.
59 E2 **Tikhoretsk** Rus. Fed.
57 D2 **Tikhvin** Rus. Fed.
57 D2 **Tikhvinskaya Gryada** *ridge* Rus. Fed.
22 C1 **Tikokino** N.Z.
49 C2 **Tikrīt** Iraq
51 J2 **Tiksi** Rus. Fed.
68 B2 **Tilburg** Neth.
120 B3 **Tilcara** Arg.
20 B1 **Tilcha** Austr.
82 C3 **Tillabéri** Niger
102 B1 **Tillamook** U.S.A.
31 A3 **Tillanchong Island** India
79 C3 **Tilos** *i.* Greece
20 B2 **Tilpa** Austr.
54 F2 **Til'tim** Rus. Fed.
57 E3 **Tim** Rus. Fed.
54 D2 **Timanskiy Kryazh** *ridge* Rus. Fed.
22 B2 **Timaru** N.Z.
59 D2 **Timashevsk** Rus. Fed.
82 B3 **Timbedgha** Maur.
18 C1 **Timber Creek** Austr.
Timbuktu Mali *see* **Tombouctou**
82 C2 **Timimoun** Alg.
78 B1 **Timiş** *r.* Romania
78 B1 **Timişoara** Romania
98 B2 **Timmins** Can.
57 E2 **Timokhino** Rus. Fed.
119 D2 **Timon** Brazil
27 C3 **Timor** *i.* Indon.
26 C3 **Timor Sea** Austr./Indon.
61 G3 **Timrå** Sweden
82 B2 **Tindouf** Alg.
43 C2 **Tingri** China
61 F4 **Tingsryd** Sweden
27 D2 **Tinian** *i.* N. Mariana Is
120 B3 **Tinogasta** Arg.
79 C3 **Tinos** Greece
79 C3 **Tinos** *i.* Greece
83 C2 **Tinrhert, Plateau du** Alg.
30 A1 **Tinsukia** India
75 D2 **Tipasa** Alg.
65 B2 **Tipperary** Rep. of Ireland
77 C2 **Tirana** Albania
Tiranë Albania *see* **Tirana**
76 B1 **Tirano** Italy
58 B2 **Tiraspol** Moldova
79 C3 **Tire** Turkey
64 A2 **Tiree** *i.* U.K.
42 B1 **Tirich Mir** *mt.* Pak.
41 B3 **Tiruchchirāppalli** India
41 B4 **Tirunelveli** India
41 B3 **Tirupati** India
41 B3 **Tiruppattur** India
41 B3 **Tiruppur** India
77 D1 **Tisa** *r.* Europe
97 D2 **Tisdale** Can.
75 D2 **Tissemsilt** Alg.
120 B2 **Titicaca, Lago** *l.* Bol./Peru
43 C2 **Titlagarh** India
78 C2 **Titu** Romania
111 D3 **Titusville** U.S.A.
67 B3 **Tiverton** U.K.
76 B2 **Tivoli** Italy
47 C2 **Ţīwī** Oman
113 D2 **Tizimín** Mex.
75 D2 **Tizi Ouzou** Alg.
82 B2 **Tiznit** Morocco
113 C3 **Tlacotalpán** Mex.
112 B2 **Tlahualilo** Mex.
113 C3 **Tlapa** Mex.
113 C3 **Tlaxcala** Mex.
113 C3 **Tlaxiaco** Mex.
82 B1 **Tlemcen** Alg.
91 C1 **Tlokweng** Botswana
96 B2 **Toad River** Can.
89 □D2 **Toamasina** Madag.
28 A1 **Toba, Danau** *l.* Indon.
42 A1 **Toba and Kakar Ranges** *mts* Pak.
115 D3 **Tobago** *i.* Trin. and Tob.
27 C2 **Tobelo** Indon.
98 B2 **Tobermory** Can.
64 A2 **Tobermory** U.K.
97 D2 **Tobin Lake** *l.* Can.
28 B2 **Toboali** Indon.
44 C1 **Tobol** *r.* Kazakh./Rus. Fed.
54 F3 **Tobol'sk** Rus. Fed.
119 D2 **Tocantinópolis** Brazil
119 D2 **Tocantins** *r.* Brazil
111 D2 **Toccoa** U.S.A.
76 A1 **Toce** *r.* Italy
120 A3 **Tocopilla** Chile
21 C3 **Tocumwal** Austr.
112 A2 **Todos Santos** Mex.
96 B3 **Tofino** Can.
64 □ **Toft** U.K.
27 C3 **Togian, Kepulauan** *is* Indon.
82 C4 **Togo** *country* Africa
42 B2 **Tohana** India
94 B2 **Tok** U.S.A.
84 B3 **Tokar** Sudan
37 E3 **Tokara-rettō** *is* Japan
59 E1 **Tokarevka** Rus. Fed.
17 G3 **Tokelau** *terr.* S. Pacific Ocean
45 D2 **Tokmak** Kyrg.
59 D2 **Tokmak** Ukr.
22 C1 **Tokoroa** N.Z.
36 B2 **Toksun** China
35 B4 **Tokushima** Japan
35 C3 **Tōkyō** Japan
89 □D3 **Tôlañaro** Madag.
122 B2 **Toledo** Brazil
74 C2 **Toledo** Spain
108 C2 **Toledo** U.S.A.
74 C2 **Toledo, Montes de** *mts* Spain
110 B2 **Toledo Bend Reservoir** U.S.A.
89 □D3 **Toliara** Madag.
26 C2 **Tolitoli** Indon.
76 B1 **Tolmezzo** Italy
72 B3 **Tolosa** Spain
113 C3 **Toluca** Mex.
55 D3 **Tol'yatti** Rus. Fed.
108 A2 **Tomah** U.S.A.
34 D2 **Tomakomai** Japan
74 B2 **Tomar** Port.
71 E1 **Tomaszów Lubelski** Pol.
71 E1 **Tomaszów Mazowiecki** Pol.
112 B3 **Tomatlán** Mex.
110 C2 **Tombigbee** *r.* U.S.A.
123 D2 **Tombos** Brazil
82 B3 **Tombouctou** Mali
88 A2 **Tombua** Angola
91 C1 **Tom Burke** S. Africa
21 C2 **Tomingley** Austr.
26 C3 **Tomini, Teluk** *g.* Indon.
77 C2 **Tomislavgrad** Bos.-Herz.
18 A2 **Tom Price** Austr.
50 G3 **Tomsk** Rus. Fed.
61 F4 **Tomtabacken** *hill* Sweden
113 C3 **Tonalá** Mex.
118 B2 **Tonantins** Brazil
27 C2 **Tondano** Indon.
17 G4 **Tonga** *country* S. Pacific Ocean
39 B3 **Tongcheng** China
38 A2 **Tongchuan** China
39 A3 **Tongdao** China
33 B2 **Tongduch'ŏn** S. Korea
68 B2 **Tongeren** Belgium
33 B2 **Tonghae** S. Korea
30 B1 **Tonghai** China
33 B1 **Tonghua** China
33 B2 **Tongjosŏn-man** *b.* N. Korea

30 B1 **Tongking, Gulf of** China/Vietnam
37 E2 **Tongliao** China
38 B2 **Tongling** China
20 B2 **Tongo** Austr.
39 A3 **Tongren** China
Tongshan China *see* **Xuzhou**
39 A4 **Tongshi** China
Tongtian He *r.* China *see* Yangtze
64 B1 **Tongue** U.K.
33 B3 **T'ongyŏng** S. Korea
37 E2 **Tongyu** China
33 A1 **Tongyuanpu** China
42 B2 **Tonk** India
31 B2 **Tônlé Sab** *l.* Cambodia
103 C3 **Tonopah** U.S.A.
61 F4 **Tønsberg** Norway
103 D2 **Tooele** U.S.A.
20 B3 **Tooleybuc** Austr.
21 D1 **Toowoomba** Austr.
105 D3 **Topeka** U.S.A.
112 B2 **Topia** Mex.
112 B2 **Topolobampo** Mex.
54 C2 **Topozero, Ozero** *l.* Rus. Fed.
102 B1 **Toppenish** U.S.A.
79 C3 **Torbalı** Turkey
44 B3 **Torbat-e Heydarīyeh** Iran
44 C3 **Torbat-e Jām** Iran
74 C1 **Tordesillas** Spain
75 C1 **Tordesilos** Spain
75 D1 **Torelló** Spain
69 F2 **Torgau** Ger.
68 A2 **Torhout** Belgium
Torino Italy *see* **Turin**
122 B1 **Torixoreu** Brazil
74 B1 **Tormes** *r.* Spain
60 H2 **Torneälven** *r.* Sweden
95 G3 **Torngat Mountains** Can.
60 H2 **Tornio** Fin.
74 B1 **Toro** Spain
98 C2 **Toronto** Can.
87 D2 **Tororo** Uganda
48 B2 **Toros Dağları** *mts* Turkey
67 B3 **Torquay** U.K.
74 B2 **Torrão** Port.
74 B1 **Torre** *mt.* Port.
75 D1 **Torreblanca** Spain
74 B1 **Torre de Moncorvo** Port.
74 C1 **Torrelavega** Spain
74 C2 **Torremolinos** Spain
20 A2 **Torrens, Lake** *salt flat* Austr.
75 C2 **Torrent** Spain
112 B2 **Torreón** Mex.
74 B2 **Torres Novas** Port.
74 B2 **Torres Vedras** Port.
75 C2 **Torrevieja** Spain
64 B2 **Torridon** U.K.
74 C2 **Torrijos** Spain
109 E2 **Torrington** *CT* U.S.A.
104 C2 **Torrington** *WY* U.S.A.
75 D1 **Torroella de Montgrí** Spain
62 B1 **Tórshavn** Faroe Is
76 A3 **Tortolì** *Sardinia* Italy
75 D1 **Tortosa** Spain
71 D1 **Toruń** Pol.
65 B1 **Tory Island** Rep. of Ireland
57 D2 **Torzhok** Rus. Fed.
90 B2 **Tosca** S. Africa
57 D2 **Tosno** Rus. Fed.
120 B3 **Tostado** Arg.
69 D1 **Tostedt** Ger.
48 B1 **Tosya** Turkey
54 D3 **Tot'ma** Rus. Fed.
35 B3 **Tottori** Japan
82 B4 **Touba** Côte d'Ivoire
82 B1 **Toubkal, Jbel** *mt.* Morocco
82 B3 **Tougan** Burkina
83 C1 **Touggourt** Alg.
73 D2 **Toul** France
39 C3 **Touliu** Taiwan
73 D3 **Toulon** France
72 C3 **Toulouse** France
30 A2 **Toungoo** Myanmar
68 A2 **Tournai** Belgium
73 C2 **Tournus** France
119 E2 **Touros** Brazil
72 C2 **Tours** France
83 D2 **Tousside, Pic** *mt.* Chad
90 B3 **Touwsrivier** S. Africa
34 D2 **Towada** Japan
102 D1 **Townsend** U.S.A.
19 D1 **Townsville** Austr.
45 E2 **Toxkan He** *r.* China
34 D2 **Tōya-ko** *l.* Japan
35 C3 **Toyama** Japan
35 C3 **Toyota** Japan
83 C1 **Tozeur** Tunisia
49 C1 **Tqvarch'eli** Georgia
48 B2 **Trâblous** Lebanon
48 B1 **Trabzon** Turkey
74 B2 **Trafalgar, Cabo** *c.* Spain
96 C3 **Trail** Can.
56 B3 **Trakai** Lith.
65 B2 **Tralee** Rep. of Ireland
65 C2 **Tramore** Rep. of Ireland
31 A3 **Trang** Thai.
27 C3 **Trangan** *i.* Indon.
23 I2 **Transantarctic Mountains** Antarctica
Transylvanian Alps *mts* Romania *see* **Carpaţii Meridionali**
76 B3 **Trapani** *Sicily* Italy
21 C3 **Traralgon** Austr.
31 B2 **Trat** Thai.
70 C2 **Traunstein** Ger.
108 B2 **Traverse City** U.S.A.
71 D2 **Třebíč** Czech Rep.
77 C2 **Trebinje** Bos.-Herz.
71 E2 **Trebišov** Slovakia
77 C1 **Trebnje** Slovenia
121 C4 **Treinta-y-Tres** Uru.
121 B5 **Trelew** Arg.
61 F4 **Trelleborg** Sweden
98 C2 **Tremblant, Mont** *hill* Can.
77 C2 **Tremiti, Isole** *is* Italy
102 D2 **Tremonton** U.S.A.
75 D1 **Tremp** Spain
71 D2 **Trenčín** Slovakia
121 B4 **Trenque Lauquén** Arg.
66 C2 **Trent** *r.* U.K.
76 B1 **Trento** Italy
109 D2 **Trenton** Can.
105 E2 **Trenton** *MO* U.S.A.
109 E2 **Trenton** *NJ* U.S.A.
99 E2 **Trepassey** Can.
121 B4 **Tres Arroyos** Arg.
123 C2 **Três Corações** Brazil
122 B2 **Três Lagoas** Brazil
121 A5 **Tres Lagos** Arg.
123 C1 **Três Marias, Represa** *resr* Brazil
123 C2 **Três Pontas** Brazil
121 B5 **Tres Puntas, Cabo** *c.* Arg.
123 D2 **Três Rios** Brazil
69 F1 **Treuenbrietzen** Ger.
76 A1 **Treviglio** Italy
76 B1 **Treviso** Italy
67 A3 **Trevose Head** *hd* U.K.
77 C3 **Tricase** Italy
41 B3 **Trichur** India
21 C2 **Trida** Austr.
68 C3 **Trier** Ger.
76 B1 **Trieste** Italy
76 B1 **Triglav** *mt.* Slovenia
27 D3 **Trikora, Puncak** *mt.* Indon.
65 C2 **Trim** Rep. of Ireland
41 C4 **Trincomalee** Sri Lanka
122 C1 **Trindade** Brazil
120 B2 **Trinidad** Bol.
115 D3 **Trinidad** *i.* Trin. and Tob.
104 C3 **Trinidad** U.S.A.
115 D3 **Trinidad and Tobago** *country* West Indies
99 E2 **Trinity Bay** Can.
79 B3 **Tripoli** Greece
Tripoli Lebanon *see* **Trâblous**
83 D1 **Tripoli** Libya
117 G5 **Tristan da Cunha** *i.* S. Atlantic Ocean
41 B4 **Trivandrum** India
76 B2 **Trivento** Italy
71 D2 **Trnava** Slovakia
77 C2 **Trogir** Croatia
77 C2 **Troia** Italy
68 C2 **Troisdorf** Ger.
99 C2 **Trois-Rivières** Can.
54 E2 **Troitsko-Pechorsk** Rus. Fed.
119 C2 **Trombetas** *r.* Brazil
91 C3 **Trompsburg** S. Africa
60 G2 **Tromsø** Norway
60 F3 **Trondheim** Norway
96 C2 **Trout Lake** Can.
96 B1 **Trout Lake** *l. N.W.T.* Can.
98 A1 **Trout Lake** *l. Ont.* Can.
67 B3 **Trowbridge** U.K.
110 C2 **Troy** *AL* U.S.A.
109 E2 **Troy** *NY* U.S.A.
73 C2 **Troyes** France
77 D2 **Trstenik** Yugo.
57 D3 **Trubchevsk** Rus. Fed.
74 B1 **Truchas** Spain
47 C2 **Trucial Coast** U.A.E.
114 B3 **Trujillo** Hond.
118 A2 **Trujillo** Peru
74 B2 **Trujillo** Spain
115 C4 **Trujillo** Venez.
110 B1 **Trumann** U.S.A.
99 D2 **Truro** Can.
67 A3 **Truro** U.K.
106 B2 **Truth or Consequences** U.S.A.
71 D1 **Trutnov** Czech Rep.
61 F3 **Trysil** Norway
71 D1 **Trzebiatów** Pol.
36 B1 **Tsagaannuur** Mongolia
89 □D2 **Tsaratanana, Massif du** *mts* Madag.
78 C2 **Tsarevo** Bulg.
90 A2 **Tses** Namibia
90 B1 **Tsetseng** Botswana
36 C1 **Tsetserleg** Mongolia
90 B2 **Tshabong** Botswana
90 B1 **Tshane** Botswana
59 D2 **Tshchikskoye Vodokhranilishche** *resr* Rus. Fed.
86 B3 **Tshela** Dem. Rep. Congo
86 C3 **Tshikapa** Dem. Rep. Congo
86 C3 **Tshikapa** *r.* Dem. Rep. Congo
91 D1 **Tshipise** S. Africa
86 C3 **Tshitanzu** Dem. Rep. Congo
86 C3 **Tshuapa** *r.* Dem. Rep. Congo
55 D4 **Tsimlyanskoye Vodokhranilishche** *resr* Rus. Fed.
Tsingtao China *see* **Qingdao**
Tsining China *see* **Jining**
89 □D2 **Tsiroanomandidy** Madag.
91 C3 **Tsomo** S. Africa
35 C4 **Tsu** Japan
35 D3 **Tsuchiura** Japan
34 D2 **Tsugarū-kaikyō** *str.* Japan
88 A2 **Tsumeb** Namibia
90 A1 **Tsumis Park** Namibia
88 B2 **Tsumkwe** Namibia
35 C3 **Tsuruga** Japan
34 C3 **Tsuruoka** Japan

35 A4 **Tsushima** *is* Japan
Tsushima-kaikyō *str.* Japan/S. Korea *see* **Korea Strait**
35 B3 **Tsuyama** Japan
91 C2 **Tswelelang** S. Africa
59 C2 **Tsyurupyns'k** Ukr.
Tthenaagoo Can. *see* **Nahanni Butte**
27 C3 **Tual** Indon.
65 B2 **Tuam** Rep. of Ireland
59 D3 **Tuapse** Rus. Fed.
22 A3 **Tuatapere** N.Z.
106 A1 **Tuba City** U.S.A.
29 C2 **Tuban** Indon.
120 D3 **Tubarão** Brazil
70 B2 **Tübingen** Ger.
83 E1 **Tubruq** Libya
112 A1 **Tubutama** Mex.
120 C2 **Tucavaca** Bol.
96 B1 **Tuchitua** Can.
106 A2 **Tucson** U.S.A.
107 C1 **Tucumcari** U.S.A.
118 B1 **Tucupita** Venez.
119 D2 **Tucuruí** Brazil
119 D2 **Tucuruí, Represa** *resr* Brazil
75 C1 **Tudela** Spain
74 B1 **Tuela** *r.* Port.
30 A1 **Tuensang** India
91 D2 **Tugela** *r.* S. Africa
32 B1 **Tuguegarao** Phil.
74 B1 **Tui** Spain
27 C3 **Tukangbesi, Kepulauan** *is* Indon.
94 C2 **Tuktoyaktuk** Can.
56 B2 **Tukums** Latvia
113 C2 **Tula** Mex.
57 E3 **Tula** Rus. Fed.
113 C2 **Tulancingo** Mex.
103 C3 **Tulare** U.S.A.
106 B2 **Tularosa** U.S.A.
78 C1 **Tulcea** Romania
58 B2 **Tul'chyn** Ukr.
97 E1 **Tulemalu Lake** Can.
107 C2 **Tulia** U.S.A.
110 C1 **Tullahoma** U.S.A.
65 C2 **Tullamore** Rep. of Ireland
72 C2 **Tulle** France
19 D1 **Tully** Austr.
107 D1 **Tulsa** U.S.A.
118 A1 **Tumaco** Col.
91 C2 **Tumahole** S. Africa
61 G4 **Tumba** Sweden
86 B3 **Tumba, Lac** *l.* Dem. Rep. Congo
21 C3 **Tumbarumba** Austr.
118 A2 **Tumbes** Peru
96 B2 **Tumbler Ridge** Can.
20 A2 **Tumby Bay** Austr.
33 B1 **Tumen** China
118 B1 **Tumereng** Guyana
32 A2 **Tumindao** *i.* Phil.
42 A2 **Tump** Pak.
119 C1 **Tumucumaque, Serra** *hills* Brazil
21 C3 **Tumut** Austr.
67 D3 **Tunbridge Wells, Royal** U.K.
48 B2 **Tunceli** Turkey
21 D2 **Tuncurry** Austr.
87 D4 **Tunduru** Tanz.
78 C2 **Tundzha** *r.* Bulg.
96 B1 **Tungsten** Can.
83 D1 **Tunis** Tunisia
83 C1 **Tunisia** *country* Africa
118 A1 **Tunja** Col.
60 F3 **Tunnsjøen** *l.* Norway
Tunxi China *see* **Huangshan**
122 B2 **Tupã** Brazil
122 C1 **Tupaciguara** Brazil
110 C2 **Tupelo** U.S.A.
120 B3 **Tupiza** Bol.
51 H2 **Tura** Rus. Fed.
54 F3 **Tura** *r.* Rus. Fed.
46 B2 **Turabah** Saudi Arabia
51 J3 **Turana, Khrebet** *mts* Rus. Fed.
22 C1 **Turangi** N.Z.
44 B2 **Turan Lowland** Asia
46 A1 **Turayf** Saudi Arabia
56 B2 **Turba** Estonia
42 A2 **Turbat** Pak.
118 A1 **Turbo** Col.
78 B1 **Turda** Romania
Turfan China *see* **Turpan**
44 C2 **Turgay** Kazakh.
79 C3 **Turgutlu** Turkey
48 B1 **Turhal** Turkey
75 C2 **Turia** *r.* Spain
76 A1 **Turin** Italy
54 F3 **Turinsk** Rus. Fed.
58 A1 **Turiys'k** Ukr.
87 D2 **Turkana, Lake** *salt l.* Eth./Kenya
45 C2 **Turkestan** Kazakh.
48 B2 **Turkey** *country* Asia
44 B2 **Turkmenbashi** Turkm.
44 B2 **Turkmenistan** *country* Asia
115 C2 **Turks and Caicos Islands** *terr.* West Indies
61 H3 **Turku** Fin.
103 B3 **Turlock** U.S.A.
22 C2 **Turnagain, Cape** N.Z.
68 B2 **Turnhout** Belgium
97 D2 **Turnor Lake** Can.
78 B2 **Turnu Măgurele** Romania
36 B2 **Turpan** China
45 D2 **Turugart Pass** China/Kyrg.
110 C2 **Tuscaloosa** U.S.A.
110 C2 **Tuskegee** U.S.A.
41 B4 **Tuticorin** India
88 B3 **Tutume** Botswana
17 F3 **Tuvalu** *country* S. Pacific Ocean
46 B2 **Tuwayq, Jabal** *hills* Saudi Arabia
46 B2 **Tuwayq, Jabal** *mts* Saudi Arabia
46 A2 **Tuwwal** Saudi Arabia
112 B2 **Tuxpan** *Nayarit* Mex.
113 C2 **Tuxpan** *Veracruz* Mex.
113 C3 **Tuxtla Gutiérrez** Mex.
30 B1 **Tuyên Quang** Vietnam
31 B2 **Tuy Hoa** Vietnam
48 B2 **Tuz Gölü** *salt l.* Turkey
49 C2 **Tuz Khurmātū** Iraq
77 C2 **Tuzla** Bos.-Herz.
59 E2 **Tuzlov** *r.* Rus. Fed.
57 E2 **Tver'** Rus. Fed.
66 B1 **Tweed** *r.* U.K.
90 A2 **Twee Rivier** Namibia
103 C4 **Twentynine Palms** U.S.A.
99 E2 **Twillingate** Can.
102 D2 **Twin Falls** U.S.A.
105 E1 **Two Harbors** U.S.A.
97 C2 **Two Hills** Can.
107 D2 **Tyler** U.S.A.
51 J3 **Tynda** Rus. Fed.
61 F3 **Tynset** Norway
79 B3 **Tyrnavos** Greece
20 B3 **Tyrrell, Lake** *dry lake* Austr.
76 B2 **Tyrrhenian Sea** France/Italy
44 B2 **Tyub-Karagan, Mys** *pt* Kazakh.
55 E3 **Tyul'gan** Rus. Fed.
54 F3 **Tyumen'** Rus. Fed.
51 J2 **Tyung** *r.* Rus. Fed.
67 A3 **Tywi** *r.* U.K.
91 D1 **Tzaneen** S. Africa

U

88 B2 **Uamanda** Angola
118 B2 **Uaupés** Brazil
123 D2 **Ubá** Brazil
123 D1 **Ubaí** Brazil
119 E3 **Ubaitaba** Brazil
86 B3 **Ubangi** *r.* C.A.R./Dem. Rep. Congo
35 B4 **Ube** Japan
74 C2 **Úbeda** Spain
122 C1 **Uberaba** Brazil
122 C1 **Uberlândia** Brazil
91 D2 **Ubombo** S. Africa
31 B2 **Ubon Ratchathani** Thai.
87 C3 **Ubundu** Dem. Rep. Congo
118 A2 **Ucayali** *r.* Peru
42 B2 **Uch** Pak.
45 E2 **Ucharal** Kazakh.
34 D2 **Uchiura-wan** *b.* Japan
51 J3 **Uchur** *r.* Rus. Fed.
96 B3 **Ucluelet** Can.
42 B2 **Udaipur** India
59 C1 **Uday** *r.* Ukr.
61 F4 **Uddevalla** Sweden
60 G2 **Uddjaure** *l.* Sweden
68 B2 **Uden** Neth.
42 B1 **Udhampur** Jammu and Kashmir
76 B1 **Udine** Italy
57 E2 **Udomlya** Rus. Fed.
30 B2 **Udon Thani** Thai.
41 B3 **Udupi** India
51 K3 **Udyl', Ozero** *l.* Rus. Fed.
35 C3 **Ueda** Japan
26 C3 **Uekuli** Indon.
86 C2 **Uele** *r.* Dem. Rep. Congo
69 E1 **Uelzen** Ger.
87 C2 **Uere** *r.* Dem. Rep. Congo
55 E3 **Ufa** Rus. Fed.
87 D3 **Ugalla** *r.* Tanz.
87 D2 **Uganda** *country* Africa
57 E2 **Uglich** Rus. Fed.
57 D2 **Uglovka** Rus. Fed.
57 D3 **Ugra** Rus. Fed.
71 D2 **Uherské Hradiště** Czech Rep.
69 E2 **Uichteritz** Ger.
64 A2 **Uig** U.K.
88 A1 **Uíge** Angola
33 B2 **Ŭijŏngbu** S. Korea
103 D2 **Uinta Mountains** U.S.A.
88 A3 **Uis Mine** Namibia
33 B2 **Ŭisŏng** S. Korea
91 C3 **Uitenhage** S. Africa
68 C1 **Uithuizen** Neth.
42 B2 **Ujjain** India
26 B3 **Ujung Pandang** Indon.
57 F3 **Ukholovo** Rus. Fed.
30 A1 **Ukhrul** India
54 E2 **Ukhta** Rus. Fed.
103 B3 **Ukiah** U.S.A.
95 H2 **Ukkusissat** Greenland
56 B2 **Ukmergė** Lith.
58 C2 **Ukraine** *country* Europe
37 D1 **Ulaanbaatar** Mongolia
36 C1 **Ulaangom** Mongolia
Ulan Bator Mongolia *see* **Ulaanbaatar**
Ulanhad China *see* **Chifeng**
37 E1 **Ulanhot** China
55 D4 **Ulan-Khol** Rus. Fed.
37 D1 **Ulan-Ude** Rus. Fed.
43 D1 **Ulan Ul Hu** *l.* China
33 B2 **Ulchin** S. Korea
Uleåborg Fin. *see* **Oulu**
56 C2 **Ülenurme** Estonia
41 B3 **Ulhasnagar** India

37 D1 **Uliastai** China
36 C1 **Uliastay** Mongolia
27 D2 **Ulithi** *atoll* Micronesia
21 D3 **Ulladulla** Austr.
64 B2 **Ullapool** U.K.
66 B1 **Ullswater** *l.* U.K.
33 C2 **Ullŭng-do** *i.* S. Korea
70 B2 **Ulm** Ger.
33 B2 **Ulsan** S. Korea
64 □ **Ulsta** U.K.
65 C1 **Ulster** *reg.* Rep. of Ireland/U.K.
20 B3 **Ultima** Austr.
79 C3 **Ulubey** Turkey
79 C2 **Uludağ** *mt.* Turkey
91 D2 **Ulundi** S. Africa
45 E2 **Ulungur Hu** *l.* China
Uluru *hill* Austr. *see* **Ayers Rock**
66 B1 **Ulverston** U.K.
55 D3 **Ul'yanovsk** Rus. Fed.
104 C3 **Ulysses** U.S.A.
58 C2 **Uman'** Ukr.
54 C2 **Umba** Rus. Fed.
27 D3 **Umboi** *i.* P.N.G.
60 H3 **Umeå** Sweden
60 H3 **Umeälven** *r.* Sweden
95 I2 **Umiiviip Kangertiva** *inlet* Greenland
94 D2 **Umingmaktok** Can.
91 D2 **Umlazi** S. Africa
85 A3 **Umm Keddada** Sudan
46 A2 **Umm Lajj** Saudi Arabia
85 B3 **Umm Ruwaba** Sudan
83 E1 **Umm Sa'ad** Libya
102 B2 **Umpqua** *r.* U.S.A.
88 A2 **Umpulo** Angola
91 C3 **Umtata** S. Africa
122 B2 **Umuarama** Brazil
77 C1 **Una** *r.* Bos.-Herz./Croatia
123 E1 **Una** Brazil
122 C1 **Unaí** Brazil
46 B2 **'Unayzah** Saudi Arabia
104 C1 **Underwood** U.S.A.
57 D3 **Unecha** Rus. Fed.
21 C2 **Ungarie** Austr.
20 A2 **Ungarra** Austr.
95 G2 **Ungava, Péninsule d'** *pen.* Can.
95 G3 **Ungava Bay** Can.
33 C1 **Unggi** N. Korea
58 B2 **Ungheni** Moldova
Unguja *i.* Tanz. *see* **Zanzibar Island**
122 B3 **União da Vitória** Brazil
118 B2 **Unini** *r.* Brazil
110 C1 **Union City** U.S.A.
90 B3 **Uniondale** S. Africa
109 D3 **Uniontown** U.S.A.
47 C2 **United Arab Emirates** *country* Asia
63 C3 **United Kingdom** *country* Europe
100 D3 **United States of America** *country* N. America
97 D2 **Unity** Can.
64 □ **Unst** *i.* U.K.
69 E2 **Unstrut** *r.* Ger.
57 E3 **Upa** *r.* Rus. Fed.
87 C3 **Upemba, Lac** *l.* Dem. Rep. Congo
90 B2 **Upington** S. Africa
102 B2 **Upper Alkali Lake** U.S.A.
96 C2 **Upper Arrow Lake** Can.
102 B2 **Upper Klamath Lake** U.S.A.
96 B1 **Upper Liard** Can.
65 C1 **Upper Lough Erne** *l.* U.K.
61 G4 **Uppsala** Sweden
46 B2 **'Uqlat aş Şuqūr** Saudi Arabia
Urad Qianqi China *see* **Xishanzui**
44 B2 **Ural** *r.* Kazakh./Rus. Fed.
21 D2 **Uralla** Austr.
Ural Mountains Rus. Fed. *see* **Ural'skiy Khrebet**
44 B1 **Ural'sk** Kazakh.
55 E3 **Ural'skiy Khrebet** *mts* Rus. Fed.
87 D3 **Urambo** Tanz.
21 C3 **Urana** Austr.
97 D2 **Uranium City** Can.
54 F2 **Uray** Rus. Fed.
66 C1 **Ure** *r.* U.K.
54 D3 **Uren'** Rus. Fed.
112 A2 **Ures** Mex.
44 C2 **Urgench** Uzbek.
68 B1 **Urk** Neth.
79 C3 **Urla** Turkey
77 D2 **Uroševac** Yugo.
112 B2 **Uruáchic** Mex.
119 D3 **Uruaçu** Brazil
112 B3 **Uruapan** Mex.
118 A3 **Urubamba** *r.* Peru
119 C2 **Urucara** Brazil
119 D2 **Uruçuí** Brazil
119 C2 **Urucurituba** Brazil
120 C3 **Uruguaiana** Brazil
121 C4 **Uruguay** *country* S. America
Urumchi China *see* **Ürümqi**
36 B2 **Ürümqi** China
21 D2 **Urunga** Austr.
78 C2 **Urziceni** Romania
35 B4 **Usa** Japan
50 E2 **Usa** *r.* Rus. Fed.
79 C3 **Uşak** Turkey
50 G1 **Ushakova, Ostrov** *i.* Rus. Fed.
45 D2 **Ushtobe** Kazakh.
121 B6 **Ushuaia** Arg.
54 E2 **Usinsk** Rus. Fed.
56 C3 **Uskhodni** Belarus
57 E3 **Usman'** Rus. Fed.
54 D2 **Usogorsk** Rus. Fed.
72 C2 **Ussel** France
34 B2 **Ussuriysk** Rus. Fed.
76 B3 **Ustica, Isola di** *i. Sicily* Italy
51 H3 **Ust'-Ilimsk** Rus. Fed.
54 E2 **Ust'-Ilych** Rus. Fed.
71 D1 **Ustka** Pol.
51 L3 **Ust'-Kamchatsk** Rus. Fed.
45 E2 **Ust'-Kamenogorsk** Kazakh.
54 F2 **Ust'-Kara** Rus. Fed.
54 E2 **Ust'-Kulom** Rus. Fed.
51 I3 **Ust'-Kut** Rus. Fed.
59 D2 **Ust'-Labinsk** Rus. Fed.
56 C2 **Ust'-Luga** Rus. Fed.
54 E2 **Ust'-Nem** Rus. Fed.
51 K2 **Ust'-Nera** Rus. Fed.
51 K2 **Ust'-Omchug** Rus. Fed.
51 H3 **Ust'-Ordynskiy** Rus. Fed.
54 E2 **Ust'-Tsil'ma** Rus. Fed.
54 D2 **Ust'-Ura** Rus. Fed.
44 B2 **Ustyurt Plateau** Kazakh./Uzbek.
57 E2 **Ustyuzhna** Rus. Fed.
57 D2 **Usvyaty** Rus. Fed.
103 D3 **Utah** *state* U.S.A.
103 D2 **Utah Lake** U.S.A.
61 I4 **Utena** Lith.
109 D2 **Utica** U.S.A.
75 C2 **Utiel** Spain
96 C2 **Utikuma Lake** Can.
68 B1 **Utrecht** Neth.
74 B2 **Utrera** Spain
60 I2 **Utsjoki** Fin.
35 C3 **Utsunomiya** Japan
55 D4 **Utta** Rus. Fed.
30 B2 **Uttaradit** Thai.
95 G1 **Uummannaq** Greenland
95 H2 **Uummannaq Fjord** *inlet* Greenland
61 H3 **Uusikaupunki** Fin.
107 D3 **Uvalde** U.S.A.
87 D3 **Uvinza** Tanz.
36 C1 **Uvs Nuur** *salt l.* Mongolia
35 B4 **Uwajima** Japan
46 A2 **'Uwayriḍ, Ḩarrat al** *lava field* Saudi Arabia
84 A2 **Uweinat, Jebel** *mt.* Sudan
51 H3 **Uyar** Rus. Fed.
83 C4 **Uyo** Nigeria
120 B3 **Uyuni, Salar de** *salt flat* Bol.
44 C2 **Uzbekistan** *country* Asia
72 C2 **Uzerche** France
73 C3 **Uzès** France
58 C1 **Uzh** *r.* Ukr.
58 A2 **Uzhhorod** Ukr.
77 C2 **Užice** Yugo.
57 E3 **Uzlovaya** Rus. Fed.
79 C3 **Üzümlü** Turkey
79 C2 **Uzunköprü** Turkey

V

91 B2 **Vaal** *r.* S. Africa
91 C2 **Vaal Dam** S. Africa
91 C1 **Vaalwater** S. Africa
60 H3 **Vaasa** Fin.
71 D2 **Vác** Hungary
120 C3 **Vacaria** Brazil
122 B2 **Vacaria, Serra** *hills* Brazil
103 B3 **Vacaville** U.S.A.
42 B2 **Vadodara** India
60 I1 **Vadsø** Norway
73 D2 **Vaduz** Liechtenstein
62 B1 **Vágar** *i.* Faroe Is
62 B1 **Vágur** Faroe Is
71 D2 **Váh** *r.* Slovakia
56 B2 **Vaida** Estonia
104 B3 **Vail** U.S.A.
47 C2 **Vakīlābād** Iran
76 B1 **Valdagno** Italy
57 D2 **Valday** Rus. Fed.
57 D2 **Valdayskaya Vozvyshennost'** *hills* Rus. Fed.
74 B2 **Valdecañas, Embalse de** *resr* Spain
61 G4 **Valdemarsvik** Sweden
73 D2 **Val-de-Meuse** France
74 C2 **Valdepeñas** Spain
121 B5 **Valdés, Península** *pen.* Arg.
121 A4 **Valdivia** Chile
98 C2 **Val-d'Or** Can.
111 D2 **Valdosta** U.S.A.
96 C2 **Valemount** Can.
73 C3 **Valence** France
75 C2 **Valencia** Spain
118 B1 **Valencia** Venez.
75 D2 **Valencia, Golfo de** *g.* Spain
68 A2 **Valenciennes** France
104 C2 **Valentine** U.S.A.
32 B1 **Valenzuela** Phil.
118 A1 **Valera** Venez.
77 C2 **Valjevo** Yugo.
56 C2 **Valka** Latvia
61 H3 **Valkeakoski** Fin.
68 B2 **Valkenswaard** Neth.
59 D2 **Valky** Ukr.
23 D2 **Valkyrie Dome** *ice feature* Antarctica
113 D2 **Valladolid** Mex.
74 C1 **Valladolid** Spain
75 C2 **Vall de Uxó** Spain
61 E4 **Valle** Norway
113 C2 **Vallecillos** Mex.

118 B1 **Valle de la Pascua** Venez.
118 A1 **Valledupar** Col.
113 C2 **Valle Hermoso** Mex.
103 B3 **Vallejo** U.S.A.
120 A3 **Vallenar** Chile
105 D1 **Valley City** U.S.A.
102 B2 **Valley Falls** U.S.A.
96 C2 **Valleyview** Can.
75 D1 **Valls** Spain
97 D3 **Val Marie** Can.
56 C2 **Valmiera** Latvia
56 C3 **Valozhyn** Belarus
122 B2 **Valparaíso** Brazil
121 A4 **Valparaíso** Chile
73 C3 **Valréas** France
27 D3 **Vals, Tanjung** *c.* Indon.
42 B2 **Valsād** India
90 B2 **Valspan** S. Africa
59 D1 **Valuyki** Rus. Fed.
74 B2 **Valverde del Camino** Spain
49 C2 **Van** Turkey
109 F1 **Van Buren** U.S.A.
96 B3 **Vancouver** Can.
102 B1 **Vancouver** U.S.A.
96 B3 **Vancouver Island** Can.
108 B3 **Vandalia** U.S.A.
91 C2 **Vanderbijlpark** S. Africa
96 B2 **Vanderhoof** Can.
56 C2 **Vändra** Estonia
61 F4 **Vänern** *l.* Sweden
61 F4 **Vänersborg** Sweden
89 □D3 **Vangaindrano** Madag.
49 C2 **Van Gölü** *salt l.* Turkey
106 C2 **Van Horn** U.S.A.
27 D3 **Vanimo** P.N.G.
51 K3 **Vanino** Rus. Fed.
72 B2 **Vannes** France
27 D3 **Van Rees, Pegunungan** *mts* Indon.
90 A3 **Vanrhynsdorp** S. Africa
61 H3 **Vantaa** Fin.
17 E4 **Vanuatu** *country* S. Pacific Ocean
108 C2 **Van Wert** U.S.A.
90 B3 **Vanwyksvlei** S. Africa
90 B2 **Van Zylsrus** S. Africa
43 C2 **Varanasi** India
60 I1 **Varangerfjorden** *sea chan.* Norway
60 I1 **Varanger Halvøya** *pen.* Norway
77 C1 **Varaždin** Croatia
61 F4 **Varberg** Sweden
77 D2 **Vardar** *r.* Macedonia
61 E4 **Varde** Denmark
60 J1 **Vardø** Norway
68 D1 **Varel** Ger.
56 B3 **Varėna** Lith.
76 A1 **Varese** Italy
123 C2 **Varginha** Brazil
61 I3 **Varkaus** Fin.
78 C2 **Varna** Bulg.
61 F4 **Värnamo** Sweden
123 D1 **Várzea da Palma** Brazil
54 C2 **Varzino** Rus. Fed.
Vasa Fin. *see* **Vaasa**
56 C2 **Vasknarva** Estonia
78 C1 **Vaslui** Romania
61 G4 **Västerås** Sweden
61 G3 **Västerdalälven** *r.* Sweden
56 A2 **Västerhaninge** Sweden
61 G4 **Västervik** Sweden
76 B2 **Vasto** Italy
58 C1 **Vasyl'kiv** Ukr.
72 C2 **Vatan** France
76 B2 **Vatican City** Europe
60 □B3 **Vatnajökull** *ice cap* Iceland
78 C1 **Vatra Dornei** Romania
61 F4 **Vättern** *l.* Sweden
106 B2 **Vaughn** U.S.A.
73 C3 **Vauvert** France
56 B3 **Vawkavysk** Belarus
61 F4 **Växjö** Sweden
54 E1 **Vaygach, Ostrov** *i.* Rus. Fed.
69 D1 **Vechta** Ger.
78 C2 **Vedea** *r.* Romania
68 C1 **Veendam** Neth.
68 B1 **Veenendaal** Neth.
97 C2 **Vegreville** Can.
74 B2 **Vejer de la Frontera** Spain
61 E4 **Vejle** Denmark
77 D2 **Velbŭzhdki Prokhod** *pass* Macedonia
68 B2 **Veldhoven** Neth.
77 B2 **Velebit** *mts* Croatia
68 C2 **Velen** Ger.
77 C1 **Velenje** Slovenia
77 D2 **Veles** Macedonia
74 C2 **Vélez-Málaga** Spain
123 D1 **Velhas** *r.* Brazil
77 D2 **Velika Plana** Yugo.
56 C2 **Velikaya** *r.* Rus. Fed.
57 D2 **Velikiye Luki** Rus. Fed.
57 D2 **Velikiy Novgorod** Rus. Fed.
54 D2 **Velikiy Ustyug** Rus. Fed.
78 C2 **Veliko Tŭrnovo** Bulg.
76 B2 **Veli Lošinj** Croatia
57 D2 **Velizh** Rus. Fed.
54 D2 **Vel'sk** Rus. Fed.
69 F1 **Velten** Ger.
59 D1 **Velykyy Burluk** Ukr.
76 B2 **Venafro** Italy
122 C2 **Venceslau Bráz** Brazil
72 C2 **Vendôme** France
57 E3 **Venev** Rus. Fed.
Venezia Italy *see* **Venice**
118 B1 **Venezuela** *country* S. America
118 A1 **Venezuela, Golfo de** *g.* Venez.
76 B1 **Venice** Italy
111 D3 **Venice** U.S.A.
76 B1 **Venice, Gulf of** Europe
68 C2 **Venlo** Neth.
68 B2 **Venray** Neth.
56 B2 **Venta** *r.* Latvia/Lith.
56 B2 **Venta** Lith.
91 C3 **Venterstad** S. Africa
67 C3 **Ventnor** U.K.
56 B2 **Ventspils** Latvia
103 C4 **Ventura** U.S.A.
107 C3 **Venustiano Carranza, Presa** *resr* Mex.
75 C2 **Vera** Spain
113 C3 **Veracruz** Mex.
42 B2 **Veraval** India
76 A1 **Verbania** Italy
76 A1 **Vercelli** Italy
73 D3 **Vercors** *reg.* France
60 F3 **Verdalsøra** Norway
122 B1 **Verde** *r. Goiás* Brazil
122 B2 **Verde** *r. Mato Grosso do Sul* Brazil
112 B2 **Verde** *r.* Mex.
106 A2 **Verde** *r.* U.S.A.
123 D1 **Verde Grande** *r.* Brazil
69 D1 **Verden (Aller)** Ger.
122 B1 **Verdinho, Serra do** *mts* Brazil
76 A2 **Verdon** *r.* France
73 D2 **Verdun** France
91 C2 **Vereeniging** S. Africa
74 B1 **Verín** Spain
59 D3 **Verkhnebakanskiy** Rus. Fed.
60 J2 **Verkhnetulomskiy** Rus. Fed.
59 E1 **Verkhniy Mamon** Rus. Fed.
57 E3 **Verkhov'ye** Rus. Fed.
58 A2 **Verkhovyna** Ukr.
51 J2 **Verkhoyanskiy Khrebet** *mts* Rus. Fed.
97 C2 **Vermilion** Can.
105 D2 **Vermillion** U.S.A.
98 A2 **Vermillion Bay** Can.
109 E2 **Vermont** *state* U.S.A.
103 E2 **Vernal** U.S.A.
96 C2 **Vernon** Can.
107 D2 **Vernon** U.S.A.
111 D3 **Vero Beach** U.S.A.
79 B2 **Veroia** Greece
76 B1 **Verona** Italy
72 C2 **Versailles** France
72 B2 **Vertou** France
91 D2 **Verulam** S. Africa
68 B2 **Verviers** Belgium
73 C2 **Vervins** France
73 D3 **Vescovato** *Corsica* France
55 E3 **Veselaya, Gora** *mt.* Rus. Fed.
59 C2 **Vesele** Ukr.
73 D2 **Vesoul** France
60 F2 **Vesterålen** *is* Norway
60 F2 **Vestfjorden** *sea chan.* Norway
62 B1 **Vestmanna** Faroe Is
61 E3 **Vestnes** Norway
76 B2 **Vesuvio** *vol.* Italy
57 E2 **Ves'yegonsk** Rus. Fed.
61 G4 **Vetlanda** Sweden
54 D3 **Vetluga** Rus. Fed.
68 A2 **Veurne** Belgium
73 D2 **Vevey** Switz.
59 D1 **Veydelevka** Rus. Fed.
48 B1 **Vezirköprü** Turkey
119 D2 **Viana** Brazil
74 B1 **Viana do Castelo** Port.
Viangchan Laos *see* Vientiane
30 B1 **Viangphoukha** Laos
122 C1 **Vianópolis** Brazil
76 B2 **Viareggio** Italy
61 E4 **Viborg** Denmark
77 C3 **Vibo Valentia** Italy
75 D1 **Vic** Spain
112 A1 **Vicente Guerrero** Mex.
76 B1 **Vicenza** Italy
73 C2 **Vichy** France
110 B2 **Vicksburg** U.S.A.
123 D2 **Viçosa** Brazil
20 A3 **Victor Harbor** Austr.
18 C1 **Victoria** *r.* Austr.
20 B3 **Victoria** *state* Austr.
96 B3 **Victoria** Can.
121 A4 **Victoria** Chile
107 D3 **Victoria** U.S.A.
87 D3 **Victoria, Lake** Africa
20 B2 **Victoria, Lake** Austr.
30 A1 **Victoria, Mount** Myanmar
27 D3 **Victoria, Mount** P.N.G.
88 B2 **Victoria Falls** *waterfall* Zambia/Zimbabwe
94 D2 **Victoria Island** Can.
18 C1 **Victoria River Downs** Austr.
90 B3 **Victoria West** S. Africa
103 C4 **Victorville** U.S.A.
78 C2 **Videle** Romania
60 □A2 **Víðidalsá** Iceland
121 B5 **Viedma** Arg.
121 A5 **Viedma, Lago** *l.* Arg.
68 B2 **Vielsalm** Belgium
69 E2 **Vienenburg** Ger.
71 D2 **Vienna** Austria
73 C2 **Vienne** France
72 C2 **Vienne** *r.* France
30 B2 **Vientiane** Laos
68 C2 **Viersen** Ger.
72 C2 **Vierzon** France
112 B2 **Viesca** Mex.
77 C2 **Vieste** Italy
30 B2 **Vietnam** *country* Asia
30 B1 **Viêt Tri** Vietnam

32 B1 **Vigan** Phil.
76 A1 **Vigevano** Italy
74 B1 **Vigo** Spain
41 C3 **Vijayawada** India
60 □B3 **Vík** Iceland
97 C2 **Viking** Can.
74 B2 **Vila Franca de Xira** Port.
74 B1 **Vilagarcía de Arousa** Spain
74 B1 **Vilalba** Spain
74 B1 **Vila Nova de Gaia** Port.
75 D1 **Vilanova i la Geltrú** Spain
74 B1 **Vila Real** Port.
74 B1 **Vilar Formoso** Port.
123 D2 **Vila Velha** Brazil
118 A3 **Vilcabamba, Cordillera** *mts* Peru
60 G3 **Vilhelmina** Sweden
118 B3 **Vilhena** Brazil
56 C2 **Viljandi** Estonia
56 B3 **Vilkaviškis** Lith.
51 H1 **Vil'kitskogo, Proliv** *str.* Rus. Fed.
112 B1 **Villa Ahumada** Mex.
74 B1 **Villablino** Spain
70 C2 **Villach** Austria
112 A2 **Villa Constitución** Mex.
112 B2 **Villa de Cos** Mex.
120 B4 **Villa Dolores** Arg.
113 C3 **Villa Flores** Mex.
113 C2 **Villagrán** Mex.
113 C3 **Villahermosa** Mex.
112 A2 **Villa Insurgentes** Mex.
120 B4 **Villa María** Arg.
120 B3 **Villa Montes** Bol.
112 B2 **Villanueva** Mex.
74 B2 **Villanueva de la Serena** Spain
74 C2 **Villanueva de los Infantes** Spain
120 C3 **Villa Ocampo** Arg.
76 A3 **Villaputzu** *Sardinia* Italy
120 C3 **Villarrica** Para.
74 C2 **Villarrobledo** Spain
120 B3 **Villa Unión** Arg.
112 B2 **Villa Unión** *Durango* Mex.
112 B2 **Villa Unión** *Sinaloa* Mex.
118 A1 **Villavicencio** Col.
120 B3 **Villazon** Bol.
75 C2 **Villena** Spain
72 C3 **Villeneuve-sur-Lot** France
110 B2 **Ville Platte** U.S.A.
73 C2 **Villeurbanne** France
70 B2 **Villingen** Ger.
56 C3 **Vilnius** Lith.
59 C2 **Vil'nohirs'k** Ukr.
59 D2 **Vil'nyans'k** Ukr.
68 B2 **Vilvoorde** Belgium
56 C3 **Vilyeyka** Belarus
51 J2 **Vilyuy** *r.* Rus. Fed.
61 G4 **Vimmerby** Sweden
121 A4 **Viña del Mar** Chile
75 D1 **Vinaròs** Spain
108 B3 **Vincennes** U.S.A.
23 G3 **Vincennes Bay** Antarctica
109 D3 **Vineland** U.S.A.
30 B2 **Vinh** Vietnam
31 B2 **Vinh Long** Vietnam
107 D1 **Vinita** U.S.A.
58 B2 **Vinnytsya** Ukr.
23 O2 **Vinson Massif** *mt.* Antarctica
61 E3 **Vinstra** Norway
76 B1 **Vipiteno** Italy
32 B1 **Virac** Phil.
97 D3 **Virden** Can.
72 B2 **Vire** France
88 A2 **Virei** Angola
123 D1 **Virgem da Lapa** Brazil
106 A1 **Virgin** *r.* U.S.A.
91 C2 **Virginia** S. Africa
105 E1 **Virginia** U.S.A.
109 D3 **Virginia** *state* U.S.A.
109 D3 **Virginia Beach** U.S.A.
103 C3 **Virginia City** U.S.A.
115 D3 **Virgin Islands (U.K.)** *terr.* West Indies
115 D3 **Virgin Islands (U.S.A.)** *terr.* West Indies
31 B2 **Virôchey** Cambodia
77 C1 **Virovitica** Croatia
68 B3 **Virton** Belgium
56 B2 **Virtsu** Estonia
77 C2 **Vis** *i.* Croatia
56 C2 **Visaginas** Lith.
103 C3 **Visalia** U.S.A.
42 B2 **Visavadar** India
32 B1 **Visayan Sea** Phil.
61 G4 **Visby** Sweden
94 D2 **Viscount Melville Sound** *sea chan.* Can.
119 D2 **Viseu** Brazil
74 B1 **Viseu** Port.
41 C3 **Vishakhapatnam** India
56 C2 **Višķi** Latvia
77 C2 **Visoko** Bos.-Herz.
76 B2 **Viterbo** Italy
51 I3 **Vitim** *r.* Rus. Fed.
123 D2 **Vitória** Brazil
119 D3 **Vitória da Conquista** Brazil
74 C1 **Vitoria-Gasteiz** Spain
72 B2 **Vitré** France
73 C2 **Vitry-le-François** France
57 D2 **Vitsyebsk** Belarus
73 D2 **Vittel** France
76 B3 **Vittoria** *Sicily* Italy
76 B1 **Vittorio Veneto** Italy
74 B1 **Viveiro** Spain
112 A2 **Vizcaíno, Sierra** *mts* Mex.
78 C2 **Vize** Turkey
41 C3 **Vizianagaram** India
68 B2 **Vlaardingen** Neth.
55 D4 **Vladikavkaz** Rus. Fed.
57 F2 **Vladimir** Rus. Fed.
34 B2 **Vladivostok** Rus. Fed.
77 D2 **Vlasotince** Yugo.
68 B1 **Vlieland** *i.* Neth.
68 A2 **Vlissingen** Neth.
77 C2 **Vlorë** Albania
70 C2 **Vöcklabruck** Austria
69 D2 **Vogelsberg** *hills* Ger.
87 D3 **Voi** Kenya
73 D2 **Voiron** France
99 D1 **Voisey Bay** Can.
77 C1 **Vojvodina** *prov.* Yugo.
60 J3 **Voknavolok** Rus. Fed.
Volcano Bay *b.* Japan *see* Uchiura-wan
57 E2 **Volga** Rus. Fed.
57 F2 **Volga** *r.* Rus. Fed.
55 D4 **Volgodonsk** Rus. Fed.
55 D4 **Volgograd** Rus. Fed.
57 D2 **Volkhov** Rus. Fed.
57 D1 **Volkhov** *r.* Rus. Fed.
69 E2 **Volkstedt** Ger.
59 D2 **Volnovakha** Ukr.
58 B2 **Volochys'k** Ukr.
59 D2 **Volodars'ke** Ukr.
58 B1 **Volodars'k-Volyns'kyy** Ukr.
58 B1 **Volodymyrets'** Ukr.
58 A1 **Volodymyr-Volyns'kyy** Ukr.
57 E2 **Vologda** Rus. Fed.
57 E2 **Volokolamsk** Rus. Fed.
59 D1 **Volokonovka** Rus. Fed.
79 B3 **Volos** Greece
56 C2 **Volosovo** Rus. Fed.
57 D2 **Volot** Rus. Fed.
57 E3 **Volovo** Rus. Fed.
55 D3 **Vol'sk** Rus. Fed.
82 C4 **Volta, Lake** *resr* Ghana
123 D2 **Volta Redonda** Brazil
55 D4 **Volzhskiy** Rus. Fed.
60 □C2 **Vopnafjörður** Iceland
56 C3 **Voranava** Belarus
54 F2 **Vorkuta** Rus. Fed.
56 B2 **Vormsi** *i.* Estonia
57 E3 **Voronezh** Rus. Fed.
57 E3 **Voronezh** *r.* Rus. Fed.
59 E1 **Vorontsovka** Rus. Fed.
59 C2 **Vorskla** *r.* Rus. Fed.
56 C2 **Võrtsjärv** *l.* Estonia
56 C2 **Võru** Estonia
90 B3 **Vosburg** S. Africa
73 D2 **Vosges** *mts* France
57 E2 **Voskresensk** Rus. Fed.
61 E3 **Voss** Norway
51 K2 **Vostochno-Sibirskoye More** *sea* Rus. Fed.
51 H3 **Vostochnyy Sayan** *mts* Rus. Fed.
54 E3 **Votkinsk** Rus. Fed.
122 C2 **Votuporanga** Brazil
68 B3 **Vouziers** France
60 J2 **Voynitsa** Rus. Fed.
59 C2 **Voznesens'k** Ukr.
61 E4 **Vrådal** Norway
34 B2 **Vrangel'** Rus. Fed.
51 N2 **Vrangelya, Ostrov** *i.* Rus. Fed.
77 D2 **Vranje** Yugo.
78 C2 **Vratnik** *pass* Bulg.
78 B2 **Vratsa** Bulg.
77 C1 **Vrbas** *r.* Bos.-Herz.
77 C1 **Vrbas** Yugo.
90 A3 **Vredenburg** S. Africa
90 A3 **Vredendal** S. Africa
68 B3 **Vresse** Belgium
77 D1 **Vršac** Yugo.
90 B2 **Vryburg** S. Africa
91 D2 **Vryheid** S. Africa
57 D1 **Vsevolozhsk** Rus. Fed.
77 D2 **Vučitrn** Yugo.
77 C1 **Vukovar** Croatia
54 E2 **Vuktyl'** Rus. Fed.
91 C2 **Vukuzakhe** S. Africa
77 B3 **Vulcano, Isola** *i.* Italy
31 B2 **Vung Tau** Vietnam
42 B2 **Vyara** India
Vyatka Rus. Fed. *see* **Kirov**
57 D2 **Vyaz'ma** Rus. Fed.
61 I3 **Vyborg** Rus. Fed.
56 C2 **Vyerkhnyadzvinsk** Belarus
55 D3 **Vyksa** Rus. Fed.
58 B2 **Vylkove** Ukr.
58 A2 **Vynohradiv** Ukr.
57 D2 **Vypolzovo** Rus. Fed.
57 D2 **Vyritsa** Rus. Fed.
59 D2 **Vyselki** Rus. Fed.
58 C1 **Vyshhorod** Ukr.
57 D2 **Vyshniy-Volochek** Rus. Fed.
71 D2 **Vyškov** Czech Rep.
54 C2 **Vytegra** Rus. Fed.

W

82 B3 **Wa** Ghana
68 B2 **Waal** *r.* Neth.
68 B2 **Waalwijk** Neth.
96 C2 **Wabasca** *r.* Can.
96 C2 **Wabasca-Desmarais** Can.
108 B3 **Wabash** *r.* U.S.A.
85 C4 **Wabē Gestro** *r.* Eth.
85 C4 **Wabē Shebelē Wenz** *r.* Eth.
97 E2 **Wabowden** Can.
111 D3 **Waccasassa Bay** U.S.A.
69 D2 **Wächtersbach** Ger.
107 D2 **Waco** U.S.A.

42 A2 **Wad** Pak.
83 D2 **Waddān** Libya
68 B1 **Waddeneilanden** *is* Neth.
68 B1 **Waddenzee** *sea chan.* Neth.
96 B2 **Waddington, Mount** Can.
68 B1 **Waddinxveen** Neth.
97 D2 **Wadena** Can.
105 D1 **Wadena** U.S.A.
84 B2 **Wadi Halfa** Sudan
84 B3 **Wad Medani** Sudan
38 C2 **Wafangdian** China
68 B2 **Wageningen** Neth.
95 F2 **Wager Bay** Can.
21 C3 **Wagga Wagga** Austr.
105 D2 **Wahoo** U.S.A.
105 D1 **Wahpeton** U.S.A.
22 B2 **Waiau** *r.* N.Z.
27 C3 **Waigeo** *i.* Indon.
26 B3 **Waikabubak** Indon.
20 A2 **Waikerie** Austr.
22 B2 **Waimate** N.Z.
43 B3 **Wainganga** *r.* India
26 C3 **Waingapu** Indon.
97 C2 **Wainwright** Can.
22 C1 **Waiouru** N.Z.
22 B2 **Waipara** N.Z.
22 C1 **Waipawa** N.Z.
22 B2 **Wairau** *r.* N.Z.
22 C1 **Wairoa** N.Z.
22 B2 **Waitaki** *r.* N.Z.
22 B1 **Waitara** N.Z.
22 B1 **Waiuku** N.Z.
87 E2 **Wajir** Kenya
35 C3 **Wakasa-wan** *b.* Japan
22 A3 **Wakatipu, Lake** N.Z.
97 D2 **Wakaw** Can.
35 C4 **Wakayama** Japan
104 D3 **WaKeeney** U.S.A.
22 B2 **Wakefield** N.Z.
66 C2 **Wakefield** U.K.
34 D1 **Wakkanai** Japan
91 D2 **Wakkerstroom** S. Africa
71 D1 **Wałbrzych** Pol.
21 D2 **Walcha** Austr.
68 C1 **Walchum** Ger.
71 D1 **Wałcz** Pol.
67 B2 **Wales** *admin. div.* U.K.
21 C2 **Walgett** Austr.
87 C3 **Walikale** Dem. Rep. Congo
103 C3 **Walker Lake** *l.* U.S.A.
20 A2 **Wallaroo** Austr.
66 B2 **Wallasey** U.K.
102 C1 **Walla Walla** U.S.A.
69 D3 **Walldürn** Ger.
21 C2 **Wallendbeen** Austr.
17 F4 **Wallis and Futuna Islands** *terr.* S. Pacific Ocean
64 □ **Walls** U.K.
66 B1 **Walney, Isle of** *i.* U.K.
67 C2 **Walsall** U.K.
104 C3 **Walsenburg** U.S.A.
69 D1 **Walsrode** Ger.
111 D2 **Walterboro** U.S.A.
88 A3 **Walvis Bay** Namibia
87 C2 **Wamba** Dem. Rep. Congo
20 B1 **Wanaaring** Austr.
22 A2 **Wanaka** N.Z.
22 A2 **Wanaka, Lake** N.Z.
98 B2 **Wanapitei Lake** Can.
122 B3 **Wanda** Arg.
34 B1 **Wanda Shan** *mts* China
30 A1 **Wanding** China
22 C1 **Wanganui** N.Z.
21 C3 **Wangaratta** Austr.
68 C1 **Wangerooge** *i.* Ger.
33 B1 **Wangqing** China
30 A1 **Wan Hsa-la** Myanmar
39 B4 **Wanning** China
68 B2 **Wanroij** Neth.
67 C3 **Wantage** U.K.
38 A2 **Wanxian** China
38 A2 **Wanyuan** China
41 B3 **Warangal** India
69 D2 **Warburg** Ger.
18 B2 **Warburton** Austr.
20 A1 **Warburton** *watercourse* Austr.
43 B2 **Wardha** India
64 C1 **Ward Hill** U.K.
96 B2 **Ware** Can.
69 F1 **Waren** Ger.
68 C2 **Warendorf** Ger.
21 D1 **Warialda** Austr.
90 A2 **Warmbad** Namibia
103 C3 **Warm Springs** U.S.A.
102 C2 **Warner Lakes** U.S.A.
111 D2 **Warner Robins** U.S.A.
120 B2 **Warnes** Bol.
20 B3 **Warracknabeal** Austr.
21 C2 **Warrego** *r.* Austr.
21 C2 **Warren** Austr.
110 B2 **Warren** *AR* U.S.A.
108 C2 **Warren** *OH* U.S.A.
109 D2 **Warren** *PA* U.S.A.
65 C1 **Warrenpoint** U.K.
105 E3 **Warrensburg** U.S.A.
90 B2 **Warrenton** S. Africa
83 C4 **Warri** Nigeria
66 B2 **Warrington** U.K.
20 B3 **Warrnambool** Austr.
71 E1 **Warsaw** Pol.
Warszawa Pol. *see* **Warsaw**
71 C1 **Warta** *r.* Pol.
21 D1 **Warwick** Austr.
67 C2 **Warwick** U.K.
103 C3 **Wasco** U.S.A.
104 C1 **Washburn** U.S.A.
109 D3 **Washington** *DC* U.S.A.
105 E2 **Washington** *IA* U.S.A.
108 B3 **Washington** *IN* U.S.A.
105 E3 **Washington** *MO* U.S.A.
111 E1 **Washington** *NC* U.S.A.
108 C2 **Washington** *PA* U.S.A.
102 B1 **Washington** *state* U.S.A.
109 E2 **Washington, Mount** U.S.A.
108 C3 **Washington Court House** U.S.A.
42 A2 **Washuk** Pak.
98 C1 **Waskaganish** Can.
97 E2 **Waskaiowaka Lake** Can.
90 A2 **Wasser** Namibia
98 C2 **Waswanipi, Lac** *l.* Can.
26 C3 **Watampone** Indon.
109 E2 **Waterbury** U.S.A.
97 D2 **Waterbury Lake** Can.
65 C2 **Waterford** Rep. of Ireland
65 C2 **Waterford Harbour** Rep. of Ireland
105 E2 **Waterloo** U.S.A.
91 C1 **Waterpoort** S. Africa
109 D2 **Watertown** *NY* U.S.A.
105 D2 **Watertown** *SD* U.S.A.
108 B2 **Watertown** *WI* U.S.A.
109 F2 **Waterville** U.S.A.
67 C3 **Watford** U.K.
104 C1 **Watford City** U.S.A.
97 D2 **Wathaman** *r.* Can.
107 D1 **Watonga** U.S.A.
97 D2 **Watrous** Can.
87 C2 **Watsa** Dem. Rep. Congo
108 B2 **Watseka** U.S.A.
86 C3 **Watsi Kengo** Dem. Rep. Congo
96 B1 **Watson Lake** Can.
103 B3 **Watsonville** U.S.A.
27 C3 **Watubela, Kepulauan** *is* Indon.
27 D3 **Wau** P.N.G.
85 A4 **Wau** Sudan
21 D2 **Wauchope** Austr.
108 B2 **Waukegan** U.S.A.
107 D2 **Waurika** U.S.A.
108 B2 **Wausau** U.S.A.
67 D2 **Waveney** *r.* U.K.
105 E2 **Waverly** U.S.A.
111 D2 **Waycross** U.S.A.
105 D2 **Wayne** U.S.A.
111 D2 **Waynesboro** *GA* U.S.A.
109 D3 **Waynesboro** *VA* U.S.A.
111 D1 **Waynesville** U.S.A.
42 B1 **Wazirabad** Pak.
66 C1 **Wear** *r.* U.K.
107 D2 **Weatherford** U.S.A.
102 B2 **Weaverville** U.S.A.
98 B1 **Webequie** Can.
85 C4 **Webi Shabeelle** *r.* Somalia
105 D1 **Webster** U.S.A.
105 E2 **Webster City** U.S.A.
23 Q2 **Weddell Sea** Antarctica
68 B2 **Weert** Neth.
21 C2 **Wee Waa** Austr.
68 C2 **Wegberg** Ger.
71 E1 **Węgorzewo** Pol.
69 F3 **Weiden in der Oberpfalz** Ger.
38 B2 **Weifang** China
38 C2 **Weihai** China
21 C1 **Weilmoringle** Austr.
69 E2 **Weimar** Ger.
38 A2 **Weinan** China
19 D1 **Weipa** Austr.
21 C1 **Weir** *r.* Austr.
108 C2 **Weirton** U.S.A.
30 B1 **Weishan** China
69 E2 **Weißenfels** Ger.
70 C2 **Weißkugel** *mt.* Austria/Italy
71 D1 **Wejherowo** Pol.
108 C3 **Welch** U.S.A.
85 B3 **Weldiya** Eth.
91 C2 **Welkom** S. Africa
67 C2 **Welland** *r.* U.K.
19 C1 **Wellesley Islands** Austr.
21 C2 **Wellington** Austr.
22 B2 **Wellington** N.Z.
90 A3 **Wellington** S. Africa
105 D3 **Wellington** U.S.A.
121 A5 **Wellington, Isla** *i.* Chile
21 C3 **Wellington, Lake** Austr.
96 B2 **Wells** Can.
67 B3 **Wells** U.K.
102 D2 **Wells** U.S.A.
18 B2 **Wells, Lake** *salt flat* Austr.
22 B1 **Wellsford** N.Z.
66 D2 **Wells-next-the-Sea** U.K.
70 C2 **Wels** Austria
67 B2 **Welshpool** U.K.
91 C2 **Wembesi** S. Africa
98 C1 **Wemindji** Can.
102 B1 **Wenatchee** U.S.A.
39 B4 **Wenchang** China
82 B4 **Wenchi** Ghana
38 C2 **Wendeng** China
69 E1 **Wendisch Evern** Ger.
85 B4 **Wendo** Eth.
102 D2 **Wendover** U.S.A.
39 B3 **Wengyuan** China
39 C3 **Wenling** China
39 A3 **Wenshan** China
20 B2 **Wentworth** Austr.
39 C3 **Wenzhou** China
91 C2 **Wepener** S. Africa
90 B2 **Werda** Botswana
69 F2 **Werdau** Ger.
69 F1 **Werder** Ger.
69 F3 **Wernberg-Köblitz** Ger.
69 E2 **Wernigerode** Ger.
69 D2 **Werra** *r.* Ger.
21 D2 **Werris Creek** Austr.
69 D3 **Wertheim** Ger.
68 C2 **Wesel** Ger.

69 E1 **Wesendorf** Ger.
69 D1 **Weser** *r.* Ger.
69 D1 **Weser** *sea chan.* Ger.
19 C1 **Wessel, Cape** Austr.
19 C1 **Wessel Islands** Austr.
91 C2 **Wesselton** S. Africa
48 B2 **West Bank** *terr.* Asia
108 B2 **West Bend** U.S.A.
67 C2 **West Bromwich** U.K.
68 C2 **Westerburg** Ger.
68 C1 **Westerholt** Ger.
18 B2 **Western Australia** *state* Austr.
90 B3 **Western Cape** *prov.* S. Africa
84 A2 **Western Desert** *des.* Egypt
41 B3 **Western Ghats** *mts* India
82 A2 **Western Sahara** *terr.* Africa
Western Samoa *country* S. Pacific Ocean *see* **Samoa**
68 A2 **Westerschelde** *est.* Neth.
68 C1 **Westerstede** Ger.
68 C2 **Westerwald** *hills* Ger.
121 B6 **West Falkland** *i.* Falkland Is
108 B3 **West Frankfort** U.S.A.
115 D2 **West Indies** N. America
64 A1 **West Loch Roag** *b.* U.K.
96 C2 **Westlock** Can.
68 B2 **Westmalle** Belgium
110 B1 **West Memphis** U.S.A.
108 C3 **Weston** U.S.A.
67 B3 **Weston-super-Mare** U.K.
111 D3 **West Palm Beach** U.S.A.
105 E3 **West Plains** U.S.A.
105 D2 **West Point** U.S.A.
22 B2 **Westport** N.Z.
65 B2 **Westport** Rep. of Ireland
97 D2 **Westray** Can.
64 C1 **Westray** *i.* U.K.
68 B1 **West-Terschelling** Neth.
65 B1 **West Town** Rep. of Ireland
108 C3 **West Virginia** *state* U.S.A.
21 C2 **West Wyalong** Austr.
102 D2 **West Yellowstone** U.S.A.
27 C3 **Wetar** *i.* Indon.
96 C2 **Wetaskiwin** Can.
69 D2 **Wetzlar** Ger.
27 D3 **Wewak** P.N.G.
65 C2 **Wexford** Rep. of Ireland
97 D2 **Weyakwin** Can.
97 D3 **Weyburn** Can.
69 D1 **Weyhe** Ger.
67 B3 **Weymouth** U.K.
22 C1 **Whakatane** N.Z.
97 E1 **Whale Cove** Can.
64 □ **Whalsay** *i.* U.K.
22 B1 **Whangarei** N.Z.
66 C2 **Wharfe** *r.* U.K.
107 D3 **Wharton** U.S.A.
96 C1 **Wha Ti** Can.
104 B2 **Wheatland** U.S.A.
106 B1 **Wheeler Peak** *NM* U.S.A.
103 D3 **Wheeler Peak** *NV* U.S.A.
108 C2 **Wheeling** U.S.A.
66 B1 **Whernside** *hill* U.K.
96 B2 **Whistler** Can.
66 C1 **Whitby** U.K.
110 B2 **White** *r.* U.S.A.
18 B2 **White, Lake** *salt flat* Austr.
99 E2 **White Bay** Can.
104 C1 **White Butte** *mt.* U.S.A.
20 B2 **White Cliffs** Austr.
96 C2 **Whitecourt** Can.
102 D1 **Whitefish** U.S.A.
66 B1 **Whitehaven** U.K.
65 D1 **Whitehead** U.K.
96 A1 **Whitehorse** Can.
110 B3 **White Lake** *l.* U.S.A.
103 C3 **White Mountain Peak** U.S.A.
84 B3 **White Nile** *r.* Sudan/Uganda
54 C2 **White Sea** Rus. Fed.
102 D1 **White Sulphur Springs** U.S.A.
111 E2 **Whiteville** U.S.A.
82 B3 **White Volta** *watercourse* Burkina/Ghana
106 B2 **Whitewater Baldy** *mt.* U.S.A.
98 B1 **Whitewater Lake** Can.
97 D2 **Whitewood** Can.
64 B3 **Whithorn** U.K.
22 C1 **Whitianga** N.Z.
103 C3 **Whitney, Mount** U.S.A.
19 D2 **Whitsunday Island** Austr.
20 A2 **Whyalla** Austr.
30 A1 **Wiang Phran** Thai.
68 A2 **Wichelen** Belgium
105 D3 **Wichita** U.S.A.
107 D2 **Wichita Falls** U.S.A.
64 C1 **Wick** U.K.
106 A2 **Wickenburg** U.S.A.
65 C2 **Wicklow** Rep. of Ireland
65 D2 **Wicklow Head** *hd* Rep. of Ireland
65 C2 **Wicklow Mountains** Rep. of Ireland
66 B2 **Widnes** U.K.
69 D1 **Wiehengebirge** *hills* Ger.
68 C2 **Wiehl** Ger.
71 D1 **Wieluń** Pol.
Wien Austria *see* **Vienna**
71 D2 **Wiener Neustadt** Austria
68 B1 **Wieringerwerf** Neth.
68 D2 **Wiesbaden** Ger.
68 C1 **Wiesmoor** Ger.
71 D1 **Wieżyca** *hill* Pol.
67 C3 **Wight, Isle of** *i.* U.K.
64 B3 **Wigtown** U.K.
68 B2 **Wijchen** Neth.
20 B2 **Wilcannia** Austr.
91 C3 **Wild Coast** S. Africa
91 C2 **Wilge** *r.* S. Africa
68 D1 **Wilhelmshaven** Ger.
109 D2 **Wilkes-Barre** U.S.A.
23 H3 **Wilkes Land** *reg.* Antarctica
97 D2 **Wilkie** Can.
106 B2 **Willcox** U.S.A.
68 B2 **Willebroek** Belgium
115 D3 **Willemstad** Neth. Antilles
20 B3 **William, Mount** Austr.
20 A1 **William Creek** Austr.
106 A1 **Williams** U.S.A.
109 D3 **Williamsburg** U.S.A.
96 B2 **Williams Lake** Can.
108 C3 **Williamson** U.S.A.
109 D2 **Williamsport** U.S.A.
111 E1 **Williamston** U.S.A.
90 B3 **Williston** S. Africa
104 C1 **Williston** U.S.A.
96 B2 **Williston Lake** Can.
103 B3 **Willits** U.S.A.
105 D1 **Willmar** U.S.A.
90 B3 **Willowmore** S. Africa
91 C3 **Willowvale** S. Africa
18 B2 **Wills, Lake** *salt flat* Austr.
20 A3 **Willunga** Austr.
20 A2 **Wilmington** Austr.
109 D3 **Wilmington** *DE* U.S.A.
111 E2 **Wilmington** *NC* U.S.A.
68 D2 **Wilnsdorf** Ger.
111 E1 **Wilson** U.S.A.
21 C3 **Wilson's Promontory** *pen.* Austr.
68 B3 **Wiltz** Lux.
18 B2 **Wiluna** Austr.
67 C3 **Winchester** U.K.
108 C3 **Winchester** *KY* U.S.A.
109 D3 **Winchester** *VA* U.S.A.
66 B1 **Windermere** *l.* U.K.
88 A3 **Windhoek** Namibia
105 D2 **Windom** U.S.A.
19 D2 **Windorah** Austr.
104 B2 **Wind River Range** *mts* U.S.A.
21 D2 **Windsor** Austr.
98 B2 **Windsor** Can.
115 D3 **Windward Islands** Caribbean Sea
115 C3 **Windward Passage** Cuba/Haiti
105 D3 **Winfield** U.S.A.
68 A2 **Wingene** Belgium
21 D2 **Wingham** Austr.
98 B1 **Winisk** Can.
98 B1 **Winisk** *r.* Can.
98 B1 **Winisk Lake** Can.
31 A2 **Winkana** Myanmar
97 E3 **Winkler** Can.
82 B4 **Winneba** Ghana
108 B2 **Winnebago, Lake** U.S.A.
102 C2 **Winnemucca** U.S.A.
104 C2 **Winner** U.S.A.
110 B2 **Winnfield** U.S.A.
105 E1 **Winnibigoshish, Lake** U.S.A.
97 E3 **Winnipeg** Can.
97 E2 **Winnipeg** *r.* Can.
97 E2 **Winnipeg, Lake** Can.
97 D2 **Winnipegosis, Lake** Can.
109 E2 **Winnipesaukee, Lake** U.S.A.
110 B2 **Winnsboro** U.S.A.
105 E2 **Winona** *MN* U.S.A.
110 C2 **Winona** *MS* U.S.A.
68 C1 **Winschoten** Neth.
69 D1 **Winsen (Aller)** Ger.
69 E1 **Winsen (Luhe)** Ger.
106 A1 **Winslow** U.S.A.
111 D1 **Winston-Salem** U.S.A.
69 D2 **Winterberg** Ger.
111 D3 **Winter Haven** U.S.A.
68 C2 **Winterswijk** Neth.
73 D2 **Winterthur** Switz.
19 D2 **Winton** Austr.
22 A3 **Winton** N.Z.
20 A2 **Wirrabara** Austr.
67 D2 **Wisbech** U.K.
108 A2 **Wisconsin** *r.* U.S.A.
108 B2 **Wisconsin** *state* U.S.A.
108 B2 **Wisconsin Rapids** U.S.A.
71 D1 **Wisła** *r.* Pol.
69 E1 **Wismar** Ger.
91 C2 **Witbank** S. Africa
66 D2 **Withernsea** U.K.
68 B1 **Witmarsum** Neth.
67 C3 **Witney** U.K.
91 D2 **Witrivier** S. Africa
69 E1 **Wittenberge** Ger.
69 E1 **Wittenburg** Ger.
69 E1 **Wittingen** Ger.
68 C3 **Wittlich** Ger.
68 C1 **Wittmund** Ger.
69 F1 **Wittstock** Ger.
88 A3 **Witvlei** Namibia
69 D2 **Witzenhausen** Ger.
71 D1 **Władysławowo** Pol.
71 D1 **Włocławek** Pol.
21 C3 **Wodonga** Austr.
27 C3 **Wokam** *i.* Indon.
67 C3 **Woking** U.K.
69 F2 **Wolfen** Ger.
69 E1 **Wolfenbüttel** Ger.
69 D2 **Wolfhagen** Ger.
104 B1 **Wolf Point** U.S.A.
69 E1 **Wolfsburg** Ger.
68 C3 **Wolfstein** Ger.
99 D2 **Wolfville** Can.
70 C1 **Wolgast** Ger.
70 C1 **Wolin** Pol.
97 D2 **Wollaston Lake** Can.
97 D2 **Wollaston Lake** *l.* Can.
94 D2 **Wollaston Peninsula** Can.
21 D2 **Wollongong** Austr.
69 E2 **Wolmirsleben** Ger.

69 E1 **Wolmirstedt** Ger.
68 C1 **Wolvega** Neth.
67 B2 **Wolverhampton** U.K.
33 B2 **Wŏnju** S. Korea
96 B2 **Wonowon** Can.
33 B2 **Wŏnsan** N. Korea
21 C3 **Wonthaggi** Austr.
20 A2 **Woocalla** Austr.
19 C1 **Woodah, Isle** *i.* Austr.
67 D2 **Woodbridge** U.K.
104 B3 **Woodland Park** U.S.A.
18 C2 **Woodroffe, Mount** Austr.
19 C1 **Woods, Lake** *salt flat* Austr.
97 E3 **Woods, Lake of the** Can./U.S.A.
21 C3 **Woods Point** Austr.
99 D2 **Woodstock** *N.B.* Can.
108 C2 **Woodstock** *Ont.* Can.
22 C2 **Woodville** N.Z.
107 D1 **Woodward** U.S.A.
21 D2 **Woolgoolga** Austr.
20 A2 **Woomera** Austr.
108 C2 **Wooster** U.S.A.
90 A3 **Worcester** S. Africa
67 B2 **Worcester** U.K.
109 E2 **Worcester** U.S.A.
70 C2 **Wörgl** Austria
66 B1 **Workington** U.K.
66 C2 **Worksop** U.K.
104 B2 **Worland** U.S.A.
69 D3 **Worms** Ger.
67 C3 **Worthing** U.K.
105 D2 **Worthington** U.S.A.
26 C3 **Wotu** Indon.
27 C3 **Wowoni** *i.* Indon.
Wrangel Island Rus. Fed. *see* **Vrangelya, Ostrov**
96 A2 **Wrangell** U.S.A.
64 B1 **Wrath, Cape** U.K.
104 C2 **Wray** U.S.A.
90 A2 **Wreck Point** S. Africa
66 B2 **Wrexham** U.K.
104 B2 **Wright** U.S.A.
31 A2 **Wrightmyo** India
96 B1 **Wrigley** Can.
71 D1 **Wrocław** Pol.
71 D1 **Września** Pol.
38 B2 **Wu'an** China
38 A2 **Wudu** China
38 A2 **Wuhai** China
38 B2 **Wuhan** China
38 B2 **Wuhu** China
Wujin China *see* **Changzhou**
83 C4 **Wukari** Nigeria
30 B1 **Wuliang Shan** *mts* China
27 C3 **Wuliaru** *i.* Indon.
39 A3 **Wumeng Shan** *mts* China
98 B1 **Wunnummin Lake** Can.
69 F2 **Wunsiedel** Ger.
69 D1 **Wunstorf** Ger.
30 A1 **Wuntho** Myanmar
68 C2 **Wuppertal** Ger.
90 A3 **Wuppertal** S. Africa
69 E2 **Wurzbach** Ger.
69 D3 **Würzburg** Ger.
69 F2 **Wurzen** Ger.
36 C2 **Wuwei** China
38 A2 **Wuxi** *Chongqing* China
38 C2 **Wuxi** *Jiangsu* China
Wuxing China *see* **Huzhou**
39 A3 **Wuxuan** China
Wuyang China *see* **Zhenyuan**
37 E1 **Wuyiling** China
39 B3 **Wuyishan** China
39 B3 **Wuyi Shan** *mts* China
38 A1 **Wuyuan** China
38 A2 **Wuzhong** China
39 B3 **Wuzhou** China
21 C2 **Wyangala Reservoir** Austr.
20 B3 **Wycheproof** Austr.
67 B3 **Wye** *r.* U.K.
18 B1 **Wyndham** Austr.
110 B1 **Wynne** U.S.A.
97 D2 **Wynyard** Can.
104 B2 **Wyoming** *state* U.S.A.
71 E1 **Wyszków** Pol.
108 C3 **Wytheville** U.S.A.

85 D3 **Xaafuun** Somalia
89 C3 **Xai-Xai** Moz.
38 A1 **Xamba** China
30 B1 **Xam Hua** Laos
88 A2 **Xangongo** Angola
49 C2 **Xankändi** Azer.
79 B2 **Xanthi** Greece
118 B3 **Xapuri** Brazil
75 C2 **Xátiva** Spain
Xiaguan China *see* **Dali**
39 B3 **Xiamen** China
38 A2 **Xi'an** China
38 A3 **Xianfeng** China
38 B2 **Xiangfan** China
Xiangjiang China *see* Huichang
30 B2 **Xiangkhoang** Laos
39 B3 **Xiangtan** China
Xiangyang China *see* Xiangfan
38 B2 **Xiantao** China
38 A2 **Xianyang** China
38 B2 **Xiaogan** China
37 E1 **Xiao Hinggan Ling** *mts* China
38 C2 **Xiaoshan** China
30 B1 **Xichang** China
113 C2 **Xicohténcatl** Mex.
38 A2 **Xifeng** *Gansu* China
39 A3 **Xifeng** *Guizhou* China
43 C2 **Xigazê** China
51 I4 **Xilinhot** China
38 B2 **Xincai** China
36 C2 **Xinghai** China
38 B2 **Xinghua** China
39 B3 **Xingning** China
38 A2 **Xingping** China
38 B2 **Xingtai** China
119 C2 **Xingu** *r.* Brazil
39 A3 **Xingyi** China
39 B3 **Xinhua** China
36 C2 **Xining** China
43 C1 **Xinjiang Uygur Zizhiqu** *aut. reg.* China
Xinjing China *see* **Jingxi**
33 A1 **Xinmin** China
39 B3 **Xinning** China
38 B2 **Xintai** China
38 B2 **Xinxiang** China
38 B2 **Xinyang** China
39 B3 **Xinyu** China
37 D2 **Xinzhou** *Shanxi* China
38 B2 **Xinzhou** *Shanxi* China
74 B1 **Xinzo de Limia** Spain
Xiongshan China *see* Zhenghe
36 C2 **Xiqing Shan** *mts* China
119 D3 **Xique Xique** Brazil
38 A1 **Xishanzui** China
39 A3 **Xiushan** China
39 B3 **Xiuying** China
38 B2 **Xixia** China
43 C1 **Xizang Zizhiqu** *aut. reg.* China
39 A3 **Xuanwei** China
38 B2 **Xuanzhou** China
38 B2 **Xuchang** China
Xucheng China *see* **Xuwen**
85 C4 **Xuddur** Somalia
Xujiang China *see* Guangchang
39 B3 **Xun Jiang** *r.* China
39 B3 **Xunwu** China
39 B3 **Xuwen** China
39 A3 **Xuyong** China
38 B2 **Xuzhou** China

85 B4 **Yabēlo** Eth.
37 D1 **Yablonovyy Khrebet** *mts* Rus. Fed.
111 D1 **Yadkin** *r.* U.S.A.
43 C2 **Yadong** China
57 E2 **Yagnitsa** Rus. Fed.
86 B1 **Yagoua** Cameroon
96 C3 **Yahk** Can.
112 B2 **Yahualica** Mex.
48 B2 **Yahyalı** Turkey
35 C4 **Yaizu** Japan
102 B1 **Yakima** U.S.A.
82 B3 **Yako** Burkina
34 D2 **Yakumo** Japan
94 B3 **Yakutat** U.S.A.
51 J2 **Yakutsk** Rus. Fed.
59 D2 **Yakymivka** Ukr.
31 B3 **Yala** Thai.
21 C3 **Yallourn** Austr.
79 C2 **Yalova** Turkey
59 C3 **Yalta** Ukr.
33 A1 **Yalu Jiang** *r.* China/N. Korea
35 D3 **Yamagata** Japan
35 B4 **Yamaguchi** Japan
54 F1 **Yamal, Poluostrov** *pen.* Rus. Fed.
21 D1 **Yamba** Austr.
85 A4 **Yambio** Sudan
78 C2 **Yambol** Bulg.
54 G2 **Yamburg** Rus. Fed.
30 A1 **Yamethin** Myanmar
82 B4 **Yamoussoukro** Côte d'Ivoire
58 B2 **Yampil'** Ukr.
43 C2 **Yamuna** *r.* India
43 D2 **Yamzho Yumco** *l.* China
51 K2 **Yana** *r.* Rus. Fed.
38 A2 **Yan'an** China
118 A3 **Yanaoca** Peru
46 A2 **Yanbu' al Baḩr** Saudi Arabia
38 C2 **Yancheng** China
18 A3 **Yanchep** Austr.
38 B2 **Yangcheng** China
39 B3 **Yangchun** China
33 B2 **Yangdok** N. Korea
39 B3 **Yangjiang** China
30 A2 **Yangôn** Myanmar
38 B2 **Yangquan** China
39 B3 **Yangshuo** China
38 C3 **Yangtze** *r.* China
37 E2 **Yangtze, Mouths of the** China
38 B2 **Yangzhou** China
33 B1 **Yanji** China
105 D2 **Yankton** U.S.A.
51 K2 **Yano-Indigirskaya Nizmennost'** *lowland* Rus. Fed.
51 K2 **Yanskiy Zaliv** *g.* Rus. Fed.
21 C1 **Yantabulla** Austr.
38 C2 **Yantai** China
86 B2 **Yaoundé** Cameroon
27 D2 **Yap** *i.* Micronesia
27 D3 **Yapen** *i.* Indon.
27 D3 **Yapen, Selat** *sea chan.* Indon.

112 A2 **Yaqui** *r.* Mex.
19 D2 **Yaraka** Austr.
54 D3 **Yaransk** Rus. Fed.
46 B3 **Yarīm** Yemen
45 D3 **Yarkant He** *r.* China
43 D2 **Yarlung Zangbo** *r.* China
99 D2 **Yarmouth** Can.
54 F2 **Yarongo** Rus. Fed.
57 E2 **Yaroslavl'** Rus. Fed.
34 B2 **Yaroslavskiy** Rus. Fed.
21 C3 **Yarram** Austr.
57 D2 **Yartsevo** Rus. Fed.
57 E3 **Yasnogorsk** Rus. Fed.
31 B2 **Yasothon** Thai.
21 C2 **Yass** Austr.
79 C3 **Yatağan** Turkey
97 E1 **Yathkyed Lake** Can.
35 B4 **Yatsushiro** Japan
118 B2 **Yavari** *r.* Brazil/Peru
43 B2 **Yavatmāl** India
58 A2 **Yavoriv** Ukr.
35 B4 **Yawatahama** Japan
49 D2 **Yazd** Iran
110 B2 **Yazoo City** U.S.A.
79 B3 **Ydra** *i.* Greece
31 A2 **Ye** Myanmar
45 D3 **Yecheng** China
112 B2 **Yécora** Mex.
57 E3 **Yefremov** Rus. Fed.
57 E2 **Yegor'yevsk** Rus. Fed.
85 B4 **Yei** Sudan
54 F3 **Yekaterinburg** Rus. Fed.
57 E3 **Yelets** Rus. Fed.
82 A3 **Yélimané** Mali
64 □ **Yell** *i.* U.K.
96 C1 **Yellowknife** Can.
Yellow River *r.* China *see* **Huang He**
37 E2 **Yellow Sea** N. Pacific Ocean
104 C1 **Yellowstone** *r.* U.S.A.
104 A2 **Yellowstone Lake** U.S.A.
56 C3 **Yel'sk** Belarus
46 B3 **Yemen** *country* Asia
54 E2 **Yemva** Rus. Fed.
59 D2 **Yenakiyeve** Ukr.
30 A1 **Yenangyaung** Myanmar
30 B1 **Yên Bai** Vietnam
82 B4 **Yendi** Ghana
79 C3 **Yenice** Turkey
79 C3 **Yenifoça** Turkey
36 C1 **Yenisey** *r.* Rus. Fed.
21 C2 **Yeoval** Austr.
67 B3 **Yeovil** U.K.
19 E2 **Yeppoon** Austr.
51 I2 **Yerbogachen** Rus. Fed.
49 C1 **Yerevan** Armenia
45 D1 **Yereymentau** Kazakh.
55 D3 **Yershov** Rus. Fed.
33 B2 **Yesan** S. Korea
45 C1 **Yesil'** Kazakh.
79 C3 **Yeşilova** Turkey
67 A3 **Yes Tor** *hill* U.K.
21 D1 **Yetman** Austr.
30 A1 **Yeu** Myanmar
72 B2 **Yeu, Île d'** *i.* France
59 C2 **Yevpatoriya** Ukr.
59 D2 **Yeysk** Rus. Fed.
56 C2 **Yezyaryshcha** Belarus
39 A3 **Yibin** China
38 B2 **Yichang** China
37 E1 **Yichun** *Heilong.* China
39 B3 **Yichun** *Jiangxi* China
78 C2 **Yıldız Dağları** *mts* Turkey
48 B2 **Yıldızeli** Turkey
38 A2 **Yinchuan** China
33 B1 **Yingchengzi** China
39 B3 **Yingde** China
38 C1 **Yingkou** China
38 B2 **Yingshan** *Hubei* China
38 A2 **Yingshan** *Sichuan* China
39 B3 **Yingtan** China
45 E2 **Yining** China
85 B4 **Yirga Alem** Eth.
38 B2 **Yishui** China
36 C2 **Yiwu** China
39 B3 **Yiyang** China
39 A3 **Yizhou** China
60 I2 **Yli-Kitka** *l.* Fin.
60 H2 **Ylitornio** Fin.
60 H3 **Ylivieska** Fin.
29 C2 **Yogyakarta** Indon.
86 B2 **Yoko** Cameroon
35 C3 **Yokohama** Japan
83 D4 **Yola** Nigeria
35 D3 **Yonezawa** Japan
39 B3 **Yong'an** China
39 B3 **Yongchun** China
38 A2 **Yongdeng** China
33 B2 **Yŏngdŏk** S. Korea
39 C3 **Yongkang** China
30 B1 **Yongsheng** China
39 B3 **Yongzhou** China
118 A1 **Yopal** Col.
18 A3 **York** Austr.
66 C2 **York** U.K.
105 D2 **York** *NE* U.S.A.
109 D3 **York** *PA* U.S.A.
19 D1 **York, Cape** Austr.
20 A3 **Yorke Peninsula** Austr.
20 A3 **Yorketown** Austr.
66 C2 **Yorkshire Wolds** *hills* U.K.
97 D2 **Yorkton** Can.
55 D3 **Yoshkar-Ola** Rus. Fed.
65 C3 **Youghal** Rep. of Ireland
21 C2 **Young** Austr.
20 A3 **Younghusband Peninsula** Austr.
108 C2 **Youngstown** U.S.A.
82 B3 **Youvarou** Mali
45 E2 **Youyi Feng** *mt.* China/Rus. Fed.
48 B2 **Yozgat** Turkey
102 B2 **Yreka** U.S.A.
Yr Wyddfa *mt.* U.K. *see* Snowdon
73 C2 **Yssingeaux** France
61 F4 **Ystad** Sweden
Ysyk-Köl Kyrg. *see* **Balykchy**
45 D2 **Ysyk-Köl** *salt l.* Kyrg.
60 □A3 **Ytri-Rangá** *r.* Iceland
30 B1 **Yuanjiang** China
30 B1 **Yuan Jiang** *r.* China
30 B1 **Yuanmou** China
103 B3 **Yuba City** U.S.A.
34 D2 **Yūbari** Japan
113 C3 **Yucatán** *pen.* Mex.
114 B2 **Yucatan Channel** Cuba/Mex.
38 B2 **Yuci** China
18 C2 **Yuendumu** Austr.
39 B3 **Yueyang** China
54 F2 **Yugorsk** Rus. Fed.
77 D2 **Yugoslavia** *country* Europe
51 L2 **Yukagirskoye Ploskogor'ye** *plat.* Rus. Fed.
94 A2 **Yukon** *r.* Can./U.S.A.
96 A1 **Yukon Territory** *admin. div.* Can.
39 B3 **Yulin** *Guangxi* China
38 A2 **Yulin** *Shaanxi* China
30 B1 **Yulongxue Shan** *mt.* China
106 A2 **Yuma** *AZ* U.S.A.
104 C2 **Yuma** *CO* U.S.A.
36 C2 **Yumen** China
48 B2 **Yunak** Turkey
38 B2 **Yuncheng** China
39 B3 **Yunfu** China
30 B1 **Yunnan** *prov.* China
20 A2 **Yunta** Austr.
38 B2 **Yunyang** China
118 A2 **Yurimaguas** Peru
45 E3 **Yurungkax He** *r.* China
39 C3 **Yü Shan** *mt.* Taiwan
36 C2 **Yushu** China
Yushuwan China *see* **Huaihua**
49 C1 **Yusufeli** Turkey
45 E3 **Yutian** China
30 B1 **Yuxi** China
34 D2 **Yuzhno-Kuril'sk** Rus. Fed.
51 K3 **Yuzhno-Sakhalinsk** Rus. Fed.
59 C2 **Yuzhnoukrayinsk** Ukr.
72 C2 **Yvetot** France

Z

68 B1 **Zaandam** Neth.
37 D1 **Zabaykal'sk** Rus. Fed.
46 B3 **Zabīd** Yemen
44 C3 **Zābol** Iran
47 D2 **Zābolī** Iran
112 B3 **Zacapu** Mex.
112 B2 **Zacatecas** Mex.
113 C3 **Zacatepec** Mex.
113 C3 **Zacatlán** Mex.
79 B3 **Zacharo** Greece
112 B2 **Zacoalco** Mex.
113 C2 **Zacualtipán** Mex.
77 C2 **Zadar** Croatia
31 A3 **Zadetkyi Kyun** *i.* Myanmar
57 E3 **Zadonsk** Rus. Fed.
74 B2 **Zafra** Spain
48 B2 **Zagazig** Egypt
77 C1 **Zagreb** Croatia
49 C2 **Zagros, Kūhhā-ye** *mts* Iran
47 D2 **Zāhedān** Iran
48 B2 **Zahlé** Lebanon
46 B3 **Ẓahrān** Saudi Arabia
Zaire *country* Africa *see* **Congo, Democratic Republic of**
77 D2 **Zaječar** Yugo.
57 E3 **Zakharovo** Rus. Fed.
49 C2 **Zākhō** Iraq
79 B3 **Zakynthos** Greece
Zakynthos *i.* Greece *see* Zante
71 D2 **Zalaegerszeg** Hungary
78 B1 **Zalău** Romania
46 B2 **Ẓalim** Saudi Arabia
85 A3 **Zalingei** Sudan
46 A2 **Ẓalmā, Jabal aẓ** *mt.* Saudi Arabia
96 C2 **Zama City** Can.
Zambeze *r.* Moz. *see* Zambezi
88 B2 **Zambezi** Zambia
88 C2 **Zambezi** *r.* Zambia
88 B2 **Zambia** *country* Africa
32 B2 **Zamboanga** Phil.
32 B2 **Zamboanga Peninsula** Phil.
74 B1 **Zamora** Spain
112 B3 **Zamora de Hidalgo** Mex.
71 E1 **Zamość** Pol.
43 B1 **Zanda** China
68 B2 **Zandvliet** Belgium
108 C3 **Zanesville** U.S.A.
49 C2 **Zanjān** Iran
79 B3 **Zante** *i.* Greece
87 D3 **Zanzibar** Tanz.
87 D3 **Zanzibar Island** Tanz.
57 E3 **Zaokskiy** Rus. Fed.
83 C2 **Zaouatallaz** Alg.
38 B2 **Zaozhuang** China
57 D2 **Zapadnaya Dvina** *r.* Europe
57 D2 **Zapadnaya Dvina** Rus. Fed.
36 B1 **Zapadnyy Sayan** *reg.* Rus. Fed.

107 D3 **Zapata** U.S.A.
60 J2 **Zapolyarnyy** Rus. Fed.
59 D2 **Zaporizhzhya** Ukr.
69 E2 **Zappendorf** Ger.
48 B2 **Zara** Turkey
113 B2 **Zaragoza** Mex.
75 C1 **Zaragoza** Spain
44 B3 **Zarand** Iran
44 C3 **Zaranj** Afgh.
56 C2 **Zarasai** Lith.
57 E3 **Zaraysk** Rus. Fed.
118 B1 **Zaraza** Venez.
83 C3 **Zaria** Nigeria
58 B1 **Zarichne** Ukr.
49 D3 **Zarqān** Iran
34 B2 **Zarubino** Rus. Fed.
71 D1 **Żary** Pol.
83 D1 **Zarzis** Tunisia
43 B1 **Zaskar Mountains** India
91 C3 **Zastron** S. Africa
37 E1 **Zavitinsk** Rus. Fed.
71 D1 **Zawiercie** Pol.
45 E2 **Zaysan, Ozero** *l.* Kazakh.
36 C3 **Zayü** China
58 B1 **Zdolbuniv** Ukr.
Zealand *i.* Denmark *see* Sjælland
68 A2 **Zeebrugge** Belgium
91 C2 **Zeerust** S. Africa
18 C2 **Zeil, Mount** Austr.
69 F2 **Zeitz** Ger.
56 B3 **Zelenogradsk** Rus. Fed.
50 E1 **Zemlya Aleksandry** *i.* Rus. Fed.
50 E1 **Zemlya Frantsa-Iosifa** *is* Rus. Fed.
50 F1 **Zemlya Vil'cheka** *i.* Rus. Fed.
75 D2 **Zemmora** Alg.
113 C3 **Zempoaltépetl, Nudo de** *mt.* Mex.
33 B1 **Zengfeng Shan** *mt.* China
73 D2 **Zermatt** Switz.
59 E2 **Zernograd** Rus. Fed.
69 E2 **Zeulenroda** Ger.
69 D1 **Zeven** Ger.
68 C2 **Zevenaar** Neth.
51 J3 **Zeya** Rus. Fed.
47 C2 **Zeydābād** Iran
51 J3 **Zeyskoye Vodokhranilishche** *resr* Rus. Fed.
71 D1 **Zgierz** Pol.
43 C2 **Zhabdün** China
56 B3 **Zhabinka** Belarus
44 A2 **Zhalpaktal** Kazakh.
45 C1 **Zhaltyr** Kazakh.
44 B2 **Zhanaozen** Kazakh.
Zhangde China *see* **Anyang**
38 B1 **Zhangjiakou** China
39 B3 **Zhangping** China
39 B3 **Zhangshu** China
36 C2 **Zhangye** China
39 B3 **Zhangzhou** China
39 B3 **Zhanjiang** China
39 B3 **Zhao'an** China
39 B3 **Zhaoqing** China
39 A3 **Zhaotong** China
43 C1 **Zhari Namco** *salt l.* China
45 E2 **Zharkent** Kazakh.
57 D2 **Zharkovskiy** Rus. Fed.
45 E2 **Zharma** Kazakh.
58 C2 **Zhashkiv** Ukr.
39 C3 **Zhejiang** *prov.* China
50 F1 **Zhelaniya, Mys** *c.* Rus. Fed.
57 E3 **Zheleznogorsk** Rus. Fed.
38 A2 **Zhenba** China
39 A3 **Zheng'an** China
39 B3 **Zhenghe** China
38 B2 **Zhengzhou** China
38 B2 **Zhenjiang** China
39 A3 **Zhenyuan** China
59 E1 **Zherdevka** Rus. Fed.
54 D2 **Zheshart** Rus. Fed.
45 C2 **Zhezkazgan** *Karagandinskaya Oblast'* Kazakh.
45 C2 **Zhezkazgan** *Karagandinskaya Oblast'* Kazakh.
51 J2 **Zhigansk** Rus. Fed.
Zhi Qu *r.* China *see* **Yangtze**
44 C1 **Zhitikara** Kazakh.
56 D3 **Zhlobin** Belarus
58 B2 **Zhmerynka** Ukr.
42 A1 **Zhob** Pak.
51 L1 **Zhokhova, Ostrov** *i.* Rus. Fed.
30 A1 **Zhongdian** China
Zhonghe China *see* **Xiushan**
38 A2 **Zhongning** China
39 B3 **Zhongshan** China
Zhongshan China *see* Lupanshui
38 A2 **Zhongwei** China
Zhongxin China *see* Zhongdian
38 B2 **Zhoukou** China
38 C2 **Zhoushan** China
59 C2 **Zhovti Vody** Ukr.
33 A2 **Zhuanghe** China
38 B2 **Zhucheng** China
57 D3 **Zhukovka** Rus. Fed.
57 E2 **Zhukovskiy** Rus. Fed.
38 B2 **Zhumadian** China
Zhuoyang China *see* **Suiping**
39 B3 **Zhuzhou** *Hunan* China
39 B3 **Zhuzhou** *Hunan* China
58 A2 **Zhydachiv** Ukr.
56 C3 **Zhytkavichy** Belarus
58 B1 **Zhytomyr** Ukr.
71 D2 **Žiar nad Hronom** Slovakia
38 B2 **Zibo** China
71 D1 **Zielona Góra** Pol.
30 A1 **Zigaing** Myanmar
39 A3 **Zigong** China
38 B2 **Zigui** China
82 A3 **Ziguinchor** Senegal
112 B3 **Zihuatanejo** Mex.
71 D2 **Žilina** Slovakia
51 H3 **Zima** Rus. Fed.
113 C2 **Zimapán** Mex.
89 C3 **Zimbabwe** *country* Africa
82 A4 **Zimmi** Sierra Leone
78 C2 **Zimnicea** Romania
83 C3 **Zinder** Niger
46 B3 **Zinjibār** Yemen
71 D2 **Zirc** Hungary
30 A1 **Ziro** India
47 C2 **Zīr Rūd** Iran
71 D2 **Zistersdorf** Austria
113 B3 **Zitácuaro** Mex.
71 C1 **Zittau** Ger.
55 E3 **Zlatoust** Rus. Fed.
71 D2 **Zlín** Czech Rep.
57 D3 **Zlynka** Rus. Fed.
57 E3 **Zmiyevka** Rus. Fed.
59 D2 **Zmiyiv** Ukr.
57 E3 **Znamenka** *Orlovskaya Oblast'* Rus. Fed.
59 E1 **Znamenka** *Tambovskaya Oblast'* Rus. Fed.
59 C2 **Znam"yanka** Ukr.
71 D2 **Znojmo** Czech Rep.
59 D1 **Zolochiv** *Kharkivs'ka Oblast'* Ukr.
58 A2 **Zolochiv** *L'vivs'ka Oblast'* Ukr.
59 C2 **Zolotonosha** Ukr.
57 E3 **Zolotukhino** Rus. Fed.
89 C2 **Zomba** Malawi
48 B1 **Zonguldak** Turkey
73 D3 **Zonza** *Corsica* France
82 B4 **Zorzor** Liberia
83 D2 **Zouar** Chad
82 A2 **Zouérat** Maur.
77 D1 **Zrenjanin** Yugo.
57 D2 **Zubtsov** Rus. Fed.
73 D2 **Zug** Switz.
49 C1 **Zugdidi** Georgia
74 B2 **Zújar** *r.* Spain
68 C2 **Zülpich** Ger.
89 C2 **Zumbo** Moz.
113 C3 **Zumpango** Mex.
106 B1 **Zuni Mountains** U.S.A.
39 A3 **Zunyi** China
77 C1 **Županja** Croatia
73 D2 **Zürich** Switz.
68 C1 **Zutphen** Neth.
83 D1 **Zuwārah** Libya
58 C2 **Zvenyhorodka** Ukr.
89 C3 **Zvishavane** Zimbabwe
71 D2 **Zvolen** Slovakia
77 C2 **Zvornik** Bos.-Herz.
82 B4 **Zwedru** Liberia
91 C3 **Zwelitsha** S. Africa
71 D2 **Zwettl** Austria
69 F2 **Zwickau** Ger.
68 C1 **Zwolle** Neth.
51 L2 **Zyryanka** Rus. Fed.

ACKNOWLEDGEMENTS

Maps designed and created by
HarperCollins Cartographic, Glasgow

The mapping is available in digital and electronic form.
For details and information on Bartholomew Mapping Solutions visit
http://www.bartholomewmaps.com
or contact
Bartholomew Mapping Solutions
Tel: +44 (0) 141 306 3155
Fax: +44 (0) 141 306 3104
E-mail:bartholomew@harpercollins.co.uk

IMAGES
pages 146-157
Continent maps: Mountain High Maps™